Study Guide and Student Solutions Manual

Organic Chemistry

NINTH EDITION

John McMurry

Prepared by

Susan McMurry

CENGAGE
Learning

Australia • Brazil • Mexico • Singapore • United Kingdom • United States

For product information and technology assistance, contact us at
**Cengage Learning Customer & Sales Support,
1-800-354-9706**.

For permission to use material from this text or product, submit all requests online at **www.cengage.com/permissions**
Further permissions questions can be emailed to
permissionrequest@cengage.com.

ISBN: 978-1-305-08214-4

Cengage Learning
20 Channel Center Street
Boston, MA 02210
USA

Cengage Learning is a leading provider of customized learning solutions with office locations around the globe, including Singapore, the United Kingdom, Australia, Mexico, Brazil, and Japan. Locate your local office at: **www.cengage.com/global**.

Cengage Learning products are represented in Canada by Nelson Education, Ltd.

To learn more about Cengage Learning Solutions, visit **www.cengage.com**.

Purchase any of our products at your local college store or at our preferred online store **www.cengagebrain.com**.

Printed in the United States of America
Print Number: 02   Print Year: 2015

Table of Contents

Appendices

Preface

What enters your mind when you hear the words "organic chemistry?" Some of you may think, "the chemistry of life," or "the chemistry of carbon." Other responses might include "pre-med, "pressure," "difficult," or "memorization." Although formally the study of the compounds of carbon, the discipline of organic chemistry encompasses many skills that are common to other areas of study. Organic chemistry is as much a liberal art as a science, and mastery of the concepts and techniques of organic chemistry can lead to improved competence in other fields.

As you work on the problems that accompany the text, you will bring to the task many problem-solving techniques. For example, planning an organic synthesis requires the skills of a chess player; you must plan your moves while looking several steps ahead, and you must keep your plan flexible. Structure-determination problems are like detective problems, in which many clues must be assembled to yield the most likely solution. Naming organic compounds is similar to the systematic naming of biological specimens; in both cases, a set of rules must be learned and then applied to the specimen or compound under study.

The problems in the text fall into two categories: drill and complex. Drill problems, which appear throughout the text and at the end of each chapter, test your knowledge of one fact or technique at a time. You may need to rely on memorization to solve these problems, which you should work on first. More complicated problems require you to recall facts from several parts of the text and then use one or more of the problem-solving techniques mentioned above. As each major type of problem—synthesis, nomenclature, or structure determination—is introduced in the text, a solution is extensively worked out in this *Solutions Manual*.

Here are several suggestions that may help you with problem solving:

1. The text is organized into chapters that describe individual functional groups. As you study each functional group, *make sure that you understand the structure and reactivity of that group*. In case your memory of a specific reaction fails you, you can rely on your general knowledge of functional groups for help.

2. *Use molecular models.* It is difficult to visualize the three-dimensional structure of an organic molecule when looking at a two-dimensional drawing. Models will help you to appreciate the structural aspects of organic chemistry and are indispensable tools for understanding stereochemistry.

3. Every effort has been made to make this *Solutions Manual* as clear, attractive, and error-free as possible. Nevertheless, you should *use the Solutions Manual in moderation*. The principal use of this book should be to check answers to problems you have already worked out. The *Solutions Manual* should not be used as a substitute for effort; at times, struggling with a problem is the only way to teach yourself.

4. *Look through the appendices at the end of the Solutions Manual.* Some of these appendices contain tables that may help you in working problems; others present information related to the history of organic chemistry.

Although the *Solutions Manual* is written to accompany *Organic Chemistry,* it contains several unique features. Each chapter of the *Solutions Manual* begins with an outline of the text that can be used for a concise review of the text material and can also serve as a reference. After every few chapters a Review Unit has been inserted. In most cases, the chapters covered in the Review Units are related to each other, and the units are planned to appear at approximately the place in the textbook where a test might be given. Each unit lists the vocabulary for the chapters covered, the skills needed to solve problems, and several important points that might need reinforcing or that restate material in the text from a slightly different point of view. Finally, the small self-test that has been included allows you to test yourself on the material from more than one chapter.

I have tried to include many types of study aids in this *Solutions Manual*. Nevertheless, this book can only serve as an adjunct to the larger and more complete textbook. If *Organic Chemistry* is the guidebook to your study of organic chemistry, then the *Solutions Manual* is the roadmap that shows you how to find what you need.

Susan McMurry

Acknowledgments

Cengage Learning would like to acknowledge KC Russell of Northern Kentucky University and James Vyvyan of Western Washington University for providing solutions to the new mechanism and spectroscopy problems. We would also like to thank Paul Adams of the University of Arkansas for his review work.

Chapter 1 – Structure and Bonding

Chapter Outline

I. Atomic Structure (Sections 1.1–1.3).
  A. Introduction to atomic structure (Section 1.1).
    1. An atom consists of a dense, positively charged nucleus surrounded by negatively charged electrons.
      a. The nucleus is made up of positively charged protons and uncharged neutrons.
      b. The nucleus contains most of the mass of the atom.
      c. Electrons move about the nucleus at a distance of about 2×10^{-10} m (200 pm).
    2. The atomic number (Z) gives the number of protons in the nucleus.
    3. The mass number (A) gives the total number of protons and neutrons.
    4. All atoms of a given element have the same value of Z.
      a. Atoms of a given element can have different values of A.
      b. Atoms of the same element with different values of A are called isotopes.
  B. Orbitals (Section 1.2).
    1. The distribution of electrons in an atom can be described by a wave equation.
      a. The solution to a wave equation is an orbital, represented by Ψ.
      b. Ψ^2 predicts the volume of space in which an electron is likely to be found.
    2. There are four different kinds of orbitals (s, p, d, f).
      a. The s orbitals are spherical.
      b. The p orbitals are dumbbell-shaped.
      c. Four of the five d orbitals are cloverleaf-shaped.
    3. An atom's electrons are organized into electron shells.
      a. The shells differ in the numbers and kinds of orbitals they contain.
      b. Electrons in different orbitals have different energies.
      c. Each orbital can hold up to a maximum of two electrons.
    4. The two lowest-energy electrons are in the $1s$ orbital.
      a. The $2s$ orbital is the next higher in energy.
      b. The next three orbitals are $2p_x$, $2p_y$ and $2p_z$, which have the same energy.
        i. Each p orbital has a region of zero density, called a node.
      c. The lobes of a p orbital have opposite algebraic signs.
  C. Electron Configuration (Section 1.3).
    1. The ground-state electron configuration of an atom is a listing of the orbitals occupied by the electrons of the atom in the lowest energy configuration.
    2. Rules for predicting the ground-state electron configuration of an atom:
      a. Orbitals with the lowest energy levels are filled first.
        i. The order of filling is $1s$, $2s$, $2p$, $3s$, $3p$, $4s$, $3d$.
      b. Only two electrons can occupy each orbital, and they must be of opposite spin.
      c. If two or more orbitals have the same energy, one electron occupies each until all are half-full (Hund's rule). Only then does a second electron occupy one of the orbitals.
        i. All of the electrons in half-filled shells have the same spin.

II. Chemical Bonding Theory (Sections 1.4–1.5).

    A. Development of chemical bonding theory (Section 1.4).

        1. Kekulé and Couper proposed that carbon has four "affinity units"; carbon is tetravalent.

        2. Kekulé suggested that carbon can form rings and chains.

        3. Van't Hoff and Le Bel proposed that the 4 atoms to which carbon forms bonds sit at the corners of a regular tetrahedron.

        4. In a drawing of a tetrahedral carbon, a wedged line represents a bond pointing toward the viewer, a dashed line points behind the plane of the page, and a solid line lies in the plane of the page.

    B. Covalent bonds.

        1. Atoms bond together because the resulting compound is more stable than the individual atoms.

            a. Atoms tend to achieve the electron configuration of the nearest noble gas.

            b. Atoms in groups 1A, 2A and 7A either lose electrons or gain electrons to form ionic compounds.

            c. Atoms in the middle of the periodic table share electrons by forming covalent bonds.

            d. The neutral collection of atoms held together by covalent bonds is a molecule.

        2. Covalent bonds can be represented two ways.

            a. In electron-dot structures, bonds are represented as pairs of dots.

            b. In line-bond structures, bonds are represented as lines drawn between two bonded atoms.

        3. The number of covalent bonds formed by an atom depends on the number of electrons it has and on the number it needs to achieve an octet.

        4. Valence electrons not used for bonding are called lone-pair (nonbonding) electrons.

            a. Lone-pair electrons are often represented as dots.

    C. Valence bond theory (Section 1.5).

        1. Covalent bonds are formed by the overlap of two atomic orbitals, each of which contains one electron. The two electrons have opposite spins.

        2. Bonds formed by the head-on overlap of two atomic orbitals are cylindrically symmetrical and are called σ bonds.

        3. Bond strength is the measure of the amount of energy needed to break a bond.

        4. Bond length is the optimum distance between nuclei.

        5. Every bond has a characteristic bond length and bond strength.

III. Hybridization (Sections 1.6–1.10).

    A. sp^3 Orbitals (Sections 1.6, 1.7).

        1. Structure of methane (Section 1.6).

            a. When carbon forms 4 bonds with hydrogen, one $2s$ orbital and three $2p$ orbitals combine to form four equivalent atomic orbitals (sp^3 hybrid orbitals).

            b. These orbitals are tetrahedrally oriented.

            c. Because these orbitals are unsymmetrical, they can form stronger bonds than unhybridized orbitals can.

    d. These bonds have a specific geometry and a bond angle of 109.5°.

  2. Structure of ethane (Section 1.7).

    a. Ethane has the same type of hybridization as occurs in methane.

    b. The C–C bond is formed by overlap of two sp^3 orbitals.

    c. Bond lengths, strengths and angles are very close to those of methane.

B. sp^2 Orbitals (Section 1.8).

  1. If one carbon $2s$ orbital combines with two carbon $2p$ orbitals, three hybrid sp^2 orbitals are formed, and one p orbital remains unchanged.

  2. The three sp^2 orbitals lie in a plane at angles of 120°, and the unhybridized p orbital is perpendicular to them.

  3. Two different types of bonds form between two carbons.

    a. A σ bond forms from the overlap of two sp^2 orbitals.

    b. A π bond forms by sideways overlap of two p orbitals.

    c. This combination is known as a carbon–carbon double bond.

  4. Ethylene is composed of a carbon–carbon double bond and four σ bonds formed between the remaining four sp^2 orbitals of carbon and the $1s$ orbitals of hydrogen.

    a. The double bond of ethylene is both shorter and stronger than the C–C bond of ethane.

C. sp Orbitals (Section 1.10).

  1. If one carbon $2s$ orbital combines with one carbon $2p$ orbital, two hybrid sp orbitals are formed, and two p orbitals are unchanged.

  2. The two sp orbitals are 180° apart, and the two p orbitals are perpendicular to them and to each other.

  3. Two different types of bonds form.

    a. A σ bond forms from the overlap of two sp orbitals.

    b. Two π bonds form by sideways overlap of four unhybridized p orbitals.

    c. This combination is known as a carbon–carbon triple bond.

  4. Acetylene is composed of a carbon–carbon triple bond and two σ bonds formed between the remaining two sp orbitals of carbon and the $1s$ orbitals of hydrogen.

    a. The triple bond of acetylene is the strongest carbon–carbon bond.

D. Hybridization of nitrogen and oxygen (Section 1.10).

  1. Covalent bonds between other elements can be described by using hybrid orbitals.

  2. Both the nitrogen atom in ammonia and the oxygen atom in water form sp^3 hybrid orbitals.

    a. The lone-pair electrons in these compounds occupy sp^3 orbitals.

  3. The bond angles between hydrogen and the central atom is often less than 109° because the lone-pair electrons take up more room than the σ bond.

  4. Because of their positions in the third row, phosphorus and sulfur can form more than the typical number of covalent bonds.

IV. Molecular orbital theory (Section 1.11).

A. Molecular orbitals arise from a mathematical combination of atomic orbitals and belong to the entire molecule.

  1. Two $1s$ orbitals can combine in two different ways.

      a.   The additive combination is a bonding MO and is lower in energy than the two hydrogen $1s$ atomic orbitals.

      b.   The subtractive combination is an antibonding MO and is higher in energy than the two hydrogen $1s$ atomic orbitals.

  2. Two p orbitals in ethylene can combine to form two π MOs.

      a.   The bonding MO has no node; the antibonding MO has one node.

  3. A node is a region between nuclei where electrons aren't found.

      a.   If a node occurs between two nuclei, the nuclei repel each other.

V. Chemical structures (Section 1.12).

  A. Drawing chemical structures.

    1. Condensed structures don't show C–H bonds and don't show the bonds between CH_3, CH_2 and CH units.

    2. Skeletal structures are simpler still.

      a.   Carbon atoms aren't usually shown.

      b.   Hydrogen atoms bonded to carbon aren't usually shown.

      c.   Other atoms (O, N, Cl, etc.) are shown.

Solutions to Problems

1.1   **(a)**   To find the ground-state electron configuration of an element, first locate its atomic number. For oxygen, the atomic number is 8; oxygen thus has 8 protons and 8 electrons. Next, assign the electrons to the proper energy levels, starting with the lowest level. Fill each level *completely* before assigning electrons to a higher energy level.

Oxygen

$2p$ ⇅ ↑ ↑

$2s$ ⇅

$1s$ ⇅

Notice that the $2p$ electrons are in different orbitals. According to *Hund's rule*, we must place one electron into each orbital of the same energy level until all orbitals are half-filled.

Remember that only two electrons can occupy the same orbital, and that they must be of opposite spin.

A different way to represent the ground-state electron configuration is to simply write down the occupied orbitals and to indicate the number of electrons in each orbital. For example, the electron configuration for oxygen is $1s^2\, 2s^2\, 2p^4$.

Nitrogen

$2p$ ↑ ↑ ↑

$2s$ ⇅

$1s$ ⇅

   (b)   Nitrogen, with an atomic number of 7, has 7 electrons. Assigning these to energy levels:
The more concise way to represent ground-state electron configuration for nitrogen: $1s^2\, 2s^2\, 2p^3$

   (c)   Sulfur has 16 electrons.

$3p$ ⇅ ↑ ↑

$3s$ ⇅

Sulfur

$2p$ ⇅ ⇅ ⇅

$2s$ ⇅

$1s$ ⇅

$1s^2\, 2s^2\, 2p^6\, 3s^6\, 3p^4$

1.2 The elements of the periodic table are organized into groups that are based on the number of outer-shell electrons each element has. For example, an element in group 1A has one outershell electron, and an element in group 5A has five outer-shell electrons. To find the number of outer-shell electrons for a given element, use the periodic table to locate its group.

    (a) Magnesium (group 2A) has two electrons in its outermost shell.

    (b) Cobalt is a transition metal, which has two electrons in the $4s$ subshell, plus seven electrons in its $3d$ subshell.

    (c) Selenium (group 6A) has six electrons in its outermost shell.

1.3 A solid line represents a bond lying in the plane of the page, a wedged bond represents a bond pointing out of the plane of the page toward the viewer, and a dashed bond represents a bond pointing behind the plane of the page.

1.4

1.5 Identify the group of the central element to predict the number of covalent bonds the element can form.

    (a) Carbon (Group 4A) has four electrons in its valence shell and forms four bonds to achieve the noble-gas configuration of neon. A likely formula is CCl_4.

	Element	Group	Likely Formula
(b)	Al	3A	AlH_3
(c)	C	4A	CH_2Cl_2
(d)	Si	4A	SiF_4
(e)	N	5A	CH_3NH_2

1.6   Start by drawing the electron-dot structure of the molecule.

(1)   Determine the number of valence, or outer-shell electrons for each atom in the molecule. For chloroform, we know that carbon has four valence electrons, hydrogen has one valence electron, and each chlorine has seven valence electrons.

$$\cdot \overset{\displaystyle\cdot}{\underset{\displaystyle\cdot}{C}} \cdot \qquad 4 \times 1 = 4$$

$$H \cdot \qquad 1 \times 1 = 1$$

$$: \overset{\displaystyle\cdot\cdot}{\underset{\displaystyle\cdot\cdot}{Cl}} \cdot \qquad 7 \times 3 = \underline{21}$$

$$\qquad\qquad\qquad 26 \quad \text{total valence electrons}$$

(2)   Next, use two electrons for each single bond.

$$\begin{array}{c} H \\ \cdot\cdot \\ Cl : C : Cl \\ \cdot\cdot \\ Cl \end{array}$$

(3)   Finally, use the remaining electrons to achieve a noble gas configuration for all atoms. For a line-bond structure, replace the electron dots between two atoms with a line.

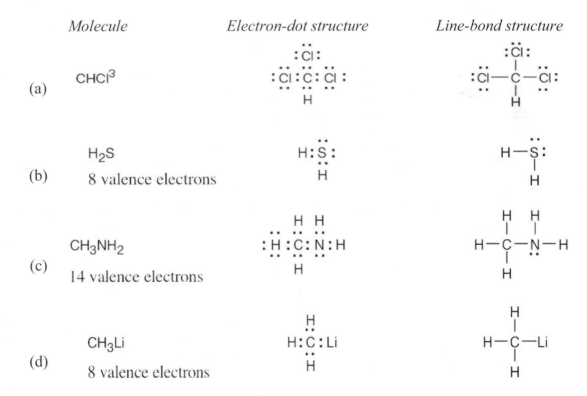

	Molecule	Electron-dot structure	Line-bond structure
(a)	$CHCl^3$		
(b)	H_2S 8 valence electrons		
(c)	CH_3NH_2 14 valence electrons		
(d)	CH_3Li 8 valence electrons		

1.7   Each of the two carbons has 4 valence electrons. Two electrons are used to form the carbon–carbon bond, and the 6 electrons that remain can form bonds with a maximum of 6 hydrogens. Thus, the formula C_2H_7 is not possible.

1.8   Connect the carbons and add hydrogens so that all carbons are bonded to four different atoms.

Propane

The geometry around all carbon atoms is tetrahedral, and all bond angles are approximately 109°.

1.9

Hexane

1.10

Propene

The C3–H bonds are σ bonds formed by overlap of an sp^3 orbital of carbon 3 with an s orbital of hydrogen.

The C2–H and C1–H bonds are σ bonds formed by overlap of an sp^2 orbital of carbon with an s orbital of hydrogen.

The C2–C3 bond is a σ bond formed by overlap of an sp^3 orbital of carbon 3 with an sp^2 orbital of carbon 2.

There are two C1–C2 bonds. One is a σ bond formed by overlap of an sp^3 orbital of carbon 1 with an sp^2 orbital of carbon 2. The other is a π bond formed by overlap of a p orbital of carbon 1 with a p orbital of carbon 2. All four atoms connected to the carbon–carbon double bond lie in the same plane, and all bond angles between these atoms are 120°. The bond angle between hydrogen and the sp^3-hybridized carbon is 109°.

1.11

All atoms lie in the same plane, and all bond angles are approximately 120°.

1,3-Butadiene

1.12

Aspirin.

All carbons are sp^2 hybridized, with the exception of the indicated carbon. All oxygen atoms have two lone pairs of electrons.

1.13

Propyne

The C3-H bonds are σ bonds formed by overlap of an sp^3 orbital of carbon 3 with an s orbital of hydrogen.

The C1-H bond is a σ bond formed by overlap of an sp orbital of carbon 1 with an s orbital of hydrogen.

The C2-C3 bond is a σ bond formed by overlap of an sp orbital of carbon 2 with an sp^3 orbital of carbon 3.

There are three C1-C2 bonds. One is a σ bond formed by overlap of an sp orbital of carbon 1 with an sp orbital of carbon 2. The other two bonds are π bonds formed by overlap of two p orbitals of carbon 1 with two p orbitals of carbon 2.

The three carbon atoms of propyne lie in a straight line: the bond angle is 180°. The H–$C_1 \equiv C_2$ bond angle is also 180°. The bond angle between hydrogen and the sp^3-hybridized carbon is 109°.

1.14

(a)

The sp^3-hybridized oxygen atom has tetrahedral geometry.

(b)

Tetrahedral geometry at nitrogen and carbon.

(c)

Like nitrogen, phosphorus has five outer-shell electrons. PH_3 has tetrahedral geometry.

(d) $H_3C\diagdown_S\diagup CH_2CH_2\underset{\underset{NH_2}{|}}{CH}\overset{\overset{O}{||}}{C}OH$ The sp^3-hybridized sulfur atom has tetrahedral geometry.

1.15 Remember that the end of a line represents a carbon atom with 3 hydrogens, a two-way intersection represents a carbon atom with 2 hydrogens, a three-way intersection represents a carbon with 1 hydrogen and a four-way intersection represents a carbon with no hydrogens.

(a)

Adrenaline – $C_9H_{13}NO_3$

(b)

Estrone – $C_{18}H_{22}O_2$

1.16 Several possible skeletal structures can satisfy each molecular formula.

(a) C_5H_{12}

(b) C_2H_7N

(c) C_3H_6O

(d) C_4H_9Cl

1.17

PABA

Visualizing Chemistry

1.18 (a)

$C_8H_{17}N$

(b)

$C_3H_7NO_2$

1.19 Citric acid ($C_6H_8O_7$) contains seven oxygen atoms, each of which has two electron lone pairs. Three of the oxygens form double bonds with carbon.

Citric acid

1.20

Acetaminophen

All carbons are sp^2 hybridized, except for the carbon indicated as sp^3. The two oxygen atoms and the nitrogen atom have lone pair electrons, as shown.

1.21

Aspartame

Additional Problems

Electron Configuration

1.22

	Element	Atomic Number	Number of valence electrons
(a)	Zinc	30	2
(b)	Iodine	53	7
(c)	Silicon	14	4
(d)	Iron	26	2 (in $4s$ nutshell),6 (in $3d$ subshell)

1.23

	Element	Atomic Number	Ground-state electron configuration
(a)	Potassium	19	$1s^2\ 2s^2\ 2p^6\ 3s^2\ 3p^6\ 4s^1$
(b)	Arsenic	33	$1s^2\ 2s^2\ 2p^6\ 3s^2\ 3p^6\ 4s^2\ 3d^{10}\ 4p^3$
(c)	Aluminum	13	$1s^2\ 2s^2\ 2p^6\ 3s^2\ 3p^1$
(d)	Germanium	32	$1s^2\ 2s^2\ 2p^6\ 3s^2\ 3p^6\ 4s^2\ 3d^{10}\ 4p^2$

Electron-Dot and Line-Bond Structures

1.24 (a) NH_2OH (b) $AlCl_3$ (c) CH_2CL_2 (d) CH_2O

1.25 (a)   The 4 valence electrons of carbon can form bonds with a maximum of 4 hydrogens. Thus, it is not possible for the compound CH_5 to exist.

  (b)   If you try to draw a molecule with the formula C_2H_6N, you will see that it I impossible for both carbons and nitrogen to have a complete octet of electrons. Therefore, C_2H_6N is unlikely to exist.

  (c)   A compound with the formula $C_3H_5Br_2$ doesn't have filled outer shells for all atoms and is thus unlikely to exist.

1.26

H
··
H:C:C:::N: Acetonitrile
··
H

In the compound acetonitrile, nitrogen has eight electrons in its outer electron shell. Six are used in the carbon-nitrogen triple bond, and two are a nonbonding electron pair.

1.27

Vinyl chloride

Vinyl chloride has 18 valence electrons. Eight electrons are used for 4 single bonds, 4 electrons are used in the carbon–carbon double bond, and 6 electrons are in the 3 lone pairs that surround chlorine.

1.28 (a)   (b)   (c)

1.29 In molecular formulas of organic molecules, carbon is listed first, followed by hydrogen. All other elements are listed in alphabetical order.

Compound		Molecular Formula
(a)	Aspirin	$C_9H_8O_4$
(b)	Vitamin C	$C_6H_8O_6$
(c)	Nicotine	$C_{10}H_{14}N_2$
(d)	Glucose	$C_6H_{12}O_6$

1.30 To work a problem of this sort, you must draw all possible structures consistent with the rules of valence. You must systematically consider all possible attachments, including those that have branches, rings and multiple bonds.

(a)

```
    H   H   H
    |   |   |
H — C — C — C — H
    |   |   |
    H   H   H
```

(b)

```
    H   H
    |   |
H — C — N — H
    |
    H
```

(c)

```
    H   H                    H       H
    |   |                    |       |
H — C — C — O — H       H — C — O — C — H
    |   |                    |       |
    H   H                    H       H
```

(d)

```
    H   H   H                   H   Br  H
    |   |   |                   |   |   |
H — C — C — C — Br         H — C — C — C — H
    |   |   |                   |   |   |
    H   H   H                   H   H   H
```

(e)

```
    H   O                    H                     O
    |   ||                   |              H.   ⁄    ＼  H
H — C — C — H          H — C = C — O — H        C — C
    |                        |              H ⁄        ＼ H
    H                        H
```

(f)

```
    H   H   H            H   H   H          H   H       H         H           H
    |   |   |            |   |   |          |   |       |         |           |
H — C — C — C — H    H — C — C — C — H  H — C — C — N — C — H  H — C — N — C — H
    |   |   |            |   |   |          |   |   |   |         |   |   |
    H   H   N           H   N   H          H   H   H   H         H   C   H
          ⁄   ＼           ⁄   ＼                                    H ⁄ | ＼ H
        H     H         H     H                                          H
```

1.31

```
    CH₃
    |
H — C ⋯ H        Ethanol
    ▲
    OH
```

1.32

```
    O   O   H   O
    ||  ||  |   ||
H — O — C — C — C — C — O — H
            |
            H
```

```
    O   H   O   O
    ||  |   ||  ||
H — C — C — C — C — O — H
        |
        H — O
```

```
    O   H   O   O
    ||  |   ||  ||
H — O — C — C — C — C — H
        |
        H — O
```

1.33 (a)   (b)   (c)   (d)

1.34

H—C—O⁻ K⁺   All other bonds are covalent.

ionic

Hybridization

1.35 The H₃C– carbon is sp^3 hybridized, and the –CN carbon is sp hybridized.

1.36 (a)   (b)   (c)   (d)

sp^3 sp^3 sp^3
CH₃CH₂CH₃

sp^2 sp^2 sp sp
H₂C=CH—C≡CH

1.37

Benzene

All carbon atoms of benzene are sp^2 hybridized, and all bond angles of benzene are 120°. Benzene is a planar molecule.

1.38 (a)   (b)   (c)

Glycine   Pyridine   Lactic acid

1.39 Examples:

(a) $CH_3CH_2CH=CH_2$ (b) $H_2C=CH-CH=CH_2$ (c) $H_2C=CH-C\equiv CH$

1.40 (a)   (b)

Procaine   Vitamin C

1.41

Pyridoxal phosphate

The bond angles formed by atoms having sp^3 hybridization are approximately 109°.
The bond angles formed by atoms having sp^2 hybridization are approximately 120°.

Skeletal Structures

1.42 (a)   (b)   (c)   (d)

1.43 (a)

$C_{10}H_{11}N$

(b)

$C_{11}H_{11}BrO_2$

(c)

$C_9H_{12}O$

1.44

Quetiapine (Seroquel) $C_{21}H_{25}N_3O_2S$

1.45

Oseltamivir
(Tamiflu)
$C_{16}H_{28}N_2O_4$

Clopidogrel
(Plavix)
$C_{16}H_{16}ClNO_2S$

General Problems

1.46 In a compound containing a carbon–carbon triple bond, atoms bonded to the *sp*-hybridized carbons must lie in a straight line. It is not possible to form a five-membered ring if four carbons must have a linear relationship.

1.47

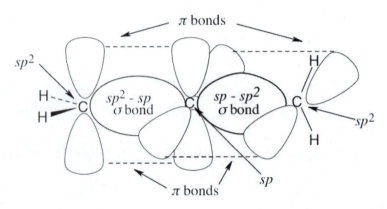

The central carbon of allene forms two σ bonds and two π bonds. The central carbon is *sp*-hybridized, and the two terminal carbons are *sp*²-hybridized. The bond angle formed by the three carbons is 180°, indicating linear geometry for the carbons of allene.

1.48

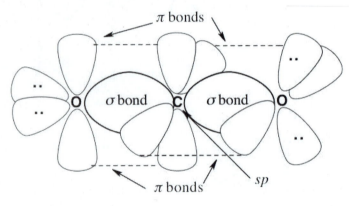

Carbon dioxide is a linear molecule.

1.49

H₃C N C O (Caffeine structure)

Caffeine

All of the indicated atoms are *sp*²-hybridized.

1.50 (a) The positively charged carbon atom is surrounded by six valence electrons; carbon has three valence electrons, and each hydrogen brings three valence electrons.

(b) The positively charged carbon is sp^2-hybridized.

(c) A carbocation is planar about the positively charged carbon.

1.51

(a) A carbanion is isoelectronic with (has the same number of electrons as) a trivalent nitrogen compound.

(b) The negatively charged carbanion carbon has eight valence electrons.

(c) The carbon atom is sp^3-hybridized.

(d) A carbanion is tetrahedral.

1.52 According to the Pauli Exclusion Principle, two electrons in the same orbital must have opposite spins. Thus, the two electrons of triplet (spin-unpaired) methylene must occupy different orbitals. In triplet methylene, sp-hybridized carbon forms one bond to each of two hydrogens. Each of the two unpaired electrons occupies a p orbital. In singlet (spin-paired) methylene the two electrons can occupy the same orbital because they have opposite spins. Including the two C–H bonds, there are a total of three occupied orbitals. We predict sp^2 hybridization and planar geometry for singlet methylene.

Triplet methylene
(linear)

Singlet methylene
(planar)

1.53

The two compounds differ in the way that the carbon atoms are connected.

1.54

$H_2C{=}CH{-}CH_3$

One compound has a double bond, and one has a ring.

1.55 CH_3CH_2OH CH_3OCH_3

The two compounds differ in the location of the oxygen atom.

1.56

$CH_3CH_2CH{=}CH_2$ $CH_3CH{=}CHCH_3$

The compounds differ in the way that the carbon atoms are connected and in the location of the double bond.

1.57

Ibuprofen

Naproxen

Acetaminophen

* = sp^3-hybridized carbon

\# = sp^2-hybridized carbon

(a), (b)

Compound	sp^3-Hybridized carbons	sp^2-Hybridized carbons
Ibuprofen	6	7
Naproxen	3	11
Acetaminophen	1	7

(c)   Each of the structures has a six-membered ring containing three double bonds, each has a methyl group, and each has a C=O group.

Chapter 2 – Polar Covalent Bonds; Acids and Bases

Chapter Outline

I. Polar covalent bonds (Sections 2.1–2.3).

    A. Electronegativity (Section 2.1).

        1. Although some bonds are totally ionic and some are totally covalent, most chemical bonds are polar covalent bonds.

            a. In these bonds, electrons are attracted to one atom more than to the other atom.

        2. Bond polarity is due to differences in electronegativity (EN).

            a. Elements on the right side of the periodic table are more electronegative than elements on the left side.

            b. Carbon has an EN of 2.5.

            c. Elements with EN > 2.5 are more electronegative than carbon.

            d. Elements with EN < 2.5 are less electronegative than carbon.

        3. The difference in EN between two elements can be used to predict the polarity of a bond.

            a. If ΔEN < 0.4, a bond is nonpolar covalent.

            b. If ΔEN is between 0.4 and 2.0, a bond is polar covalent.

            c. If ΔEN > 2.0, a bond is ionic.

            d. The symbols Δ+ and Δ– are used to indicate partial charges.

            e. A crossed arrow is used to indicate bond polarity.

                i. The tail of the arrow is electron-poor, and the head of the arrow is electron-rich.

        4. Electrostatic potential maps are also used to show electron-rich (red) and electron-poor (blue) regions of molecules.

        5. An inductive effect is an atom's ability to polarize a bond.

    B. Dipole moment (Section 2.2).

        1. Dipole moment is the measure of a molecule's overall polarity.

        2. Dipole moment (μ) = Q x r, where Q = charge and r = distance between charges.

            a. Dipole moment is measured in debyes (D).

        3. Dipole moment can be used to measure charge separation.

        4. Water and ammonia have large values of D; methane and ethane have D = 0.

    C. Formal charge (Section 2.3).

        1. Formal charge (FC) indicates electron "ownership" in a molecule.

        2. $(FC) = \begin{bmatrix} \text{\# of valence} \\ \text{electrons} \end{bmatrix} - \begin{bmatrix} \dfrac{\text{\# of bonding electrons}}{2} \end{bmatrix} - \begin{bmatrix} \text{\# nonbonding} \\ \text{electrons} \end{bmatrix}$

II. Resonance (Sections 2.4–2.6).

    A. Chemical structures and resonance (Section 2.4).

        1. Some molecules (acetate ion, for example) can be drawn as two (or more) different electron-dot structures.

            a. These structures are called resonance structures.

    b. The true structure of the molecule is intermediate between the resonance structures.

    c. The true structure is called a resonance hybrid.

  2. Resonance structures differ only in the placement of π and nonbonding electrons.

    a. All atoms occupy the same positions.

  3. Resonance is an important concept in organic chemistry.

  B. Rules for resonance forms (Section 2.5).

    1. Individual resonance forms are imaginary, not real.

    2. Resonance forms differ only in the placement of their π or nonbonding electrons.

      a. A curved arrow is used to indicate the movement of electrons, not atoms.

    3. Different resonance forms of a molecule don't have to be equivalent.

      a. If resonance forms are nonequivalent, the structure of the actual molecule resembles the more stable resonance form(s).

    4. Resonance forms must obey normal rules of valency.

    5. The resonance hybrid is more stable than any individual resonance form.

  C. A useful technique for drawing resonance forms (Section 2.6).

    1. Any three-atom grouping with a multiple bond adjacent to a nonbonding p orbital has two resonance forms.

    2. One atom in the grouping has a lone electron pair, a vacant orbital or a single electron.

    3. By recognizing these three-atom pieces, resonance forms can be generated.

III. Acids and bases (Sections 2.7–2.11).

  A. Brønsted–Lowry definition (Section 2.7).

    1. A Brønsted–Lowry acid donates an H^+ ion; a Brønsted–Lowry base accepts H^+.

    2. The product that results when a base gains H^+ is the conjugate acid of the base; the product that results when an acid loses H^+ is the conjugate base of the acid.

    3. Water can act either as an acid or as a base.

  B. Acid and base strength (Section 2.8–2.10).

    1. A strong acid reacts almost completely with water (Section 2.8).

    2. The strength of an acid in water is indicated by K_a, the acidity constant.

    3. Strong acids have large acidity constants, and weaker acids have smaller acidity constants.

    4. The pK_a is normally used to express acid strength.

      a. pK_a = –log K_a

      b. A strong acid has a small pK_a, and a weak acid has a large pK_a.

      c. The conjugate base of a strong acid is a weak base, and the conjugate base of a weak acid is a strong base.

    5. Predicting acid–base reactions from pK_a (Section 2.9).

      a. An acid with a low pK_a (stronger acid) reacts with the conjugate base of an acid with a high pK_a (stronger base).

      b. In other words, the products of an acid–base reaction are more stable than the reactants.

6. Organic acids and organic bases (Section 2.10).
    a. There are two main types of organic acids:
        i.   Acids that contain hydrogen bonded to oxygen.
        ii. Acids that have hydrogen bonded to the carbon next to a C=O group.
    b. The main type of organic base contains a nitrogen atom with a lone electron pair.
C. Lewis acids and bases (Section 2.11).
    1. A Lewis acid accepts an electron pair.
        a. A Lewis acid may have either a vacant low-energy orbital or a polar bond to hydrogen.
        b. Examples include metal cations, halogen acids, group 3 compounds and transition-metal compounds.
    2. A Lewis base has a pair of nonbonding electrons.
        a. Most oxygen- and nitrogen-containing organic compounds are Lewis bases.
        b. Many organic Lewis bases have more than one basic site.
    3. A curved arrow shows the movement of electrons from a Lewis base to a Lewis acid.
IV. Noncovalent interactions in molecules (Section 2.12).
    A. Dipole–dipole interactions occur between polar molecules as a result of electrostatic interactions among dipoles.
        1. These interactions may be either attractive or repulsive.
        2. The attractive geometry is lower in energy and predominates.
    B. Dispersion forces result from the constantly changing electron distribution within molecules.
        1. These forces are transient and weak, but their cumulative effect may be important.
    C. Hydrogen bonds.
        1. Hydrogen bonds form between a hydrogen bonded to an electronegative atom and an unshared electron pair on another electronegative atom.
        2. Hydrogen bonds are extremely important in living organisms.
        3. Hydrophilic substances dissolve in water because they are capable of forming hydrogen bonds.
        4. Hydrophobic substances don't form hydrogen bonds and usually don't dissolve in water.

Solutions to Problems

2.1 After solving this problem, use Figure 2.2 to check your answers. The larger the number, the more electronegative the element.

More electronegative		*Less electronegative*	
(a) H	(2.1)	Li	(1.0)
(b) Br	(2.8)	B	(2.0)
(c) Cl	(3.0)	I	(2.5)
(d) C	(2.5)	H	(2.1)

Carbon is slightly more electronegative than hydrogen.

2.2 As in Problem 2.1, use Figure 2.2. The partial negative charge is placed on the more electronegative atom, and the partial positive charge is placed on the less electronegative atom.

(a)

$$\overset{\delta+ \quad \delta-}{H_3C-Cl}$$

(b)

$$\overset{\delta+ \quad \delta-}{H_3C-NH_2}$$

(c)

$$\overset{\delta- \quad \delta+}{H_2N-H}$$

(d)

$$H_3C-SH$$

Carbon and sulfur have identical electronegativies.

(e)

$$\overset{\delta- \quad \delta+}{H_3C-MgBr}$$

(f)

$$\overset{\delta+ \quad \delta-}{H_3C-F}$$

2.3 Use Figure 2.2 to find the electronegativities of each element. Calculate ΔEN and rank the answers in order of increasing ΔEN.

Carbon:	EN = 2.5	Carbon:	EN = 2.5	Fluorine:	EN = 4.0
Lithium:	EN = 1.0	Potassium:	EN = 0.8	Carbon:	EN = 2.5
	$\Delta EN = 1.5$		$\Delta EN = 1.7$		$\Delta EN = 1.5$

Carbon:	EN = 2.5	Oxygen:	EN = 3.5
Magnesium:	EN = 1.2	Carbon:	EN = 2.5
	$\Delta EN = 1.3$		$\Delta EN = 1.0$

The most polar bond has the largest ΔEN. Thus, in order of increasing bond polarity:

$$H_3C - OH < H_3C - MgBr < H_3C - Li, \ H_3C - F < H_3C - K$$

2.4   In an electrostatic potential map, the color red indicates regions of a molecule that are electron-rich. The map shows that chlorine is the most electronegative atom in chloromethane, and the direction of polarity of the C–Cl bond is:

Chloromethane

2.5

Ethylene glycol

The dipole moment of ethylene glycol is zero because the bond polarities of the two carbon–oxygen bonds cancel.

2.6   For each bond, identify the more electronegative element, and draw an arrow that points from the less electronegative element to the more electronegative element. Estimate the sum of the individual dipole moments to arrive at the dipole moment for the entire molecule.

(a)   0 dipole moment

(b)   net dipole moment

(c)   net dipole moment

(d)   net dipole moment

2.7   To find the formal charge of an atom in a molecule, follow these two steps:

(1)   Draw an electron-dot structure of the molecule.

(2)   Use the formula in Section 2.3 (shown below) to determine formal charge for each atom. The periodic table shows the number of valence electrons of the element, and the electron-dot structure shows the number of bonding and nonbonding electrons.

$$\text{Formal charge } (FC) = \left[\begin{array}{c} \# \text{ of valence} \\ \text{electrons} \end{array}\right] - \left[\begin{array}{c} \# \text{ of bonding electrons} \\ \hline 2 \end{array}\right] - \left[\begin{array}{c} \# \text{ nonbonding} \\ \text{electrons} \end{array}\right]$$

(a)

$$H_2C = N = \ddot{N}: \quad = \quad H:\overset{H}{\underset{..}{C}}::\overset{1}{N}::\overset{2}{\underset{..}{N}}:$$

For carbon: $FC = 4 - \dfrac{8}{2} - 0 = 0$

For nitrogen 1: $FC = 5 - \dfrac{8}{2} - 0 = +1$

For nitrogen 2: $FC = 5 - \dfrac{4}{2} - 4 = -1$

Remember: *Valence electrons* are the electrons characteristic of a specific element. *Bonding electrons* are those electrons involved in bonding to other atoms. *Nonbonding electrons* are those electrons in lone pairs.

(b)

$$H_3C - C \equiv N - \ddot{\underset{..}{O}}: \quad = \quad H:\overset{\overset{H}{..}}{\underset{\underset{H}{..}}{C}}:\overset{1}{C}:::\overset{2}{N}:\overset{..}{\underset{..}{O}}:$$

For carbon 1: $FC = 4 - \dfrac{8}{2} - 0 = 0$

For carbon 2: $FC = 4 - \dfrac{8}{2} - 0 = 0$

For nitrogen : $FC = 5 - \dfrac{8}{2} - 0 = +1$

For oxygen: $FC = 6 - \dfrac{2}{2} - 6 = -1$

(c)

$$H_3C - N \equiv C: \quad = \quad H:\overset{\overset{H}{..}}{\underset{\underset{H}{..}}{C}}:\overset{1}{N}:::\overset{2}{C}:$$

For carbon 1: $FC = 4 - \dfrac{8}{2} - 0 = 0$

For carbon 2: $FC = 4 - \dfrac{6}{2} - 2 = -1$

For nitrogen : $FC = 5 - \dfrac{8}{2} - 0 = +1$

2.8

$$\text{Formal charge (FC)} = \left[\begin{array}{c} \text{\# of valence} \\ \text{electrons} \end{array}\right] - \left[\begin{array}{c} \text{\# of bonding electrons} \\ \hline 2 \end{array}\right] - \left[\begin{array}{c} \text{\# nonbonding} \\ \text{electrons} \end{array}\right]$$

Methyl phosphate

For oxygen 1: $FC = 6 - \dfrac{4}{2} - 4 = 0$

For oxygen 2: $FC = 6 - \dfrac{4}{2} - 4 = 0$

For oxygen 3 : $FC = 6 - \dfrac{2}{2} - 6 = -1$

For oxygen 4: $FC = 6 - \dfrac{2}{2} - 6 = -1$

Oxygen atoms 3 and 4 each have a formal charge of –1, and oxygen atoms 1 and 2 have a formal charge of 0.

2.9 Try to locate the three-atom groupings that are present in resonance forms.

(a) These two structures represent resonance forms. The three-atom grouping (C–C double bond and an adjacent vacant *p* orbital) is pictured on the right.

(b) These two structures represent different compounds, not resonance structures.

2.10 Look for three-atom groupings that contain a multiple bond next to an atom with a *p* orbital. Exchange the positions of the bond and the electrons in the *p* orbital to draw the resonance form of each grouping.

(a) Methyl phosphate anion has 3 three-atom groupings and thus has 3 resonance forms.

Recall from Chapter 1 that phosphorus, a third-row element, can form more than four covalent bonds

(b)

(c)

(d)

2.11 When an acid loses a proton, the product is the conjugate base of the acid. When a base gains a proton, the product is the conjugate acid of the base.

$$H-NO_3 \;+\; :NH_3 \;\rightleftharpoons\; NO_3^- \;+\; NH_4^+$$

Acid   Base   Conjugate   Conjugate
                                  base   acid

2.12 Recall from Section 2.8 that a stronger acid has a smaller pK_a and a weaker acid has a larger pK_a. Accordingly, phenylalanine (pK_a = 1.83) is a stronger acid than tryptophan (pK_a = 2.83).

2.13 HO–H is a stronger acid than H$_2$N–H. Since H$_2$N$^-$ is a stronger base than HO$^-$, the conjugate acid of H$_2$N$^-$ (H$_2$N–H) is a weaker acid than the conjugate acid of HO$^-$ (HO–H).

2.14 Use Table 2.3 to find the strength of each acid. A reaction takes place as written if the stronger acid is the reactant.

(a)

$$H-CN \;+\; CH_3CO_2^-\,Na^+ \;\xrightarrow{\;?\;}\; Na^+\,{}^-CN \;+\; CH_3CO_2H$$

pK_a = 9.3   pK_a = 4.7
Weaker acid   Stronger acid

Remember that the lower the pK_a, the stronger the acid. Thus CH$_3$CO$_2$H, not HCN, is the stronger acid, and the above reaction will not take place to a significant extent in them direction written.

(b)

$$CH_3CH_2O-H + Na^+ \; ^-CN \xrightarrow{\;?\;} CH_3CH_2O^- \; Na^+ + HCN$$

$pK_a = 16$

Weaker acid

$pK_a = 9.3$

Stronger acid

Using the same reasoning as in part (a), we can see that the above reaction will not occur to a significant extent.

2.15

$pK_a = 19$

Stronger acid

$pK_a = 36$

Weaker acid

As written, the above reaction will take place to virtual completion due to the large difference in pK_a values.

2.16 Enter −9.31 into a calculator and use the INV LOG function to arrive at the answer $K_a = 4.9 \times 10^{-10}$.

2.17 Locate the electron pair(s) of the Lewis base and draw a curved arrow from the electron pair to the Lewis acid. The electron pair moves from the atom at the tail of the arrow (Lewis base) to the atom at the point of the arrow (Lewis acid).(Note: electron dots have been omitted from Cl⁻ to reduce clutter.)

(a)

(b)

2.18 (a)   The nitrogen on the left is more electron-rich and more basic. The indicated hydrogen is most electron-poor (bluest) and is most acidic.

more basic (red)

Imidazole

most acidic (blue)

(b)

2.19

only one –OH group

several –OH groups

* = polar group

Vitamin A

Vitamin C

Vitamin C is water-soluble (hydrophilic) because it has several polar –OH groups that can form hydrogen bonds with water. Vitamin A is fat-soluble (hydrophobic) because most of its atoms can't form hydrogen bonds with water.

Visualizing Chemistry

2.20 Naphthalene has three resonance forms.

2.21

HO—C(=O)

Ibuprofen

2.22 Electrostatic potential maps show that the electron-rich regions of the cis isomer lie on the same side of the double bond, leading to a net dipole moment. Because the electron-rich regions of the trans isomer are symmetrical about the double bond, the individual bond dipole moments cancel, and the isomer has no overall dipole moment.

net dipole moment   zero dipole moment

cis-1,2-Dichloroethylene   *trans*-1,2-Dichloroethylene

2.23 (a)   (b)

Adenine   Cytosine

Mechanism Problems

2.24

(a)

(b)

(c)

2.25

(a)

(b)

(c)

(d)

2.26

(a)

(b)

(c)

2.27

(a)

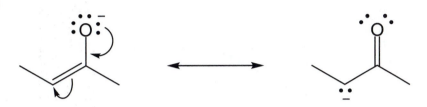

(b)

(c)

Additional Problems

Electronegativity and Dipole Moments

2.28 Use Figure 2.2 if you need help. The most electronegative element is starred.

(a) CH₂F*Cl   (b) *FCH₂CH₂CH₂Br   (c) H*OCH₂CH₂NH₂   (d) CH₃*OCH₂Li

2.29   *More polar*   *Less polar*

(a)   H₃C══Cl   Cl──Cl
          δ+   δ−

(b)   H══Cl   H₃C══H
          δ+   δ−   δ−   δ+

(c)   HO──CH₃   (CH₃)₃Si══CH₃
          δ−   δ+   δ+   δ−

(d)   Li══OH   H₃C══Li
          δ+   δ−   δ−   δ+

2.30 (a)   **(b)**   **(c)**   **(d)**

no dipole moment

2.31 (a)   In Section 2.2, we found that $\mu = Q \times r$. For a proton and an electron separated by 100 pm, $\mu = 4.80$ D. If the two charges are separated by 136 pm, $\mu = 6.53$ D.

(b)   Since the observed dipole moment is 1.08 D, the H–Cl bond has (1.08 D / 6.53 D) × 100 % = 16.5 % ionic character.

2.32   In phosgene, the individual bond polarities tend to cancel, but in formaldehyde, the bond polarities add to each other. Thus, phosgene has a smaller dipole moment than form aldehyde.

Phosgene   Formaldehyde

2.33   The magnitude of a dipole moment depends on both charge and distance between atoms. Fluorine is more electronegative than chlorine, but a C–F bond is shorter than a C–Cl bond. Thus, the dipole moment of CH₃F is smaller than that of CH₃Cl.

2.34   The observed dipole moment is due to the lone pair electrons on sulfur.

H₃C──S̈H

Formal Charges

2.35 To save space, molecules are shown as line-bond structures with lone pairs, rather than as electron-dot structures.

(a)

$(CH_3)_2\overset{..}{O}-BF_3$

Oxygen: $FC = 6 - \dfrac{6}{2} - 2 = +1$

Boron: $FC = 3 - \dfrac{8}{2} - 0 = -1$

(b)

$H_2\overset{..}{C}-\overset{1}{\overset{..}{N}}\equiv\overset{2}{N}:$

Carbon: $FC = 4 - \dfrac{6}{2} - 2 = -1$

Nitrogen 1: $FC = 5 - \dfrac{8}{2} - 0 = +1$

Nitrogen 2: $FC = 5 - \dfrac{6}{2} - 2 = 0$

(c)

$\begin{array}{c} CH_3 \\ | \\ H_2\overset{..}{C}-P-CH_3 \\ | \\ CH_3 \end{array}$

Carbon: $FC = 4 - \dfrac{6}{2} - 2 = -1$

Phosphorus: $FC = 5 - \dfrac{8}{2} - 0 = +1$

(d)

$H_2C=\overset{1}{N}=\overset{2}{\overset{..}{N}}:$

Carbon: $FC = 4 - \dfrac{8}{2} - 0 = 0$

Nitrogen 1: $FC = 5 - \dfrac{8}{2} - 0 = +1$

Nitrogen 2: $FC = 5 - \dfrac{4}{2} - 4 = -1$

(e)

$:\overset{1}{\overset{.}{O}}=\overset{2}{\overset{..}{O}}-\overset{3}{\overset{..}{O}}:$

Oxygen 1: $FC = 6 - \dfrac{4}{2} - 4 = 0$

Oxygen 2: $FC = 6 - \dfrac{6}{2} - 2 = +1$

Oxygen 3: $FC = 6 - \dfrac{2}{2} - 6 = -1$

(f)

Nitrogen: $FC = 5 - \dfrac{8}{2} - 0 = +1$

Oxygen: $FC = 6 - \dfrac{2}{2} - 6 = -1$

2.36 As in Problem 2.31, molecules are shown as line-bond structures with lone-pair electrons indicated. Only calculations for atoms with non-zero formal charge are shown.

(a)

Oxygen: $FC = 6 - \dfrac{2}{2} - 6 = -1$

Nitrogen: $FC = 5 - \dfrac{8}{2} - 0 = +1$

(b)

Nitrogen 1: $FC = 5 - \dfrac{4}{2} - 4 = -1$

Nitrogen 2: $FC = 5 - \dfrac{8}{2} - 0 = +1$

Nitrogen 3: $FC = 5 - \dfrac{6}{2} - 2 = 0$

(c)

Nitrogen 1: $FC = 5 - \dfrac{6}{2} - 2 = 0$

Nitrogen 2: $FC = 5 - \dfrac{8}{2} - 0 = +1$

Nitrogen 3: $FC = 5 - \dfrac{4}{2} - 4 = -1$

Resonance

2.37 Resonance forms do not differ in the position of nuclei. The two structures in (a) are not resonance forms because the positions of the carbon and hydrogen atoms outside the ring are different in the two forms.

The pairs of structures in parts (b), (c), and (d) represent resonance forms.

2.38

(a)

(b)

(c)

The last resonance structure is a minor contributor because its carbon lacks a complete electron octet.

(d)

(e)

2.39 The two structures are not resonance forms because the positions of the carbon atoms are different in the two forms.

Acids and Bases

2.40

2.41

The O–H hydrogen of acetic acid is more acidic than the C–H hydrogens. The –OH oxygen is electronegative, and, consequently, the –O–H bond is more strongly polarized than the –C–H bonds. In addition, the acetate anion is stabilized by resonance.

2.42

(a)

(b)

(c)

(d)

$$H\!:\!\overset{..}{\underset{..}{F}}\!:$$

(e)

$$H\!:\!\overset{H}{\underset{H}{C}}\!:\!\overset{..}{\underset{..}{S}}\!:\!\overset{H}{\underset{H}{C}}\!:\!H$$

(f)

$$:\!\overset{:\overset{..}{Cl}:}{\underset{:\overset{..}{Cl}:}{\underset{..}{Cl}:Ti:\overset{..}{Cl}}}\!:$$

The Lewis acids shown below can accept an electron pair either because they have a vacant orbital or because they can donate H^+. The Lewis bases have nonbonding electron pairs.

Lewis acids: $AlBr_3$, BH_3, HF, $TiCl_4$

Lewis bases: $CH_3CH_2\overset{..}{N}H_2$, $H_3C\!-\!\overset{..}{\underset{..}{S}}\!-\!CH_3$

2.43

(a) CH_3OH + H_2SO_4 ⇌ $CH_3OH_2^+$ + HSO_4^-
        stronger   stronger   weaker   weaker
          base   acid   acid   base

(b) CH_3OH + $NaNH_2$ ⇌ $CH_3O^-\,Na^+$ + NH_3
        stronger   stronger   weaker   weaker
          acid   base   base   acid

(c) $CH_3NH_3^+\,Cl^-$ + NaOH ⇌ CH_3NH_2 + H_2O + NaCl
        stronger   stronger   weaker   weaker
          acid   base   base   acid

2.44 The substances with the largest values of pK_a are the least acidic.

Least acidic ──────────────────────────▶ *Most acidic*

$$\underset{CH_3\overset{\overset{\displaystyle O}{\|}}{C}CH_3}{} \;<\; \text{⬡—OH} \;<\; CH_3\overset{\overset{\displaystyle O}{\|}}{C}CH_2\overset{\overset{\displaystyle O}{\|}}{C}CH_3 \;<\; CH_3\overset{\overset{\displaystyle O}{\|}}{C}OH$$

pK_a = 19.3 pK_a = 9.9 pK_a = 9 pK_a = 4.76

2.45 To react completely (> 99.9%) with NaOH, an acid must have a pK_a at least 3 units smaller than the pK_a of H_2O. Thus, all substances in the previous problem except acetone react completely with NaOH.

2.46 The stronger the acid (smaller pK_a), the weaker its conjugate base. Since NH_4^+ is a stronger acid than $CH_3NH_3^+$, CH_3NH_2 is a stronger base than NH_3.

2.47

$$pK_a = 15.7 \qquad pK_a \approx 18$$
stronger acid   weaker acid

The reaction takes place as written because water is a stronger acid than *tert*-butyl alcohol. Thus, a solution of potassium *tert*-butoxide in water can't be prepared.

2.48

2.49 (a)   Acetone: $K_a = 5 \times 10^{-20}$ (b)   Formic acid: $K_a = 1.8 \times 10^{-4}$

2.50 (a)   Nitromethane: $pK_a = 10.30$ (b)   Acrylic acid: $pK_a = 4.25$

2.51

$$\text{Formic acid} + H_2O \xrightleftharpoons{K_a} \text{Formate}^- + H_3O^+$$
$$[\,0.050\ M] \qquad\qquad\qquad [x] \qquad\qquad [x]$$

$$K_a = 1.8 \times 10^{-4} = \frac{x^2}{0.050 - x}$$

If you let $0.050 - x = 0.050$, then $x = 3.0 \times 10^{-3}$ and $pH = 2.52$. If you calculate x exactly using the quadratic equation, then $x = 2.9 \times 10^{-3}$ and $pH = 2.54$.

2.52 Only acetic acid will react with sodium bicarbonate. Acetic acid is the only substance in Problem 2.40 that is a stronger acid than carbonic acid.

General Problems

2.53 In maleic acid, the individual dipole moments add to produce a net dipole moment for the whole molecule. The individual dipole moments in fumaric acid cancel, resulting in a zero dipole moment.

Maleic acid   Fumaric acid

2.54 Sodium bicarbonate reacts with acetic acid to produce carbonic acid, which breaks down to form CO_2. Thus, bubbles of CO_2 indicate the presence of an acid stronger than carbonic acid, in this case acetic acid, as the pK_a values indicate. Phenol does not react with sodium bicarbonate.

2.55 Reactions (a) and (c) are reactions between Brønsted–Lowry acids and bases; the stronger acid and stronger base are identified. Reactions (b) and (d) occur between Lewis acids and bases.

(a)
$$CH_3OH \; + \; H^+ \longrightarrow CH_3\overset{+}{O}H_2$$
base   acid

(b)

$$CH_3CCH_3 \; + \; TiCl_4 \longrightarrow CH_3CCH_3$$
base   acid

(c)

acid   base

(d)

base   acid

2.56 Pairs (a) and (d) represent resonance structures; pairs (b) and (c) do not. For two structures to be resonance forms, all atoms must be in the same positions in all resonance forms.

2.57

(a)

$$H_3C-\overset{+}{N}\overset{:O:}{\diagup} \quad \longleftrightarrow \quad H_3C-\overset{+}{N}\overset{:O:^-}{\diagup}$$

(b)

$$\overset{..}{O}=\overset{+}{\overset{..}{O}}-\overset{..}{O}:^- \quad \longleftrightarrow \quad ^-:\overset{..}{O}-\overset{+}{\overset{..}{O}}=\overset{..}{O}$$

(c)

$$H_2C=\overset{+}{N}=\overset{..}{N}:^- \quad \longleftrightarrow \quad H_2\overset{..}{C}-\overset{+}{N}\equiv N:$$

2.58 The cation pictured can be represented by two resonance forms. Reaction with water can occur at either positively charged carbon, resulting in two products.

2.59

(a) $\delta- O$ $\delta+ C ^+$

(b) $\delta+ C$ $\delta- OH$

(c) $\delta- O$ $\delta+ C ^+ \delta- NH_2$

(d) $-C\equiv N$ $\delta+ \quad \delta-$

2.60

2.61

When phenol loses a proton, the resulting anion is stabilized by resonance. The methanol anion is not stabilized by resonance.

2.62

2.63 (a)   The central carbon of carbonate ion is sp^2 and trigonal planar. The three resonance forms contribute equally to the overall resonance hybrid. Thus, the molecule has no dipole.

(b)   The oxygen is sp³ hybridized and tetrahedral. Therefore, there is a dipole pointing towards the lone pairs.

(c)   The carbocation is sp² hybridized and trigonal planar. Thus, there is no dipole.

$\overset{+}{C}(CH_3)_3$

2.64 The equilibrium always favors the weaker acid/base pair (acid with the higher pK_a)

(a)

pK$_a$ 9.9   pK$_a$ 4.2

(b)

$CH_3CH_2CH_2OH$ + $\overset{\ominus}{N}H_2$ $\longrightarrow$ $CH_3CH_2CH_2\overset{\ominus}{O}$ + NH_3

pK$_a$ 16.1   pK$_a$ 36

The "a" in pKa here should be subscript.

(c)

$\overset{\ominus}{C}H_3$ + CH_3NO_2 $\rightleftharpoons$ CH_4 + $\overset{\ominus}{C}H_2NO_2$

pK$_a$ 10.3   pK$_a$ 60

The "a" in pKa here should be subscript.

2.65 (a)   London dispersion forces

(b)   Hydrogen bonding

(c)   Hydrogen bonding causes the carboxylic acids interact with one another and that force is stronger than the dispersion forces between acetic acid and oil.

2.66

(a)

(b)

(c)

(d)

2.67 Being more electronegative than carbon, the three chlorine atoms inductively remove electron density from the carbon atom to which they are all attached. This effect propagates down the chain, ultimately reducing the electron density on the oxygen. This makes the oxygen-hydrogen bond weaker and the molecule more acidic.

Major Topics Covered (with vocabulary:)

Atomic Structure:
atomic number   mass number   wave equation   orbital   shell   node   electron configuration

Chemical Bonding Theory:
covalent bond   Lewis structure   lone-pair electrons   line-bond structure   valence-bond theory
sigma (σ) bond   bond strength   bond length   molecular orbital theory   bonding MO
antibonding MO

Hybridization:
sp^3 hybrid orbital bond angle sp^2 hybrid orbital   pi (π) bond sp hybrid orbital

Polar covalent bonds:
polar covalent bond   electronegativity (EN)   electrostatic potential maps   inductive effect
dipole moment   formal charge   dipolar molecule

Resonance:
resonance form   resonance hybrid

Acids and Bases:
Brønsted-Lowry acid   Brønsted-Lowry base   conjugate acid   conjugate base   acidity constant
K_a pK_a organic acid   organic base   Lewis acid   Lewis base

Chemical Structures:
condensed structure   skeletal structure   space-filling models   ball-and-stick models

Types of Problems:

After studying these chapters you should be able to:

- Predict the ground state electronic configuration of atoms.
- Draw Lewis electron-dot structures of simple compounds.
- Predict and describe the hybridization of bonds in simple compounds.
- Predict bond angles and shapes of molecules.
- Predict the direction of polarity of a chemical bond, and predict the dipole moment of a simple compound.
- Calculate formal charge for atoms in a molecule.
- Draw resonance forms of molecules.
- Predict the relative acid/base strengths of Brønsted acids and bases.
- Predict the direction of Brønsted acid/base reactions.
- Calculate: pK_a from K_a, and vice versa.
              pH of a solution of a weak acid.
- Identify Lewis acids and bases.
- Draw chemical structures from molecular formulas, and vice versa.

Points to Remember:

*   In order for carbon, with valence shell electron configuration of $2s^2 2p^2$, to form four sp^3 hybrid orbitals, it is necessary that one electron be promoted from the $2s$ subshell to the $2p$ subshell. Although this promotion requires energy, the resulting hybrid orbitals are able to form stronger bonds, and compounds containing these bonds are more stable.

*   Assigning formal charge to atoms in a molecule is helpful in showing where the electrons in a bond are located. Even if a bond is polar covalent, in some molecules the electrons "belong" more to one of the atoms than the other. This "ownership" is useful for predicting the outcomes of chemical reactions, as we will see in later chapters.

*   Resonance structures are representations of the distribution of π and nonbonding electrons in a molecule. Electrons don't move around in the molecule, and the molecule doesn't change back and forth, from structure to structure. Rather, resonance structures are an attempt to show, by conventional line-bond drawings, the electron distribution of a molecule that can't be represented by any one structure.

*   As in general chemistry, acid-base reactions are of fundamental importance in organic chemistry. Organic acids and bases, as well as inorganic acids and bases, occur frequently in reactions, and large numbers of reactions are catalyzed by Brønsted acids and bases and Lewis acids and bases.

Self-Test:

A	B	C
Ricinine	**Oxaflozane**	**1,3,4-Oxadiazole**
(a toxic component of castor beans)	(an antidepressant)	

For **A** (ricinine) and **B** (oxaflozane): Add all missing electron lone pairs. Identify the hybridization of all carbons. Indicate the direction of bond polarity for all bonds with $\Delta EN \geq 0.5$. In each compound, which bond is the most polar? Convert **A** and **B** to molecular formulas.

Draw a resonance structure for **B**. Which atom (or atoms) of **B** can act as a Lewis base?

Add missing electron lone pairs to **C**. Is it possible to draw resonance forms for **C**? If so, draw at least one resonance form, and describe it.

Multiple Choice:

1. Which element has $4s^2 4p^2$ as its valence shell electronic configuration?
    (a) Ca (b) C   (c) Al   (d) Ge

2. Which compound (or group of atoms) has an oxygen with a +1 formal charge?
    (a) NO_3- (b) O_3 (c) acetone anion   (d) acetate anion

    The following questions involve these acids: (i) HW ($pK_a = 2$);   (ii) HX ($pK_a = 6$);
    (iii) HY ($pK_a = 10$);   (iv) HZ ($pK_a = 20$).

3. Which of the above acids react almost completely with water to form hydroxide ion?
    (a) none of them   (b) all of them   (c) HY and HZ   (d) HZ

4. The conjugate bases of which of the above acids react almost completely with water to
    form hydroxide ion?
    (a) none of them   (b) all of them   (c) HZ   (d) HY and HZ

5. If you want to convert HX to X^-, which bases can you use?
    (a) W^- (b) Y^- (c) Z^- (d) Y^- or Z^-

6. If you add equimolar amounts of HW, X^- and HY to a solution, what are the principal
    species in the resulting solution?
    (a) HW, HX, HY   (b) W^-, HX, HY   (c) HW, X^-, HY   (d) HW, HX, Y^-

7. What is the approximate pH difference between a solution of 1 M HX and a solution of
    1 M HY?
    (a) 2 (b) 3 (c) 4 (d) 6

8. If you wanted to write the structure of a molecule that shows carbon and hydrogen atoms
    as groups, without indicating many of the carbon-hydrogen bonds, you would draw a:
    (a) molecular formula (b) Kekulé structure (c) skeletal structure (d) condensed structure

9. Which of the following molecules has zero net dipole moment?

10. In which of the following bonds is carbon the more electronegative element?
    (a) C — Br   (b) C — I   (c) C — P   (d) C — S

Chapter 3 – Organic Compounds: Alkanes and Their Stereochemistry

Chapter Outline

I. Functional Groups (Section 3.1).

    A. Functional groups are groups of atoms within a molecule that have a characteristic chemical behavior.

    B. The chemistry of every organic molecule is determined by its functional groups.

    C. Functional groups described in this text can be grouped into three categories:

        1. Functional groups with carbon–carbon multiple bonds.

        2. Groups in which carbon forms a single bond to an electronegative atom.

        3. Groups with a carbon–oxygen double bond.

II. Alkanes (Sections 3.2–3.5).

    A. Alkanes and alkane isomers (Section 3.2).

        1. Alkanes are formed by overlap of carbon sp^3 orbitals.

        2. Alkanes are described as saturated hydrocarbons.

            a. They are hydrocarbons because they contain only carbon and hydrogen.

            b. They are saturated because all bonds are single bonds.

            c. The general formula for alkanes is C_nH_{2n+2}.

        3. For alkanes with four or more carbons, the carbons can be connected in more than one way.

            a. If the carbons are in a row, the alkane is a straight-chain alkane.

            b. If the carbon chain has a branch, the alkane is a branched-chain alkane.

        4. Alkanes with the same molecular formula can exist in different forms known as isomers.

            a. Isomers whose atoms are connected differently are constitutional isomers.

                i. Constitutional isomers are always different compounds with different properties but with the same molecular formula.

            b. Most alkanes can be drawn in many ways.

        5. Straight-chain alkanes are named according to the number of carbons in their chain.

    B. Alkyl groups (Section 3.3).

        1. An alkyl group is the partial structure that results from the removal of a hydrogen atom from an alkane.

            a. Alkyl groups are named by replacing the -ane of an alkane name by -yl.

            b. *n*-Alkyl groups are formed by removal of an end hydrogen atom of a straight chain alkane.

            c. Branched-chain alkyl groups are formed by removal of a hydrogen atom from an internal carbon.

                i. The prefixes *sec-* and *tert-* refer to the degree of substitution at the branching carbon atom.

        2. There are four possible degrees of alkyl substitution for carbon.

            a. A primary carbon is bonded to one other carbon.

            b. A secondary carbon is bonded to two other carbons.

  c. A tertiary carbon is bonded to three other carbons.

  d. A quaternary carbon is bonded to four other carbons.

  e. The symbol **R** refers to the rest of the molecule.

 3. Hydrogens are also described as primary, secondary and tertiary.

  a. Primary hydrogens are bonded to primary carbons (RCH_3).

  b. Secondary hydrogens are bonded to secondary carbons (R_2CH_2).

  c. Tertiary hydrogens are bonded to tertiary carbons (R_3CH).

C. Naming alkanes (Section 3.4).

 1. The system of nomenclature used in this book is the IUPAC system.
In this system, a chemical name has a locant, a prefix, a parent and a suffix.

  i. The locant shows the location of substituents and functional groups.

  ii. The prefix indicates the type of substituent or functional group.

  iii. The parent shows the number of carbons in the principal chain.

  iv. The suffix identifies the functional group family.

 2. Naming an alkane:

  a. Find the parent hydrocarbon.

   i. Find the longest continuous chain of carbons, and use its name as the parent name.

   ii. If two chains have the same number of carbons, choose the one with more branch points.

  b. Number the atoms in the parent chain.

   i. Start numbering at the end nearer the first branch point.

   ii. If branching occurs an equal distance from both ends, begin numbering at the end nearer the second branch point.

  c. Identify and number the substituents.

   i. Give each substituent a number that corresponds to its position on the parent chain.

   ii. Two substituents on the same carbon receive the same number.

  d. Write the name as a single word.

   i. Use hyphens to separate prefixes and commas to separate numbers.

   ii. Use the prefixes, *di-*, *tri-*, *tetra-* if necessary, but don't use them for alphabetizing.

  e. Name a complex substituent as if it were a compound, and set it off within parentheses.

   i. Some simple branched-chain alkyl groups have common names.

   ii. The prefix *iso* is used for alphabetizing, but *sec-* and *tert-* are not.

D. Properties of alkanes (Section 3.5).

 1. Alkanes are chemically inert to most laboratory reagents.

 2. Alkanes react with O_2 (combustion) and Cl_2 (substitution).

3. The boiling points and melting points of alkanes increase with increasing molecular weight.
    a. This effect is due to weak dispersion forces.
    b. The strength of these forces increases with increasing molecular weight.
4. Increased branching lowers an alkane's boiling point.

III. Conformations of straight-chain alkanes (Sections 3.6–3.7).
    A. Conformations of ethane (Section 3.6).
        1. Rotation about a single bond produces isomers that differ in conformation.
            a. These isomers (conformers) have the same connections of atoms and can't be isolated.
        2. These isomers can be represented in two ways:
            a. Sawhorse representations view the C–C bond from an oblique angle.
            b. Newman projections view the C–C bond end-on and represent the two carbons as a circle.
        3. There is a barrier to rotation that makes some conformers of lower energy than others.
            a. The lowest energy conformer (staggered conformation) occurs when all C–H bonds are as far from each other as possible.
            b. The highest energy conformer (eclipsed conformation) occurs when all C–H bonds are as close to each other as possible.
            c. Between these two conformations lie an infinite number of other conformations.
        4. The staggered conformation is 12 kJ/mol lower in energy than the eclipsed conformation.
            a. This energy difference is due to torsional strain from interactions between C–H bonding orbitals on one carbon and C–H antibonding orbitals on an adjacent carbon, which stabilize the staggered conformer.
            b. The torsional strain resulting from a single C–H interaction is 4.0 kJ/mol.
            c. The barrier to rotation can be represented on a graph of potential energy vs. angle of rotation (dihedral angle).
    B. Conformations of other alkanes (Section 3.7).
        1. Conformations of propane.
            a. Propane also shows a barrier to rotation that is 14 kJ/mol.
            b. The eclipsing interaction between a C–C bond and a C–H bond is 6.0 kJ/mol.
        2. Conformations of butane.
            a. Not all staggered conformations of butane have the same energy; not all eclipsed conformations have the same energy.
                i. In the lowest energy conformation (anti) the two large methyl groups are as far from each other as possible.
                ii. The eclipsed conformation that has two methyl–hydrogen interactions and a H–H interaction is 16 kJ/mol higher in energy than the anti conformation.
                iii. The conformation with two methyl groups 60° apart (gauche conformation) is 3.8 kJ/mol higher in energy than the anti conformation.

(a). This energy difference is due to steric strain – the repulsive interaction that results from forcing atoms to be closer together than their atomic radii allow.

iv. The highest energy conformations occur when the two methyl groups are eclipsed.

(a). This conformation is 19 kJ/mol less stable than the anti conformation. The value of a methyl–methyl eclipsing interaction is 11 kJ/mol.

b. The most favored conformation for any straight-chain alkane has carbon–carbon bonds in staggered arrangements and large substituents anti to each other.

c. At room temperature, bond rotation occurs rapidly, but a majority of molecules adopt the most stable conformation.

Solutions to Problems

3.1 Notice that certain functional groups have different designations if other functional groups are also present in a molecule. For example, a molecule containing a carbon–carbon double bond and no other functional group is an alkene; if other groups are present, the group is referred to as a carbon–carbon double bond. Similarly, a compound containing a benzene ring, and only carbon- and hydrogen-containing substituents, is an arene; if other groups present, the ring is labeled an aromatic ring.

(a)

sulfide   carboxylic acid

$CH_3SCH_2CH_2CHCOH$

NH$_2$ ← amine

Methionine

(b)

CO$_2$H

carboxylic acid

aromatic ring

Ibuprofen

(c)

ether   amide   C–C double bond

H_3C—O

CH$_3$

alcohol

HO

aromatic ring   Capsaicin

CH$_3$

3.2

(a) CH_3OH
Methanol

(b)   CH$_3$

Toluene

(c)   O
CH_3COH
Acetic acid

(d) CH_3NH_2
Methylamine

(e)   O
$CH_3CCH_2NH_2$
Aminoacetone

(f)

1,3-Butadiene

3.3

amine   H H   O—CH$_3$

H_3C—N—C—C=O   ester

H—C   C   H

H   C

H H   C–C double bond

Arecoline $C_8H_{13}NO_2$

3.4 We know that carbon forms four bonds and hydrogen forms one bond. Thus, draw all possible six-carbon skeletons and add hydrogens so that all carbons have four bonds. To draw all possible skeletons in this problem: (1) Draw the six-carbon straight-chain skeleton. (2) Draw a five-carbon chain, identify the different types of carbon atoms on the chain, and add a –CH₃ group to each of the different types of carbons, generating two skeletons. (3) Repeat the process with the four-carbon chain to give rise to the last two skeletons. Add hydrogens to the remaining carbons to complete the structures.

$$CH_3CH_2CH_2CH_2CH_2CH_3 \qquad CH_3CH_2CH_2\overset{\overset{\displaystyle CH_3}{|}}{C}HCH_3 \qquad CH_3CH_2\overset{\overset{\displaystyle CH_3}{|}}{C}HCH_2CH_3$$

$$CH_3CH_2\overset{\overset{\displaystyle CH_3}{|}}{\underset{\underset{\displaystyle CH_3}{|}}{C}}CH_3 \qquad CH_3\overset{\overset{\displaystyle CH_3}{|}}{C}H\overset{\overset{\displaystyle }{}}{C}H CH_3 \atop \underset{\displaystyle CH_3}{|}$$

3.5 **(a)** Nine isomeric esters of formula $C_5H_{10}O_2$ can be drawn. The procedure is described in Problem 3.4.

$$CH_3CH_2CH_2\overset{\overset{\displaystyle O}{||}}{C}OCH_3 \qquad CH_3\overset{\overset{\displaystyle O}{||}}{\underset{\underset{\displaystyle CH_3}{|}}{C}}HCOCH_3 \qquad CH_3CH_2\overset{\overset{\displaystyle O}{||}}{C}OCH_2CH_3$$

$$CH_3\overset{\overset{\displaystyle O}{||}}{C}OCH_2CH_2CH_3 \qquad CH_3\overset{\overset{\displaystyle O}{||}}{C}OCHCH_3 \atop \underset{\displaystyle CH_3}{|} \qquad HCOCH_2CH_2CH_2CH_3$$

$$HCOCHCH_2CH_3 \qquad HCOCH_2CHCH_3 \qquad HCOCCH_3$$

(b) Two isomers can be drawn.

$$CH_3CH_2CH_2C≡N \quad \text{and} \quad CH_3\overset{\overset{\displaystyle CH_3}{|}}{C}HC≡N$$

(c) Three isomers can be drawn.

$$CH_3CH_2SSCH_2CH_3 \qquad CH_3SSCH_2CH_2CH_3 \qquad CH_3SS\overset{}{C}HCH_3 \atop \underset{\displaystyle CH_3}{|}$$

3.6   **(a)**   Two alcohols have the formula C_3H_8O.

$$CH_3CH_2CH_2OH \quad \text{and} \quad CH_3\overset{\overset{\displaystyle OH}{|}}{C}HCH_3$$

(b)   Four bromoalkanes have the formula C_4H_9Br.

$$CH_3CH_2CH_2CH_2Br \quad CH_3CH_2\overset{\overset{\displaystyle Br}{|}}{C}HCH_3 \quad CH_3\overset{\overset{\displaystyle CH_3}{|}}{C}HCH_2Br \quad CH_3\overset{\overset{\displaystyle Br}{|}}{\underset{\underset{\displaystyle CH_3}{|}}{C}}CH_3$$

(c)   Four thioesters have the formula C_4H_8OS.

$$CH_3CH_2\overset{\overset{\displaystyle O}{||}}{C}SCH_3 \quad CH_3CH_2\overset{\overset{\displaystyle O}{||}}{S}CCH_3 \quad H\overset{\overset{\displaystyle O}{||}}{C}S\overset{\overset{\displaystyle CH_3}{|}}{C}HCH_3 \quad H\overset{\overset{\displaystyle O}{||}}{C}SCH_2CH_2CH_3$$

3.7

$$CH_3CH_2CH_2CH_2CH_2\text{---}\xi \quad CH_3CH_2CH_2\overset{}{C}H\text{---}\xi \quad CH_3CH_2\overset{}{C}H\text{---}\xi \quad CH_3CH_2\overset{}{C}HCH_2\text{---}\xi$$
$$\qquad\qquad\qquad\qquad\quad CH_3 \qquad\qquad\quad CH_2CH_3 \qquad\qquad\quad CH_3$$

$$CH_3\overset{}{C}HCH_2CH_2\text{---}\xi \quad CH_3CH_2\overset{\overset{\displaystyle CH_3}{|}}{C}\text{---}\xi \quad CH_3\overset{\overset{\displaystyle CH_3}{|}}{C}H\overset{}{C}H\text{---}\xi \quad CH_3\overset{\overset{\displaystyle CH_3}{|}}{C}CH_2\text{---}\xi$$
$$\quad CH_3 \qquad\qquad\qquad CH_3 \qquad\qquad\qquad CH_3 \qquad\qquad\qquad CH_3$$

3.8

(a)
$$\quad\;\; p$$
$$\quad\;\; CH_3$$
$$\quad\;\; |$$
$$CH_3CHCH_2CH_2CH_3$$
$$p \quad t \quad s \quad s \quad p$$

(b)
$$\quad\; p \quad t \quad p$$
$$\quad CH_3CHCH_3$$
$$\qquad\quad |$$
$$CH_3CH_2CHCH_2CH_3$$
$$p \quad s \quad t \quad s \quad p$$

(c)
$$\quad\; p \qquad\quad p$$
$$\quad CH_3 \qquad CH_3$$
$$\quad\; | \qquad\qquad |$$
$$CH_3CHCH_2\text{--}C\text{--}CH_3$$
$$p \quad t \quad s \quad |q \quad p$$
$$\qquad\qquad\quad CH_3$$
$$\qquad\qquad\qquad p$$

3.9   The carbons and the attached hydrogens have the same classification.

(a)
$$\quad\;\; p$$
$$\quad\;\; CH_3$$
$$\quad\;\; |$$
$$CH_3CHCH_2CH_2CH_3$$
$$p \quad t \quad s \quad s \quad p$$

(b)
$$\quad\; p \quad t \quad p$$
$$\quad CH_3CHCH_3$$
$$\qquad\quad |$$
$$CH_3CH_2CHCH_2CH_3$$
$$p \quad s \quad t \quad s \quad p$$

(c)
$$\quad\; p \qquad\quad p$$
$$\quad CH_3 \qquad CH_3$$
$$\quad\; | \qquad\qquad |$$
$$CH_3CHCH_2\text{--}C\text{--}CH_3$$
$$p \quad t \quad s \qquad | \quad p$$
$$\qquad\qquad\qquad CH_3$$
$$\qquad\qquad\qquad p$$

3.10

(a)
$$\qquad\quad CH_3$$
$$\quad\; t \;\; |$$
$$CH_3CHCHCH_3$$
$$\quad\; | \;\; t$$
$$\quad CH_3$$

(b)
$$\boxed{CH_3CHCH_3}$$
$$\qquad\quad |$$
$$CH_3CH_2CHCH_2CH_3$$

(c)
$$q \searrow CH_3$$
$$\qquad\quad |$$
$$CH_3CH_2CCH_3$$
$$\quad s \qquad |$$
$$\qquad\qquad CH_3$$

3.11 (a)

CH₃CH₂CH₂CH₂CH₃

$$\underset{\text{CH}_3}{\text{CH}_3\text{CH}_2\overset{|}{\text{C}}\text{HCH}_3}$$

$$\text{CH}_3\overset{\overset{\displaystyle\text{CH}_3}{|}}{\underset{\underset{\displaystyle\text{CH}_3}{|}}{\text{C}}}\text{CH}_3$$

Pentane   2-Methylbutane   2,2-Dimethylpropane

(b) *Step 1*: Find the longest continuous carbon chain and use it as the parent name. In (b), the chain is a pentane.
Step 2: Identify the substituents. In (b), both substituents are methyl groups.
Step 3: Number the substituents. In (b), the methyl groups are in the 2- and 3-positions.
Step 4: Name the compound. Remember that the prefix *di-* must be used when two are the same. The IUPAC name is 2,3-dimethylpentane.

$$\underset{\underset{\displaystyle\text{CH}_3}{|}}{\overset{\overset{\displaystyle\text{CH}_3}{|}}{\underset{5\ \ \ 4\ \ \ 3}{\text{CH}_3\text{CH}_2\text{CH}}\underset{2}{\text{CH}}}}-\underset{1}{\text{CH}_3} \qquad \text{2,3-Dimethylpentane}$$

(c)   (d)

$$\underset{\underset{\displaystyle\text{CH}_3}{|}}{(\text{CH}_3)_2\text{CHCH}_2\text{CHCH}_3}$$

$$\underset{\underset{\displaystyle\text{CH}_3}{|}}{(\text{CH}_3)_3\text{CCH}_2\text{CH}_2\text{CH}}$$

2,4-Dimethylpentane   2,2,5-Trimethylhexane

3.12 When you are asked to draw the structure corresponding to a given name, draw the parent carbon chain, attach the specified groups to the proper carbons, and fill in the remaining hydrogens.

(a)

$$\underset{\underset{\displaystyle\text{CH}_3}{|}}{\overset{\overset{\displaystyle\text{CH}_3}{|}}{\text{CH}_3\text{CH}_2\text{CH}_2\text{CH}_2\text{CH}_2\text{CHCH}}}\text{CH}_2\text{CH}_3$$

(b)

$$\underset{\underset{\displaystyle\text{CH}_3\ \ \text{CH}_2\text{CH}_3}{|\ \ \ \ \ |}}{\overset{\overset{\displaystyle\text{CH}_3}{|}}{\text{CH}_3\text{CH}_2\text{CH}_2\text{C}}}-\text{CHCH}_2\text{CH}_3$$

3,4-Dimethylnonane   3-Ethyl-4,4-dimethylheptane

(c)

$$\underset{\underset{\displaystyle\text{CH}_2\text{CH}_2\text{CH}_3}{|}}{\text{CH}_3\text{CH}_2\text{CH}_2\text{CH}_2\text{CHCH}_2\text{C(CH}_3)_3}$$

(d)

$$\underset{\underset{\displaystyle\text{CH}_3}{|}}{\overset{\overset{\displaystyle\text{CH}_3\ \ \ \text{CH}_3}{|\ \ \ \ \ |}}{\text{CH}_3\text{CHCH}_2\text{CCH}_3}}$$

2,2-Dimethyl-4-propyloctane   2,2,4-Trimethylpentane

3.13

CH₃CH₂CH₂CH₂CH₂—   CH₃CH₂CH₂CH—   CH₃CH₂CHCH₂—   CH₃CHCH₂CH₂—
                                    | | |
                                   CH₃   CH₃   CH₃

Pentyl   1-Methylbutyl   2-Methylbutyl   3-Methylbutyl

            CH₃   CH₃   CH₃
             | | |
CH₃CH₂C—   CH₃CHCH—   CH₃CCH₂—   CH₃CH₂CH—
             | | | |
            CH₃   CH₃   CH₃   CH₂CH₃

1,1-Dimethylpropyl   1,2-Dimethylpropyl   2,2-Dimethylpropyl   1-Ethylpropyl

3.14

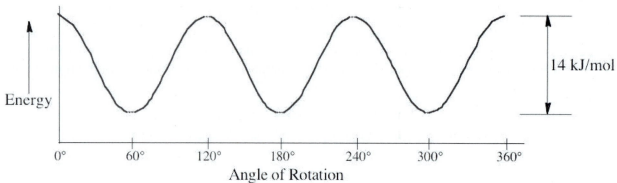

3,3,4,5-Tetramethylheptane

3.15 The graph shows the energy of a conformation as a function of angle of rotation.

14 kJ/mol

Energy

0°   60°   120°   180°   240°   300°   360°

Angle of Rotation

3.16

(a)

The *most stable* conformation is staggered and occurs at 60°, 180°, and 300°.

(b)

The *least stable* conformation is eclipsed and occurs at 0°, 120°, 240°, and 360°.

(c), (d)

3.17

This conformation of 2,3-dimethylbutane is the most stable because it is staggered and has the fewest $CH_3 \leftrightarrow CH_3$ gauche interactions.

3.18 The conformation is a staggered conformation in which the hydrogens on carbons 2 and 3 are 60° apart. Draw the Newman projection.

The Newman projection shows three gauche interactions, each of which has an energy cost of 3.8 kJ/mol. The total strain energy is 11.4 kJ/mol (3 × 3.8 kJ/mol).

Visualizing Chemistry

3.19

(a)

aromatic
ring

amine

Phenylalanine $C_9H_{11}NO_2$

(b)

amide

amine

aromatic
ring

Lidocaine $C_{14}H_{22}N_2O$

3.20

(a)

3,3,5-Trimethylheptane

(b)

3-Ethyl-2-methylpentane

(c)

2,2,4-Trimethylpentane

(d)

4-Isopropyl-2-methylheptane

3.21

In this conformation, all groups are staggered and the two methyl groups are 180° apart.

Additional Problems

Functional Groups

3.22

(a)

CH₂OH
↑
alcohol

NHCH₃

aromatic ring   amine

(b)

ketone

C–C double bond

(c)

amide

H
|
N
C=O
|
CH₃

aromatic ring

(d)

carboxylic acid

O
‖
CH₃CHCOH
|
NH₂
↑
amine

(e)

C–C double bond

ketone

O

C–C double bond

(f)

carboxylic acid chloride

Cl

C–C triple bond

C
‖
O

3.23 Different answers to this problem and to Problem 3.24 are acceptable.

(a)

O
‖
CH₃CH₂CCH₂CH₃

(b)

O
‖
CH₃CNHCH₂CH₃

(c)

O
‖
CH₃COCH₂CH₂CH₃

(d)

O
‖
C
H

(e)

O   O
‖   ‖
CH₃CCH₂COCH₂CH₃

(f) H₂NCH₂CH₂OH

3.24 For (a) and (h), only one structure is possible.

(a)

$$C_4H_8O: \quad CH_3\overset{\overset{\displaystyle O}{\|}}{C}CH_2CH_3$$

(b) C_5H_9N: $CH_3CH_2CH_2CH_2C{\equiv}N$

(c)

$$C_4H_6O_2: \quad H\overset{\overset{\displaystyle O}{\|}}{C}CH_2CH_2\overset{\overset{\displaystyle O}{\|}}{C}H$$

(d) $C_6H_{11}Br$: $CH_3CH_2CH{=}CHCH_2CH_2Br$

(e) C_6H_{14}: $CH_3CH_2CH_2CH_2CH_2CH_3$

(f)

C_6H_{12}:

(g) C_5H_8: $CH_3CH{=}CHCH{=}CH_2$

(h)

$$C_5H_8O: \quad H_2C{=}CH\overset{\overset{\displaystyle O}{\|}}{C}CH_2CH_3$$

3.25

(a)

$$\overset{\overset{\displaystyle O}{\|}}{\underset{}{C}}{\leftarrow}sp^2$$

(b) sp

$$\downarrow$$
$$-C{\equiv}N$$

(c)

$$\overset{\overset{\displaystyle O}{\|}}{\underset{OH}{C}}{\leftarrow}sp^2$$

3.26 (a)   Although it is stated that biacetyl contains no rings or carbon–carbon double bonds, it is obvious from the formula for biacetyl that some sort of multiple bond must be present. The structure for biacetyl contains two carbon–oxygen double bonds.

(b)   Ethyleneimine contains a three-membered ring.

(c)   Glycerol contains no multiple bonds or rings.

(a)

$$CH_3\overset{\overset{\displaystyle O}{\|}}{C}{-}\overset{\overset{\displaystyle O}{\|}}{C}CH_3$$
Biacetyl

$\left(\text{also possible: } H\overset{\overset{\displaystyle O}{\|}}{C}CH_2CH_2\overset{\overset{\displaystyle O}{\|}}{C}H \quad H\overset{\overset{\displaystyle O}{\|}}{C}CH_2\overset{\overset{\displaystyle O}{\|}}{C}CH_3\right)$

(b)

$$\overset{\displaystyle H}{\underset{}{\overset{|}{N}}}$$
$$H_2C{-}CH_2$$
Ethyleneimine

(c)

$$HOCH_2\overset{\overset{\displaystyle OH}{|}}{C}HCH_2OH$$
Glycerol

Isomers

3.27 (a)   Eighteen isomers have the formula C_8H_{18}. Three are pictured.

$$CH_3CH_2CH_2CH_2CH_2CH_2CH_2CH_3 \qquad CH_3CH_2CH_2\overset{\overset{\displaystyle CH_2CH_3}{|}}{C}HCH_2CH_3 \qquad CH_3\overset{\overset{\displaystyle CH_3}{|}}{\underset{\underset{\displaystyle CH_3}{|}}{C}}\!-\!\overset{\overset{\displaystyle CH_3}{|}}{\underset{\underset{\displaystyle CH_3}{|}}{C}}CH_3$$

(b)   Structures with the formula $C_4H_8O_2$ may represent esters, carboxylic acids or many other complicated molecules. Three possibilities:

$$CH_3CH_2CH_2\overset{\overset{\displaystyle O}{\|}}{C}OH \qquad CH_3CH_2\overset{\overset{\displaystyle O}{\|}}{C}OCH_3 \qquad HOCH_2CH\!=\!CHCH_2OH$$

3.28

$$CH_3CH_2CH_2CH_2CH_2CH_2CH_3$$
Heptane

$$CH_3CH_2CH_2CH_2\overset{\overset{\displaystyle CH_3}{|}}{C}HCH_3$$
2-Methylhexane

$$CH_3CH_2CH_2\overset{\overset{\displaystyle CH_3}{|}}{C}HCH_2CH_3$$
3-Methylhexane

$$CH_3CH_2CH_2\overset{\overset{\displaystyle CH_3}{|}}{\underset{\underset{\displaystyle CH_3}{|}}{C}}CH_3$$
2,2-Dimethylpentane

$$CH_3CH_2\overset{\overset{\displaystyle CH_3}{|}}{C}H\overset{\overset{\displaystyle CH_3}{|}}{\underset{\underset{\displaystyle CH_3}{|}}{C}}HCH_3$$
2,3-Dimethylpentane

$$CH_3\overset{\overset{\displaystyle CH_3}{|}}{C}HCH_2\overset{\overset{\displaystyle CH_3}{|}}{C}HCH_3$$
2,4-Dimethylpentane

$$CH_3CH_2\overset{\overset{\displaystyle CH_3}{|}}{\underset{\underset{\displaystyle CH_3}{|}}{C}}CH_2CH_3$$
3,3-Dimethylpentane

$$CH_3CH_2\overset{\overset{\displaystyle CH_2CH_3}{|}}{\underset{}{C}}HCH_2CH_3$$
3-Ethylpentane

$$CH_3\overset{\overset{\displaystyle CH_3}{|}}{C}H\!-\!\overset{\overset{\displaystyle CH_3}{|}}{\underset{\underset{\displaystyle CH_3}{|}}{C}}CH_3$$
2,2,3-Trimethylbutane

3.29 (a)

$$CH_3\overset{\overset{\displaystyle Br}{|}}{C}H\overset{}{\underset{\underset{\displaystyle CH_3}{|}}{C}}HCH_3$$
same

$$CH_3\overset{\overset{\displaystyle CH_3}{|}}{C}H\overset{}{\underset{\underset{\displaystyle Br}{|}}{C}}HCH_3$$
same

$$CH_3\overset{\overset{\displaystyle CH_3}{|}}{C}H\overset{}{\underset{\underset{\displaystyle Br}{|}}{C}}HCH_3$$
same

(b)

$$\underset{\substack{|\\ ^1CH_2OH\\ \text{different}}}{CH_3CH_2\overset{\overset{\displaystyle CH_3}{|}}{C}HCH_2\overset{\overset{\displaystyle }{}}{C}HCH_3}$$

$$\underset{\substack{|\\ CH_3\\ \text{same}}}{HOCH_2\overset{\overset{\displaystyle CH_2CH_3}{|}}{C}HCH_2\overset{\overset{\displaystyle }{}}{C}HCH_3}$$

$$\underset{\substack{\\ \text{same}}}{CH_3CH_2\overset{\overset{\displaystyle CH_3}{|}}{C}HCH_2\overset{\overset{\displaystyle CH_3}{|}}{C}HCH_2OH}$$

(c)

same   same   different

Give the number "1" to the carbon bonded to –OH, and count to find the longest chain containing the –OH group.

3.30 The isomers may be either alcohols or ethers.

$$CH_3CH_2CH_2CH_2-OH \qquad CH_3CH_2\overset{\overset{\displaystyle OH}{|}}{C}HCH_3 \qquad CH_3\overset{\overset{\displaystyle CH_3}{|}}{C}HCH_2-OH \qquad CH_3\overset{\overset{\displaystyle CH_3}{|}}{\underset{\underset{\displaystyle OH}{|}}{C}}CH_3$$

$$CH_3CH_2CH_2-OCH_3 \qquad CH_3\overset{\overset{\displaystyle OCH_3}{|}}{C}HCH_3 \qquad CH_3CH_2-OCH_2CH_3$$

3.31 First, draw all straight-chain isomers. Then proceed to the simplest branched structure.

(a)   There are four alcohol isomers with the formula $C_4H_{10}O$.

$$CH_3CH_2CH_2CH_2OH \qquad CH_3CH_2\overset{\overset{\displaystyle OH}{|}}{C}HCH_3 \qquad CH_3\overset{\overset{\displaystyle CH_3}{|}}{C}HCH_2OH \qquad CH_3\overset{\overset{\displaystyle OH}{|}}{\underset{\underset{\displaystyle CH_3}{|}}{C}}CH_3$$

(b)   There are 17 isomers of $C_5H_{13}N$. Nitrogen can be bonded to one, two or three alkyl groups.

CH₃CH₂CH₂CH₂CH₂NH₂

$$\underset{}{CH_3CH_2CH_2\overset{\overset{\displaystyle NH_2}{|}}{C}HCH_3}$$

$$CH_3CH_2\overset{\overset{\displaystyle NH_2}{|}}{C}HCH_2CH_3$$

(c)   There are 3 ketone isomers with the formula $C_5H_{10}O$.

(d)   There are 4 isomeric aldehydes with the formula $C_5H_{10}O$. Remember that the aldehyde functional group can occur only at the end of a chain.

(e)   There are 4 esters with the formula $C_4H_8O_2$.

(f)   There are 3 ethers with the formula $C_4H_{10}O$.

$CH_3CH_2OCH_2CH_3$ $CH_3OCH_2CH_2CH_3$ CH_3OCHCH_3
$\overset{\displaystyle CH_3}{|}$ (above the last)

3.32

(a) CH_3CH_2OH

(b) $CH_3\overset{\overset{\displaystyle CH_3}{|}}{\underset{\underset{\displaystyle CH_3}{|}}{C}}C\equiv N$

(c) $CH_3\overset{\overset{\displaystyle SH}{|}}{C}HCH_3$

(d) $CH_3\overset{\overset{\displaystyle OH}{|}}{C}HCH_2OH$

(e) $CH_3\overset{\overset{\displaystyle CH_3}{|}}{C}HOCH_3$

(f) $CH_3\overset{\overset{\displaystyle CH_3}{|}}{\underset{\underset{\displaystyle CH_3}{|}}{C}}CH_3$

Naming Compounds

3.33

$CH_3CH_2CH_2CH_2CH_2Br$

1-Bromopentane

$CH_3CH_2CH_2\overset{\overset{\displaystyle Br}{|}}{C}HCH_3$

2-Bromopentane

$CH_3CH_2\overset{\overset{\displaystyle Br}{|}}{C}HCH_2CH_3$

3-Bromopentane

3.34

$CH_3\overset{\overset{\displaystyle CH_3}{|}}{C}HCH_2CH_2\overset{\overset{\displaystyle CH_3}{|}}{C}HCH_2Cl$

1-Chloro-2,5-dimethylhexane

$CH_3\overset{\overset{\displaystyle CH_3}{|}}{C}HCH_2CH_2\overset{\overset{\overset{\displaystyle CH_3}{|}}{C}}{\underset{\underset{\displaystyle Cl}{|}}{}}CH_3$

2-Chloro-2,5-dimethylhexane

$CH_3\overset{\overset{\displaystyle CH_3}{|}}{C}HCH_2\overset{\overset{\displaystyle CH_3}{|}}{C}H\overset{}{\underset{\underset{\displaystyle Cl}{|}}{C}}HCH_3$

3-Chloro-2,5-dimethyl-
hexane

3.35

(a) $CH_3CH_2CH_2CH_2CH_2\overset{\overset{\displaystyle CH_3}{|}}{C}HCH_3$

2-Methylheptane

(b) $CH_3CH_2\overset{}{\underset{\underset{\displaystyle CH_2CH_3}{|}}{C}}H\!-\!CH_2\overset{\overset{\displaystyle CH_3}{|}}{\underset{\underset{\displaystyle CH_3}{|}}{C}}CH_3$

4-Ethyl-2,2-dimethylhexane

(c) $CH_3CH_2CH_2CH_2\overset{\overset{\displaystyle CH_3}{|}}{\underset{\underset{\displaystyle CH_2CH_3}{|}}{C}}\!-\!\overset{\overset{\displaystyle CH_3}{|}}{C}HCH_2CH_3$

4-Ethyl-3,4-dimethyloctane

(d) $CH_3CH_2CH_2\overset{\overset{\displaystyle CH_3}{|}}{\underset{\underset{\displaystyle CH_3}{|}}{C}}CH_2\overset{\overset{\displaystyle CH_3}{|}}{C}HCH_3$

2,4,4-Trimethylheptane

(e)

$$CH_3CH_2CH_2CH_2CHCH_2C \underset{\underset{CH_2CH_3}{|}}{\overset{\overset{CH_3}{|}}{}} \underset{}{\overset{\overset{CH_2CH_3\ CH_3}{|\ \ \ \ |}}{}} CHCH_3$$

3,3-Diethyl-2,5-dimethylnonane

(f)

$$CH_3CH_2CH_2CHCHCH_2CH_3 \overset{\overset{CH_3CHCH_3}{|}}{\underset{\underset{CH_3}{|}}{}}$$

4-Isopropyl-3-methylheptane

3.36

(a)

$$CH_3CHCH_3 \overset{\overset{CH_3}{|}}{}$$

2-Methylpropane

(b)

$$CH_3CCH_3 \overset{\overset{CH_3}{|}}{\underset{\underset{CH_3}{|}}{}}$$

2,2-Dimethylpropane

(c) $CH_3CH_2CH_2CH_2CH_2CH_3$

Hexane

3.37

(a)

$$CH_3CHCH_3 \overset{\overset{CH_3}{|}}{}$$

2-Methylpropane

(b) CH_3CH_3

Ethane

3.38

(a)

$$CH_3CHCH_2CH_2CH_3 \overset{\overset{CH_3}{|}}{}$$

2-Methylpentane

(b)

$$CH_3CH_2CCH_3 \overset{\overset{CH_3}{|}}{\underset{\underset{CH_3}{|}}{}}$$

2,2-Dimethylbutane

(c)

$$CH_3CHCCH_2CH_2CH_3 \overset{\overset{H_3C\ \ CH_3}{|\ \ \ \ |}}{\underset{\underset{CH_3}{|}}{}}$$

2,3,3-Trimethylhexane

(d)

$$CH_3CH_2CHCH_2CH_2CHCH_3 \overset{\overset{CH_2CH_3\ \ \ CH_3}{|\ \ \ \ \ \ \ \ \ \ |}}{}$$

5-Ethyl-2-methylheptane

(e)

$$CH_3CH_2CH_2CHCH_2CCH_3 \overset{\overset{CH_3\ \ \ \ CH_2CH_3}{|\ \ \ \ \ \ \ \ \ |}}{\underset{\underset{CH_3}{|}}{}}$$

3,3,5-Trimethyloctane

(f)

$$CH_3C - CCH_2CH_2CH_3 \overset{\overset{H_3C\ \ \ CH_3}{|\ \ \ \ \ |}}{\underset{\underset{H_3C\ \ \ CH_3}{|\ \ \ \ \ |}}{}}$$

2,2,3,3-Tetramethylhexane

3.39

$$CH_3CH_2CH_2CH_2CH_2CH_3$$
Hexane

$$\overset{\displaystyle CH_3}{\underset{\displaystyle |}{CH_3CH_2CH_2CHCH_3}}$$
2-Methylpentane

$$\overset{\displaystyle CH_3}{\underset{\displaystyle |}{CH_3CH_2CHCH_2CH_3}}$$
3-Methylpentane

$$\overset{\displaystyle CH_3}{\underset{\displaystyle |}{\underset{\displaystyle CH_3}{\underset{\displaystyle |}{CH_3CH_2CCH_3}}}}$$
2,2-Dimethylbutane

$$\overset{\displaystyle CH_3}{\underset{\displaystyle |}{\underset{\displaystyle CH_3}{\underset{\displaystyle |}{CH_3CHCHCH_3}}}}$$
2,3-Dimethylbutane

3.40

	Structure and Correct Name	Error
(a)	CH₂CH₃ ⁸ ... CH₃ CH₃CHCH₂CH₂CH₂CCH₃ |1 CH₃ Correct name: 2,2,6-Trimethyloctane	The longest chain is an octane and has only methyl branches.
(b)	CH₃ CH₃CHCHCH₂CH₂CH₃ 1 ... 6 CH₂CH₃ Correct name: 3-Ethyl-2-methylhexane	The longest chain is a hexane. Numbering should start from the opposite end of the carbon chain, nearer the first branch.
(c)	CH₃ CH₃CH₂C—CHCH₂CH₃ 1 ... 6 CH₃ CH₂CH₃ Correct name: 4-Ethyl-3,3-dimethylhexane	Numbering should start from the opposite end of the carbon chain. See step 2(b) in Section 3.4.
(d)	CH₃ CH₃ CH₃CH₂CH—CCH₂CH₂CH₂CH₃ 1 ... 8 CH₃ Correct name: 3,4,4-Trimethyloctane	Numbering should start from the opposite end of the carbon chain.
(e)	CH₃ CH₃CH₂CH₂CHCH₂CHCH₃ 8 ... CH₃CHCH₃ 1 Correct name: 2,3,5-Trimethyloctane	The longest chain is an octane.

3.41

(a)

$$CH_3CCH_2CCH_2CH_3$$

with substituents CH_3, CH_2CH_3 on top and CH_3, CH_2CH_3 on bottom

4,4-Diethyl-2,2-dimethylhexane

(b)

$$CH_3CH_2CH_2CH_2CH_2CHCH_2CH_2CH_2CH_2CH_3$$

with substituent $CH_2CH_2CHCH_3$ and CH_3

6-(3-Methylbutyl)-undecane

Remember that you must choose an alkane whose principal chain is long enough so that the substituent does not become part of the principal chain.

Conformations

3.42 (a), (b)

$$CH_3CH_2 - CHCH_3$$

with substituent CH_3, arrow pointing to C2–C3 bond

2-Methylbutane

3.8 kJ/mol — Most stable conformation

6.0 kJ/mol, 4.0 kJ/mol, 11.0 kJ/mol — Least stable comformation

The energy difference between the two conformations is (11.0 + 6.0 + 4.0) kJ/mol − 3.8 kJ/mol = 17.2 kJ/mol.

(c)   Consider the least stable conformation to be at zero degrees. Keeping the front of the projection unchanged, rotate the back by 60° to obtain each conformation.

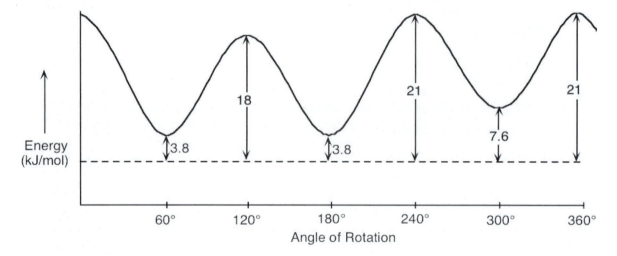

at 60°: energy = 3.8 kJ/mol   at 120°: energy = 18.0 kJ/mol   at 180°: energy = 3.8 kJ/mol

at 240°: energy = 21.0 kJ/mol   at 300°: energy = 7.6 kJ/mol

Use the lowest energy conformation as the energy minimum. The highest energy conformation is 17.2 kJ/mol higher in energy than the lowest energy conformation.

3.43 Each CH₃↔CH₃ gauche interaction has a value of 3.8 kJ/mol.

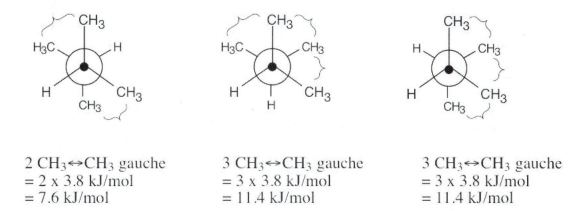

2 CH₃↔CH₃ gauche
= 2 x 3.8 kJ/mol
= 7.6 kJ/mol

3 CH₃↔CH₃ gauche
= 3 x 3.8 kJ/mol
= 11.4 kJ/mol

3 CH₃↔CH₃ gauche
= 3 x 3.8 kJ/mol
= 11.4 kJ/mol

3.44 Since we are not told the values of the interactions for 1,2-dibromoethane, the diagram can only be qualitative.

The anti conformation is at 180°.
The gauche conformations are at 60°, 300°.

3.45 The eclipsed conformation at 0° rotation has the largest dipole moment but is a high energy that is present in low abundance. The anti conformation has no net dipole because the polarities of the individual bonds cancel. The gauche conformation, however, has a dipole moment. Because the observed dipole moment is 1.0 D at room temperature, a mixture of gauche and anti conformations must be present.

3.46 The best way to draw pentane is to make a model and to copy it onto the page. A model shows the relationship among atoms, and its drawing shows how these relationships in two dimensions. From your model, you should be able to see that all atoms are staggered in the drawing.

3.47

General Problems

3.48

(a) $CH_3CH_2CH_2CH_2CH_2Br$ (b)

(c)
$$CH_3CHC\equiv N$$
with CH_3 above

(d)

(e)   There are no aldehyde isomers. However, the structure below is a ketone isomer.

$$CH_3CCH_3$$
with O double bonded above

(f)

3.49 (a)   Because malic acid has two $-CO_2H$ groups, the formula for the rest of the molecule is C_2H_4O. Possible structures for malic acid are:

primary alcohol   secondary alcohol   tertiary alcohol

(b)   Because only one of these compounds (the second one) is also a secondary alcohol, it must be malic acid.

3.50 To solve this type of problem, read the problem carefully, word for word. Then try to interpret parts of the problem. For example:

1)   Formaldehyde is an aldehyde,

2)   It trimerizes – that is, 3 formaldehydes come together to form a compound $C_3H_6O_3$. Because no atoms are eliminated, all of the original atoms are still present.

3)   There are no carbonyl groups. This means that trioxane cannot contain any –C=O functional groups. If you look back to Table 3.1, you can see that the only oxygen-containing functional groups that can be present are either ethers or alcohols.

4)   A monobromo derivative is a compound in which one of the –H's has been replaced by a –Br. Because only one monobromo derivative is possible, we know that there can only be one type of hydrogen in trioxane. The only possibility for trioxane is:

Trioxane

3.51 The highest energy conformation of bromoethane has a strain energy of 15 kJ/mol. Because this includes two H–H eclipsing interactions of 4.0 kJ/mol each, the value of an H–Br eclipsing interaction is 15 kJ/mol – 2(4.0 kJ/mol) = 7.0 kJ/mol.

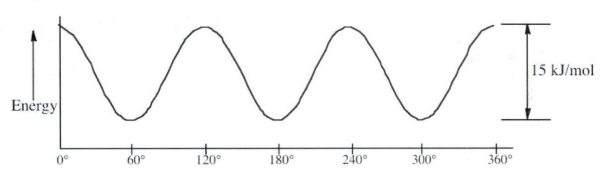

3.52

	Most stable	Strain enegy	Least stable	Strain energy

(a)

* 3.8 kJ/mol

21 kJ/mol

(b)

7.6 kJ/mol

23 kJ/mol

(c)

7.6 kJ/mol

26 kJ/mol

(d)
15.2 kJ/mol

** 28 kJ/mol

*most stable overall (least strain)
**least stable overall (most strain)

3.53

Zocor   C$_{25}$H$_{38}$O$_5$ Pravachol   C$_{23}$H$_{36}$O$_7$

A	alcohol	**B**	ester
C	carboxylic acid	**D**	C–C double bond (alkene)

The carboxylic acid and alcohol groups in Pravachol are an ester (lactone) group in Zocor. The alcohol at the bottom left of Pravachol is a methyl group in Zocor.

3.54 A puckered ring allows all the bonds in the ring to have a nearly tetrahedral bond angle. (If the ring were flat, C–C–C bond angles would be 120°.) Also, a puckered conformation relieves strain due to eclipsed hydrogens.

3.55

In one of the 1, 2-dimethylcyclohexanes the two methyl groups are on the same side of the ring, and in the other isomer the methyl groups are on opposite sides.

Chapter 4 – Organic Compounds: Cycloalkanes and Their Stereochemistry

Chapter Outline

I. Cycloalkanes – alicyclic compounds – (Sections 4.1–4.2).
    A. Cycloalkanes have the general formula C_nH_{2n}, if they have one ring.
    B. Naming cycloalkanes (Section 4.1).
        1. Find the parent.
            a. If the number of carbon atoms in the ring is larger than the number in the largest substituent, the compound is named as an alkyl-substituted cycloalkane.
            b. If the number of carbon atoms in the ring is smaller than the number in the largest substituent, the compound is named as a cycloalkyl-substituted alkane.
        2. Number the substituents.
            a. Start at a point of attachment and number the substituents so that the second substituent has the lowest possible number.
            b. If necessary, proceed to the next substituent until a point of difference is found.
            c. If two or more substituents might potentially receive the same number, number them by alphabetical priority.
            d. Halogens are treated in the same way as alkyl groups.
    C. Cis–trans isomerism in cycloalkanes (Section 4.2).
        1. Unlike open-chain alkanes, cycloalkanes have much less rotational freedom.
            a. Very small rings are rigid.
            b. Large rings have more rotational freedom.
        2. Cycloalkanes have a "top" side and a "bottom" side.
            a. If two substituents are on the same side of a ring, the ring is cis-disubstituted.
            b. If two substituents are on opposite sides of a ring, the ring is trans-disubstituted.
        3. Substituents in the two types of disubstituted cycloalkanes are connected in the same order but differ in spatial orientation.
            a. These cycloalkenes are stereoisomers that are known as cis–trans isomers.
            b. Cis–trans isomers are stable compounds that can't be interconverted.
II. Conformations of cycloalkanes (Sections 4.3–4.9).
    A. General principles (Section 4.3).
        1. Ring strain.
            a. A. von Baeyer suggested that rings other than those of 5 or 6 carbons were too strained to exist.
            b. This concept of angle strain is true for smaller rings, but larger rings can be easily prepared.
        2. Heats of combustion of cycloalkanes.
            a. To measure strain, it is necessary to measure the total energy of a compound and compare it to a strain-free reference compound.
            b. Heat of combustion measures the amount of heat released when a compound is completely burned in oxygen.
               i. The more strained the compound, the higher the heat of combustion.
              ii. Strain per CH_2 unit can be calculated and plotted as a function of ring size.

      c. Graphs show that only small rings have serious strain.

   3. The nature of ring strain.

      a. Rings tend to adopt puckered conformations.

      b. Several factors account for ring strain.

         i. Angle strain occurs when bond angles are distorted from their normal values.

        ii. Torsional strain is due to eclipsing of bonds.

       iii. Steric strain results when atoms approach too closely.

B. Conformations of small rings (Section 4.4).

   1. Cyclopropane.

      a. Cyclopropane has bent bonds.

      b. Because of bent bonds, cyclopropane is more reactive than other cycloalkanes.

   2. Cyclobutane.

      a. Cyclobutane has less angle strain than cyclopropane but has more torsional strain.

      b. Cyclobutane has almost the same total strain as cyclopropane.

      c. Cyclobutane is slightly bent to relieve torsional strain, but this increases angle strain.

   3. Cyclopentane

      a. Cyclopentane has little angle strain but considerable torsional strain.

      b. To relieve torsional strain, cyclopentane adopts a puckered conformation.

         i. In this conformation, one carbon is bent out of plane; hydrogens are nearly staggered.

C. Conformations of cyclohexane (Sections 4.5–4.8).

   1. Chair cyclohexane (Section 4.5).

      a. The chair conformation of cyclohexane is strain-free.

      b. In a standard drawing of cyclohexane, the lower bond is in front, and the upper bond is in back.

      c. The twist-boat conformation of cyclohexane has little angle strain but experiences both steric strain and torsional strain.

   2. Axial and equatorial bonds in cyclohexane (Section 4.6).

      a. There are two kinds of positions on a cyclohexane ring.

         i. Six axial hydrogens are perpendicular to the plane of the ring.

        ii. Six equatorial hydrogens are roughly in the plane of the ring.

      b. Each carbon has one axial hydrogen and one equatorial hydrogen.

      c. Each side of the ring has alternating axial and equatorial hydrogens.

      d. All hydrogens on the same face of the ring are cis.

   3. Conformational mobility of cyclohexanes.

      a. Different chair conformations of cyclohexanes interconvert by a ring-flip.

      b. After a ring-flip, an axial bond becomes an equatorial bond, and vice versa.

      c. The energy barrier to interconversion is 45 kJ/mol, making interconversion rapid at room temperature.

4. Conformations of monosubstituted cyclohexanes (Section 4.7).
    a. Both conformations aren't equally stable at room temperature.
        i.   In methylcyclohexane, 95% of molecules have the methyl group in the equatorial position.
    b. The energy difference is due to 1, 3-diaxial interactions.
        i.   These interactions, between an axial group and a ring hydrogen two carbons away, are due to steric strain.
        ii. They are the same interactions as occur in gauche butane.
    c. Axial methylcyclohexane has two gauche interactions that cause it to be 7.6 kJ/mol less stable than equatorial methylcyclohexane.
    d. All substituents are more stable in the equatorial position.
        i.   The size of the strain depends on the nature and size of the group.
5. Conformations of disubstituted cyclohexanes (Section 4.8).
    a. In *cis*-1,2-dimethylcyclohexane, one methyl group is axial and one is equatorial in both chair conformations, which are of equal energy.
    b. In *trans*-1,2-dimethylcyclohexane, both methyl groups are either both axial or both equatorial.
        i.   The conformation with both methyl groups axial is 15.2 kJ/mol less stable than the conformation with both groups equatorial due to 1,3 diaxial interactions.
        ii. The trans isomer exists almost exclusively in the diequatorial conformation.
    c. This type of conformational analysis can be carried out for most substituted cyclohexanes.
D. Conformations of polycyclic (fused-ring) molecules (Section 4.9).
    1. Decalin has two rings that can be either cis-fused or trans-fused.
        a. The two decalins are nonconvertible.
    2. Steroids have four fused rings.
    3. Bicyclic ring systems have rings that are connected by bridges.
        a. In norbornane, the six-membered ring is locked into a boat conformation.

Solutions to Problems

4.1 The steps for naming a cycloalkane are very similar to the steps used for naming an open-chain alkane.

Step 1: Name the parent cycloalkane. In (a), the parent is cyclohexane. If the compound has an alkyl substituent with more carbons than the ring size, the compound is named as a cycloalkyl-substituted alkane, as in (c).

Step 2: Identify the substituents. In (a), both substituents are methyl groups.

Step 3: Number the substituents so that the second substituent receives the lowest possible number. In (a), the substituents are in the 1- and 4- positions.

Step 4: Name the compound. If two different alkyl groups are present, cite them alphabetically. Halogen substituents follow the same rules as alkyl substituents.

(a)

CH₃

CH₃

1,4-Dimethylcyclohexane

(b)

CH₂CH₂CH₃

CH₃

1-Methyl-3-propylcyclopentane

(c)

3-Cyclobutylpentane

(d)

CH₂CH₃

Br

1-Bromo-4-ethylcyclodecane

(e)

CH₃

CH(CH₃)₂

1-Isopropyl-2-methylcyclohexane

(f)

Br

CH₃

C(CH₃)₃

4-Bromo-1-*tert*-butyl-
2-methylcycloheptane

4.2 To draw a substituted cycloalkane, simply draw the ring and attach substituents in the specified positions. The structure in (b) is named as a cyclobutyl-substituted alkane because the alkyl chain has more carbons than the ring.

(a)

CH₃

CH₃

1,1-Dimethylcyclooctane

(b)

3-Cyclobutylhexane

(c)

(d)

1,2-Dichlorocyclopentane

1,3-Dibromo-5-methylcyclohexane

4.3

3-Ethyl-1,1-dimethylcyclopentane

4.4   Two substituents are cis if they both have either dashed or wedged bonds. The substituents are trans if one has a wedged bond and the other has a dashed bond.

(a)

trans-1-Chloro-4-methylcyclohexane

(b)

cis-1-Ethyl-3-methylcycloheptane

4.5

(a)

trans-1-Bromo-3-methylcyclohexane

(b)

cis-1,2-Dimethylcyclobutane

(c)

trans-1-*tert*-Butyl-2-ethylcyclohexane

4.6

 CO$_2$H

CH$_3$

 Prostaglandin F$_{2\alpha}$

The two hydroxyl groups are cis because they both point behind the plane of the page (both dashed bonds). The carbon chains have a trans relationship (one is dashed and the other is wedged).

4.7

(a)

cis-1,2-Dimethylcyclopentane

(b)

cis-1-Bromo-3-methylcyclobutane

4.8

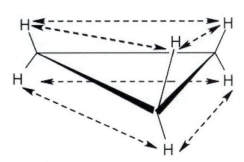

All hydrogen atoms on the same side of the cyclopropane ring are eclipsed by neighboring hydrogens. If we draw each hydrogen–hydrogen interaction, we count six eclipsing interactions, three on each side of the ring. Since each of these interactions costs 4.0 kJ/mol, all six cost 24.0 kJ/mol. 24 kJ/mol ÷115 kJ/mol x 100% = 21% of the total strain energy of cyclopropane is due to torsional strain.

4.9

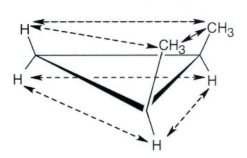

Eclipsing interaction	Energy cost (kJ/mol)	*cis isomer*		*trans isomer*	
		# of interactions	Total energy cost (kJ/mol)	# of interactions	Total energy cost (kJ/mol)
H ↔ H	4.0	3	12.0	2	8.0
H ↔ CH₃	6.0	2	12.0	4	24.0
CH₃ ↔ CH₃	11	1	11	0	0
			35		32

The added energy cost of eclipsing interactions causes *cis*-1,2-dimethylcyclopropane to be of higher energy and to be less stable than the trans isomer.

4.10

If cyclopentane were planar, it would have ten hydrogen–hydrogen interactions with a total energy cost of 40 kJ/mol (10 x 4.0 kJ/mol) . The measured total strain energy of 26 kJ/mol indicates that 14 kJ/mol of eclipsing strain in cyclopentane (35%) has been relieved by puckering.

4.11

(a)   (b)

The methyl groups are farther apart in the more stable conformation of *cis*-1,3-dimethylcyclobutane.

4.12 Use the technique in Section 4.5 to draw the cyclohexane ring. Figure 4.10 shows how to attach axial (a) and equatorial (e) bonds.

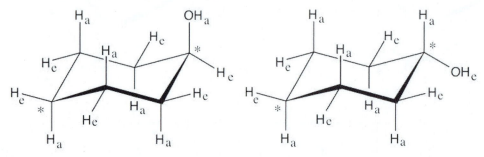

The conformation with –OH in the equatorial position is more stable.
Note: The starred ring carbons lie in the plane of the paper.

4.13 In trans-1,4-disubstituted cyclohexanes, the methyl substituents are either both axial or both equatorial.

trans-1,4-Dimethylcyclohexane

4.14 In a ring-flip, an axial substituent becomes equatorial, and an equatorial substituent becomes axial.

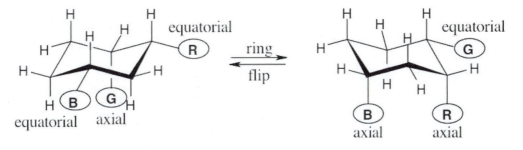

4.15 Table 4.1 shows that an axial hydroxyl group causes 2 x 2.1 kJ/mol of steric strain. Thus, the energy difference between axial and equatorial cyclohexanol is 4.2 kJ/mol.

4.16 There is very little energy difference between an axial and an equatorial cyano group because the small linear cyano group takes up very little room and produces practically no 1,3-diaxial interactions.

Cyclohexanecarbonitrile

4.17 Table 4.1 shows that an axial bromine causes 2 × 1.0 kJ/mol of steric strain. Thus, the energy difference between axial and equatorial bromocyclohexane is 2.0 kJ/mol. According to Figure 4.12, this energy difference corresponds approximately to a 75:25 ratio of more stable:less stable conformer. Thus, 75% of bromocyclohexane molecules are in the equatorial conformation, and 25% are in the axial conformation at any given moment.

4.18 Draw the two chair conformations of each molecule, and look for gauche and 1,3-diaxial interactions. Use Table 4.1 to estimate the values of the interactions. Calculate the total strain; the conformation with the smaller value for strain energy is more stable.

(a)

trans-1-Chloro-3-methylcyclohexane

2 (H ↔ CH₃) = 7.6 kJ/mol   2 (H ↔ Cl) = 2.0 kJ/mol

The second conformation is more stable than the first.

(b)

cis-1-Ethyl-2-methylcyclohexane

one CH₃ ↔ CH₂CH₃ gauche   one CH₃ ↔ CH₂CH₃ gauche
interaction = 3.8 kJ/mol   interaction = 3.8 kJ/mol
2 (H–CH₂CH₃) = 8.0 kJ/mol   2 (H–CH₃) = 7.6 kJ/mol
_____   _____
Total = 11.8 kJ/mol   Total = 11.4 kJ/mol

The second conformation is slightly more stable than the first.

(c)

cis-1-Bromo-4-ethylcyclohexane

2 (H ↔ CH₂CH₃) = 8.0 kJ/mol   2 (H ↔ Br) = 2.0 kJ/mol

The second conformation is more stable than the first.

(d)

cis-1-*tert*-Butyl-4-ethylcyclohexane

$$2 \, [H \leftrightarrow C(CH_3)_3] = 22.8 \; kJ/mol \qquad\qquad 2 \, (H \leftrightarrow CH_2CH_3) = 8.0 \; kJ/mol$$

The second conformation is more stable than the first.

4.19 The three substituents have the orientations shown in the first structure. To decide if the conformation shown is the more stable conformation or the less stable conformation, perform a ring-flip on the illustrated conformation and do a calculation of the total strain in each structure as in the previous problem. Notice that each conformation has a Cl – CH₃ gauche interaction, but we don't need to know its energy cost because it is present in both conformations.

$$
\begin{aligned}
2 \, [H \leftrightarrow CH_3] &= 7.6 \; kJ/mol \\
\underline{2 \, [H \leftrightarrow Cl]} \;\; &= 2.0 \; kJ/mol \\
&= 9.6 \; kJ/mol
\end{aligned}
\qquad\qquad
2 \, [H \leftrightarrow CH_3] = 7.6 \; kJ/mol
$$

The illustrated conformation is the less stable chair form.

4.20

Trans-decalin is more stable than *cis*-decalin because of three 1,3-diaxial interactions present in the cis isomer. You can recognize these interactions if you think of the circled parts of *cis*-decalin as similar to axial methyl groups. The gauche interactions that occur with axial methyl groups also occur in *cis*-decalin.

4.21 Both ring fusions are trans because the bridgehead groups are on opposite faces of the fused ring system.

Visualizing Chemistry

4.22

(a)

cis-1-Ethyl-3-methylcyclopentane

(b)

1,1,4-Trimethylcyclohexane

4.23

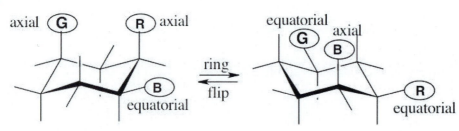

trans-1-Chloro-3-methylcyclohexane

2 (H ↔ CH₃) = 7.6 kJ/mol   2 (H ↔ Cl) = 2.0 kJ/mol

The conformation shown (the left structure) is the less stable conformation.

4.24

4.25 The green substituent is axial, and the red and blue substituents are equatorial.

cis relationship: blue–green
trans relationship: green–red, blue–red

4.26

α-Glucose   β-Glucose

The only difference between α-glucose and β-glucose is in the orientation of the –OH group at carbon 1: the –OH group is axial in α-glucose, and it is equatorial in β-glucose. You would expect β-glucose to be more stable because all of its substituents are in the equatorial position.

Additional Problems

Cycloalkane Isomers

4.27

The last two structures are cis-trans isomers.

4.28   Constitutional isomers differ in the way that atoms are connected.

cis-1,2-Dibromo-   cis-1,3-Dibromo-   trans-1,3-Dibromo-
cyclopentane   cyclopentane   cyclopentane

constitutional isomers of cis-1,2-dibromocyclopentane

4.29   Stereoisomers have different three-dimensional geometry.

trans-1,3-Dimethylcyclobutane   cis-1,3-Dimethylcyclobutane

4.30

(a)

cis-1,3-Dibromocyclohexane   trans-1,4-Dibromocyclohexane

constitutional isomers

(b)

2,3-Dimethylhexane   2,3,3-Trimethylpentane

constitutional isomers

(c)

identical

4.31 Stereoisomers:

trans-1,2-Dichlorocyclobutane   cis-1,2-Dichlorocyclobutane

Constitutional isomers of trans-1,2-dichlorocyclobutane:

1,1-Dichlorocyclobutane   cis-1,3-Dichlorocyclobutane

trans-1,3-Dichlorocyclobutane is also a constitutional isomer.

4.32

CH_2OH

OH red

OH green

OH black

OH blue

cis relationship: red–green, blue–black

trans relationship: red–blue, green–blue
                    red–black, green–black

4.33

Two cis–trans isomers of 1,3,5-trimethylcyclohexane are possible. In one isomer (**A**), all methyl groups are cis; in **B**, one methyl group is trans to the other two.

Cycloalkane Conformation and Stability

4.34

4.35 Make a model of *cis*-1,2-dichlorocyclohexane. Notice that all cis substituents are on the same side of the ring and that two adjacent cis substituents have an axial–equatorial relationship. Now, perform a ring-flip on the cyclohexane.

After the ring-flip, the relationship of the two substituents is still axial–equatorial. No two adjacent cis substituents can be converted to being both axial or both equatorial by a ring-flip. Don't forget that there are only two chair conformations of any given cyclohexane.

4.36 For a *trans*-1,2-disubstituted cyclohexane, two adjacent substituents must be either both axial or both equatorial.

A ring flip converts two adjacent axial substituents to equatorial substituents, and vice versa. As in Problem 4.35, no two adjacent trans substituents can have an axial-equatorial relationship.

4.37

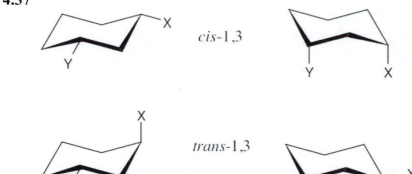

A cis-1,3-disubstituted isomer exists almost exclusively in the diequatorial conformation, which has no 1,3-diaxial interactions. The trans isomer must have one group axial, leading to 1,3-diaxial interactions. Thus, the trans isomer is less stable than the cis isomer. When a molecule has two conformations available, the molecule exists mainly in the lower energy conformation.

4.38

The *trans*-1,4-isomer is more stable because both substituents can be equatorial.

4.39

less stable
cis

more stable
trans

Two types of interaction are present in *cis*-1,2-dimethylcyclobutane. One interaction occurs between the two methyl groups, which are almost eclipsed. The other is an across-the-ring interaction between methyl group at position 1 of the ring and a hydrogen at Position 3. Because neither of these interactions are present in trans isomer, it is more stable than the cis isomer.

more stable
cis

less stable
trans

In *trans*-1,3-dimethylcyclobutane, an across-the-ring interaction occurs between the methyl group at position 3 of the ring and a hydrogen at position 1. Because no interactions are present in the cis isomer, it is more stable than the trans isomer.

4.40   To solve this problem: (1) Find the energy cost of a 1,3-diaxial interaction by using Table 4.1. (2) Convert this energy difference into a percent by using Figure 4.12.

(a)

H ⟵---⟶ CH(CH₃)₂

$2 (H \leftrightarrow CH(CH_3)_2) = 9.2$ kJ/mol
% equatorial = 97
% axial = 3

(b)

H ⟵---⟶ F

$2 (H \leftrightarrow F) = 1.0$ kJ/mol
% equatorial = 63
% axial = 37

(c)

H ⟵---⟶ CN

$2 (H \leftrightarrow CN) = 0.8$ kJ/mol
% equatorial = 60
% axial = 40

4.41 Make sure you know the difference between axial–equatorial and cis–trans. Axial
substituents are parallel to the axis of the ring; equatorial substituents lie around the
equator of the ring. Cis substituents are on the same side of the ring; trans substituents are
on opposite side of the ring.

(a)   1,3-trans

   axial, equatorial   equatorial, axial

(b)   1,4-cis

   axial, equatorial   equatorial, axial

(c)   1,3-cis

   axial, axial   equatorial, equatorial

(d)   1,5-trans is the same as 1,3-trans

(e)   1,5-cis is the same as 1,3-cis

(f)

   axial, axial   equatorial, equatorial

   1,6-trans is the same as 1,2-trans.

Cyclohexane Conformational Analysis

4.42

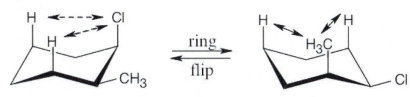

cis-1-Chloro-2-methylcyclohexane

Use Table 4.1 to find the values of 1,3-diaxial interactions. For the first conformation, the steric strain is 2×1.0 kJ/mol $= 2.0$ kJ/mol. The steric strain in the second conformation is 2×3.8 kJ/mol, or 7.6 kJ/mol. The first conformation is more stable than the second conformation by 5.6 kJ/mol.

4.43

no 1,3-diaxial
interactions

2×3.8 kJ/mol $= 7.6$ kJ/mol
2×1.0 kJ/mol $= 2.0$ kJ/mol
$= 9.6$ kJ/mol

trans-1-Chloro-2-methylcyclohexane

Use Table 4.1 to find the values of 1,3-diaxial interactions. The first conformation is more stable than the second conformation by a maximum of 9.6 kJ/mol. (A gauche interaction between the two substituents in the diequatorial conformation reduces the value of the energy difference, but its value can't be determined with the given data.)

4.44

Galactose

In this conformation, all substituents, except for one hydroxyl group, are equatorial.

4.45 From the flat-ring drawing you can see that the methyl group and the –OH group have a *cis* relationship, and the isopropyl group has a *trans* relationship to both of these groups. Draw a chair cyclohexane ring and attach the groups with the correct relationship.

In this conformation, all substituents are equatorial. Now, perform a ring flip.

The second conformation is less stable because all substituents are axial.

4.46

Menthol   cis–trans isomers of menthol

The substituents on the ring have the following relationships:

	Menthol	Isomer A	Isomer B	Isomer C
—CH(CH₃)₂, —CH₃	trans	trans	cis	cis
—CH(CH₃)₂, —OH	trans	cis	trans	cis
—CH₃, —OH	cis	trans	trans	cis

The substituents on the ring have the following relationships:

4.47

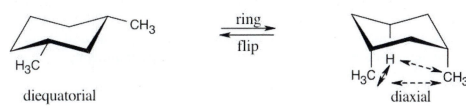

diequatorial   diaxial

The large energy difference between conformations is due to the severe 1,3 diaxial interaction between the two methyl groups.

4.48 Diaxial cis-1,3-dimethylcyclohexane contains three 1,3-diaxial interactions – two H ↔ CH_3 interactions of 3.8 kJ/mol each, and one CH_3 ↔ CH_3 interaction. If the diaxial conformation is 23 kJ/mol less stable than the diequatorial, 23 kJ/mol – 2(3.8 kJ/mol) ≈ 15 kJ/mol of this strain energy must be due to the CH_3 ↔ CH_3 interaction.

4.49

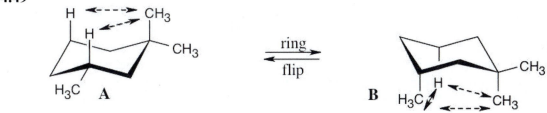

2 H ↔ CH_3 interactions = 7.6 kJ/mol

2 H ↔ CH_3 interactions = 7.6 kJ/mol
1 CH_3 ↔ CH_3 interaction = 15 kJ/mol
                                    ≈ 23 kJ/mol

Conformation **A** is favored.

4.50

Conformation **A** of *cis*-1-chloro-3-methylcyclohexane has no 1,3-diaxial interactions and is the more stable conformation. Steric strain in **B** is due to one CH_3 ↔ H interaction (3.8 kJ/mol), one Cl ↔ H interaction (1.0 kJ/mol) and one CH_3 ↔ Cl interaction. Since the total-strain energy of B is 15.5 kJ/mol, 15.5 kJ/mol - 3.8 kJ/mol - 1.0 kJ/mol = 10.7 kJ/mol of strain is caused by a CH_3 ↔ Cl interaction.

General Problems

4.51 Note: In working with decalins, it is essential to use models. Many structural features of decalins that are obvious with models are not easily visualized with drawings.

trans-Decalin   *cis*-Decalin

No 1,3-diaxial interactions are present in *trans*-decalin.

At the ring junction of *cis*-decalin, one ring acts as an axial substituent of the other (see circled bonds). The circled part of ring **B** has two 1,3-diaxial interactions with ring **A** (indicated by arrows). Similarly, the circled part of ring **A** has two 1,3-diaxial interactions with ring **B**; one of these interactions is the same as an interaction of part of the **B** ring with ring **A**. These three 1,3-diaxial interactions have a total energy cost of 3 x 3.8 kJ/ mol = 11.4 kJ/mol. *Cis*-decalin is therefore less stable than *trans*-decalin by 11.4 kJ/mol.

4.52 A ring-flip converts an axial substituent into an equatorial substituent and vice versa. At the ring junction of *trans*-decalin, each ring is a trans–trans diequatorial substituent of the other. If a ring-flip were to occur, the two rings would become axial substituents of each other. You can see with models that a diaxial ring junction is impossibly strained. Consequently, *trans*-decalin does not ring-flip.

The rings of *cis*-decalin are joined by an axial bond and an equatorial bond. After a ring-flip, the rings are still linked by an equatorial and an axial bond. No additional strain or interaction is introduced by a ring-flip of *cis*-decalin.

4.53 Build models to see the stability difference between the two [4.1.0] ring systems. In both cases, fusing a three-membered ring to a six-membered ring distorts the bond angles of both rings, causing angle strain. This strain is much more severe in the *trans* isomer than in the *cis* isomer.

cis-Bicyclo[4.1.0]heptane   *trans*-Bicyclo[4.1.0]heptane

4.54

Simvastatin (Zocor)

Pravastatin (Pravachol)

Atorvastatin (Lipitor)

(a)   The indicated bonds on simvastatin are trans.

(b)   The –H bond and the –OH bond have a cis relationship. The third bond is trans to both of them.

(c)   The three indicated bonds on atorvastatin are attached to sp^2-hybridized carbons of a planar ring and lie in the same plane.

4.55 In the flat-ring structure shown, all –OH groups have a alternating relationship except for the starred group (below). If all of the groups had a trans relationship, the most stable conformation would have all –OH groups in the equatorial position. We expect that the most stable conformation of this structure has one group in the axial position.

Draw both rings and add –OH groups having the indicated relationships. Perform a ring-flip on the structure you have drawn to arrive at the other conformation.

myo-Inositol   more stable

4.56

There are eight cis–trans stereoisomers of *myo*-inositol. The first isomer is the most stable because all hydroxyl groups can assume an equatorial conformation.

4.57

1-Norbornene

If you build a model of 1-norbornene, you will find that it is almost impossible to form the bridgehead double bond. *sp²*-Hybridization at the double bond requires all carbons bonded to the starred carbons to lie in a common plane in order for the *p* orbitals to overlap to form the π bond. The bicyclic ring system forces these atoms out of plane, and the bridgehead double bond can't form.

4.58 A steroid ring system is fused, and ring-flips don't occur. Thus, substituents such as the methyl groups shown remain axial. Substituents on the same side of the ring system as the methyl groups are in alternating axial and equatorial positions. Thus, an "up" substituent at C3 (a) is equatorial.

Substituents on the bottom side of the ring system also alternate axial and equatorial positions. A substituent at C7 (b) is axial, and one at C11 (c) is equatorial

4.59

Amantadine

4.60

mirror

The two *trans*-1,2-dimethylcyclopentanes are mirror IMAGEs.

4.61 Draw the four possible isomers of 4-*tert*-butylcyclohexane-1,3-diol. Make models of these isomers also. The bulky *tert*-butyl group determines the stable conformation because of its strong preference for the equatorial position.

Only when the two hydroxyl groups are cis diaxial (structure **1**) can the acetal ring form. In any other conformation, the oxygen atoms are too far apart to be incorporated into a six-membered ring.

4.62

cis

trans

All four conformations of the two isomers are illustrated. The second conformation of each pair has a high degree of steric strain, and thus each isomer adopts the first conformation. Since only the cis isomer has the hydroxyl group in the necessary axial position, it oxidizes faster than the trans isomer.

Chapter 5 – Stereochemistry at Tetrahedral Centers

Chapter Outline

I. Handedness (Sections 5.1–5.4).
   A. Enantiomers and tetrahedral carbon (Section 5.1).
      1. When four different groups are bonded to a carbon atom, two different arrangements are possible.
         a. These arrangements are mirror images.
         b. The two mirror-image molecules are enantiomers.
   B. The reason for handedness in molecules: chirality (Section 5.2).
      1. Molecules that are not superimposable on their mirror-images are chiral.
         a. A molecule is not chiral if it contains a plane of symmetry.
         b. A molecule with no plane of symmetry is chiral.
      2. A carbon bonded to four different groups is a chirality center.
      3. It is sometimes difficult to locate a chirality center in a complex molecule.
      4. The groups $-CH_2-$, $-CH_3$, C=O, C=C, and C≡C can't be chirality centers.
   C. Optical activity (Section 5.3).
      1. Solutions of certain substances rotate the plane of plane-polarized light.
         a. These substances are said to be optically active.
      2. The angle of rotation can be measured with a polarimeter.
      3. The direction of rotation can also be measured.
         a. A compound whose solution rotates plane-polarized light to the right is termed dextrorotatory.
         b. A compound whose solution rotates plane-polarized light to the left is termed levorotatory.
      4. Specific rotation.
         a. The extent of rotation depends on concentration, path length and wavelength.
         b. Specific rotation is the observed rotation of a sample with concentration = 1 g/mL, sample path length of 1 dm, and light of wavelength = 589 nm.
         c. Specific rotation is a physical constant characteristic of a given optically active compound.
   D. Pasteur's discovery of enantiomerism (Section 5.4).
      1. Pasteur discovered two different types of crystals in a solution that he was evaporating.
      2. The crystals were mirror images.
      3. Solutions of each of the two types of crystals were optically active, and their specific rotations were equal in magnitude but opposite in sign.
      4. Pasteur postulated that some molecules are handed and thus discovered the phenomenon of enantiomerism.
II. Stereoisomers and configurations (Sections 5.5–5.8).
   A. Specification of configurations of stereoisomers (Section 5.5).
      1. Rules for assigning configurations at a chirality center:

      a.   Assign priorities to each group bonded to the carbon by using Cahn–Ingold–Prelog rules.

          i.   Rank each atom by atomic number.

              (a). An atom with a higher atomic number receives a higher priority than an atom with a lower atomic number.

          ii. If a decision can't be reached based on the first atom, look at the second or third atom until a difference is found.

          iii. Multiple-bonded atoms are equivalent to the same number of single-bonded atoms.

      b.   Orient the molecule so that the group of lowest priority is pointing to the rear.

      c.   Draw a curved arrow from group 1 to group 2 to group 3.

      d.   If the arrow rotates clockwise, the chirality center is *R*, and if the arrow rotates counterclockwise, the chirality center is *S*.

    2. The sign of optical rotation is not related to *R,S* designation.

    3. X-ray experiments have proven *R,S* conventions to be correct.

  B. Diastereomers (Section 5.6).

    1. A molecule with two chirality centers can have four possible stereoisomers.

      a.   The stereoisomers group into two pairs of enantiomers.

      b.   A stereoisomer from one pair is the diastereomer of a stereoisomer from the other pair.

    2. Diastereomers are stereoisomers that are not mirror images.

    3. Epimers are diastereomers whose configuration differs at only one chirality center.

  C. Meso compounds (Section 5.7).

    1. A meso compound occurs when a compound with two chirality centers possesses a plane of symmetry.

    2. A meso compound is achiral despite having two chirality centers.

    3. The physical properties of meso compounds, diastereomers and racemic mixtures differ from each other and from the properties of enantiomers.

  D. Racemic mixtures and the resolution of enantiomers (Section 5.8).

    1. A racemic mixture (racemate) is a 50:50 mixture of two enantiomers.

      a.   Racemic mixtures show zero optical rotation.

    2. Some racemic mixtures can be resolved into their component enantiomers.

      a.   If a racemic mixture of a carboxylic acid reacts with a chiral amine, the product ammonium salts are diastereomers.

      b.   The diastereomeric salts differ in chemical and physical properties and can be separated.

      c.   The original enantiomers can be recovered by acidification.

III. A review of isomerism (Section 5.9).

  A. Constitutional isomers differ in connections between atoms.

    1. Skeletal isomers have different carbon skeletons.

    2. Functional isomers contain different functional groups.

    3. Positional isomers have functional groups in different positions.

B. Stereoisomers have the same connections between atoms, but different geometry.

1. Enantiomers have a mirror-image relationship.

2. Diastereomers are non-mirror-image stereoisomers.

a. Configurational diastereomers.

b. Cis–trans isomers differ in the arrangement of substituents on a ring or a double bond.

IV. Chirality at atoms other than carbon (Section 5.10).

A. Other elements with tetrahedral atoms can be chirality centers.

B. Trivalent nitrogen can, theoretically, be chiral, but rapid inversion of the nitrogen lone pair interconverts the enantiomers.

C. Chiral phosphines and trivalent sulfur compounds can be isolated because their rate of inversion is slower.

V. Prochirality (Section 5.11).

A. A molecule is prochiral if it can be converted from achiral to chiral in a single chemical step.

B. Identifying prochirality.

1. For sp^2 carbon, draw the plane that includes the atoms bonded to the sp^2 carbon.

a. Assign priorities to the groups bonded to the carbon.

b. Draw an curved arrow from group 1 to group 2 to group 3.

c. The face of the plane on which the curved arrow rotates clockwise is the *Re* face.

d. The face on which the arrow rotates counterclockwise is the *Si* face.

2. An atom that is sp^3-hybridized may have a prochirality center if, when one of its attached groups is replaced, it becomes a chirality center.

a. For −CH₂X, imagine a replacement of one hydrogen with deuterium.

b. Rank the groups, including the deuterium.

c. If the replacement leads to *R* chirality, the atom is *pro-R*.

d. If the replacement leads to *S* chirality, the atom is *pro-S*.

C. Many biochemical reactions involve prochiral compounds.

VI. Chirality in nature and chiral environments (Section 5.12).

A. Different enantiomers of a chiral molecule have different properties in nature.

1. (+)-Limonene has the odor of oranges, and (−)-limonene has the odor of lemons.

2. Racemic fluoxetine is an antidepressant, but the *S* enantiomer is effective against migraine.

B. In nature, a molecule must fit into a chiral receptor, and only one enantiomer usually fits.

Solutions to Problems

5.1   Objects having a plane of symmetry are achiral.
Chiral: screw, shoe.
Achiral: soda can, screwdriver.

5.2   Use the following rules to locate centers that are *not* chirality centers, then examine the remaining centers to find a carbon with four different groups attached.

1.   All −CH₃ and −CX₃ carbons are not chirality centers.

2.   All −CH₂− and −CX₂− carbons are not chirality centers.

3.   All $-\overset{|}{C}=\overset{|}{C}-$ and −C≡C− carbons are not chirality centers.
By rule 3, all aromatic ring carbons are not chirality centers.

(a)

Coniine

(b)

Menthol

(c)

Dextromethorphan

5.3   Refer to Problem 5.2 if you need help.

Alanine

5.4

(a)

Threose

(b)

Enflurane

5.5   By convention, a (−) rotation indicates rotation to the left, and thus cocaine is levorotatory.

5.6

$$\text{Use the formula } [\alpha]_D = \frac{\alpha}{l \times C}, \text{ where}$$

$$
\begin{array}{rcl}
[\alpha]_D &=& \text{specific rotation} \\
\alpha &=& \text{observed rotation} \\
l &=& \text{path length of cell (in dm)} \\
C &=& \text{concentration (in g/mL)}
\end{array}
$$

$$
\text{In this problem:} \quad
\begin{array}{rcl}
\alpha &=& 1.21° \\
l &=& 5.00 \text{ cm} = 0.500 \text{ dm} \\
C &=& 1.50 \text{ g}/10.0 \text{ mL} = 0.150 \text{ g/mL}
\end{array}
$$

$$[\alpha]_D = \frac{+1.21°}{0.500 \text{ dm} \times 0.150 \text{ g/mL}} = +16.1°$$

5.7   Review the sequence rules presented in Section 5.5. A summary:

Rule 1:   An atom with a higher atomic number has priority over an atom with a lower atomic number.

Rule 2:   If a decision can't be reached by using Rule 1, look at the second, third, or fourth atom away from the double-bond carbon until a decision can be made.

Rule 3:   Multiple-bonded atoms are equivalent to the same number of single-bonded atoms.

	Higher	*Lower*	*Rule*			*Higher*	*Lower*	*Rule*
(a)	$-Br$	$-H$	1		(b)	$-Br$	$-Cl$	1
(c)	$-CH_2CH_3$	$-CH_3$	2		(d)	$-OH$	$-NH_2$	1
(e)	$-CH_2OH$	$-CH_3$	2		(f)	$-CH=O$	$-CH_2OH$	3

5.8   Use the sequence rules in Section 5.5.

(a)   By Rule 1, $-H$ is of lowest priority, and $-OH$ is of highest priority. By Rule 2, $-CH_2CH_2OH$ is of higher priority than $-CH_2CH_3$.

Highest $\longrightarrow$ *Lowest*
$-OH, -CH_2CH_2OH, -CH_2CH_3, -H$

(b)   By Rule 3, $-CO_2H$ is considered as $-\overset{\overset{\displaystyle -O\quad O-}{\diagdown\diagup}}{C}-$ OH. Because 3 oxygens are attached to a $-CO_2H$ carbon and only one oxygen is attached to $-CH_2OH$, $-CO_2H$ is of higher priority than $-CH_2OH$. $-CO_2CH_3$ is of higher priority than $-CO_2H$ by Rule 2, and $-OH$ is of highest priority by Rule 1.

Highest $\longrightarrow$ *Lowest*

(b) $-OH, -CO_2CH_3. -CO_2H, -CH_2OH$

(c) $-NH_2, -CN, -CH_2NHCH_3, -CH_2NH_2$

(d) $-SSCH_3, -SH, -CH_2SCH_3, -CH_3$

5.9 All stereochemistry problems are easier if you use models. Part (a) will be solved by two methods – with models and without models.

(a) *With models:* Build a model of (a). Orient the model so that group 4 is pointing to the rear. Note the direction of rotation of arrows that go from group 1 to group 2 to group 3. The arrows point counterclockwise, and the configuration is *S*.

Without models: Imagine yourself looking at the molecule, with the group of lowest priority pointing to the back. Your viewpoint would be at the upper right of the molecule, and you would see group 1 on the left, group 3 on the right and group 2 at the bottom. The arrow of rotation travels counterclockwise, and the configuration is *S*.

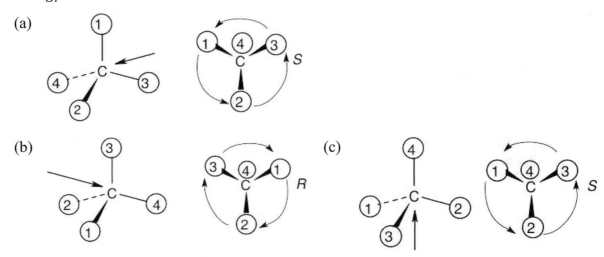

5.10 *Step 1.* For each chirality center, rank substituents by the Cahn–Ingold–Prelog system; give the number 4 to the lowest priority substituent. For part (a):

Substituent	Priority
$-SH$	1
$-CO_2H$	2
$-CH_3$	3
$-H$	4

Step 2. As in the previous problem, orient yourself so that you are 180° from the lowest priority group (indicated by the arrow in the drawing). From that viewpoint, draw the molecule as it looks when you face it. Draw the arrow that travels from group 1 to group 2 to group 3, and note its direction of rotation. The molecule in (a) has *S* configuration.

(a)

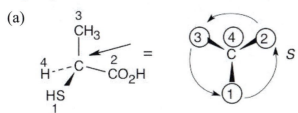

(b)

(c)

In (b), the observer is behind the page, looking out and down toward the right. In (c), the observer is behind the page looking out and up to the left.

5.11

$CH_3CH_2CH_2\overset{*}{C}CH_3$ (S)-2-Pentanol

Substituent	Priority
—OH	1
—CH₂CH₂CH₃	2
—CH₃	3
—H	4

5.12 Fortunately, methionine is shown in the correct orientation.

$CH_3SCH_2CH_2$ (S)-Methionine

5.13 For (a): (Note: the phosphate group is represented as **P.**)

(a) R,R   (b) S,R   (c) R,S   (d) S,S

a, d are enantiomers and are diastereomeric with b, c.
b, c are enantiomers and are diastereomeric with a, d.
Structure (a) is D-erythrose 4-phosphate, structure (d) is its enantiomer, and structures (b) and (c) are its diastereomers.

5.14

Morphine

Morphine has five chirality centers and, in principle, can have $2^5 = 32$ stereoisomers. Most of these stereoisomers are too strained to exist.

5.15

Isoleucine

5.16 To decide if a structure represents a meso compound, try to locate a plane of symmetry that divides the molecule into two halves that are mirror images. Molecular models are always helpful.

(a)

meso

(b) and (c) are not meso structures.

(d)

plane of symmetry

meso

5.17 For a molecule to exist as a meso form, it must possess a plane of symmetry. 2,3-Butanediol can exist as a pair of enantiomers *or* as a meso compound, depending on the configurations at carbons 2 and 3.

(a)

not meso

(b) 2,3-Pentanediol has no symmetry plane and thus can't exist in a meso form.

(c) 2,4-Pentanediol can exist in a meso form.

2,4-Pentanediol can also exist as a pair of enantiomers (2R,4R) and (2S,4S) that are not meso compounds.

5.18 The molecule represents a meso compound. The symmetry plane passes through the carbon bearing the −OH group and between thr two ring carbons that are bonded to methyl groups.

5.19

Acetic acid   (S)-2-Butanol   sec-Butyl acetate

The product is the pure S-ester. No new chirality centers are formed during the reaction, and the configuration at the chirality center of (S)-2-butanol is unchanged.

5.20

The two product salts have the configurations (*R*,*S*) and (*S*,*S*) and are diastereomers.

5.21 (a)

(*S*)-5-Chloro-2-hexene   Chlorocyclohexane

These two compounds are constitutional isomers (skeletal isomers).

(b)   The two dibromopentane stereoisomers are diastereomers.

5.22 For each molecule, replace the left hydrogen with ^{2}H. Give priorities to the groups and assign *R*,*S* configuration to the chirality center. If the configuration is *R*, the replaced hydrogen is *pro-R*, and if the configuration is *S*, the replaced hydrogen is *pro-S*.

(a)

(*S*)-Glyceraldehyde

(b)

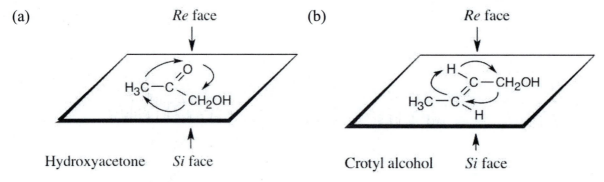

5.23 Draw the plane that includes the sp^2 carbon and its substituents, and rank the substituents. For the upper face, draw the arrow that proceeds from group 1 to group 2 to group 3. If the direction of rotation is clockwise, the face is the *Re* face; if rotation is counterclockwise, the face is the *Si* face.

(a)

Re face

H₃C—C (=O) CH₂OH

Hydroxyacetone   *Si* face

(b)

Re face

H—C—CH₂OH / H₃C—C—H

Crotyl alcohol   *Si* face

5.24 Use the strategy in the previous problem to identify the faces of the plane that contains the sp^2 carbon. Draw the product that results from reaction at the *Re* face, and assign configuration to the chirality center.

Re face

⁻H:

H₃C—C (=O) CO₂⁻

Pyruvate   *Si* face

HO—C(S)—CO₂⁻ with H and H₃C

(S)-Lactate

5.25 Addition of −OH takes place on the *Re* face of C2 of aconitate. Addition of −H occurs on the *Re* face of C3 to yield (2*R*,3*S*)-isocitrate. H and OH add from opposite sides of the double bond.

cis-Aconitate

(2*R*,3*S*)-Isocitrate

Visualizing Chemistry

5.26 Structures (a), (b), and (d) are identical (*R* enantiomer), and (c) represents the *S* enantiomer.

5.27

(a)

(*S*)-Serine

(b)

(*R*)-Adrenaline

R = ring

5.28 Locate the plane of symmetry that identifies the structure as a meso compound.

(a) meso   (b) meso   (c) not meso

5.29

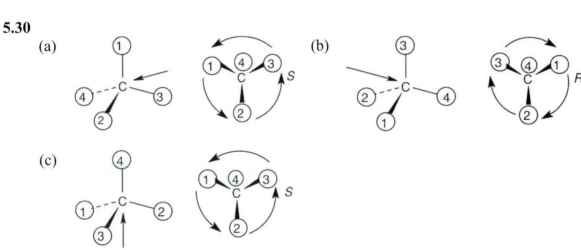

Pseudoephedrine

5.30

(a)

(b)

(c)

Additional Problems

Chirality and Optical Activity

5.31 Chiral: (d) golf club, (e) spiral staircase
Achiral: (a) basketball, (b) fork, (c) wine glass, (f) snowflake.

5.32

(a)

$$CH_3CH_2CH_2CHCH_2CHCH_3$$

2,4-Dimethylheptane has one chirality center.

(b)

$$CH_3CH_2CCH_2CHCH_2CH_3$$

5-Ethyl-3,3-dimethylheptane is achiral.

(c)

Cl ▬◀ ⬡ ▶■ Cl

cis-1,4-Dichlorocyclohexane is achiral. Note the plane of symmetry that passes through the –Cl groups.

5.33

(a)

$$\underset{Cl}{\overset{Cl}{|}}$$
CH₃CH₂CH₂CHCH₃
*
2-Chloropentane

(b)

OH
|
CH₃CH₂CH₂CH₂CHCH₃
*
2-Hexanol

(c)

CH₃
|
CH₃CH₂CHCH=CH₂
*
3-Methyl-1-pentene

(d)

CH₃
|
CH₃CH₂CH₂CH₂CHCH₂CH₃
*
3-Methylheptane

5.34

CH₃CH₂CH₂CH₂CH₂OH
achiral

OH
|
CH₃CH₂CH₂CHCH₃
*
chiral

OH
|
CH₃CH₂CHCH₂CH₃
achiral

CH₃
|
CH₃CCH₂OH
|
CH₃
achiral

CH₃
|
CH₃CH₂CHCH₂OH
*
chiral

OH
|
CH₃CH₂CCH₃
|
CH₃
achiral

OH
|
CH₃CHCHCH₃
| *
CH₃
chiral

CH₃
|
HOCH₂CH₂CHCH₃
achiral

5.35

(a)

OH
|
CH₃CH₂CHCH₃
*

(b)

CH₃
|
CH₃CH₂CHCO₂H
*

(c)

Br OH
| |
CH₃CHCHCH₃
*   *

(d)

Br   O
| ‖
CH₃CHC
*   \
       H

5.36

Erythronolide B

Erythronolide B has ten chirality centers.

Assigning Configuration to Chirality Centers

5.37 Identical molecules: b (*S* enantiomer), c (*R* enantiomer), d (*S* enantiomer). Pair of enantiomers: a

5.38 The specific rotation of (2*R*,3*R*)-dichloropentane is equal in magnitude and opposite in sign to the specific rotation of (2*S*,3*S*)-dichloropentane because the compounds are enantiomers. There is no predictable relationship between the specific rotations of the (2*R*,3*S*) and (2*R*,3*R*) isomers because they are diastereomers.

5.39–5.40

The (2*R*,4*S*) stereoisomer is the enantiomer of the (2*S*,4*R*) stereoisomer.

The (2*S*,4*S*) and (2*R*,4*R*) stereoisomers are diastereomers of the (2*S*,4*R*) stereoisomer.

5.41

(a)

(b)

(c)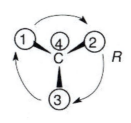

5.42

Highest ⟶ *Lowest*

(a) —C(CH₃)₃, —CH=CH₂, —CH(CH₃)₂, —CH₂CH₃

(b) ⟍⬡ , —C≡CH, —C(CH₃)₃, —CH=CH₂

(c) —CO₂CH₃, —COCH₃, —CH₂OCH₃, —CH₂CH₃

(d) —Br, —CH₂Br, —CN, —CH₂CH₂Br

5.43

(a) (b) (c)

5.44

(a) (b) (c)

5.45

(a)

Biotin

(b)

Prostaglandin E$_1$

5.46

(a)

(b)

5.47

(a)

(b)

5.48

Ascorbic acid

5.49

(a)

(b)

5.50

(+)-Xylose

Meso Compounds

5.51

(a)

CH₂CH₃
H ◣ C ◢ CH₃
← symmetry plane
H ◥ C ◤ CH₃
CH₂CH₃

(b)

CH₂CH₃
H ◣ C ◢ CH₃
symmetry plane ↘
CH₂
H ◥ C ◤ CH₃
CH₂CH₃

(c)

CH₃
H ◣ S C ◢ OH
symmetry plane →
H ◥ C R ◤ OH
CH₃

This compound is
also a meso compound.

5.52

(a)

(b)

(c)

5.53 Both of the diastereomers shown below are meso compounds with three chirality centers. Each is a meso compound because it has a symmetry plane, and in each structure the central carbon is bonded to four different groups (a group with R configuration, a group with S configuration, $-OH$, and $-H$).

5.54 (a)–(c)

Ribose   Enantiomer of ribose

Ribose has three chirality centers, which give rise to eight (2^3) stereoisomers.

(d)   Ribose has six diastereomers.

5.55 Ribitol is an optically inactive meso compound. Catalytic hydrogenation converts the aldehyde functional group into a hydroxyl group and makes the two halves of ribitol mirror images of each other.

Ribose   Ribitol

Prochirality

5.56

(a)   Malic acid   (b)   Methionine   (c)   Cysteine

5.57

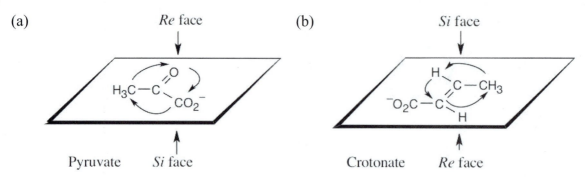

(a) Re face / Pyruvate / Si face   (b) Si face / Crotonate / Re face

5.58 Remember that *each sp²* carbon has a *Re* face and a *Si* face.

(S)-3-Hydroxybutyrate

5.59 If you perform the "replacement test" to assign *pro-R/pro-S* prochirality, you will see that the right "arm" of citrate is *pro-R* and the product pictured on the right is formed. The *pro-S* arm is unchanged.

5.60

5.61

(R)-3-Hydroxybutyryl ACP   *trans*-Crotonyl ACP
The reaction removes the *pro-R* hydrogen.

General Problems

5.62

B and **C** are enantiomers and are optically active. Compound **A** is their diastereomer and is a *meso* compound, which is not optically active.

The two isomeric cyclobutane-1,3-dicarboxylic acids are achiral and are optically inactive.

5.63

5.64

Cystine has the (*S,S*) configuration and is optically active.

5.65

(a)

(2*S*,3*R*)-2,3-Dibromopentane

(b)

meso-3,5-Heptanediol

5.66 All chirality centers of Cephalexin have an (*R*) configuration.

Cephalexin

5.67

Chloramphenicol

5.68 Mycomycin contains no chiral carbon atoms, yet is chiral. To see why, make a model of mycomycin. For simplicity, call $-CH=CHCH=CHCH_2CO_2H$ "A" and $-C{\equiv}CC{\equiv}CH$ "B". The carbon atoms of an allene have a linear relationship and that the π bonds formed are perpendicular to each other. Attach substituents at the sp^2 carbons.

Notice that the substituents A, H_a, and all carbon atoms lie in a plane that is perpendicular to the plane that contains B, H_b, and all carbon atoms.

Now, make another model identical to the first, except for an exchange of A and H$_a$. This new allene is not superimposable on the original allene. The two allenes are enantiomers and are chiral because they possess no plane of symmetry.

5.69   4-Methylcyclohexylideneacetic acid is chiral for the same reason that mycomycin (Problem 5.68) is chiral: It possesses no plane of symmetry and is not superimposable on its mirror image. As in the case of allenes, the two groups at one end of the molecule lie in a plane perpendicular to the plane that contains the two groups at the other end.

5.70   (a)

(S)-1-Chloro-
2-methylbutane

(S)-1,4-Dichloro-
2-methylbutane

(R)-1,2-Dichloro-
2-methylbutane

(S)-1,2-Dichloro-
2-methylbutane

+ other products

1:1 mixture

(b)   Chlorination at carbon 4 yields an optically active product because the chirality center at C2 is not affected. Chlorination at carbon 2 yields an optically inactive racemic product.

5.71

A   B   C   D

There are four stereoisomers of 2,4-dibromo-3-chloropentane. **C** and **D** are enantiomers and are optically active. **A** and **B** are optically inactive meso compounds and are diastereomers.

5.72

cis-1,4-Dimethylcyclohexane   *trans*-1,4-Dimethylcyclohexane

(a)   There is only one stereoisomer of each of the 1,4-dimethylcyclohexanes.

(b)   Neither 1,4-dimethylcyclohexane is chiral.

(c)   The two 1,4-dimethylcyclohexanes are diastereomers.

5.73

cis-1,3-Dimethylcyclohexane   *trans*-1,3-Dimethylcyclohexane

(a)   There is one stereoisomer of *cis*-1,3-dimethylcyclohexane, and there are two
       stereoisomers of *trans*-1,3-dimethylcyclohexane.

(b)   *cis*-1,3-Dimethylcyclohexane is an achiral meso compound; *trans*-1,3-
       dimethylcyclohexane exists as a pair of chiral enantiomers.

(c)   The two trans stereoisomers are enantiomers, and both are diastereomers of the cis
       stereoisomer.

5.74

ring flip   ≡

cis-1,2-Dimethylcyclohexane

The two *cis*-1,2-dimethylcyclohexane enantiomers rapidly interconvert by a ring flip,
leading to an optically inactive 1:1 mixture.

5.75

The product is (*R*)-2-butanethiol.

5.76 The reaction proceeds by addition of acetylide anion to the carbonyl group and occurs with equal probability from either face of the planar ketone carbon.

(a) The product is an optically inactive racemic mixture.

(b) The two enantiomers are formed in a 50:50 ratio.

5.77

(3*R*,4*R*)-4-Phenyl-
1-pentyn-3-ol

(3*S*,4*R*)-4-Phenyl-
1-pentyn-3-ol

(a) Reaction of sodium acetylide with a chiral aldehyde yields chiral products; the product mixture is optically active.

(b) The two products are a mixture of the (3*R*,4*R*) and (3*S*,4*R*) diastereomers of 4-phenyl-1-pentyn-3-ol. The product ratio can't be predicted, but it is not 50:50.

Review Unit 2: Alkanes and Stereochemistry

Major Topics Covered (with vocabulary):

Functional Groups.

Alkanes:

saturated   aliphatic   straight-chain alkane   branched-chain alkane   isomer
constitutional isomer   alkyl group   primary, secondary, tertiary, quaternary carbon
IUPAC system of nomenclature   primary, secondary, tertiary hydrogen   paraffin   cycloalkane
cis-trans isomer   stereoisomer

Alkane Stereochemistry:

conformer   sawhorse representation   Newman projection   staggered conformation
eclipsed conformation   torsional strain   dihedral angle   anti conformation
gauche conformation   steric strain   angle strain   heat of combustion   chair conformation
axial group   equatorial group ring-flip   1,3-diaxial interaction   conformational analysis
boat conformation   twist-boat   conformation   polycyclic molecules   bicycloalkane

Handedness:

stereoisomer   enantiomer   chiral   plane of symmetry   achiral   chirality center
plane-polarized light   optical activity   levorotatory   dextrorotatory   specific rotation

Stereoisomers and configuration:

configuration   Cahn–Ingold–Prelog rules   absolute configuration   diastereomer   meso
compound racemate   resolution   prochirality   *Re* face   *Si* face   prochirality center
pro-R   *pro-S*

Types of Problems:

After studying these chapters, you should be able to:

- Identify functional groups, and draw molecules containing a given functional group.
- Draw all isomers of a given molecular formula.
- Name and draw alkanes and alkyl groups.
- Identify carbons and hydrogens as being primary, secondary or tertiary.
- Draw energy vs. angle of rotation graphs for single bond conformations.
- Draw Newman projections of bond conformations and predict their relative stability.
- Understand the geometry of, and predict the stability of, cycloalkanes having fewer than 6 carbons.
- Draw and name substituted cyclohexanes, indicating cis/trans geometry.
- Predict the stability of substituted cyclohexanes by estimating steric interactions.
- Calculate the specific rotation of an optically active compound.
- Locate chirality centers, assign priorities to substituents, and assign *R,S* designations to chirality centers.
- Given a stereoisomer, draw its enantiomer and/or diastereomers.
- Locate the symmetry plane of a meso compound.
- Assign *Pro-R* and *Pro-S* designations to prochiral groups.
- Identify the face of an sp^2-hybridized carbon as *pro-R* or *pro-S*.

Points to Remember:

*   In identifying the functional groups in a compound, some groups have different designations that depend on the number and importance of other groups in the molecule. For example, a compound containing an −OH group and few other groups is probably named as an alcohol, but when several other groups are present , the −OH group is referred to as a hydroxyl group. There is a priority list of functional groups in the Appendix of the textbook, and this priority order will become more apparent as you progress through the text.

*   It is surprising how many errors can be made in naming compounds as simple as alkanes. Why is this? Often the problem is a result of just not paying attention. It is very easy to undercount or overcount the −CH₂− groups in a chain and to misnumber substituents. Let's work through a problem, using the rules in Section 3.4.

$$
\begin{array}{c}
\quad\;\; CH_3 \quad CH_2CH_3 \\
\quad\;\; | \qquad\; | \\
CH_3CCH_2CHCH_2CH_3 \\
\quad\;\; | \\
\quad\;\; CH_3
\end{array}
$$

    <u>Find</u> the longest chain. In the above compound, the longest chain is a hexane (Try all possibilities; there are two different six-carbon chains in the compound.) <u>Identify</u> the substituents. The compound has two methyl groups and an ethyl group. It's a good idea to list these groups to keep track of them. <u>Number</u> the chain and the groups. Try both possible sets of numbers, and see which results in the lower combination of numbers. The compound might be named either as a 2,2,4-trisubstituted hexane or a 3,5,5-trisubstituted hexane, but the first name has a lower combination of numbers. <u>Name</u> the compound, remembering the prefix *di-* and remembering to list substituents in alphabetical order. The correct name for the above compound is 4-ethyl-2,2-dimethylhexane. The acronym FINN (from the first letters of each step listed above) may be helpful.

*   When performing a ring-flip on a cyclohexane ring, keep track of the positions on the ring.

*   A helpful strategy for assigning *R,S* designations: Using models, build two enantiomers by adding four groups to each of two tetrahedral carbons. Number the groups 1–4, to represent priorities of groups at a tetrahedral carbon, and assign a configuration to each carbon. Attach a label that indicates the configuration of each enantiomer. Keep these two enantiomers, and use them to check your answer every time that you need to assign *R,S* configurations to a chiral atom.

*   When assigning *pro-R* or *pro-S* designations to a hydrogen, mentally replace the hydrogen that points out of the plane of the page. The other hydrogen is then positioned for prochirality assignment without manipulating the molecule. If the designation is *R*, the replaced hydrogen is *pro-R*; if the designation is *S*, the replaced hydrogen is *pro-S*.

Self-test

$$CH_3\text{-}CHCH_2CH_2\underset{\underset{CH_3}{|}}{C}CH_2\underset{\underset{CH_2CH_3}{|}}{CH}CH_2CH_3$$

with CH₃ on first carbon and CH(CH₃)₂ on central carbon

A

$$CH_3\text{-}CHCH_2CH_2CH_2CHNHCH(CH_3)_2$$

with CH₃ substituents

B

Metron S (an antihistamine)

Name **A**, and identify carbons as primary, secondary, tertiary or quaternary.

B is an amine with two alkyl substituents. Name these groups and identify alkyl hydrogens as primary, secondary or tertiary.

C   Metalaxyl (a fungicide)

D

Identify all functional groups of **C** (metalaxyl).

Name **D** and indicate the cis/trans relationship of the substituents. Draw both possible chair conformations, and calculate the energy difference between them.

E
Ubenimex
(an antitumor drug)

F
Epiandosterone
(an androgen)

Assign *R,S* designations to the chiral carbons in **E**. Label the circled hydrogen as *pro-R* or *pro-S*. Indicate the chirality centers in **F**. How many stereoisomers of **F** are possible?

Multiple Choice

1.   Which of the following functional groups doesn't contain a carbonyl group?
(a) aldehyde   (b) ester   (c) ether   (d) ketone

2.   Which of the following compounds contains primary, secondary, tertiary and quaternary carbons?
(a) 2,2,4-Trimethylhexane   (b) Ethylcyclohexane   (c) 2-Methyl-4-ethylcyclohexane
(d) 2,2-Dimethylcyclohexane

3.   How many isomers of the formula $C_4H_8Br_2$ are there?
(a) 4   (b) 6   (c) 8   (d) 9

4.   The lowest energy conformation of 2-methylbutane occurs:
(a) when all methyl groups are anti   (b) when all methyl groups are gauche
(c) when two methyl groups are anti   (d) when two methyl groups are eclipsed

5.   The strain in a cyclopentane ring is due to:
(a) angle strain (b) torsional strain (c) steric stain (d) angle strain and torsional strain

6.   In which molecule do the substituents in the more stable conformation have a diequatorial relationship?
(a) cis-1,2 disubstituted   (b) cis-1,3 disubstituted   (c) trans-1,3-disubstituted
(d) cis-1,4 disubstituted

7.   Which group is of lower priority than $-CH=CH_2$?
(a) $-CH(CH_3)_2$ (b) $-CH=C(CH_3)_2$ (c) $-C\equiv CH$ (d) $-C(CH_3)_3$

8.   A meso compound and a racemate are identical in all respects except:
(a) molecular formula   (b) degree of rotation of plane-polarized light
(c) connectivity of atoms   (d) physical properties

9.   Which of the following projections represents an R enantiomer?

(a) CO_2H (b)   Br   (c) HO_2C (d) CH_3

10.   How many prochirality centers does 1-bromobutane have?
(a) none   (b) 1   (c) 2   (d) 3

Chapter 6 – An Overview of Organic Reactions

Chapter Outline

I. Organic Reactions (Sections 6.1–6.6).

  A. Kinds of organic reactions (Section 6.1).

    1. Addition reactions occur when two reactants add to form one product, with no atoms left over.

    2. Elimination reactions occur when a single reactant splits into two products.

    3. Substitution reactions occur when two reactants exchange parts to yield two new products.

    4. Rearrangement reactions occur when a single product undergoes a rearrangement of bonds to yield an isomeric product.

  B. Reaction mechanisms - general information (Section 6.2).

    1. A reaction mechanism describes the bonds broken and formed in a chemical reaction, and accounts for all reactants and products.

    2. Bond breaking and formation in chemical reactions.

      a. Bond breaking is symmetrical (homolytic) if one electron remains with each fragment.

      b. Bond breaking is unsymmetrical (heterolytic) if both electrons remain with one fragment and the other fragment has a vacant orbital.

      c. Bond formation is symmetrical if one electron in a covalent bond comes from each reactant.

      d. Bond formation is unsymmetrical if both electrons in a covalent bond come from one reactant.

    3. Types of reactions.

      a. Radical reactions involve symmetrical bond breaking and bond formation.

      b. Polar reactions involve unsymmetrical bond breaking and bond formation.

      c. Pericyclic reactions will be studied later.

  C. Radical reactions (Section 6.3).

    1. Radicals are highly reactive because they contain an atom with an unpaired electron.

    2. A substitution reaction occurs when a radical abstracts an atom and a bonding electron from another molecule.

    3. An addition reaction occurs when a radical adds to a double bond.

    4. Steps in a radical reaction.

      a. The *initiation step* produces radicals by the symmetrical cleavage of a bond.

      b. The *propagation steps* occur when a radical abstracts an atom to produce a new radical and a stable molecule.

        i. This sequence of steps is a chain reaction.

      c. A *termination step* occurs when two radicals combine.

    5. In radical reactions, all bonds are broken and formed by reactions of species with odd numbers of electrons.

  D. Polar reactions (Sections 6.4–6.6).

    1. Characteristics of polar reactions (Section 6.4).

      a. Polar reactions occur as a result of differences in bond polarities within molecules.

      b. These polarities are usually due to electronegativity differences between atoms.

          i. Differences may also be due to interactions of functional groups with solvents, as well as with Lewis acids or bases.

         ii. Some bonds in which one atom is polarizable may also behave as polar bonds.

      c. In polar reactions, electron-rich sites in one molecule react with electron-poor sites in another molecule.

      d. The movement of an electron pair in a polar reaction is shown by a curved, fullheaded arrow.

         i. An electron pair moves from an atom at the tail of the arrow to a second atom at the head of the arrow.

      e. The reacting species:

         i. A nucleophile is a compound with an electron-rich atom.

         ii. An electrophile is a compound with an electron-poor atom.

         iii. Some compounds can behave as both nucleophiles and as electrophiles

      f. Many polar reactions can be explained in terms of acid-base reactions.

  2. An example of a polar reaction: addition of HBr to ethylene (Section 6.5).

      a. This reaction is known as an electrophilic addition.

      b. The π electrons in ethylene behave as a nucleophile.

      c. The reaction begins by the attack of the π electrons on the electrophile H^+.

      d. The resulting intermediate carbocation reacts with Br^- to form bromoethane.

  3. Rules for using curved arrows in polar reaction mechanisms (Section 6.6).

      a. Electrons must move from a nucleophilic source to an electrophilic sink.

      b. The nucleophile can be either negatively charged or neutral.

      c. The electrophile can be either positively charged or neutral.

      d. The octet rule must be followed.

II. Describing a reaction (Sections 6.7–6.10).

  A. Equilibria, rates, and energy changes (Section 6.7).

    1. All chemical reactions are equilibria that can be expressed by an equilibrium constant K_{eq} that shows the ratio of products to reactants.

      a. If $K_{eq} > 1$, [products] > [reactants].

      b. If $K_{eq} < 1$, [reactants] > [products].

    2. For a reaction to proceed as written, the energy of the products must be lower than the energy of the reactants.

      a. The energy change that occurs during a reaction is described by $\Delta G°$, the Gibbs free-energy change.

      b. Favorable reactions have negative $\Delta G°$ and are exergonic.

      c. Unfavorable reactions have positive $\Delta G°$ and are endergonic.

      d. $\Delta G° = -RT \ln K_{eq}$.

    3. $\Delta G°$ is composed of two terms $-\Delta H°$, and $\Delta S°$, which is temperature-dependent.

      a. $\Delta H°$ is a measure of the change in total bonding energy during a reaction.

        i.   If $\Delta H°$ is negative, a reaction is exothermic.

       ii.   If $\Delta H°$ is positive, a reaction is endothermic.

   b. $\Delta S°$ (entropy) is a measure of the freedom of motion of a reaction.

        i.   A reaction that produces two product molecules from one reactant molecule has positive entropy.

       ii.   A reaction that produces one product molecule from two reactant molecules has negative entropy.

   c. $\Delta G° = \Delta H° - T\,\Delta S°$.

  4. None of these expressions predict the rate of a reaction.

B. Bond dissociation energies (Section 6.8).

  1. The bond dissociation energy (D) measures the heat needed to break a bond to produce two radical fragments.

  2. Each bond has a characteristic strength.

  3. In exothermic reactions, the bonds formed are stronger than the bonds broken.

C. Energy diagrams and transition states (Section 6.9).

  1. Reaction energy diagrams show the energy changes that occur during a reaction.

   a. The vertical axis represents energy changes, and the horizontal axis (reaction coordinate) represents the progress of a reaction.

  2. The transition state is the highest-energy species in this reaction.

   a. It is possible for a reaction to have more than one transition state.

   b. The difference in energy between the reactants and the transition state is the energy of activation $\Delta G^{\ddagger}$.

   c. Values of $\Delta G^{\ddagger}$ range from $40 - 150$ kJ/mol.

  3. After reaching the transition state, the reaction can go on to form products or can revert to starting material.

  4. Every reaction has its own energy profile.

D. Intermediates (Section 6.10).

  1. In a reaction of at least two steps, an intermediate is the species that lies at the energy minimum between two transition states.

  2. Even though an intermediate lies at an energy minimum between two transition states, it is a high-energy species and usually can't be isolated.

  3. Each step of a reaction has its own $\Delta G^{\ddagger}$ and $\Delta G°$, but the total reaction has an overall $\Delta G°$.

  4. Biological reactions take place in several small steps, each of which has a small value of $\Delta G^{\ddagger}$.

III. A Comparison of biological and laboratory reactions (Section 6.11).

A. Laboratory reactions are carried out in organic solvents; biological reactions occur in aqueous medium.

B. Laboratory reactions take place over a wide variety of temperatures; biological reactions take place at the temperature of the organism, usually within narrow limits.

C. Laboratory reactions are uncatalyzed, or use simple catalysts; biological reactions are enzyme-catalyzed.

D. Laboratory reagents are usually small and simple; biological reactions involve large, complex coenzymes.

E. Biological reactions have high specificity for substrate, whereas laboratory reactions are relatively nonspecific.

Solutions to Problems

6.1 (a) $CH_3Br + KOH \longrightarrow CH_3OH + KBr$ substitution

(b) $CH_3CH_2Br \longrightarrow H_2C{=}CH_2 + HBr$ elimination

(c) $H_2C{=}CH_2 + H_2 \longrightarrow CH_3CH_3$ addition

6.2

$$CH_3CH_2CH_2\overset{\overset{\displaystyle CH_3}{|}}{C}HCH_2Cl \quad + \quad CH_3CH_2CH_2\overset{\overset{\displaystyle CH_3}{|}}{\underset{\underset{\displaystyle Cl}{|}}{C}}CH_3 \quad +$$

1-Chloro-2-methylpentane   2-Chloro-2-methylpentane

$$CH_3CH_2CH_2\overset{\overset{\displaystyle CH_3}{|}}{C}HCH_3 \xrightarrow[h\nu]{Cl_2} CH_3CH_2\overset{\overset{\displaystyle CH_3}{|}}{\underset{\underset{\displaystyle Cl}{|}}{C}}HCHCH_3 \quad + \quad CH_3\overset{\overset{\displaystyle CH_3}{|}}{\underset{\underset{\displaystyle Cl}{|}}{C}}HCH_2CHCH_3$$

2-Methylpentane   3-Chloro-2-methylpentane   2-Chloro-4-methylpentane

$$ClCH_2CH_2CH_2\overset{\overset{\displaystyle CH_3}{|}}{C}HCH_3$$

1-Chloro-4-methylpentane

6.3   Even though this molecule is complex, concentrate on the bonds formed and the bonds broken. The tails of the arrows show the location of the bond to be broken, and the heads show where the electrons are moving. In radical reactions, the arrow is a fishhook (half-headed).

The reaction is a radical addition to a double bond and is a rearrangement.

6.4 Keep in mind:

(1) An electrophile is electron-poor, either because it is positively charged, because it has a functional group that is positively polarized, or because it has a vacant orbital.

(2) A nucleophile is electron-rich, either because it has a negative charge, because it has a functional group containing a lone electron pair, or because it has a functional group that is negatively polarized.

(3) Some molecules can act as both nucleophiles and electrophiles, depending on the reaction conditions.

(a) The electron-poor carbon acts as an electrophile.

(b) CH_3S^- is a nucleophile because of the sulfur lone-pair electrons and because it is negatively charged.

(c) $C_4H_6N_2$ is a nucleophile because of the lone-pair electrons of nitrogen. (Only one of the nitrogens is nucleophilic, for reasons that will be explained in a later chapter.)

(d) CH_3CHO is both a nucleophile and an electrophile because of its polar C=O bond.

6.5 BF_3 is likely to be an electrophile because the electrostatic potential map indicates that it is electron-poor (blue). The electron-dot structure shows that BF_3 lacks a complete electron octet and can accept an electron pair from a nucleophile.

6.6 Reaction of cyclohexene with HCl or HBr is an electrophilic addition reaction in which halogen acid adds to a double bond to produce a haloalkane.

6.7   The mechanism is pictured in Figure 6.3. The steps: (1) Attack of the π electrons of the double bond on HBr, forming a carbocation; (2) Formation of a C−Br bond by electron pair donation from Br⁻ to form the neutral addition product.

carbocation   2-Bromo-2-methylpropane

6.8   For curved arrow problems, follow these steps:

(1)   Locate the bonding changes. In (a), a bond from nitrogen to chlorine has formed, and a Cl Cl bond has broken.

(2)   Identify the nucleophile and electrophile (in (a), the nucleophile is ammonia and the electrophile is one Cl in the Cl₂ molecule), and draw a curved arrow whose tail is near the nucleophile and whose head is near the electrophile.

(3)   Check to see that all bonding changes are accounted for. In (a), we must draw a second arrow to show the unsymmetrical bond-breaking of Cl₂ to form Cl⁻.

(a)

(b)

A bond has formed between oxygen and the carbon of bromomethane. The bond between carbon and bromine has broken. CH₃O⁻ is the nucleophile and bromomethane is the electrophile.

(c)

nucleophilic

A double bond has formed between oxygen and carbon, and a carbon–chlorine bond has broken. Electrons move from oxygen to form the double bond and from carbon to chlorine.

6.9 This mechanism will be studied in a later chapter.

6.10 A negative value of $\Delta G°$ indicates that a reaction is favorable. Thus, a reaction with $\Delta G° = -44$ kJ/mol is more favorable than a reaction with $\Delta G° = +44$ kJ/mol.

6.11 From the expression $\Delta G° = -RT \ln K_{eq}$, we can see that a large K_{eq} is related to a large negative $\Delta G°$ and a favorable reaction. Consequently, a reaction with $K_{eq} = 1000$ is more exergonic than a reaction with $K_{eq} = 0.001$.

6.12 A reaction with $\Delta G^{\ddagger} = 45$ kJ/mol is faster than a reaction with $\Delta G^{\ddagger} = 70$ kJ/mol because a larger value for $\Delta G^{\ddagger}$ indicates a slower reaction.

6.13

Visualizing Chemistry

6.14

$$CH_3CH_2CH_2CH=CH_2 + HBr \longrightarrow$$

or

$$CH_3CH_2CH=CHCH_3 + HBr \longrightarrow$$

$$CH_3CH_2CH_2\overset{\overset{\displaystyle Br}{|}}{C}HCH_3$$

6.15

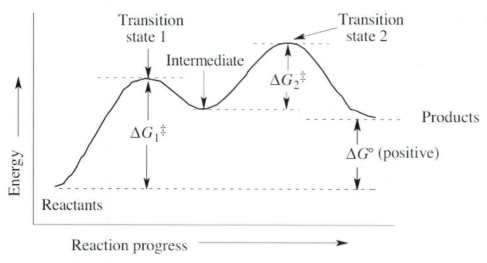

6.16 (a)   The electrostatic potential map shows that the formaldehyde oxygen is electron-rich, and the carbon–oxygen bond is polarized. The carbon atom is thus relatively electron-poor and is likely to be electrophilic.

(b)   The sulfur atom is more electron-rich than the other atoms of methanethiol and is likely to be nucleophilic.

6.17

Transition
state 1

Transition
state 2

Intermediate

$\Delta G_2^{\ddagger}$

$\Delta G_1^{\ddagger}$

Products

$\Delta G°$ (positive)

Energy

Reactants

Reaction progress

(a) $\Delta G°$ is positive.

(b)   There are two steps in the reaction.

(c)   There are two transition states, as indicated on the diagram.

6.18

(a) The reaction involves four steps, noted above.
(b) Step 1 is the most exergonic because the energy difference between reactant and product ($\Delta G°$) is greatest.
(c) Step 2 is slowest because it has the largest value of $\Delta G^{\ddagger}$.

Energy Diagrams and Reaction Mechanisms

6.19 A transition state represents a structure occurring at an energy maximum. An intermediate occurs at an energy minimum between two transition states. Even though an intermediate may be of such high energy that it cannot be isolated, it is still of lower energy than the transition states surrounding it.

6.20

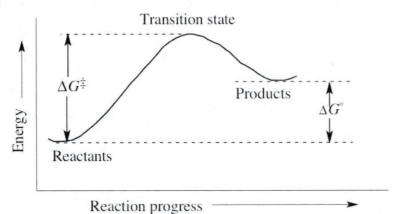

$\Delta G°$ is positive because $K_{eq} < 1$.

6.21

$\Delta G°$ is negative because Keq > 1.

6.22 Problem 6.29 shows a reaction energy diagram of a two-step exergonic reaction. Step 2 is faster than step 1 because $\Delta G^{\ddagger}_2 < \Delta G^{\ddagger}_1$.

6.23

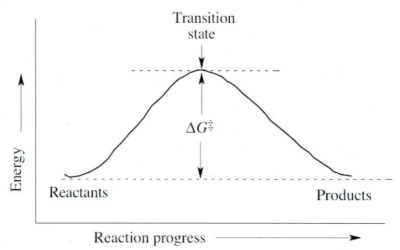

A reaction with $K_{eq} = 1$ has $\Delta G° = 0$.

6.24 (a)   The reaction is exothermic because the sign of $\Delta H°$ is negative.

(b) $\Delta G° = \Delta H° - T\Delta S°$
$= -44 \text{ kJ/mol} - (298 \text{ K}) [-0.12 \text{ kJ/(K·mol)}]$
$= -44 \text{ kJ/mol} + 36 \text{ kJ/mol}$
$= -8 \text{ kJ/mol}$

The reaction is favorable because the sign of $\Delta G°$ is negative.

6.25 (a) $K_{eq} = \dfrac{[\text{Products}]}{[\text{Reactants}]} = \dfrac{0.70}{0.30} = 2.3$

(b)   Section 6.9 states that reactions that occur spontaneously have $\Delta G^{\ddagger}$ of less than 80 kJ/mol at room temperature. Since this reaction proceeds slowly at room temperature, $\Delta G^{\ddagger}$ is probably close to 80 kJ/mol.

(c)

6.26

6.27 (a)

Initiation

Propagation

Termination

(b)

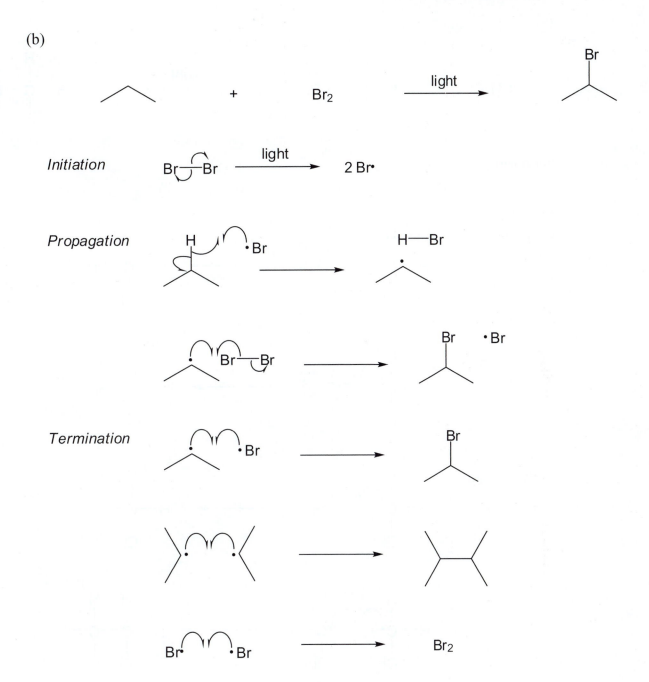

(c)

$$CH_3Cl \quad + \quad Br_2 \quad \xrightarrow{\text{light}} \quad$$

Initiation

$$Br{-}Br \xrightarrow{\text{light}} 2\,Br\cdot$$

Propagation

Termination

6.28

(a) Mechanism:

(b) Mechanism:

(c) Mechanism:

6.29

(a)

Mechanism:

(b)

Mechanism:

(c)

Mechanism:

Additional Problems

Polar Reactions

6.30

(a)

$$\overset{\delta+\quad\delta-}{CH_3CH_2C\equiv N}$$

↑
nitrile

(b)

ether

(c)

ketone $\overset{\delta-}{\underset{O}{\|}}$ $\overset{\delta-}{\underset{O}{\|}}$ ester

$$CH_3\overset{}{C}CH_2\overset{}{C}-OCH_3$$
$$\quad\quad\delta+\quad\delta+\ \delta-$$

(d) carbon–carbon
double
bonds

(e)

carbon–carbon
double bond

(f)

aromatic ring

6.31 (a)   The reaction between bromoethane and sodium cyanide is a substitution because two reagents exchange parts.

(b)   This reaction is an elimination because two products (cyclohexene and H₂O) are produced from one reactant

(c)   Two reactants form one product in this addition reaction.

(d)   This is a substitution reaction.

6.32 e = electrophilic site   n = nucleophilic site

(a)

Testosterone

(b)

Amphetamine

6.33 Nucleophiles are the electron donors in carbon bond forming reactions while the electron acceptors are the electrophiles.

(a)

$N_3^{\ominus}$ + CH_3Cl $\longrightarrow$ CH_3N_3 + $Cl^{\ominus}$

Nucleophile   Electrophile

(b)

Nucleophile   Electrophile

(c)

Electrophile   Nucleophile

6.34

(a)

(b)

6.35

(a)

(b)

(c)

(d)

Radical Reactions

6.36 Irradiation initiates the chlorination reaction by producing chlorine radicals. For every chlorine radical consumed in the propagation steps, a new Cl· radical is formed to carry on the reaction. After irradiation stops, chlorine radicals are still present to carry on the propagation steps, but, as time goes on, radicals combine in termination reactions that remove them from the reaction mixture. Because the number of radicals decreases, fewer propagation cycles occur, and the reaction gradually slows down and stops.

6.37 Pentane has three types of hydrogen atoms, $C\overset{a}{H_3}C\overset{b}{H_2}C\overset{c}{H_2}C\overset{b}{H_2}C\overset{a}{H_3}$. Although monochlorination produces $CH_3CH_2CH_2CH_2CH_2Cl$, it is not possible to avoid producing $CH_3CH_2CH_2CH(Cl)CH_3$ and $CH_3CH_2CH(Cl)CH_2CH_3$ as well. Since neopentane has only one type of hydrogen, monochlorination yields a single product.

6.38 The following compounds yield single monohalogenation products because each has only one kind of hydrogen atom.

(a) CH_3CH_3 (c) ⬡ (e) $CH_3C{\equiv}CCH_3$ (f)

6.39 To solve these problems sequentially replace one hydrogen on each carbon with a chlorine. If the compound has a different name it is a different product.

(a)   There are two products.

$\xrightarrow[\text{Light}]{Cl_2}$

1-chlorobutane   2-chlorobutane

(b)   There are four products.

$\xrightarrow[\text{Light}]{Cl_2}$

1-chloro-2-methylbutane   2-chloro-2-methylbutane   1-chloro-3-methylbutane

2-chloro-3-methylbutane

(c)   There are three products.

$\xrightarrow[\text{Light}]{Cl_2}$

1-chloro-1-methycyclopentane   1-chloro-2-methycyclopentane   1-chloro-3-methycyclopentane

6.40 To answer this question sequentially replace each hydrogen in the molecule with a hydrogen and assign the configuration to any stereocenters. If the molecule has a different name it is a different product.

(a)   There are three products.

$\xrightarrow[\text{Light}]{Cl_2}$

1-chlorobutane   (R)-2-chlorobutane   (S)-2-chlorobutane

(b)   There are seven products

(R)-2-chloro-3-methylbutane   (S)-2-chloro-3-methylbutane   1-chloro-3-methylbutane

(c)   There are nine products.

1-chloro-1-methycyclopentane

(1R,2R)-1-chloro-2-methycyclopentane   (1R,2S)-1-chloro-2-methycyclopentane   (1S,2R)-1-chloro-2-methycyclopentane   (1S,2R)-1-chloro-2-methycyclopentane

(1R,2R)-1-chloro-3-methycyclopentane   (1R,2R)-1-chloro-3-methycyclopentane   (1R,2R)-1-chloro-3-methycyclopentane   (1R,2R)-1-chloro-3-methycyclopentane

6.41

(a)

(b)

(c)

General Problems

6.42

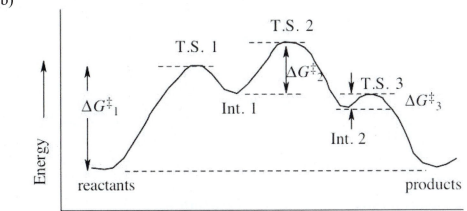

(a) $\Delta G^{\ddagger}$ for the first step is approximately 80 kJ/mol because the reaction takes place slowly at room temperature. $\Delta G^{\ddagger}$ values for the second and third steps are smaller - perhaps 60 kJ/mol for Step 2, and 40 kJ/mol for Step 3. $\Delta G°$ is approximately zero because K_{eq} is close to 1.

(b)

6.43

6.44 $\Delta G° = \Delta H° - T\Delta S°$

$= -75$ kJ/mol $- (298$ K$) (0.054$ kJ/K·mol$)$

$= -75$ kJ/mol $- 16$ kJ/mol

$= -91$ kJ/mol

The reaction is exothermic because $\Delta H°$ is negative, and it is exergonic because $\Delta G°$ is negative.

6.45

6.46 Each arrow represents either the formation of a bond or the breaking of a bond. The numbers over the arrows identify the bonds broken and formed.

	Bonds formed	Bonds broken
Step 1:	C–N (1)	C–O (2)
Step 2:	C–O (1)	C–Cl (2)
Step 3:	N–H (1)	N–H (2)

6.47

α-Terpineol

Step 1:   Attack of the double bond π electrons on the carbocation to form the isomeric carbocation.

Step 2:   Addition of water to the intermediate carbocation.

Step 3:   Deprotonation.

6.48

(a)

(b)

(c)

6.42

$$CH_3C{=}CH_2 \;+\; HBr \;\longrightarrow\; CH_3CCH_2Br \;+\; CH_3CCH_3$$

(with CH_3 substituent and H on the central carbon for the first product; CH_3 substituent and Br on central carbon for the second product)

2-Methylpropene	1-Bromo-2-methylpropane	2-Bromo-2-methylpropane

6.43

$$CH_3C{=}CH_2 \;+\; HBr \;\longrightarrow\; \left[\; CH_3\overset{+}{C}{-}CH_2 \;+\; CH_3\overset{+}{C}{-}CH_3 \;\right]$$

(first carbocation bears CH_3 substituent and H; the positive charge on the CH_2; second carbocation bears CH_3 substituent with positive charge on central carbon)

The second carbocation is more stable because more alkyl substituents are bonded to the positively charged carbon.

Chapter 7 – Alkenes: Structure and Reactivity

Chapter Outline

I. Introduction to alkene chemistry (Sections 7.1–7.7).

   A. Industrial preparation and use of alkenes (Section 7.1).

      1. Ethylene and propylene are the two most important organic chemicals produced industrially.

      2. Ethylene, propylene and butene are synthesized by thermal cracking.

         a. Thermal cracking involves homolytic breaking of C−H and C−C bonds.

         b. Thermal cracking reactions are dominated by entropy.

   B. Calculating a molecule's degree of unsaturation (Section 7.2).

      1. The degree of unsaturation of a molecule describes the number of multiple bonds and/or rings in a molecule.

      2. To calculate degree of unsaturation of a compound, first determine the equivalent hydrocarbon formula of the compound.

         a. Add the number of halogens to the number of hydrogens.

         b. Subtract one hydrogen for every nitrogen.

         c. Ignore the number of oxygens.

      3. Calculate the number of pairs of hydrogens that would be present in an alkane C_nH_{2n+2} that has the same number of carbons as the equivalent hydrocarbon of the compound of interest. The difference is the degree of unsaturation.

   C. Naming alkenes (Section 7.3).

      1. Find the longest chain containing the double bond, and name it, using "ene" as a suffix.

      2. Number the carbon atoms in the chain, beginning at the end nearer the double bond.

      3. Number the substituents and write the name.

         a. Name the substituents alphabetically.

         b. Indicate the position of the double bond.

         c. Use the suffixes -diene, -triene, etc. if more than one double bond is present.

      4. A newer IUPAC naming system places the number locant of the double bond immediately before the *-ene* suffix (not used in this book).

      5. For cycloalkenes, the double bond is between C1 and C2, and substituents receive the lowest possible numbers.

      6. A −CH₂− substituent is a methylene group, a H₂C=CH− group is a vinyl group, and a H₂C=CHCH₂− group is an allyl group.

   D. Double bond geometry (Sections 7.4–7.5).

      1. Electronic structure of alkenes (Section 7.4).

         a. Carbon atoms in a double bond are sp^2-hybridized.

         b. The two carbons in a double bond form one σ bond and one π bond.

         c. Free rotation doesn't occur around double bonds.

         d. 350 kJ/mol of energy is required to break a π bond.

      2. Cis–trans isomerism.

    a. A disubstituted alkene can have substituents either on the same side of the double bond (cis) or on opposite sides (trans).

    b. These isomers don't interconvert because free rotation about a double bond isn't possible.

    c. Cis–trans isomerism doesn't occur if one carbon in the double bond is bonded to identical substituents.

3. *E,Z* isomerism (Section 7.5).

    a. The *E,Z* system is used to describe the arrangement of substituents around a double bond that can't be described by the cis–trans system.

    b. Sequence rules for *E,Z* isomers:

      i. For each double bond carbon, rank its substituents by atomic number.

        (a). An atom with a higher atomic number receives a higher priority than an atom with a lower atomic number.

      ii. If a decision can't be reached based on the first atom, look at the second or third atom until a difference is found.

      iii. Multiple-bonded atoms are equivalent to the same number of single-bonded atoms.

    c. If the higher-ranked groups are on the same side of the double bond, the alkene has *Z* geometry.

    d. If the higher-ranked groups are on opposite sides of the double bond, the alkene has *E* geometry.

E. Stability of alkenes (Section 7.6).

1. Cis alkenes are less stable than trans alkenes because of steric strain between double bond substituents.

2. Stabilities of alkenes can be determined experimentally by measuring:

    a. Cis–trans equilibrium constants.

    b. Heats of hydrogenation – the most useful method.

3. The heat of hydrogenation of a cis isomer is a larger negative number than the heat of hydrogenation of a trans isomer.

    a. This indicates that a cis isomer is of higher energy and is less stable than a trans isomer.

4. Alkene double bonds become more stable with increasing substitution for two reasons:

    a. Hyperconjugation – a stabilizing interaction between the antibonding π orbital of the C–C bond and a filled C–H σ orbital on an adjacent substituent.

    b. More substituted double bonds have more of the stronger sp^2–sp^3 bonds.

II. Electrophilic addition reactions (Sections 7.7–7.11).

A. Addition of H–X to alkenes (Sections 7.7–7.8).

1. Mechanism of addition (Section 7.7).

    a. The electrons of the nucleophilic π bond attack the H atom of the electrophile H–X (X = Cl, Br, I, OH).

    b. Two electrons from the π bond form a new σ bond between –H and an alkene carbon.

     c. The carbocation intermediate reacts with X^- to form a C–X bond.

   2. The energy diagram has two peaks separated by a valley (carbocation intermediate).

     a. The reaction is exergonic.

     b. The first step is slower than the second step.

   3. Organic reactions are often written in different ways to emphasize different points.

   4. Orientation of addition: Markovnikov's rule (Section 7.8).

     a. In the addition of HX to a double bond, H attaches to the carbon with fewer substituents, and X attaches to the carbon with more substituents (regiospeciific).

     b. If the carbons have the same number of substituents, a mixture of products results.

B. Carbocation structure and stability (Section 7.9).

   1. Carbocations are planar; the unoccupied *p* orbital extends above and below the plane containing the cation.

   2. The stability of carbocations increases with increasing substitution.

     a. Carbocation stability can be measured by studying gas-phase dissociation enthalpies.

     b. Carbocations can be stabilized by inductive effects of neighboring alkyl groups.

     c. Carbocation can be stabilized by hyperconjugation: The more alkyl groups on the carbocation, the more opportunities there are for hyperconjugation.

C. The Hammond postulate (Section 7.10).

   1. The transition state for an endergonic reaction step resembles the product of that step because it is closer in energy.

   2. The transition state for an exergonic reaction step resembles the reactant for that step because it is closer in energy.

   3. In an electrophilic addition reaction, the transition state for alkene protonation resembles the carbocation intermediate.

   4. More stable carbocations form faster because their transition states are also stabilized.

D. Carbocation rearrangements (Section 7.11).

   1. In some electrophilic addition reactions, products from carbocation rearrangements are formed.

   2. The appearance of these products supports the two-step electrophilic addition mechanism, in which an intermediate carbocation is formed.

   3. Intermediate carbocations can rearrange to more stable carbocations by either a hydride shift (H with its electron pair) or by an alkyl shift (alkyl group with its electron pair).

   4. In both cases a group moves to an adjacent positively charged carbon, taking its bonding electron pair with it.

Solutions to Problems

7.1 Because two hydrogens must be removed from a saturated compound to introduce an unsaturation, a compound's degree of unsaturation refers to the number of pairs of hydrogens by which its formula differs from that of the corresponding saturated compound. For example, a saturated alkane with four carbons has the formula C_4H_{10}. The compound in (a), C_4H_8, which has two fewer (or one pair fewer) hydrogens, may have a double bond or a ring. C_8H_{14} thus has a degree of unsaturation of 1.

Compound	Degree of Unsaturation	Structures
(a) C_4H_8	1	$CH_3CH{=}CHCH_3$ $CH_3CH_2CH{=}CH_2$ $(CH_3)_2C{=}CH_2$

Compound	Degree of Unsaturation	Structures
(b) C_4H_6	2	$CH_2{=}CHCH{=}CH_2$ $CH_3CH{=}C{=}CH_2$ $CH_3C{\equiv}CCH_3$

Compound	Degree of Unsaturation	Structures
(c) C_3H_4	2	$H_2C{=}C{=}CH_2$ $CH_3C{\equiv}CH$

7.2 Unlike the hydrocarbons in the previous problems, the compounds in this problem contain additional elements. Review the rules for these elements.

(a) Subtract one hydrogen for each nitrogen present to find the formula of the equivalent hydrocarbon $-C_6H_4$. Compared to the alkane C_6H_{14}, the compound of formula C_6H_4 has 10 fewer hydrogens, or 5 fewer hydrogen pairs, and has a degree of unsaturation of 5.

(b) $C_6H_5NO_2$ also has 5 degrees of unsaturation because oxygen doesn't affect the equivalent hydrocarbon formula of a compound.

(c) A halogen atom is equivalent to a hydrogen atom in calculating the equivalent hydrocarbon formula. For $C_8H_9Cl_3$, the equivalent hydrocarbon formula is C_8H_{12}, and the degree of unsaturation is 3.

(d) $C_9H_{16}Br_2$ – one degree of unsaturation.

(e) $C_{10}H_{12}N_2O_3$ – 6 degrees of unsaturation.

(f) $C_{20}H_{32}ClN$ – 5 degrees of unsaturation.

7.3 A C_{16} hydrocarbon with 11 degrees of unsaturation (three rings and eight double bonds) has a formula $C_{16}H_{34} - H_{22} = C_{16}H_{12}$. Adding two hydrogens (because of the two nitrogens) and subtracting one hydrogen (because of the chlorine), gives the formula $C_{16}H_{13}ClN_2O$ for Diazepam.

Diazepam

7.4 (1) Find the longest chain containing the double bond and name it. In (a), the longest chain is a pentene.

(2) Identify the substituents. There are three methyl groups in (a).

(3) Number the substituents, remembering that the double bond receives the lowest possible number. The methyl groups are attached to C3 and C4 (two methyl groups).

(4) Name the compound, remembering to use the prefix "tri-" before "methyl" and remembering to use a number to signify the location of the double bond. The name of the compound in (a) is 3,4,4-trimethyl-1-pentene.

(a)

$$H_2C=CHCH-CCH_3$$

3,4,4-Trimethyl-1-pentene

(b)

$$CH_3CH_2CH=CCH_2CH_3$$

3-Methyl-3-hexene

(c)

$$CH_3CH=CHCHCH=CHCHCH_3$$

4,7-Dimethyl-2,5-octadiene

(d)

$$CH_3CH_2CH_2CH=CHCHCH_2CH_3$$

6-Ethyl-7-methyl-4-nonene

7.5 It's much easier to draw a structure from a given name than it is to name a structure. First, draw the carbon chain, placing the double bond or bonds in the designated locations. Then attach the cited groups in the proper positions.

(a)

$$CH_3$$

$$H_2C=CHCH_2CH_2C=CH_2$$

2-Methyl-1,5-hexadiene

(b)

$$CH_2CH_3$$

$$CH_3CH_2CH_2CH=CC(CH_3)_3$$

3-Ethyl-2,2-dimethyl-3-heptene

(c)

$$CH_3 \quad CH_3$$

$$CH_3CH=CHCH=CHC-C=CH_2$$

$$CH_3$$

2,3,3-Trimethyl-1,4,6-octatriene

(d)

$$CH_3 \qquad CH_3$$

$$CH_3CH \qquad CHCH_3$$

$$C=C$$

$$CH_3CH \qquad CHCH_3$$

$$CH_3 \qquad CH_3$$

3,4-Diisopropyl-2,5-dimethyl-3-hexene

7.6

(a)

1,2-Dimethylcyclohexene

(b)

4,4-Dimethylcycloheptene

(c)

3-Isopropylcyclopentene

7.7 In the new naming system, the bond locant appears directly before −ene or −diene.

(a)

$$CH_3 \qquad CH_3$$

$$CH_3CCH_2CH=CCH_3$$

$$CH_3$$

Old: 2,5,5-Trimethyl-2-hexene
New: 2,5,5-Trimethylhex-2-ene

(b)

Old: 2,3-Dimethyl-1,3-cyclohexadiene
New: 2,2-Dimethylcyclohexa-1,3-diene

7.8

cis-9-Tricosene

7.9   Compounds (c), (e), and (f) can exist as cis–trans isomers.

(c)

$CH_3CH_2CH=CHCH_3$

(e)

$ClCH=ClCH$

(f)

$BrCH=CHCl$

7.10

(a)

cis-4,5-Dimethyl-2-hexene

(b)

trans-6-Methyl-3-heptene

7.11   Review the sequence rules presented in Section 7.5. A summary:

Rule 1:   An atom with a higher atomic number has priority over an atom with a lower atomic number.

Rule 2:   If a decision can't be reached by using Rule 1, look at the second, third, or fourth atom away from the double-bond carbon until a decision can be made.

Rule 3:   Multiple-bonded atoms are equivalent to the same number of single-bonded atoms.

	Higher	*Lower*	*Rule*		*Higher*	*Lower*	*Rule*
(a)	$-CH_3$	$-H$	1	(b)	$-Cl$	$-CH_2Cl$	1
(c)	$-CH=CH_2$	$-CH_2CH_2Br$	3	(d)	$-OCH_3$	$-NHCH_3$	1
(e)	$-CH=O$	$-CH_2OH$	3	(f)	$-CH=O$	$-CH_2OCH_3$	3

7.12   *Highest priority* ⟶ *Lowest Priority*

(a) $-Cl$, $-OH$, $-CH_3$, $-H$

(b) $-CH_2OH$, $-CH=CH_2$, $-CH_2CH_3$, $-CH_3$

(c) $-CO_2H$, $-CH_2OH$, $-C\equiv N$, $-CH_2NH_2$

(d) $-CH_2OCH_3$, $-C\equiv N$, $-C\equiv CH$, $-CH_2CH_3$

7.13 (a)

Low   H₃C   CH₂OH   Low

C=C   *Z*

High   CH₃CH₂   Cl   High

First, consider the substituents on the right side of the double bond. −Cl ranks higher than −CH₂OH by Rule 1 of the Cahn–Ingold–Prelog rules. On the left side of the double bond, −CH₂CH₃ ranks higher than −CH₃. The isomer is *Z* when the two higher priority groups lie on the same side of the double bond. Otherwise, the isomer is *E*,

(b)

High   Cl   CH₂CH₃   Low

C=C   *E*

Low   CH₃O   CH₂CH₂CH₃   High

(c)

High   H₃C   CO₂H   High

C=C   *Z*

Low   CH₂OH   Low

Notice that the upper substituent on the left side of the double bond is of higher priority because of the methyl group attached to the ring.

(d)

Low   H   CN   High

C=C   *E*

High   H₃C   CH₂NH₂   Low

7.14

High   CO₂CH₃   High

                                      Z

Low   CH₂OH   Low

7.15   *More stable*   *Less stable*

(a)

2-Methylpropene   1-Butene
disubstituted double bond   monosubstituted double bond

(b)

trans-2-Hexene   cis-2-Hexene

no steric strain   steric strain of groups on the
                                                same side of the double bond

(c)

1-Methylcyclohexene   3-Methylcyclohexene
trisubstituted double bond   disubstituted double bond

7.16 All of these reactions are electrophilic additions of HX to an alkene. Use Markovnikov's rule to predict orientation.

(a)

+ HCl ⟶   Chlorocyclohexane

(b)

$(CH_3)_2C=CHCH_2CH_3$ + HBr ⟶ $(CH_3)_2CCH_2CH_2CH_3$

2-Bromo-2-methylpentane

In accordance with Markovnikov's rule, H forms a bond to the carbon with fewer substituents, and Br forms a bond to the carbon with more substituents.

(c)

$CH_3CHCH_2CH=CH_2$ $\xrightarrow[H_2SO_4]{H_2O}$ $CH_3CHCH_2CHCH_3$

(d)

1-Bromo-1-methylcyclohexane

7.17 Think backward in choosing the alkene starting material for synthesis of the desired haloalkanes. Remember that halogen is bonded to one end of the double bond and that more than one starting material can give rise to the desired product.

(a)

Cyclopentene

(b)

(c)

$$CH_3CH_2CH\!=\!CHCH_2CH_3 \;+\; HBr \longrightarrow CH_3CH_2CH_2\overset{\displaystyle Br}{\underset{\displaystyle |}{C}}HCH_2CH_3$$

3-Hexene

(d)

7.18 The more stable carbocation is formed

(a)

(b)

7.19 Two representations of the secondary carbocation are shown on the left below. This secondary carbocation can experience hyperconjugative overlap with two hydrogens under normal circumstances. However, in the alignment shown in the drawing, only one hydrogen (circled) is in the correct position for hyperconjugative overlap with the carbocation carbon.

Because there is rotation about the carbon–carbon bonds, all of the hydrogens starred in the representation on the far right can be involved in hyperconjugation at some time.

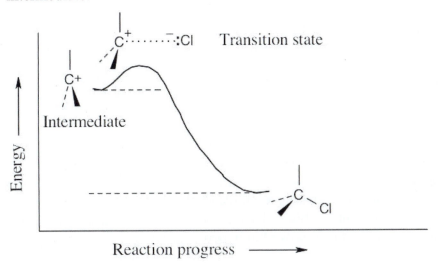

7.20 The second step in the electrophilic addition of HCl to an alkene is exergonic. According to the Hammond postulate, the transition state should resemble the carbocation intermediate.

7.21

Step 1:   Electrophilic addition of H⁺ to double bond.

Step 2:   Hydride shift that forms a more stable tertiary carbocation.

Step 3:   Reaction of carbocation with Br⁻.

Visualizing Chemistry

7.22

(a)

2,4,5-Trimethyl-2-hexene

(b)

1-Ethyl-3,3-dimethylcyclohexene

7.23

(a)

High Cl   Low

E

Low   C=O   High

        H

(b)   High   OCH$_3$ High

                      C=O

    Low   OH   Low

Z

7.24

=CHCH$_3$

or

       $\xrightarrow{\text{HCl}}$

—CH$_2$CH$_3$

$\left[\begin{array}{c} {}^{+}\text{—CH}_2\text{CH}_3 \end{array} \right] \longrightarrow$

Cl

CH$_2$CH$_3$

Either of the two compounds shown can form the illustrated tertiary carbocation when they react with HCl. In the conformation shown, the three circled hydrogens are aligned for maximum overlap with the vacant *p* orbital. Because of conformational mobility, the three starred hydrogens are also able to be involved in hyperconjugation.

7.25

Mechanism Problems

7.26 (a)

Mechanism:

(b)

Mechanism:

(c)

Mechanism:

7.27 (a)

Mechanism:

(b)

Mechanism:

(c)

Mechanism:

or

7.28 Note: Because the carbocation intermediate is resonance stabilized there are two locations where the bromine can add.

Mechanism:

7.29

methyl vinyl ether

unstable

If the oxygen is protonated resonance would result in a very unstable resonance form with a +2 charge on the oxygen.

If C1 is protonated there is no way to stabilize the positive charge

very stable

The proton adds to C2. When the proton adds to this carbon the carbocation is stabilized by formation of a extremely stable resonance form containing an extra pi-bond.

7.30

3°
carbocation

hydride
shift

3°
carbocation

7.31

bond
shift
(alkyl shift)

7.32

Step 2, in which the double bond electrons add to the carbocation, is an alkene electrophilic addition.

7.33

epi-Aristolochene

Steps 1 and 2 are alkene electrophilic additions, and steps 3 and 4 involve carbocation rearrangements.

Additional Problems

Calculating a Degree of Unsaturation

7.34 The purpose of this problem is to give you experience in calculating the number of double bonds and/or rings in a formula. Additionally, you will learn to draw structures containing various functional groups. Remember that any formula that satisfies the rules of valency is acceptable. Try to identify functional groups in the structures that you draw. Many structures are acceptable for each part of this problem.

(a) $C_{10}H_{16}$ – 3 degrees of unsaturation. Examples:

$$CH_3CH_2CH=CHCH=CHCH=CHCH_2CH_3$$

(b) C_8H_8O. The equivalent hydrocarbon is C_8H_8, which has 5 degrees of unsaturation.

(c) $C_7H_{10}Cl_2$ has C_7H_{12} as its equivalent hydrocarbon formula. $C_7H_{10}Cl_2$ has two degrees of unsaturation.

(d) $C_{10}H_{16}O_2$ – 3 degrees of unsaturation ($C_{10}H_{16}$ = equivalent hydrocarbon formula).

double bonds   ester   ketones

$H_2C{=}CHCH{=}CHCH_2CH_2CH_2CH_2COCH_3$

alcohol   ketone   aldehyde   alcohols

ether

(e) $C_5H_9NO_2$ – 2 degrees of unsaturation (C_5H_8 = equivalent hydrocarbon formula).

ketone   amide   double bond   nitro group

$CH_3CH_2CCH_2CNH_2$ $H_2C{=}CHCH_2CH_2CH_2N^{+}{-}O^{-}$

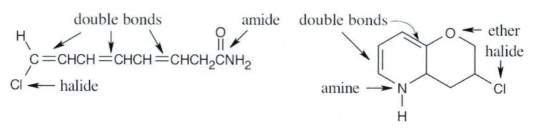

double bond   carboxylic acid   amide   ketone

$H_2C{=}CHCH_2NHCH_2COH$ amine

amine   alcohol   amine   ether

(f) $C_8H_{10}ClNO$ – 4 degrees of unsaturation (C_8H_{10} = equivalent hydrocarbon formula).

double bonds   amide   double bonds   ether   halide

$C{=}CHCH{=}CHCH{=}CHCH_2CNH_2$

Cl ← halide   amine → N   Cl

aromatic ring CH_3 halide → Cl   double bond   H   N ← amine

ketone →

HO   amine CH_2NH_2 amine   ketone

alcohol   Cl ← halide H_2N O   halide → Cl

7.35

	Compound	Equivalent hydrocarbon formula	Degree of unsaturation	Complete formula
(a)	$C_8H_?O_2$	C_8H_{18}	3	$C_8H_{12}O_2$
(b)	$C_7H_?N$	C_7H_{16}	2	$C_7H_{13}N$
(c)	$C_9H_?NO$	C_9H_{20}	4	$C_9H_{13}NO$

7.36 Solve this problem in the same way as we solved problems 7.3 and 7.27. A C_{22} hydrocarbon with 12 degrees of unsaturation (four rings and eight double bonds) has a formula $C_{22}H_{46} - H_{24} = C_{22}H_{22}$. Adding two hydrogens (because of the two nitrogens) and subtracting one hydrogen (because of the chlorine), gives the formula $C_{22}H_{23}ClN_2O_2$ for Loratadine.

Loratadine

Naming Alkenes

7.37

(a) (E)-4-Methyl-2-hexene

(b) (Z)-4-Ethyl-3,7-dimethyl-2-octene

(c) 2-Ethyl-1-butene

(d) (5E)-3,4-Dimethyl-1,5-heptadiene

(e) (2Z,4E)-4,5-Dimethyl-2,4-octadiene

(f) 1,2-Butadiene

7.38

(a)

(4E)-2,4-Dimethyl-1,4-hexadiene

(b)

cis-3,3-Dimethyl-4-propyl-1,5-octadiene

(c)

CH₃CHCH=C=CH₂
(with CH₃ branch on the first carbon)

4-Methyl-1,2-pentadiene

(d)

(3E,5Z)-2,6-Dimethyl-1,3,5,7-octatetraene

(e)

CH₃CH₂CH₂CH₂   H
            C=C
CH₃CH₂CH₂CH₂   CH₃

3-Butyl-2-heptene

(f)

(CH₃)₃C   H
        C=C
    H   C(CH₃)₃

trans-2,2,5,5-Tetramethyl-3-hexene

7.39

(a)

3-Methylcyclohexene

(b)

1,5-Dimethylcyclopentene

(c)

Ethyl-1,3-cyclobutadiene

(d)

1,2-Dimethyl-1,4-cyclohexadiene

(e)

5-Methyl-1,3-cyclohexadiene

(f)

1,5-Cyclooctadiene

7.40 Because the longest carbon chain contains 8 carbons and 3 double bonds, ocimene is an *octatriene*. Start numbering at the end that will give the lower number to the first double bond (1,3,6 is lower than 2,5,7). Number the methyl substituents and, finally, name the compound.

Ocimene
(3*E*)-3,7-Dimethylocta-1,3,6-triene

7.41

α-Farnesene

(3*E*,6*E*)-3,7,11-Trimethyl-1,3,6,10-dodecatetraene

7.42

Menthene

7.43

$CH_3CH_2CH_2CH{=}CH_2$

1-Pentene

(*Z*)-2-Pentene

(*E*)-2-Pentene

2-Methyl-1-butene

3-Methyl-1-butene

2-Methyl-2-butene

7.44 Start with 1-hexene and continue on until all hexenes are named, making sure that *E,Z* designations have been made when necessary. Then move on to all 1-pentenes, 2-pentenes, etc.

CH₃CH₂CH₂CH₂CH=CH₂

1-Hexene

(*Z*)-2-Hexene

(*E*)-2-Hexene

(*Z*)-3-Hexene

(*E*)-3-Hexene

2-Methyl-1-pentene

3-Methyl-1-pentene

4-Methyl-1-pentene

2-Methyl-2-pentene

(*Z*)-3-Methyl-2-pentene

(*E*)-3-Methyl-2-pentene

(*Z*)-4-Methyl-2-pentene

(*E*)-4-Methyl-2-pentene

2,3-Dimethyl-1-butene

3,3-Dimethyl-1-butene

2-Ethyl-1-butene

2,3-Dimethyl-2-butene

Alkene Isomers and Their Stability

7.45 *Highest priority* ⟶ *Lowest Priority*

(a)   −I, −Br, −CH₃, −H

$-I, -Br, -CH_3, -H$

(b)   −OCH₃, −OH, −CO₂H, −H

$-OCH_3, -OH, -CO_2H, -H$

(c)   −CO₂CH₃, −CO₂H, −CH₂OH, −CH₃

$-CO_2CH_3, -CO_2H, -CH_2OH, -CH_3$

(d)   −COCH₃, −CH₂CH₂OH, −CH₂CH₃, −CH₃

$-COCH_3, -CH_2CH_2OH, -CH_2CH_3, -CH_3$

(e)   −CH₂Br, −C ≡ N, −CH₂NH₂, −CH=CH₂

$-CH_2Br, -C \equiv N, -CH_2NH_2, -CH=CH_2$

(f)   −CH₂OCH₃, −CH₂OH, −CH = CH₂, CH₂CH₃

$-CH_2OCH_3, -CH_2OH, -CH = CH_2, CH_2CH_3$

7.46

(a)

(b)

(c)

(d)

7.47

(a)

(b)

(c)

(d)

High   NC   CH₃   Low

                 C=C   *E*   (correct)

Low   (CH₃)₂NCH₂   CH₂CH₃   High

(e)

High   Br

              C=C   This compound doesn't show *E-Z* isomerism.

Low   H

(f)

Low   HOCH₂   CO₂H   High

                 C=C   *E*   (correct)

High   H₃COCH₂   COCH₃   Low

7.48   The trend for alkene stability is below.

monosubstituted   cis-   terminal   trans-   trisubstituted   tetrasubstituted
                       disubstituted   disubstituted   disubstituted

(a)

terminal-   trisubstituted   tetrasubstituted
disubstituted

(b)

cis-   trisubstituted   tetrasubstituted
disubstituted

(c)

monosubstituted   terminal-   trisubstituted
                           disubstituted

7.49 As expected, the two trans compounds are more stable than their cis counterparts. The cis–trans difference is much more pronounced for the tetramethyl compound, however. Build a model of *cis*-2,2,5,5-tetramethyl-3-hexene and notice the extreme crowding of the methyl groups. Steric strain makes the cis isomer much less stable than the trans isomer and causes cis $\Delta H°_{hydrog}$ to have a much larger negative value than trans $\Delta H°_{hydrog}$ for the hexene isomers.

cis isomer   trans isomer

7.50 A model of cyclohexene shows that a six-membered ring is too small to contain a trans double bond without causing severe strain to the ring. A ten-membered ring is flexible enough to accommodate either a cis or a trans double bond, although the cis isomer has less strain than the trans isomer.

7.51 Build models of the two cyclooctenes and notice the large amount of torsional strain in *trans*-cyclooctene relative to *cis*-cyclooctene. This torsional strain, in addition to angle strain, causes the trans isomer to be of higher energy and to have a $\Delta H°_{hydrog}$ larger than the $\Delta H°_{hydrog}$ of the cis isomer.

7.52 Models show that the difference in strain between the two cyclononene isomers is smaller than the difference between the two cyclooctene isomers. This reduced strain is due to a combination of less angle strain and more puckering to relieve torsional strain and is reflected in the fact that the values of $\Delta H°_{hydrog}$ for the two cyclononene isomers are relatively close. Nevertheless, the trans isomer is still more strained than the cis isomer.

7.53

Tamoxifen   Clomiphene

Carbocations and Electrophilic Addition Reactions

7.54 The order of carbocation stability is below.

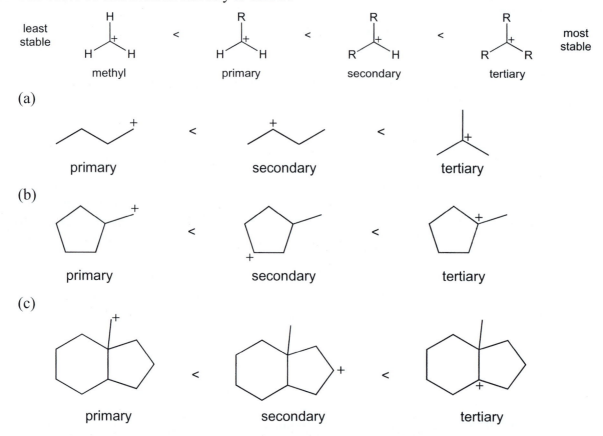

7.55 Note: Because carbocation formation is endergonic, the transition state structure will more closely resemble the intermediate carbocation that it will the starting material. Thus, a more stable carbocation will form more rapidly than a less stable one. In each example below, a tertiary carbocation will from more quickly than a secondary carbocation because the resulting transition state structure is lower in energy.

(a)

(b)

(c)

secondary carbocation
slower

vs.

tertiary carbocation
faster

7.56 Carbocations can be resonance stabilized. Draw all the resonance forms that would stabilize each carbocation.

(a)

(b)

(c)

7.57

(a)

$$CH_3CH_2CH=\underset{\underset{CH_3}{|}}{C}CH_2CH_3 \quad \xrightarrow[\text{H}_2\text{SO}_4]{\text{H}_2\text{O}} \quad CH_3CH_2CH_2\underset{\underset{\underset{OH}{|}}{\overset{\overset{CH_3}{|}}{C}}}{}CH_2CH_3$$

(b)

(c)

The third product results from rearrangement to a more stable tertiary carbocation.

(d)

$$H_2C=CHCH_2CH_2CH_2CH=CH_2 \; + \; 2 \; HCl \; \longrightarrow \; CH_3\underset{\underset{Cl}{|}}{C}HCH_2CH_2CH_2\underset{\underset{Cl}{|}}{C}HCH_3$$

7.58

(a)

$$\text{(methylenecyclohexane)} + \text{HBr} \longrightarrow \text{(1-bromo-1-methylcyclohexane)}$$

(b)

$$+ \quad \text{HBr} \longrightarrow$$

(c)

$$\underset{\overset{|}{\text{CH}_3}}{\text{CH}_3\text{CH}=\text{CHCHCH}_3} + \text{HBr} \longrightarrow \underset{\overset{|}{\text{Br}}\;\overset{|}{\text{CH}_3}}{\text{CH}_3\text{CHCH}_2\text{CHCH}_3} + \underset{\overset{|}{\text{Br}}\;\overset{|}{\text{CH}_3}}{\text{CH}_3\text{CH}_2\text{CHCHCH}_3}$$

$$+ \quad \underset{\underset{\text{Br}}{\overset{\text{CH}_3}{|}}}{\text{CH}_3\text{CH}_2\text{CH}_2\text{CCH}_3}$$

The third product results from rearrangement to a more stable tertiary carbocation.

7.59

(a)

$$\underset{\overset{|}{\text{CH}_3}}{\text{CH}_3\text{CH}_2\text{C}=\text{CHCH}_3} + \text{H}_2\text{O} \xrightarrow[\text{catalyst}]{\text{acid}} \underset{\underset{\text{OH}}{\overset{\text{CH}_3}{|}}}{\text{CH}_3\text{CH}_2\text{CCH}_2\text{CH}_3}$$

(b)

$$+ \quad \text{H}_2\text{O} \xrightarrow[\text{catalyst}]{\text{acid}}$$

(c)

$$\underset{\overset{|}{\text{CH}_3}}{\text{CH}_3\text{CHCH}_2\text{CH}=\text{CH}_2} + \text{H}_2\text{O} \xrightarrow[\text{catalyst}]{\text{acid}} \underset{\overset{|}{\text{CH}_3}\;\overset{|}{\text{OH}}}{\text{CH}_3\text{CHCH}_2\text{CHCH}_3}$$

7.60 (a)

$$\text{CH}_3\text{CH}_2\overset{\text{H}}{\underset{}{\text{CHCH}_2}}{}^+ \longrightarrow \text{CH}_3\text{CH}_2\overset{+}{\text{CHCH}_3}$$

primary carbocation   secondary carbocation

The primary carbocation rearranges to the more stable secondary carbocation by a hydride shift.

(b)

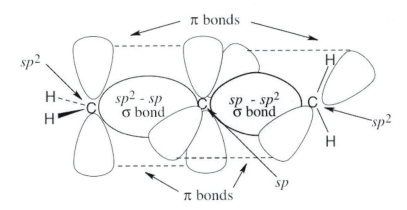

CH₃C—CHCH₃ ⟶ CH₃C⁺—CH₂CH₃

secondary carbocation   tertiary carbocation

This hydride shift produces a tertiary carbocation from rearrangement of a secondary carbocation.

(c)

primary carbocation   tertiary carbocation

An alkyl shift forms a tertiary carbocation from a primary carbocation. In this example, rearrangement involves migration of the electrons from one of the cyclobutane ring bonds to form a cyclopentane ring.

General Problems

7.61 The central carbon of allene forms two σ bonds and two π bonds. The central carbon is *sp*-hybridized, and the carbon–carbon bond angle is 180°, indicating linear geometry for the carbons of allene. The hydrogen atoms on one terminal =CH₂ units are perpendicular to the pair of hydrogen atoms on the other =CH₂ group.

7.62 The heat of hydrogenation for a typical diene is 2 x ($\Delta H°_{hydrog}$ of an alkene) = −252 kJ/mol. Thus, allene, with $\Delta H°_{hydrog}$ = −295 kJ/mol is 43 kJ/mol higher in energy than a typical diene and is less stable.

7.63

Retin A

Retin A contains five carbon–carbon double bonds. Since the ring double bond can't isomerize, Retin A can have $2^4 = 16$ isomers.

7.64

(a)

Fucoserraten

(3E, 5Z)-1,3,5-Octatriene

(b)

Ectocarpene

6-[(Z)-1-Butenyl]-1,4-cycloheptadiene

7.65   Treatment of the *tert*-butyl ester with trifluoroacetic acid cleaves the $-OC(CH_3)_3$ group and replaces it with an $-OH$ group, which has a lower priority than the $-OCH_3$ group on the upper carbon and the $-OC(CH_3)_3$ group that was removed. The result is a change in the *E,Z* designation around the double bond without breaking any of the bonds attached to the double-bond carbons.

7.66

Attack of the π electrons of the double bond on H^+ yields the carbocation pictured on the far right. A bond shift (alkyl shift) produces the bracketed intermediate, which reacts with Br^- to yield 1-bromo-2-methylcyclobutane.

7.67 (a) C₂₇H₄₆O   5 degrees of unsaturation

(b) C₁₄H₉Cl₅   8 degrees of unsaturation

(c) C₂₀H₃₄O₅   4 degrees of unsaturation

(d) C₈H₁₀N₄O₂   6 degrees of unsaturation

(e) C₂₁H₂₈O₅   8 degrees of unsaturation

(f) C₁₇H₂₃NO₃   7 degrees of unsaturation

7.68 The reaction is exergonic because it is spontaneous. According to the Hammond postulate, the transition state should resemble the isobutyl cation.

7.69

7.70

Transition State #1   *Transition State #2*

2-Bromopentane path

1-Bromopentane path

The first step (carbocation formation) is endergonic for both reaction paths, and both transition states resemble the carbocation intermediates. Transition states for the exergonic second step also resemble the carbocation intermediate. Transition state #1 fo 2-bromopentane.

7.71 Reaction of 1-chloropropane with the Lewis acid $AlCl_3$ forms a carbocation. The less stable propyl carbocation (primary undergoes a hydride shift to produce the more stable isopropyl carbocation (secondary), which reacts with benzene to give isopropylbenzene.

7.72

2,3-Dimethyl-1-butene   2-Bromo-2,3-   2,3-Dimethyl-2-butene
                            dimethylbutane

The product, 2,3-dimethyl-2-butene, is formed by elimination of HBr from 2-bromo-2,3-dimethylbutane. This product forms because it has the more substituted double bond.

Chapter 8 – Alkenes: Reactions and Synthesis

Chapter Outline

I. Preparation of alkenes (Section 8.1).
  A. Dehydrohalogenation.
     1. Reaction of an alkyl halide with a strong base forms an alkene, with loss of HX.
  B. Dehydration.
     1. Treatment of an alcohol with a strong acid forms an alkene, with loss of H_2O.
II. Addition reactions of alkenes (Sections 8.2–8.6).
  A. Addition of halogens (halogenation) (Section 8.2).
     1. Br_2 and Cl_2 react with alkenes to yield 1,2-dihaloalkanes.
     2. Reaction occurs with anti stereochemistry: Both halogens come from opposite sides of the molecule.
     3. The reaction intermediate is a cyclic halonium intermediate that is formed in a single step by interaction of an alkene with Br^+ or Cl^+.
  B. Addition of hypohalous acids (Section 8.3).
     1. Alkenes add HO−X (X = Br or Cl), forming halohydrins, when they react with halogens in the presence of H_2O.
     2. The added nucleophile (H_2O) intercepts the halonium ion to yield a halohydrin.
     3. Bromohydrin formation is usually achieved by NBS in aqueous DMSO.
     4. Aromatic rings are inert to halohydrin reagents.
  C. Addition of water to alkenes (Section 8.4).
     1. Hydration.
        a. Water adds to alkenes to yield alcohols in the presence of a strong acid catalyst.
        b. Although this reaction is important industrially, reaction conditions are too severe for most molecules.
     2. Oxymercuration.
        a. Addition of $Hg(OAc)_2$, followed by $NaBH_4$, converts an alkene to an alcohol.
        b. The mechanism of addition proceeds through a mercurinium ion.
        c. The reaction follows Markovnikov regiochemistry.
  D. Addition of water to alkenes: hydroboration/oxidation (Section 8.5).
     1. BH_3 adds to an alkene to produce an organoborane.
        a. Three molecules of alkene add to BH_3 to produce a trialkylborane.
     2. Treatment of the trialkylborane with H_2O_2 forms 3 molecules of an alcohol.
     3. Addition occurs with syn stereochemistry.
     4. Addition occurs with non-Markovnikov regiochemistry.
        a. Hydroboration is complementary to oxymercuration/reduction.
     5. The mechanism of hydroboration involves a four-center, cyclic transition state.
        a. This transition state explains syn addition.
        b. Attachment of boron to the less sterically crowded carbon atom of the alkene also explains non-Markovnikov regiochemistry.
III. Reduction and oxidation of alkenes (Sections 8.6–8.8).
  A. Reduction of alkenes (Section 8.6).

1. In organic chemistry, reduction increases electron density on carbon either by forming C−H bonds or by breaking C−O, C−N, or C−X bonds.
2. Catalytic hydrogenation reduces alkenes to saturated hydrocarbons.
   a. The catalysts most frequently used are Pt and Pd.
   b. Catalytic hydrogenation is a heterogeneous process that takes place on the surface of the catalyst.
   c. Hydrogenation occurs with syn stereochemistry.
   d. The reaction is sensitive to the steric environment around the double bond.
3. Alkenes are much more reactive than other functional groups.

B. Oxidation of alkenes (Sections 8.7–8.8).
1. In organic chemistry, oxidation decreases electron density on carbon either by forming C−O, C−N, or C−X bonds or by breaking C−H bonds.
2. Epoxidation (Section 8.7).
   a. Epoxides can be prepared by reaction of an alkene with a peroxyacid RCO_3H.
      i. The reaction occurs in one step with syn stereochemistry.
   b. Epoxides are also formed when halohydrins are treated with base.
   c. Acid-catalyzed reaction of an epoxide ring with water yields a 1,2-diol (glycol).
      i. Ring opening takes place by back-side attack of a nucleophile on the protonated epoxide ring.
      ii. A *trans*-1,2-diol is formed from an epoxycycloalkane.
3. Hydroxylation.
   a. OsO_4 causes the addition of two −OH groups to an alkene to form a diol.
      i. Hydroxylation occurs through a cyclic osmate intermediate.
   b. A safer reaction uses a catalytic amount of OsO_4 and the oxidant NMO.
   c. The reaction occurs with syn stereochemistry.
4. Cleavage to carbonyl compounds (Section 8.8).
   a. O_3 (ozone)causes cleavage of an alkene to produce aldehyde and/or ketone fragments.
      i. The reaction proceeds through a cyclic molozonide, which rearranges to an ozonide that is reduced by Zn.
   b. $KMnO_4$ in neutral or acidic solution cleaves alkenes to yield ketones, carboxylic acids or CO_2.
   c. Diols can be cleaved with HIO_4(periodic acid)to produce carbonyl compounds.

IV. Addition of carbenes (Section 8.9).
   A. A carbene (R_2C:) adds to an alkene to give a cyclopropane.
   B. The reaction occurs in a single step, without intermediates.
   C. Treatment of $HCCl_3$ with KOH forms dichlorocarbene.
      1. Addition of dichlorocarbene to a double bond is stereospecific, and only cisdichlorocyclopropanes are formed.
   D. The Simmons-Smith reaction (CH_2I_2, Zn−Cu) produces a nonhalogenated cyclopropane via a carbenoid reagent.

V. Radical additions to alkenes: chain-growth polymers (Section 8.10).

   A. Many types of polymers can be formed by radical polymerization of alkene monomers.

      1. There are 3 steps in a chain-growth polymerization reaction.

         a. *Initiation* involves cleavage of a weak bond to form a radical

            i.   The radical adds to an alkene to generate an alkyl radical.

         b. The alkyl radical adds to another alkene molecule (*propagation*) to yield a second radical.

            i.   This step is repeated many, many times.

         c. *Termination* occurs when two radical fragments combine.

      2. Mechanisms of radical reactions are shown by using fishhook arrows.

      3. As in electrophilic addition reactions, the more stable radical (more substituted) is formed in preference to the less stable radical.

   B. Biological additions of radicals to alkenes (Section 8.11).

      1. Biochemical radical reactions are more controlled than laboratory radical reactions.

VI. Stereochemistry of reactions (Sections 8.12–8.13).

   A. Addition of H_2O to an achiral alkene (Section 8.12).

      1. When H_2O adds to an achiral alkene, a racemic mixture of products is formed.

      2. The achiral cationic intermediate can react from either side to produce a racemic mixture.

      3. Alternatively, the transition states for top side reaction and bottom side reaction are enantiomers and have the same energy.

      4. Enzyme-catalyzed reactions give a single enantiomer, even when the substrate is achiral.

   B. Addition of H_2O to a chiral alkene (Section 8.13).

      1. When H^+ adds to a chiral alkene, the intermediate carbocation is chiral.

      2. The original chirality center is unaffected by the reaction.

      3. Reaction of H_2O with the carbocation doesn't occur with equal probability from either side, and the resulting product is an optically active mixture of diastereomeric alcohols.

      4. Reaction of a chiral reactant with an achiral reactant leads to unequal amounts of diastereomeric products.

Solutions to Problems

8.1

Dehydrobromination may occur in either of two directions to yield a mixture of products.

8.2

(Z)-3-Methyl-3-hexene   (E)-3-Methyl-3-hexene

(Z)-3-Methyl-2-hexene   (E)-3-Methyl-2-hexene

2-Ethyl-1-pentene

Five alkene products, including *E*, *Z* isomers, might be obtained by dehydration of 3-methyl-3-hexanol.

8.3

methyl-3-hexanol.

1,2-Dimethylcyclohexene   *trans*-1,2-Dichloro-1,2-dimethylcyclohexane

The chlorines are trans to one another in the product, as are the methyl groups.

8.4

1,2-Dimethylcyclohexene

Addition of hydrogen halides involves formation of an open carbocation, not a cyclic halonium ion intermediate. The carbocation, which is sp^2-hybridized and planar, can be attacked by chloride from either top or bottom, yielding products in which the two methyl groups can be either cis or trans to each other.

8.5

–Br and –OH are trans in the product.

8.6   Reaction of the alkene with Br_2 (formed from NBS) produces a cyclic bromonium ion. When this bromonium ion is opened by water, a partial positive charge develops at the carbon whose bond to bromine is being cleaved.

vs

less favorable

Since a secondary carbon can stabilize this charge better than a primary carbon, opening of the bromonium ion occurs at the secondary carbon to yield the Markovnikov product.

8.7   Keep in mind that oxymercuration is equivalent to Markovnikov addition of H_2O to an alkene.

(a)

$$CH_3CH_2CH_2CH{=}CH_2 \xrightarrow[\text{2. NaBH}_4]{\text{1. Hg(OAc)}_2,\ H_2O} CH_3CH_2CH_2\overset{\overset{\displaystyle OH}{|}}{C}HCH_3$$

(b)

$$CH_3CH_2CH{=}CCH_3 \xrightarrow[\text{2. NaBH}_4]{\text{1. Hg(OAc)}_2,\ H_2O} CH_3CH_2CH_2CCH_3$$

(with CH_3 substituent on the alkene carbon, and product bearing CH_3 and OH groups)

8.8   Think backwards to select the possible alkene starting materials for the alcohols pictured.

(a) $CH_3C{=}CHCH_2CH_2CH_3$ (with CH_3 substituent)   *or*

$H_2C{=}CCH_2CH_2CH_2CH_3$ (with CH_3 substituent)

$$\xrightarrow[\text{2. NaBH}_4]{\text{1. Hg(OAc)}_2,\ H_2O}$$

$CH_3CCH_2CH_2CH_2CH_3$ (with OH and CH_3 substituents)

(b)

$$\xrightarrow[\text{2. NaBH}_4]{\text{1. Hg(OAc)}_2,\ H_2O}$$

(product: cyclohexyl-CH(OH)CH$_3$)

Oxymercuration occurs with Markovnikov orientation.

8.9   Hydroboration/oxidation occurs with non-Markovnikov regiochemistry to give products in which $-OH$ is bonded to the less highly substituted carbon.

(a) $CH_3C{=}CHCH_2CH_3$ (with CH_3 substituent)

less substituted

$$\xrightarrow[\text{2. H}_2O_2,\ ^-OH]{\text{1. BH}_3,\ THF}$$

$CH_3C{-}CHCH_2CH_3$ (with CH_3, H, and OH substituents)

(b)   *less substituted*

$$\xrightarrow[\text{2. H}_2O_2,\ ^-OH]{\text{1. BH}_3,\ THF}$$

(product: cyclohexyl-CH(OH)CH$_3$)

8.10 As described in Worked Example 8.2, the strategy in this sort of problem begins with a look backward. In more complicated syntheses this approach is essential, but even in problems in which the functional group(s) in the starting material and the reagents are known, this approach is effective.

All the products in this problem result from hydroboration/oxidation of a double bond. The $-OH$ group is bonded to the less substituted carbon of the double bond in the starting material.

(a)

$$CH_3CHCH=CH_2 \quad \xrightarrow[\text{2. } H_2O_2, \ ^-OH]{\text{1. } BH_3, \ THF} \quad CH_3CHCH_2CH_2OH$$

with CH$_3$ substituents on the second carbon.

(b)

$$CH_3C=CHCH_3 \quad \xrightarrow[\text{2. } H_2O_2, \ ^-OH]{\text{1. } BH_3, \ THF} \quad CH_3CHCHCH_3$$

with CH$_3$ and OH substituents.

This product can also result from oxymercuration of the starting material in (a).

(c)

$$\xrightarrow[\text{2. } H_2O_2, \ ^-OH]{\text{1. } BH_3, \ THF}$$

8.11 The drawings below show the transition states resulting from addition of BH$_3$ to the double bond of the cycloalkene. Addition can occur on either side of the double bond.

Reaction of the two neutral alkylborane adducts with hydrogen peroxide gives two alcohol isomers. In one isomer, the two methyl groups have a cis relationship, and in the other isomer they have a trans relationship.

8.12 Catalytic hydrogenation produces alkanes from alkenes.

(a)

$$CH_3C{=}CHCH_2CH_3 \xrightarrow[\text{Pd/C in ethanol}]{H_2} CH_3CHCH_2CH_2CH_3$$

2-Methyl-2-pentene   2-Methylpentane

(b)

3,3-Dimethylcyclopentene $\xrightarrow[\text{Pd/C in ethanol}]{H_2}$ 1,1-Dimethylcyclopentane

(c)

3-*tert*-Butylcyclohexene $\xrightarrow[\text{Pd/C in ethanol}]{H_2}$ *tert*-Butylcyclohexane

8.13 Epoxidation using *m*-chloroperoxybenzoic acid (RCO₃H) is a syn addition of oxygen to a double bond. The original bond stereochemistry is retained.

cis-2-Butene $\xrightarrow{RCO_3H}$ *cis*-2,3-Epoxybutane

In the epoxide product, as in the alkene starting material, the methyl groups are cis.

8.14 Reaction of an alkene with a catalytic amount of OsO₄, in the presence of *N*-morpholine *N*oxide (NMO), yields a diol product. To pick a starting material for these products, choose an alkene that has a double bond between the diol carbons. The products in (b) and (c) can also be formed by ring opening of an epoxide formed either from a peroxyacid or from a halohydrin.

(a)

1-Methylcyclohexene $\xrightarrow[\text{OsO}_4]{\text{catalytic}}$ $\xrightarrow{\text{NMO}}$

(b)

$$CH_3CH_2CH{=}C(CH_3)_2 \xrightarrow[\text{NMO}]{\text{OsO}_4} CH_3CH_2CH{-}CCH_3$$

2-Methyl-2-pentene

(c)

$$CH_2\!=\!CHCH\!=\!CH_2 \xrightarrow[\text{NMO}]{\text{OsO}_4} \begin{array}{c} \text{HO OH} \\ | \ | \\ HOCH_2CHCHCH_2OH \end{array}$$

1,3-Butadiene

8.15 Both sets of reactants cleave double bonds. Aqueous KMnO₄ produces a carboxylic acid from a double bond carbon that is monosubstituted and a ketone from a double bond carbon that is disubstituted. Ozone produces an aldehyde from a double bond carbon that is monosubstituted and a ketone from a double bond carbon that is disubstituted. If the double bond is part of a ring, both carbonyl groups occur in the same product molecule.

(a)

(b)

8.16 Orient the fragments so that the oxygens point toward each other. Remove the oxygens, and draw a double bond between the remaining carbons.

(a)

$$(CH_3)_2C\!=\!CH_2 \xrightarrow[\text{2. Zn, H}_3\text{O}^+]{\text{1. O}_3} (CH_3)_2C\!=\!O \ + \ O\!=\!CH_2$$

(b)

$$CH_3CH_2CH\!=\!CHCH_2CH_3 \xrightarrow[\text{2. Zn, H}_3\text{O}^+]{\text{1. O}_3} CH_3CH_2CH\!=\!O \ + \ O\!=\!CHCH_2CH_3$$

8.17 Reaction of a double bond with chloroform under basic conditions gives a product with a cyclopropane ring in which one of the carbons has two chlorine atoms bonded to it. Reaction of a double bond with CH₂I₂ yields a product with a cyclopropane ring that has a −CH₂− group.

(a)

(b)

$$\begin{array}{c} CH_3 \\ | \\ CH_3CHCH_2CH\!=\!CHCH_3 \end{array} + \ CH_2I_2 \xrightarrow{\text{Zn(Cu)}} \begin{array}{c} CH_3 \quad CH_2 \\ | \quad \diagup \diagdown \\ CH_3CHCH_2CH\!-\!CHCH_3 \end{array}$$

Depending on the stereochemistry of the double bond of the alkene in (b), two different isomers can be formed.

8.18 Find the smallest repeating unit in each polymer and add a double bond. This is the monomer unit.

Monomer   Polymer

(a)

$H_2C=CHOCH_3$

(b)

$ClHC=CHCl$

8.19 One radical abstracts a hydrogen atom from a second radical, and the remaining two electrons create a double bond.

8.20 Look back to Figure 8.12, which shows the reaction of (R)-4-methyl-1-hexene with H_3O^+. In a similar way, we can write a reaction mechanism for the reaction of H_3O^+ with (S)-4-methyl-1-hexene.

(2S,4S)-4-Methyl-2-hexanol   (2R,4S)-4-Methyl-2-hexanol

The products shown above are diastereomers and are formed in unequal amounts. The (2S,4S) stereoisomer is the enantiomer of the (2R,4R) isomer (shown in Figure 8.12), and the transition states leading to the formation of these two isomers are enantiomeric and of equal energy. Thus, the (2*S*,4*S*) and (2*R*,4*R*) enantiomers are formed in equal amounts. A similar argument can be used to show that the (2*R*,4*S*)and (2*S*,4*R*) isomers are formed in equal amounts. The product mixture is optically inactive.

8.21

Two enantiomeric carbocations are formed. Each carbocation can react with H_2O from either the top or the bottom to yield a total of four stereoisomers. The same argument used in Problem 8.20 can be used to show that the (1*S*,3*R*) and (1*R*,3*S*) enantiomers are formed in equal amounts, and the (1*S*,3*S*) and (1*R*,3*R*) isomers are formed in equal amounts. The result is a non-50:50 mixture of two racemic pairs.

Visualizing Chemistry

8.22

(a)

$$\xrightarrow{\text{RCO}_3\text{H}}$$

$$\underset{\overset{|}{\text{CH}_3}}{\text{CH}_3\text{C}}\overset{\overset{\text{CH}_3}{|}}{—}\text{CHCH}_2\text{CHCH}_2\text{CH}_3$$

with epoxide O

$$\underset{\text{2,5-Dimethyl-2-heptene}}{\overset{\overset{\text{CH}_3}{|}\overset{\text{CH}_3}{|}}{\text{CH}_3\text{C}=\text{CHCH}_2\text{CHCH}_2\text{CH}_3}} \quad \xrightarrow[\text{H}_3\text{O}^+]{\text{KMnO}_4} \quad \underset{\text{H}_3\text{C}}{\overset{\text{H}_3\text{C}}{>}}\text{C}=\text{O} \; + \; \text{O}=\overset{\overset{\text{OH}\text{CH}_3}{||}}{\text{CCH}_2\text{CHCH}_2\text{CH}_3}$$

$$\xrightarrow[\text{2. Zn, CH}_3\text{CO}_2\text{H}]{\text{1. O}_3} \quad \underset{\text{H}_3\text{C}}{\overset{\text{H}_3\text{C}}{>}}\text{C}=\text{O} \; + \; \text{O}=\overset{\overset{\text{CH}_3}{|}}{\text{CHCH}_2\text{CHCH}_2\text{CH}_3}$$

(b)

$$\xrightarrow{\text{RCO}_3\text{H}}$$

3,3-Dimethylcyclopentene

$$\xrightarrow[\text{H}_3\text{O}^+]{\text{KMnO}_4}$$

$$\xrightarrow[\text{2. Zn, CH}_3\text{CO}_2\text{H}]{\text{1. O}_3}$$

8.23 (a)

$$\underset{\text{2-Ethyl-3-methyl-1-butene}}{\overset{\overset{\text{CH}_3}{|}}{\underset{\overset{|}{\text{CH}_2}}{\text{CH}_3\text{CHCCH}_2\text{CH}_3}}} \quad \xrightarrow[\text{2. NaBH}_4]{\text{1. Hg(OAc)}_2, \text{H}_2\text{O}} \quad \underset{\overset{|}{\text{H}}\overset{|}{\text{CH}_3}}{\overset{\overset{\text{CH}_3\text{OH}}{||}}{\text{CH}_3\text{C}—\text{CCH}_2\text{CH}_3}}$$

or

$$\underset{\text{3,4-Dimethyl-2-pentene}}{\overset{\overset{\text{CH}_3}{|}}{\underset{\overset{|}{\text{CH}_3}}{\text{CH}_3\text{CHC}=\text{CHCH}_3}}} \quad \xrightarrow[\text{2. NaBH}_4]{\text{1. Hg(OAc)}_2, \text{H}_2\text{O}} \quad \underset{\overset{|}{\text{H}}\overset{|}{\text{CH}_3}}{\overset{\overset{\text{CH}_3\text{OH}}{||}}{\text{CH}_3\text{C}—\text{CCH}_2\text{CH}_3}}$$

Only oxymercuration/reduction can be used to produce an alcohol that has −OH bonded to the more substituted carbon. A third alkene, 2,3-dimethyl-2-pentene, gives a mixture of tertiary alcohols when treated with either BH_3 or $Hg(OAc)_2$.

(b)

4,4-Dimethylcyclopentene

Both hydroboration/oxidation and oxymercuration yield the same alcohol product from the symmetrical alkene starting material.

8.24

Two possible alcohols might be formed by hydroboration/oxidation of the alkene shown. One product results from addition of BH_3 to the top face of the double bond (not formed), and the other product results from addition to the bottom face of the double bond (formed). Addition from the top face does not occur because a methyl group on the bridge of the bicyclic ring system blocks approach of the borane.

8.25

4-Methylcyclohexene

RCO_3H = *meta*-Chloroperoxybenzoic acid

Since the hydroxyl groups in the diol product have a trans relationship, the product can only be formed by epoxide hydrolysis. (Treatment of the alkene with OsO_4 yields a product in which the two −OH groups have a cis relationship.)

Mechanism Problems

8.26 (a)

Mechanism:

(b)

Mechanism:

(c)

Mechanism:

8.27 (a)

(b)

(c)

8.28 (a)

Mechanism:

(b)

BH$_3$

Mechanism:

(c)

Mechanism:

8.29 (a)

+   CF$_3$CO$_3$H $\longrightarrow$ + enantiomer

Mechanism:

(b)

Mechanism:

8.30 (a)

+ enantiomer

Mechanism:

(b)

+ enantiomer

Mechanism:

(c)

Mechanism:

molecule reoriented
with epoxide in the
plane of the paper

8.31

8.32 (a)   The product of this reaction is *achiral* (no chiral centers). Therefore, it will not rotate plane polarized light.

H₂, Pd/C

(b)   The product of this reaction is a *meso* compound. Therefore, it does not rotate plane polarized light.

(c) There are two products to this reaction. The first product is a *single enantiomer* that would rotate plane polarized light, and the second product is *achiral* (no chiral centers). Therefore, the product mixture **would rotate** plane polarized light.

8.33

Step 1: Protonation of the double bond.

Step 2: Nucleophilic attack of methanol on the carbocation.

Step 3: Loss of proton.

The above mechanism is the same as the mechanism shown in Section 8.4 with one exception: In this problem, methanol, rather than water, is the nucleophile, and an ether, rather than an alcohol, is the observed product.

8.34 (a)

(b)

$$(FC) = \left[\begin{array}{c} \# \text{ of valence} \\ \text{electrons} \end{array} \right] - \left[\begin{array}{c} \# \text{ of bonding electrons} \\ \hline 2 \end{array} \right] - \left[\begin{array}{c} \# \text{ nonbonding} \\ \text{electrons} \end{array} \right]$$

	I	N1	N2	N3
A	0	0	+1	−1
B	0	−1	+1	0

Formal charge calculations show a partial negative charge on N1.

(c)   Addition of IN₃ to the alkene yields a product in which —I is bonded to the primary carbon and —N₃ is bonded to the secondary carbon. If addition occurs with Markovnikov orientation, I⁺ must be the electrophile, and the reaction must proceed through an iodonium ion intermediate. Opening of the iodonium ion gives Markovnikov product for the reasons discussed in Problem 8.6. The bond polarity of iodine azide is:

8.35

γ-Bisabolene

bromonium ion

10-Bromo-α-chamigrene

cyclic carbocation

8.36

Lauirediol → (Bromo-peroxidase) → bromonium ion → carbocation → Prelaureatin

8.37

In step 1, carbon dioxide is lost from the trichloroacetate anion. In step 2, elimination of chloride anion produces dichlorocarbene. Step 2 is the same for both the above reaction and the base-induced elimination of HCl from chloroform, and both reactions proceed through the trichloromethanide anion intermediate.

8.38

Cyclohexyl methyl ether

8.39

The reaction mechanism involves the following steps:

Step 1: Addition of Hg(OAc)₂ to one of the double bonds to form a cyclic mercurinium ion.

Step 2: Reaction of a second double bond with the mercurinium ion to form a six-membered ring and a different carbocation.

Step 3: A second cyclization forms the other ring and yields another carbocation.

Step 4: Removal of −H gives a double bond.

8.40

Step 1: Formation of a cyclic bromonium ion.

Step 2: Nucleophilic attack of −OH on the bromonium ion.

Step 3: Loss of H⁺.

The above mechanism is the same as that for halohydrin formation, shown in Section 8.3. In this case, the nucleophile is the hydroxyl group of 4-penten-1-ol.

8.41 Hydroboration of 2-methyl-2-pentene at 160°C is reversible. The initial organoborane intermediate can eliminate BH_3 in either of two ways, yielding either 2-methyl-2-pentene or 4-methyl-2-pentene, which in turn can undergo reversible hydroboration to yield either 4-methyl-2-pentene or 4-methyl-1-pentene. The effect of these reversible reactions is to migrate the double bond along the carbon chain. A final hydroboration then yields the most stable (primary) organoborane, which is oxidized to form 4-methyl-1-pentanol.

Additional Problems

Reactions of Alkenes

8.42

(a)

H_2/Pd

(b)

Br_2

(c)

OsO_4

NMO

(d)

Cl_2

H_2O

(e)

CH_2I_2, Zn/Cu

(f)

RCO_3H

RCO_3H = *meta*-Chloroperoxybenzoic acid

8.43

(a)

$$
\begin{array}{l}
\underset{\text{CH}_3}{\overset{\text{CH}_3}{\text{CH}_3\text{CH}_2\text{CH}_2\text{CH}_2\text{C}=\text{CH}_2}} \qquad \text{2-Methyl-1-hexene} \\[2pt]
\text{CH}_3\text{CH}_2\text{CH}_2\text{CH}=\text{C}(\text{CH}_3)_2 \qquad \text{2-Methyl-2-hexene} \\[2pt]
\text{CH}_3\text{CH}_2\text{CH}=\text{CHCH}(\text{CH}_3)_2 \qquad \text{2-Methyl-3-hexene} \\[2pt]
\text{CH}_3\text{CH}=\text{CHCH}_2\text{CH}(\text{CH}_3)_2 \qquad \text{5-Methyl-2-hexene} \\[2pt]
\text{H}_2\text{C}=\text{CHCH}_2\text{CH}_2\text{CH}(\text{CH}_3)_2 \qquad \text{5-Methyl-1-hexene}
\end{array}
$$

$\xrightarrow{\text{H}_2/\text{Pd}}$

$$\underset{\text{2-Methylhexane}}{\overset{\text{CH}_3}{\text{CH}_3\text{CH}_2\text{CH}_2\text{CH}_2\text{CHCH}_3}}$$

(b)

3,3-Dimethylcyclohexene

4,4-Dimethylcyclohexene

$\xrightarrow{\text{H}_2/\text{Pd}}$

1,1-Dimethylcyclohexane

(c)

$$\text{CH}_3\text{CH}=\text{CHCH}_2\text{CH}(\text{CH}_3)_2 \xrightarrow{\text{Br}_2} \underset{\text{2,3-Dibromo-5-methylhexane}}{\overset{\overset{\text{Br}\quad\text{Br}}{|\qquad|}}{\text{CH}_3\text{CH}-\text{CHCH}_2\text{CH}(\text{CH}_3)_2}}$$

5-Methyl-2-hexene

(d)

$$\underset{\underset{\text{CH}_3}{|}}{\text{H}_2\text{C}=\text{CHCHCH}_2\text{CH}_2\text{CH}_2\text{CH}_3} \xrightarrow{\text{HCl}} \underset{\underset{\text{CH}_3}{|}}{\overset{\overset{\text{Cl}}{|}}{\text{CH}_3\text{CHCHCH}_2\text{CH}_2\text{CH}_2\text{CH}_3}}$$

3-Methyl-1-heptene

2-Chloro-3-methylheptane

(e)

$$\text{CH}_3\text{CH}_2\text{CH}_2\text{CH}=\text{CH}_2 \xrightarrow[\text{2. NaBH}_4]{\text{1. Hg(OAc)}_2,\ \text{H}_2\text{O}} \underset{\text{2-Pentanol}}{\overset{\overset{\text{OH}}{|}}{\text{CH}_3\text{CH}_2\text{CH}_2\text{CHCH}_3}}$$

1-Pentene

(f)

$=\text{CH}_2 \xrightarrow{\text{CH}_2\text{I}_2,\ \text{Zn/Cu}}$

8.44

(a)

(b)

(c)

Remember that −H and −OH add syn across the double bond.

(d)

8.45

secondary carbocation
less stable

tertiary carbocation
more stable

Remember from Section 7.10 that a reaction that forms a more stable carbocation intermediate is faster than a comparable reaction that forms a less stable carbocation intermediate. Thus, the reaction of 1-methylcyclohexene with HBr is faster than the reaction of cyclohexene with HBr.

8.46 Recall the mechanism of hydroboration and note that the hydrogen added to the double bond comes from borane. The product of hydroboration with BD_3 has deuterium bonded to the more substituted carbon; $-D$ and $-OH$ are cis to one another.

8.47

cis-2-Butene   *cis*-1,2-Dimethylcyclopropane

trans-2-Butene   *trans*-1,2-Dimethylcyclopropane

The Simmons–Smith reaction occurs with syn stereochemistry. Only *cis*-1,2-dimethylcyclopropane is produced from *cis*-2-butene, and only *trans*-1,2-dimethylcyclopropane is produced from *trans*-2-butene.

8.48

8.49

Conjugation with the oxygen lone pair electrons makes the double bond more nucleophilic.

Reaction with HCl yields a cation intermediate that can be stabilized by the oxygen electrons.

Addition of Cl⁻ leads to the observed product.

There are two reasons why the other regioisomer is not formed: (1) Carbon 1 is less nucleophilic than carbon 2; (2) The cation intermediate that would result from protonation at carbon 1 can't be stabilized by the oxygen electrons.

Synthesis Using Alkenes

8.50

(a)

(b)

Acid-catalyzed hydration and hydroboration/oxidation are both additional routes to this product.

(c)

(d)

(e)

(f)

8.51 Because ozonolysis gives only one product, we can assume that the alkene is symmetrical.

2,3-Dimethyl-2-butene

8.52 Remember that alkenes can give ketones, carboxylic acids, and CO₂ on oxidative cleavage with KMnO₄ in acidic solution.

(a)

$$CH_3CH_2CH=CH_2 \xrightarrow[\text{H}_3\text{O}^+]{\text{KMnO}_4} CH_3CH_2CO_2H + CO_2$$

(b)

$$CH_3CH_2CH_2CH=C(CH_3)_2 \xrightarrow[\text{H}_3\text{O}^+]{\text{KMnO}_4} CH_3CH_2CH_2CO_2H + (CH_3)_2C=O$$

(c)

(d)

8.53 (a) Addition of HI occurs with Markovnikov regiochemistry – iodine adds to the more substituted carbon.

(b) Hydroxylation of double bonds produces cis, not trans, diols.

(c) Ozone reacts with both double bonds of 1,4-cyclohexadiene.

(d) Because hydroboration is a syn addition, the –H and the –OH added to the double bond must be cis to each other.

8.54 (a) This alcohol can't be synthesized selectively by hydroboration/oxidation. Consider the two possible starting materials.

1.

$$CH_3CH_2CH_2CH=CH_2 \xrightarrow[\text{2. H}_2\text{O}_2, \text{ }^-\text{OH}]{\text{1. BH}_3, \text{ THF}} CH_3CH_2CH_2CH_2CH_2OH$$

1-Pentene yields only the primary alcohol.

2.

$$CH_3CH_2CH=CHCH_3 \xrightarrow[\text{2. H}_2\text{O}_2, \text{ }^-\text{OH}]{\text{1. BH}_3, \text{ THF}} \underset{\text{OH}}{CH_3CH_2CHCH_2CH_3} + \underset{\text{OH}}{CH_3CH_2CH_2CHCH_3}$$

2-Pentene yields a mixture of alcohols.

(b)

$$(CH_3)_2C=C(CH_3)_2 \xrightarrow[\text{2. H}_2\text{O}_2, \text{ }^-\text{OH}]{\text{1. BH}_3, \text{ THF}} \underset{\text{OH}}{(CH_3)_2CHC(CH_3)_2}$$

2,3-Dimethyl-2-butene yields the desired alcohol exclusively.

(c)   This alcohol can't be formed cleanly by a hydroboration reaction. The –H and –OH added to a double bond must be cis to each other.

(d)   The product shown is not a hydroboration product; hydroboration yields an alcohol in which ⁻OH is bonded to the less substituted carbon.

Polymers

8.55

8.56

N-Vinylpyrrolidone   Poly(vinyl pyrrolidone)

8.57

Vinylidene   Vinyl
chloride   chloride

Saran

The dashed lines cross the bonds formed during the polymerization reaction. The structural fragments that lie between the dashed lines are the monomer units. Saran is a copolymer of vinylidene chloride and vinyl chloride.

General Problems

8.58 (a)   Compound **A** has three degrees of unsaturation. Because compound **A** contains only one double bond, the other two degrees of unsaturation must be rings.

(b), (c)

Other compounds containing two fused rings and a shared double bond also yield symmetrical diketone products.

8.59 (1)   Hydrocarbon **A** (C_6H_{12}) has one double bond or ring.

(2)   Because **A** reacts with one equivalent of H_2, it has one double bond and no ring.

(3)   Compound **A** forms a diol (**B**) when reacted with OsO_4.

(4)   When alkenes are oxidized with $KMnO_4$ they give either carboxylic acids or ketones, depending on the substitution pattern of the double bond.

    (a)   A ketone is produced from what was originally a disubstituted carbon in the double bond.

    (b)   A carboxylic acid is produced from what was originally a monosubstituted carbon in the double bond.

(5)   One fragment from $KMnO_4$ oxidation is a carboxylic acid, $CH_3CH_2CO_2H$.

    (a)   This fragment was $CH_3CH_2CH=$ (a monosubstituted double bond) in compound **A**.

    (b)   It contains three of the six carbons of compound **A**.

(6)   (a)   The other fragment contains three carbons.

    (b)   It forms ketone **C** on oxidation.

    (c)   The only three carbon ketone is acetone, $O=C(CH_3)_2$.

    (d)   This fragment was $=C(CH_3)_2$ in compound **A**.

(7)   If we join the fragment in 5(a) with the one in 6(d), we get:

$$CH_3CH_2CH=C(CH_3)_2 \qquad C_6H_{12}$$
$$\mathbf{A}$$

The complete scheme:

8.60 The oxidative cleavage reaction of alkenes with O_3, followed by Zn in acid, produces aldehyde and ketone functional groups at sites where double bonds used to be. On ozonolysis, these two dienes yield only aldehydes because all double bonds are monosubstituted.

Because the other diene is symmetrical, only one dialdehyde, OCHCH₂CHO, is produced.

8.61 Try to solve this problem phrase by phrase.

(1) $C_{10}H_{18}O$ has two double bonds and/or rings.

(2) $C_{10}H_{18}O$ must be an alcohol because it undergoes reaction with H_2SO_4 to yield an alkene.

(3)   When $C_{10}H_{18}O$ is treated with dilute H_2SO_4, a mixture of alkenes of the formula $C_{10}H_{16}$ is produced.

(4)   Since the major alkene product **B** yields only cyclopentanone, C_5H_8O, on ozonolysis, **B** and **A** contain two rings. **A** therefore has no double bonds.

8.62

Cyclooctane   1,5-Cyclooctadiene

8.63

Focus on the stereochemistry of the three-membered ring. Simmons–Smith reaction of 1,1-diiodoethane with the double bond occurs with syn stereochemistry and can produce two isomers. In one of these isomers (**A**), the methyl group is on the same side of the three-membered ring as the cyclohexane ring carbons. In **B**, the methyl group is on the side of the three-membered ring opposite to the cyclohexane ring carbons.

8.64

$$CH_3(CH_2)_{12}CH{=}CH(CH_2)_7CH_3 \xrightarrow[H_3O^+]{KMnO_4} CH_3(CH_2)_{12}CO_2H + CH_3(CH_2)_7CO_2H$$

8.65 C_8H_8 has five double bonds and/or rings. One of these double bonds reacts with H_2/Pd. Stronger conditions cause the uptake of four equivalents of H_2. C_8H_8 thus contains four double bonds, three of which are in an aromatic ring, and one $C = C$ double bond. A good guess for C_8H_8 at this point is:

Reaction of a double bond with $KMnO_4$ yields cleavage products of the highest possible degree of oxidation. In this case, the products are $CO_2 + C_6H_5CO_2H$.

8.66 (a) Bromine dissolved in CH_2Cl_2 has a reddish-brown color. When an alkene such as cyclopentene is added to the bromine solution, the double bond reacts with bromine, and the color disappears. This test distinguishes cyclopentene from cyclopentane, which does not react with Br_2. Alternatively, each compound can be treated with H_2/Pd. The alkene takes up H_2, and the alkane is unreactive.

(b) An aromatic compound such as benzene is unreactive to the Br_2/CH_2Cl_2 reagent and can be distinguished from 2-hexene, which decolorizes Br_2/CH_2Cl_2. Also, an aromatic compound doesn't take up H_2 under reaction conditions used for hydrogenation of alkenes.

8.67 (a) α-Terpinene, $C_{10}H_{16}$, has three degrees of unsaturation.

(b) Hydrogenation removes only two degrees of saturation, producing a hydrocarbon $C_{10}H_{20}$, that has one ring. α-Terpinene thus has two double bonds and one ring.

(c)

α-Terpinene   Glyoxal   6-Methyl-2,5-
                                                heptanedione

8.68 The models of the cis and trans diols show that it is much easier to form a five-membered cyclic periodate from the cis diol A than from the trans diol B. The cis periodate intermediate is of lower energy than the trans periodate intermediate because of the lack of strain in the cis periodate ring. Because any factor that lowers the energy of a transition state or intermediate also lowers $\Delta G^{\ddagger}$ and increases the rate of reaction, diol cleavage should proceed more slowly for trans diols than for cis diols.

8.69

trans-1-Bromo-3-   *cis*-1-Bromo-3-
methylcyclohexane   methylcyclohexane

trans-1-Bromo-2-   *cis*-1-Bromo-2-
methylcyclohexane   methylcyclohexane

In the reaction of 3-methylcyclohexene with HBr, two intermediate carbocations of approximately equal stability are formed. Both react with bromide ion from top and bottom faces to give four different products.

The most stable cation intermediate from protonation of 3-bromocyclohexene is a cyclic bromonium ion, which is attacked by Br⁻ from the opposite side to yield trans product.

8.70

Addition of one equivalent of HX or X_2 to a triple bond occurs with Markovnikov regiochemistry to yield a product in which the two added atoms usually have a transrelationship across the double bond.

8.71

Formation of the cyclic osmate, which occurs with syn stereochemistry, retains the cistrans stereochemistry of the double bond because osmate formation is a single-step reaction. Oxidation of the osmate does not affect the stereochemistry of the carbon–oxygen bond, and the diol produced from *cis*-2-butene is a stereoisomer of the diol produced from *trans*-2-butene.

8.65 **A** has four multiple bonds/rings.

2-Phenyl-3-pentanol is also an acceptable answer.

Review Unit 3: Organic Reactions, Alkenes

Major Topics Covered (with vocabulary):

Organic Reactions:
addition reaction elimination reaction substitution reaction rearrangement reaction
reaction mechanism hemolytic heterolytic homogenic heterogenic radical reaction
polar reaction initiation propagation termination electronegativity polarizability
curved arrow electrophile nucleophile carbocation

Describing a Reaction:
K_{eq} $\Delta G°$ exergonic endergonic enthalpy entropy heat of reaction exothermic
endothermic bond dissociation energy reaction energy diagram transition state activation
energy reaction intermediate

Introduction to alkenes:
degree of unsaturation methylene group vinyl group allyl group cis–trans isomerism
E,Z isomerism heat of hydrogenation hyperconjugation

Electrophilic addition reactions:
electrophilic addition reaction regiospecific Markovnikov's rule Hammond Postulate
carbocation rearrangement hydride shift

Other reactions of alkenes:
dehydrohalogenation dehydration anti stereochemistry bromonium ion halohydrin
hydration oxymercuration hydroboration syn stereochemistry carbine stereospecific
Simmons–Smith reaction hydrogenation hydroxylation diol osmate molozonide ozonide

Polymerization reactions:
polymer monomer chain branching radical polymerization cationic polymerization

Types of Problems:

After studying these chapters you should be able to:

- Identify reactions as polar, radical, substitution, elimination, addition, or rearrangement reactions.
- Understand the mechanism of radical reactions.
- Identify reagents as electrophiles or nucleophiles.
- Use curved arrows to draw reaction mechanisms.
- Understand the concepts of equilibrium and rate.
- Calculate K_{eq} and $\Delta G°$ of reactions, and use bond dissociation energies to calculate $\Delta H°$ of reactions.
- Draw reaction energy diagrams and label them properly.
- Calculate the degree of unsaturation of any compound, including those containing N, O, and halogen.
- Name acyclic and cyclic alkenes, and draw structures corresponding to names.
- Assign *E,Z* priorities to groups.
- Assign cis-trans and *E,Z* designations to double bonds.
- Predict the relative stability of alkene double bonds.
- Formulate mechanisms of electrophilic addition reactions.

—   Predict the products of reactions involving alkenes.

—   Choose the correct alkene starting material to yield a given product.

—   Deduce the structure of an alkene from its molecular formula and products of cleavage.

—   Carry out syntheses involving alkenes.

Points to Remember:

*   In virtually all cases, a compound is of lower energy than the free elements of which it is composed. Thus, energy is released when a compound is formed from its component elements, and energy is required when bonds are broken. Entropy decreases when a compound is formed from its component elements (because disorder decreases). For two compounds of similar structure, less energy is required to break all bonds of the higher energy compound than is required to break all bonds of the lower energy compound.

*   Calculating the degree of unsaturation is an absolutely essential technique in the structure determination of <u>all</u> organic compounds. It is the starting point for deciding which functional groups are or aren't present in a given compound, and eliminates many possibilities. When a structure determination problem is given, always calculate the degree of unsaturation first.

*   All cis–trans isomers can also be described by the *E,Z* designation, but not all *E,Z* isomers can be described by the cis–trans designation.

*   Bond dissociation energies, described in Chapter 6, measure the energy required to homolytically break a bond. They are not the same as dissociation enthalpies, which measure the ability of a compound to dissociate heterolytically. Bond dissociation energies can be used to calculate dissociation enthalpies in the gas phase if other quantities are also known.

*   Not all hydrogens bonded to carbons adjacent to a carbocation can take part in hyperconjugation at the same time. At any given instant, some of the hydrogens have C-H bonds that lie in the plane of the carbocation and are not suitably oriented for hyperconjugative overlap.

Self-Test:

A

What type of reaction is occurring in **A**? Would you expect that the reaction occurs by a polar or a radical mechanism? If K_{eq} for the reaction at 298 K is 10^{-3}, what sign do you expect for $\Delta G°$? Would you expect $\Delta S°$ to be negative or positive? What about $\Delta H°$?

B Abscisic acid (a plant hormone)   **C**

Give *E,Z* configurations for the double bonds in **B**. Provide a name for **C** (include bond stereochemistry). Predict the products of reaction of **C** with (a) 1 equiv HBr   (b) H_2, Pd/C
(c) BH_3, THF, then H_2O_2, HO^- (d) O_3, then Zn, H_3O^+.

Two isomeric compounds **D** and **E** have the formula $C_{10}H_{16}$. On hydrogenation, each compound reacts with two molar equivalents of H_2. Ozonolysis of each compound yields the following fragments:

How many rings/double bonds do **D** and **E** have? What are the structures of **D** and **E**?

Multiple Choice:

1. Which of the following molecules is not a nucleophile?
    (a) BH_3 (b) NH_3 (c) HO^- (d) $H_2C = CH_2$

2. Which of the following reactions probably has the greatest entropy increase?
    (a) addition reaction   (b) elimination reaction   (c) substitution reaction   (d) rearrangement

3. At a specific temperature T, a reaction has negative $\Delta S°$ and $K_{eq} > 1$. What can you say
    about $\Delta G°$ and $\Delta H°$?
    (a) $\Delta G°$ is negative and $\Delta H°$ is positive
    (b) $\Delta G°$ and $\Delta H°$ are both positive
    (c) $\Delta G°$ and $\Delta H°$ are both negative
    (d) $\Delta G°$ is negative but you can't predict the sign of $\Delta H°$.

4. In which of the following situations is $\Delta G^{\ddagger}$ likely to be smallest?
    (a) a slow exergonic reaction
    (b) a fast exergonic reaction
    (c) a fast endergonic reaction
    (d) a slow endergonic reaction

5. What is the degree of unsaturation of a compound whose molecular formula is $C_{11}H_{13}N$?
    (a) 4   (b) 5   (c) 6   (d) 7

6. Two equivalents of H_2 are needed to hydrogenate a hydrocarbon. It is also known that the
    compound contains two rings and has 15 carbons. What is its molecular formula?
    (a) $C_{15}H_{22}$ (b) $C_{15}H_{24}$ (c) $C_{15}H_{28}$ (d) $C_{15}H_{32}$

7. What is the usual relationship between the heats of hydrogenation of a pair of cis/trans
    alkene isomers?
    (a) Both have positive heats of hydrogenation
    (b) Both have negative heats of hydrogenation, and ΔH_{hydrog} for the cis isomer has a
    greater negative value
    (c) Both have negative heats of hydrogenation, and ΔH_{hydrog} for the trans isomer has a
    greater negative value
    (d) Both have negative heats of hydrogenation, but the relationship between the two
    values of ΔH_{hydrog} can't be predicted.

8. In a two-step exergonic reaction, what is the relationship of the two transition states?
    (a) both resemble the intermediate
    (b) the first resembles the starting material, and the second resembles the product
    (c) the first resembles the intermediate and the second resembles the product
    (d) there is no predictable relationship between the two transition states

9.   For synthesis of an alcohol, acid-catalyzed hydration of an alkene is useful in all of the
     following instances except:
     (a) when an alkene has no acid-sensitive groups
     (b) when an alkene is symmetrical
     (c) when a large amount of the alcohol is needed
     (d) when two possible carbocationintermediates are of similar stability.

10. A reaction that produces a diol from an alcohol is a:
     (a) hydration   (b) hydrogenation   (c) hydroboration   (d) hydroxylation

Chapter 9 – Alkynes: An Introduction to Organic Synthesis

Chapter Outline

I. Introduction to alkynes (Section 9.1–9.2).
   A. Naming alkynes (Section 9.1).
      1. The rules for naming alkynes are like the rules for alkenes (Sec. 7.3), with a few exceptions.
         a. The suffix -*yne* is used for an alkyne.
         b. Compounds with both double bonds and triple bonds are *enynes*.
         c. When there is a choice in numbering, double bonds receive lower numbers than triple bonds.
         d. Compounds can also contain alkynyl groups.
   B. Preparation of alkynes (Section 9.2).
      1. Alkynes can be prepared by elimination reactions of 1,2-dihalides, using a strong base.
      2. The dihalides are formed by addition of X_2 to alkenes.
      3. Vinylic halides give alkynes when treated with a strong base.
II. Reactions of alkynes (Sections 9.3–9.6).
   A. General principles (Section 9.3).
      1. Alkyne triple bonds result from the overlap of two *sp*-hybridized carbon atoms.
         a. One σ bond and two π bonds are formed.
      2. The length (120 pm) and strength (965 kJ/mol) of a $-C\equiv C-$ bond make it the strongest carbon–carbon bond.
      3. Alkynes are somewhat less reactive than alkenes in electrophilic addition reactions
   B. Addition of X_2 and HX.
      1. HX adds to alkynes by an electrophilic addition mechanism.
         a. Addition of two equivalents of HX occurs if the acid is in excess.
         b. Addition occurs with Markovnikov regiochemistry and with trans stereochemistry.
      2. X_2 also adds in the same manner, and trans stereochemistry is observed.
      3. The intermediate in addition reactions is a vinylic carbocation, which forms less readily than an alkyl carbocation.
      4. Mechanisms of some alkyne addition reactions are complex.
   C. Hydration reactions of alkynes (Section 9.4).
      1. Hg(II)-catalyzed additions.
         a. The $-OH$ group adds to the more substituted carbon to give Markovnikov product.
         b. The intermediate enol product tautomerizes to a ketone.
         c. The mechanism is similar to that of addition to alkenes, but no $NaBH_4$ is necessary for removal of Hg.
         d. A mixture of products is formed from an internal alkyne, but a terminal alkyne yields a methyl ketone.
      2. Hydroboration/oxidation of alkynes.

     a. Hydroboration/oxidation of alkynes gives an intermediate enol product that tautomerizes to a carbonyl product.

        i. Hydroboration of a terminal alkyne gives an aldehyde.

        ii. Hydroboration of an internal alkyne gives a ketone.

     b. Hydroboration/ oxidation is complementary to Hg(II)-catalyzed hydration.

  D. Reduction of alkynes (Section 9.5).

    1. Complete reduction to an alkane occurs when H_2/Pd is used.

    2. Partial reduction to a cis alkene occurs with H_2 and a Lindlar catalyst.

    3. Partial reduction with Li in NH_3 produces a trans alkene.

     a. The reaction proceeds through an anion radical → vinylic radical → vinylic anion.

     b. The more stable trans vinylic anion is formed.

  E. Oxidative cleavage of alkynes (Section 9.6).

    1. O_3 or $KMnO_4$ cleave alkyne bonds to produce carboxylic acids or CO_2 (terminal alkyne).

    2. Oxidative cleavage reactions were formerly used for structure determinations.

III. Alkyne acidity (Sections 9.7–9.8).

  A. Formation of acetylide anions (Section 9.7).

    1. Terminal alkynes are weakly acidic (pK_a = 25).

    2. Very strong bases (⁻NH_2) can deprotonate a terminal alkyne, yielding an acetylide anion.

    3. Acetylide anions are stabilized by the large amount of "*s* character" of the orbital that holds the electron.

  B. Alkylation of acetylide anions (Section 9.8).

    1. Acetylide anions are strongly nucleophilic.

    2. Acetylide anions can react with haloalkanes to form substitution products.

     a. The nucleophilic acetylide anion attacks the electrophilic carbon of a haloalkane to produce a new alkyne.

     b. This reaction is called an alkylation reaction.

     c. Any terminal alkyne can form an alkylation product.

    3. Acetylide alkylations are limited to primary alkyl bromides and iodides.

     a. Acetylide ions cause dehydrohalogenation reactions with secondary and tertiary halides.

IV. Organic synthesis (Section 9.9).

  A. Reasons for the study of organic synthesis.

    1. In the pharmaceutical and chemical industries, synthesis produces new molecules, or better routes to important molecules.

    2. In academic laboratories, synthesis is done for creative reasons.

    3. In the classroom, synthesis is a tool for teaching the logic of organic chemistry.

  B. Strategies for organic synthesis.

    1. Work backward from the structure of the product, but −

    2. Keep the structure of the starting material in mind.

Solutions to Problems

9.1   The rules for naming alkynes are almost the same as the rules for naming alkenes. The suffix *-yne* is used, and compounds containing both double bonds and triple bonds are *-enynes*, with the double bond taking numerical precedence.

(a)

$$CH_3CHC{\equiv}CCHCH_3$$

with CH_3 groups on the 2nd and 5th carbons

2,5-Dimethyl-3-hexyne

(b)

$$HC{\equiv}CCCH_3$$

with two CH_3 groups

3,3-Dimethyl-1-butyne

(c)

$$CH_3CH_2CC{\equiv}CCH_2CH_2CH_3$$

with two CH_3 groups

3,3-Dimethyl-4-octyne

(d)

$$CH_3CH_2CC{\equiv}CCHCH_3$$

with CH_3 groups

2,5,5-Trimethyl-3-heptyne

(e)

6-Isopropylcyclodecyne

(f) $CH_3CH{=}CHCH{=}CHC{\equiv}CCH_3$

2,4-Octadien-6-yne
(not 4,6-Octadien-2-yne)

9.2

$CH_3CH_2CH_2CH_2C{\equiv}CH$
1-Hexyne

$CH_3CH_2CH_2C{\equiv}CCH_3$
2-Hexyne

$CH_3CH_2C{\equiv}CCH_2CH_3$
3-Hexyne

$CH_3CH_2CHC{\equiv}CH$ with CH_3
3-Methyl-1-pentyne

$CH_3CHCH_2C{\equiv}CH$ with CH_3
4-Methyl-1-pentyne

$CH_3CHC{\equiv}CCH_3$ with CH_3
4-Methyl-2-pentyne

$CH_3CC{\equiv}CH$ with two CH_3
3,3-Dimethyl-1-butyne

9.3   Markovnikov addition is observed with alkynes as well as with alkenes.

(a)

$$CH_3CH_2CH_2C\equiv CH \; + \; 2\;Cl_2 \;\longrightarrow\; CH_3CH_2CH_2CCl_2CHCl_2$$

(b)

(c)

Two products result from addition to an internal alkyne.

9.4

This symmetrical internal alkyne yields only one product.

Two ketone products result from hydration of 2-methyl-4-octyne.

9.5   (a)

(b)

The desired ketone can be prepared only as part of a product mixture.

9.6 Remember that hydroboration yields aldehydes from terminal alkynes and ketones from internal alkynes.

(a)

(b)

9.7

(a)

(b)

9.8 The correct reducing reagent gives a double bond with the desired geometry.

(a)

$CH_3CH_2CH_2CH_2CH_2C\equiv CCH_3$ $\xrightarrow{Li/NH_3}$

2-Octyne

trans-2-Octene

(b)

$CH_3CH_2CH_2C\equiv CCH_2CH_3$ $\xrightarrow[Lindlar]{H_2}$

3-Heptyne

cis-3-Heptene

(c)

3-Methyl-1-pentyne

Li/NH$_3$

or

H$_2$ Lindlar

$CH_3CH_2CHCH=CH_2$ (CH$_3$)

3-Methyl-1-pentene

9.9   A base that is strong enough to deprotonate acetone must be the conjugate base of an acid weaker than acetone. In this problem, only $Na^+ \, ^-C{\equiv}CH$ is a base strong enough to deprotonate acetone.

9.10 Remember that the alkyne must be a terminal alkyne and the halide must be primary. More than one combination of terminal alkyne and halide may be possible.

Alkyne	*R'X (X=Br or I)*	*Product*
(a) $CH_3CH_2CH_2C{\equiv}CH$ *or* $HC{\equiv}CCH_3$	CH_3X $CH_3CH_2CH_2X$	$CH_3CH_2CH_2C{\equiv}CCH_3$ 2-Hexyne
(b) $(CH_3)_2CHC{\equiv}CH$	CH_3CH_2X	$(CH_3)_2CHC{\equiv}CCH_2CH_3$ 2-Methyl-3-hexyne
(c) [cyclohexyl]$-C{\equiv}CH$	CH_3X	[cyclohexyl]$-C{\equiv}CCH_3$

Products (b) and (c) can be synthesized by only one route because only primary halides can be used for acetylide alkylations.

9.11 The cis double bond can be formed by hydrogenation of an alkyne, which can be synthesized by an alkylation reaction of a terminal alkyne.

$$CH_3C{\equiv}CH \xrightarrow[\text{2. CH}_3\text{Br, THF}]{\text{1. NaNH}_2,\ \text{NH}_3} CH_3C{\equiv}CCH_3 \xrightarrow[\substack{\text{Lindlar}\\\text{catalyst}}]{H_2} \underset{\text{cis-2-Butene}}{\overset{H \qquad\quad H}{\underset{H_3C \qquad\quad CH_3}{C{=}C}}}$$

9.12 The starting material is $CH_3CH_2CH_2C{\equiv}CCH_2CH_2CH_3$. Look at the functional groups in the target molecule and work backward to 4-octyne.

(a)   To reduce a triple bond to a double bond with *cis* stereochemistry use H_2 with Lindlar catalyst.

$$CH_3CH_2CH_2C{\equiv}CCH_2CH_2CH_3 \xrightarrow[\substack{\text{Lindlar}\\\text{catalyst}}]{H_2} \underset{\text{cis-4-Octene}}{\overset{CH_3CH_2CH_2 \qquad CH_2CH_2CH_3}{\underset{H \qquad\qquad H}{C{=}C}}}$$

(b)   An aldehyde is the product of double-bond cleavage of an alkene with O_3. The starting material can be either *cis*-4-octene or *trans*-4-octene.

$$CH_3CH_2CH_2 \quad CH_2CH_2CH_3$$
$$C=C$$
$$H \qquad H$$
from (a)

$\xrightarrow[\text{2. Zn, } CH_3CO_2H]{\text{1. } O_3}$

$$CH_3CH_2CH_2C \overset{O}{\underset{H}{\big\|}}$$

Butanal

(c)   Addition of HBr to *cis*-4-octene [part (a)] yields 4-bromooctane.

$$CH_3CH_2CH_2 \quad CH_2CH_2CH_3$$
$$C=C$$
$$H \qquad H$$
from (a)

$\xrightarrow{\text{HBr}}$

$$\overset{Br}{\underset{|}{}}$$
$$CH_3CH_2CH_2CHCH_2CH_2CH_2CH_3$$

4-Bromooctane

Alternatively, lithium/ammonia reduction of 4-octyne, followed by addition of HBr, gives 4-bromooctane.

(d)   Hydration or hydroboration/oxidation of *cis*-4-octene [part (a)] yields 4-hydroxyoctane (4-octanol).

$$CH_3CH_2CH_2 \quad CH_2CH_2CH_3$$
$$C=C$$
$$H \qquad H$$
from (a)

$\xrightarrow[\text{2. } NaBH_4]{\text{1. } Hg(OAc)_2, H_2O}$

or

$\xrightarrow[\text{2. } H_2O_2, \ ^-OH]{\text{1. } BH_3, THF}$

$$\overset{OH}{\underset{|}{}}$$
$$CH_3CH_2CH_2CHCH_2CH_2CH_2CH_3$$

4-Hydroxyoctane
(4-Octanol)

(e)   Addition of Cl_2 to 4-octene [part (a)] yields 4,5-dichlorooctane.

$$CH_3CH_2CH_2 \quad CH_2CH_2CH_3$$
$$C=C$$
$$H \qquad H$$
from (a)

$\xrightarrow{Cl_2}$

$$\overset{Cl \ Cl}{\underset{| \ |}{}}$$
$$CH_3CH_2CH_2CHCHCH_2CH_2CH_3$$

4,5-Dichlorooctane

(f) $KMnO_4$ cleaves 4-octyne into two four-carbon fragments.

$$CH_3CH_2CH_2C\equiv CCH_2CH_2CH_3$$

$\xrightarrow[H_3O^+]{KMnO_4}$

$$2 \ CH_3CH_2CH_2CO_2H$$

Butanoic acid

9.13 The following syntheses are explained in detail in order to illustrate retrosynthetic logic – the system of planning syntheses by working backwards.

(a)   1. An immediate precursor to $CH_3CH_2CH_2CH_2CH_2CH_2CH_2CH_2CH_2CH_3$ might be an alkene or alkyne. Try $C_8H_{17}C{\equiv}CH$, which can be reduced to decane by H_2/Pd.

2. The alkyne $C_8H_{17}C{\equiv}CH$ can be formed by alkylation of $HC{\equiv}C{:}^-Na^+$ by $C_8H_{17}Br$, 1-bromooctane.

3. $HC{\equiv}C{:}^-Na^+$ can be formed by treatment of $HC{\equiv}CH$ with $NaNH_2$, NH_3.

$$HC{\equiv}CH \xrightarrow[NH_3]{NaNH_2} HC{\equiv}C{:}^-Na^+ \xrightarrow[THF]{C_8H_{17}Br} C_8H_{17}C{\equiv}CH \xrightarrow{H_2/Pd} Decane$$

$C_8H_{17}Br$ = 1-Bromooctane

(b)   1. An immediate precursor to $CH_3CH_2CH_2CH_2C(CH_3)_3$ might be $HC{\equiv}CCH_2CH_2C(CH_3)_3$, which, when hydrogenated, yields 2,2-dimethylhexane.

2. $HC{\equiv}CCH_2CH_2C(CH_3)_3$ can be formed by alkylation of $HC{\equiv}C{:}^-Na^+$ (from a.) with $BrCH_2CH_2C(CH_3)_3$.

$$HC{\equiv}CH \xrightarrow[NH_3]{NaNH_2} HC{\equiv}C{:}^-Na^+$$

$$BrCH_2CH_2C(CH_3)_3 \xrightarrow[THF]{HC{\equiv}C{:}^-Na^+} HC{\equiv}CCH_2CH_2C(CH_3)_3 \xrightarrow[Pd]{2\ H_2} CH_3CH_2CH_2CH_2C(CH_3)_3$$

2,2-Dimethylhexane

(c)   1. $CH_3CH_2CH_2CH_2CH_2CHO$ can be made by treating $CH_3CH_2CH_2CH_2C{\equiv}CH$ with borane, followed by H_2O_2.

2. $CH_3CH_2CH_2CH_2C{\equiv}CH$ can be synthesized from $CH_3CH_2CH_2CH_2Br$ and $HC{\equiv}C{:}^-Na^+$.

$$HC{\equiv}CH \xrightarrow[NH_3]{NaNH_2} HC{\equiv}C{:}^-Na^+$$

$$\begin{array}{c} CH_3CH_2CH_2CH_2Br \\ + \\ HC{\equiv}C{:}^-Na^+ \end{array} \xrightarrow{THF} CH_3CH_2CH_2CH_2C{\equiv}CH \xrightarrow[2.\ H_2O_2,\ ^-OH]{1.\ BH_3,\ THF} CH_3CH_2CH_2CH_2CH_2\overset{\displaystyle O}{\overset{\displaystyle \|}{C}}H$$

Hexanal

(d)　1. The desired ketone can be formed by mercuric-ion-catalyzed hydration of 1-heptyne.

2. 1-Heptyne can be synthesized by an alkylation of sodium acetylide by 1-bromopentane.

$$HC\equiv CH \xrightarrow[NH_3]{NaNH_2} HC\equiv C:^- Na^+$$

$$HC\equiv C:^- Na^+ + CH_3CH_2CH_2CH_2CH_2Br \xrightarrow{THF} CH_3CH_2CH_2CH_2CH_2C\equiv CH$$

$$CH_3CH_2CH_2CH_2CH_2C\equiv CH \xrightarrow[HgSO_4]{H_2SO_4,\ H_2O} CH_3CH_2CH_2CH_2CH_2CCH_3$$

2-Heptanone

Visualizing Chemistry

9.14 (a)

(i) H₂, Lindlar catalyst

(ii) H₃O⁺, HgSO₄

4,4-Dimethyl-1-hexyne

(b)

(i) H₂, Lindlar catalyst

(ii) H₃O⁺, HgSO₄

2,7-Dimethyl-4-octyne

9.15 (a)

1. BH₃, THF
2. H₂O₂, ⁻OH

4-Methyl-1-pentyne

An aldehyde is formed by reacting a terminal alkyne with borane, followed by oxidation.

(b)

9.16 First, draw the structure of each target compound. Then, analyze the structures for a synthetic route.

(a)

$$CH_3CHCH_2CH-CH_2$$
with OH and CH₂ groups

(b)

$$H_2C=CHCH_2CH_2CCH_3$$
with O group

(a) The left side and the right side might have double bonds as immediate precursors; the right side may result from a Simmons-Smith carbenoid addition to an alkene, and the left side may result from hydration of an alkene. Let's start with 3-bromo-1-propene.

(b) The right side can result from Hg-catalyzed addition of H_2O to a terminal alkyne.

9.17 It's not possible to form a small ring containing a triple bond because the angle strain that would result from bending the bonds of an *sp*-hybridized carbon to form a small ring is too great.

Mechanism Problems

9.18 (a)

(b)

(c)

9.19 Assuming that strong acids add to akynes in the same manner that they add to alkenes, propose a mechanism for each reaction below.

(a)

(b)

(c)

9.20 The mercury catalyzed hydration of alkynes involves the formation of an organomercury enol intermediate. Provide the electron pushing mechanism to show how each intermediate below is formed.

(a)

Mechanism:

(b)

Mechanism:

(c)

Mechanism:

9.21 The final step in the hydration of an alkyne under acidic conditions is the tautomerism of an enol intermediate to the corresponding ketone. The mechaism involves a protonation followed by a deprotonation. Show the mechanish each tautomerism below.

(a)

(b)

Mechanism:

(c)

Mechanism:

9.22 Predict the product(s) and show the complete electron pushing mechanism for each dissolving metal reduction below.

(a)

Mechanism:

(b)

Mechanism:

(c)

Mechanism:

9.23 Identify the mechanisms for the reactions below as being either *polar*, *radical* or *both*.

(a) *Both*. The addition of single electrons is a radical process, while adding protons is polar.

$$CH_3CH_2C\equiv CCH_2CH_3 \xrightarrow[\text{NH}_3]{\text{Li}}$$

(b) Halogenation is *polar* reaction.

$$CH_3CH_2C\equiv CCH_2CH_3 \xrightarrow{\text{Br}_2}$$

(c) Hydrohalogenation is a *polar* reaction.

$$CH_3CH_2C\equiv CCH_2CH_3 \xrightarrow{\text{2 HBr}}$$

9.24 Predict the product and provide the complete electron pushing mechanism for the two-step synthetic processes below.

(a)

Mechanism:

(b)

Mechanism:

(c)

CH₃CH₂C≡CH →[1. NaNH₂][2. PhCH₂Br]→ CH₃CH₂—C≡C—CH₂Ph

Mechanism:

CH₃CH₂—C≡C—H   ⁻NH₂   →   CH₃CH₂—C≡C:⁻   PhCH₂—Br   →   CH₃CH₂—C≡C—CH₂Ph

9.25

Repeating this process five more times replaces all hydrogen atoms with deuterium atoms. The first line represents the mechanism for acid-catalyzed tautomerization of a ketone

Additional Problems

Naming Alkynes

9.26

(a)

CH₃CH₂C≡CCCH₃ (with CH₃ above and CH₃ below the final carbon)

2,2-Dimethyl-3-hexyne

(b) CH₃C≡CCH₂C≡CCH₂CH₃

2,5-Octadiyne

(c)

CH₃CH=CC≡CCHCH₃ (with CH₃ above each central carbon)

3,6-Dimethyl-2-hepten-4-yne

(d)

HC≡CCCH₂C≡CH (with CH₃ above and CH₃ below)

3,3-Dimethyl-1,5-hexadiyne

(e) $H_2C=CHCH=CHC\equiv CH$ (f)

1,3-Hexadien-5-yne

$$CH_3CH_2\underset{\underset{CH_2CH_3}{|}}{C}HC\equiv C\underset{\underset{CH_3}{|}}{C}H\overset{\overset{CH_2CH_3}{|}}{C}HCHCH_3$$

3,6-Diethyl-2-methyl-4-octyne

9.27

(a)
$$CH_3CH_2CH_2C\equiv C\underset{\underset{CH_3}{|}}{\overset{\overset{CH_3}{|}}{C}}CH_2CH_3$$

3,3-Dimethyl-4-octyne

(b)
$$CH_3C\equiv CC\equiv C\underset{}{C}HCH_2\underset{\underset{CH_2CH_3}{|}}{\overset{\overset{CH_3}{|}}{C}}HC\equiv CH$$

3-Ethyl-5-methyl-1,6,8-decatriyne

(c)
$$CH_3\underset{\underset{CH_3}{|}}{\overset{\overset{CH_3}{|}}{C}}C\equiv C\underset{\underset{CH_3}{|}}{\overset{\overset{CH_3}{|}}{C}}CH_3$$

2,2,5,5-Tetramethyl-3-hexyne

(d)

3,4-Dimethylcyclodecyne

(e) $CH_3CH=CHCH=CHC\equiv CH$ (f)

3,5-Heptadien-1-yne

$$CH_3CH_2C\equiv CCH_2\underset{\underset{CH_3}{|}}{\overset{\overset{CH_3}{|}}{C}}-\underset{}{C}HCH=CH_2$$
with Cl on the CH

3-Chloro-4,4-dimethyl-1-nonen-6-yne

(g)
$$CH_3CH_2CH_2CH_2\underset{\underset{CH_3CHCH_2CH_3}{|}}{C}HC\equiv CH$$

3-*sec*-Butyl-1-heptyne

(h)
$$CH_3CH_2CH_2\underset{\underset{C(CH_3)_3}{|}}{C}HC\equiv C\underset{\underset{CH_3}{|}}{C}HCH_3$$

5-*tert*-Butyl-2-methyl-3-octyne

9.28 (a) $CH_3CH=CHC\equiv CC\equiv CCH=CHCH=CHCH=CH_2$.

1,3,5,11-Tridecatetraen-7,9-diyne

Using *E–Z* notation: (3*E*,5*E*,11*E*)-1,3,5,11-Tridecatetraen-7,9-diyne

The parent alkane of this hydrocarbon is tridecane.

(b) $CH_3C \equiv CC \equiv CC \equiv CC \equiv CCH = CH_2$. 1-Tridecen-3,5,7,9,11-pentayne
      This hydrocarbon also belongs to the tridecane family.

Reactions of Alkynes

9.29 This reaction mechanism is similar to the mechanism of halohydrin formation.

Step 1: Attack of π electrons on Br_2.
Step 2: Opening of cyclic cation by H_2O.
Step 3: Deprotonation.
Step 4: Tautomerization (for mechanism, see Problem 9.48 and Section 9.4).

9.30

9.31

(a)

$$CH_3CH_2CH_2CH_2C \equiv CH \xrightarrow[\text{HBr}]{\text{1 equiv}}$$

$$CH_3CH_2CH_2CH_2\overset{\overset{\displaystyle Br}{|}}{C} = CH_2$$

(b)

$$CH_3CH_2CH_2CH_2C \equiv CH \xrightarrow[\text{Cl}_2]{\text{1 equiv}}$$

$$\underset{\underset{\displaystyle Cl}{}}{CH_3CH_2CH_2CH_2} \diagdown \underset{}{C} = \underset{\underset{\displaystyle H}{}}{C} \diagup Cl$$

(c)

$$CH_3CH_2CH_2CH_2C \equiv CH \xrightarrow[\substack{\text{Lindlar} \\ \text{catalyst}}]{\text{H}_2}$$

$$CH_3CH_2CH_2CH_2CH = CH_2$$

(d)

$$CH_3CH_2CH_2CH_2C \equiv CH \xrightarrow[\text{2. CH}_3\text{Br}]{\text{1. NaNH}_2,\ \text{NH}_3}$$

$$CH_3CH_2CH_2CH_2C \equiv CCH_3$$

(e)

$$CH_3CH_2CH_2CH_2C \equiv CH \xrightarrow[\text{HgSO}_4]{\text{H}_2\text{O, H}_2\text{SO}_4}$$

$$CH_3CH_2CH_2CH_2\overset{\overset{\displaystyle O}{\|}}{C}CH_3$$

(f)

$$CH_3CH_2CH_2CH_2C \equiv CH \xrightarrow[\text{HCl}]{\text{2 equiv}}$$

$$CH_3CH_2CH_2CH_2\overset{\overset{\displaystyle Cl}{|}}{\underset{\underset{\displaystyle Cl}{|}}{C}}CH_3$$

9.32

(a)

$$CH_3(CH_2)_3C \equiv C(CH_2)_3CH_3 \xrightarrow[\substack{\text{Lindlar} \\ \text{catalyst}}]{\text{H}_2}$$

$$\underset{\underset{\displaystyle H}{}}{CH_3CH_2CH_2CH_2} \diagdown \underset{}{C} = \underset{\underset{\displaystyle H}{}}{C} \diagup CH_2CH_2CH_2CH_3$$

(b)

$$CH_3(CH_2)_3C \equiv C(CH_2)_3CH_3 \xrightarrow{\text{Li in NH}_3}$$

$$\underset{\underset{\displaystyle H}{}}{CH_3CH_2CH_2CH_2} \diagdown \underset{}{C} = \underset{\underset{\displaystyle CH_2CH_2CH_2CH_3}{}}{C} \diagup H$$

(c)

$$CH_3(CH_2)_3C \equiv C(CH_2)_3CH_3 \xrightarrow[\text{Br}_2]{\text{1 equiv}}$$

$$\underset{\underset{\displaystyle Br}{}}{CH_3CH_2CH_2CH_2} \diagdown \underset{}{C} = \underset{\underset{\displaystyle CH_2CH_2CH_2CH_3}{}}{C} \diagup Br$$

(d)

$$CH_3(CH_2)_3C \equiv C(CH_2)_3CH_3 \xrightarrow[\text{2. H}_2\text{O}_2,\ ^-\text{OH}]{\text{1. BH}_3,\ \text{THF}}$$

$$CH_3CH_2CH_2CH_2CH_2\overset{\overset{\displaystyle O}{\|}}{C}CH_2CH_2CH_2CH_3$$

(e)

$$CH_3(CH_2)_3C\equiv C(CH_2)_3CH_3 \xrightarrow[\text{HgSO}_4]{\text{H}_2\text{O, H}_2\text{SO}_4} CH_3CH_2CH_2CH_2CH_2\overset{\overset{\displaystyle O}{\|}}{C}CH_2CH_2CH_2CH_3$$

(f)

$$CH_3(CH_2)_3C\equiv C(CH_2)_3CH_3 \xrightarrow[\text{Pd/C}]{\text{excess H}_2} CH_3(CH_2)_8CH_3$$

9.33 Mixtures of products are sometimes formed since the alkynes are unsymmetrical.

(a)

$$CH_3CH_2CH_2C\equiv CCH_3 \xrightarrow[\text{Br}_2]{\text{2 equiv}} CH_3CH_2CH_2C(Br_2)C(Br_2)CH_3$$

(b)

$$CH_3CH_2CH_2C\equiv CCH_3 \xrightarrow[\text{HBr}]{\text{1 equiv}}$$

CH$_3$CH$_2$CH$_2$ \ / H
    C=C
Br / \ CH$_3$

+

CH$_3$CH$_2$CH$_2$ \ / Br
    C=C
H / \ CH$_3$

(c)

$$CH_3CH_2CH_2C\equiv CCH_3 \xrightarrow[\text{HBr}]{\text{excess}} CH_3CH_2CH_2C(Br_2)CH_2CH_3$$

$$+ \quad CH_3CH_2CH_2CH_2C(Br_2)CH_3$$

(d)

$$CH_3CH_2CH_2C\equiv CCH_3 \xrightarrow{\text{Li in NH}_3}$$

CH$_3$CH$_2$CH$_2$ \ / H
    C=C
H / \ CH$_3$

(e)

$$CH_3CH_2CH_2C\equiv CCH_3 \xrightarrow[\text{HgSO}_4]{\text{H}_2\text{O, H}_2\text{SO}_4} CH_3CH_2CH_2\overset{\overset{\displaystyle O}{\|}}{C}CH_2CH_3$$

$$+ \quad CH_3CH_2CH_2CH_2\overset{\overset{\displaystyle O}{\|}}{C}CH_3$$

9.34 Both KMnO$_4$ and O$_3$ oxidation of alkynes yield carboxylic acids; terminal alkynes give CO$_2$ also. In (a), (b), and (c), the observed products can also be formed by KMnO$_4$ oxidation of the corresponding alkenes.

(a)

$$CH_3(CH_2)_5C\equiv CH \xrightarrow[\text{H}_3\text{O}^+]{\text{KMnO}_4} CH_3(CH_2)_5CO_2H \ + \ CO_2$$

(b)

$$\xrightarrow[\text{H}_3\text{O}^+]{\text{KMnO}_4}$$

$$+ \quad CH_3CO_2H$$

(c)

$$\xrightarrow[\text{H}_3\text{O}^+]{\text{KMnO}_4}$$

$$\text{HO}_2\text{C(CH}_2)_8\text{CO}_2\text{H}$$

Since only one cleavage product is formed, the parent hydrocarbon must have contained a triple bond as part of a ring.

(d)

$$\overset{\text{CH}_3}{\underset{|}{\text{CH}_3\text{CH}}} = \text{CCH}_2\text{CH}_2\text{C} \equiv \text{CH} \quad \xrightarrow[\text{2. Zn, H}_3\text{O}^+]{\text{1. O}_3} \quad \text{CH}_3\text{CHO} + \text{CH}_3\overset{\text{O}}{\overset{||}{\text{C}}}\text{CH}_2\text{CH}_2\text{CO}_2\text{H} + \text{CO}_2$$

Notice that the products of this ozonolysis contain aldehyde and ketone functional groups, as well as a carboxylic acid and CO_2. The parent hydrocarbon must thus contain a double and a triple bond.

(e)

$$\xrightarrow[\text{2. Zn, H}_3\text{O}^+]{\text{1. O}_3} \quad \overset{\text{O}}{\overset{||}{\text{H}}}\text{CCH}_2\text{CH}_2\text{CH}_2\text{CH}_2\overset{\text{O}}{\overset{||}{\text{C}}}\text{CO}_2\text{H} + \text{CO}_2$$

9.35

Organic Synthesis

9.36

9.37

(a)

$$CH_3CH_2C{\equiv}CH \xrightarrow[\text{HgSO}_4]{\text{H}_2\text{O, H}_2\text{SO}_4} CH_3CH_2\overset{\overset{\displaystyle O}{\|}}{C}CH_3$$

(b)

$$CH_3CH_2C{\equiv}CH \xrightarrow[\text{2. H}_2\text{O}_2,\ ^-\text{OH}]{\text{1. BH}_3,\ \text{THF}} CH_3CH_2CH_2CHO$$

(c)

$$\xrightarrow[\text{2. CH}_3\text{Br}]{\text{1. NaNH}_2,\ \text{NH}_3}$$

(d)

$$\xrightarrow[\text{Lindlar catalyst}]{\text{H}_2}$$

(e)

$$CH_3CH_2C{\equiv}CH \xrightarrow[\text{H}_3\text{O}^+]{\text{KMnO}_4} CH_3CH_2CO_2H \ + \ CO_2$$

(f) $CH_3CH_2CH_2CH_2CH=CH_2$ $\xrightarrow[CH_2Cl_2]{Br_2}$ $CH_3CH_2CH_2CH_2CH(Br)CH_2Br$

$\downarrow$ 1. 2 $NaNH_2$, NH_3
2. H_3O^+

$CH_3CH_2CH_2CH_2C\equiv CH$

9.38

(a) $CH_3CH_2CH_2C\equiv CH$ $\xrightarrow[\substack{\text{Lindlar}\\\text{catalyst}}]{H_2}$ $CH_3CH_2CH_2CH=CH_2$ $\xrightarrow[\substack{2.\ Zn,\ H_3O^+}]{1.\ O_3}$ $CH_3CH_2CH_2CHO$
$+$
CH_2O

(b) $(CH_3)_2CHCH_2C\equiv CH$ $\xrightarrow[\substack{2.\ CH_3CH_2Br}]{1.\ NaNH_2,\ NH_3}$ $(CH_3)_2CHCH_2C\equiv CCH_2CH_3$

$\downarrow$ Li in NH_3

$$\underset{(CH_3)_2CHCH_2}{H}\diagdown\underset{}{}C=C\diagup\overset{CH_2CH_3}{\underset{H}{}}$$

9.39 The product contains a cis-disubstituted cyclopropane ring, which can be formed from a Simmons–Smith reaction of CH_2I_2 with a cis alkene. The alkene with a cis bond can be produced from an alkyne by hydrogenation using a Lindlar catalyst. The needed alkyne can be formed from the starting material shown by an alkylation using bromomethane.

$CH_3CH_2CH_2CH_2C\equiv CH$ $\xrightarrow[\substack{2.\ CH_3Br}]{1.\ NaNH_2,\ NH_3}$ $CH_3CH_2CH_2CH_2C\equiv CCH_3$

$\downarrow$ H_2
Lindlar catalyst

$$\underset{CH_3CH_2CH_2CH_2}{H}\diagdown C=C\diagup\underset{CH_3}{H}$$

$\xleftarrow[\text{Zn (Cu)}]{CH_2I_2}$

cyclopropane ring structure with $CH_3CH_2CH_2CH_2$ and CH_3 substituents

9.40

The trans double bond in the second target molecule is a product of reduction of a triple bond with Li in NH₃. The alkyne was formed by an alkylation of a terminal alkyne with bromomethane. The terminal alkyne was synthesized from the starting alkene by bromination, followed by dehydrohalogenation.

9.41

(a)

$$CH_3CH_2C{\equiv}CH \xrightarrow[CH_2Cl_2]{2\ Cl_2} CH_3CH_2C(Cl_2)CHCl_2$$

1,1,2,2-Tetrachlorobutane

(b)

$$CH_3CH_2C{\equiv}CH \xrightarrow[\substack{Lindlar \\ catalyst}]{H_2} CH_3CH_2CH{=}CH_2 \xrightarrow[KOH]{CHCl_3} CH_3CH_2CH{-}CH_2$$

1,1-Dichloro-2-ethyl-cyclopropane

9.42 In all of these problems, an acetylide ion (or an anion of a terminal alkyne) is alkylated by a haloalkane.

(a)

$$HC{\equiv}CH \xrightarrow[\substack{2.\ CH_3CH_2CH_2Br}]{1.\ NaNH_2,\ NH_3} CH_3CH_2CH_2C{\equiv}CH$$

(b)

$$HC{\equiv}CH \xrightarrow[\substack{2.\ CH_3CH_2Br}]{1.\ NaNH_2,\ NH_3} CH_3CH_2C{\equiv}CH \xrightarrow[\substack{2.\ CH_3CH_2Br}]{1.\ NaNH_2,\ NH_3} CH_3CH_2C{\equiv}CCH_2CH_3$$

(c)

$$HC{\equiv}CH \xrightarrow[\substack{2.\ (CH_3)_2CHCH_2Br}]{1.\ NaNH_2,\ NH_3} (CH_3)_2CHCH_2C{\equiv}CH$$

$$\downarrow \substack{H_2,\ Lindlar\ catalyst \\ or \quad Li\ in\ NH_3}$$

$$(CH_3)_2CHCH_2CH{=}CH_2$$

(d)

$$CH_3CH_2CH_2C\equiv CH \xrightarrow[\text{2. } CH_3CH_2CH_2Br]{\text{1. } NaNH_2, NH_3} CH_3CH_2CH_2C\equiv CCH_2CH_2CH_3$$

from (a)

$$\downarrow \text{H}_2\text{O, H}_2\text{SO}_4 \quad \text{HgSO}_4$$

$$CH_3CH_2CH_2\overset{\overset{\displaystyle O}{\|}}{C}CH_2CH_2CH_2CH_3$$

Hydroboration/oxidation can also be used to form the ketone from 4-octyne.

(e)

$$HC\equiv CH \xrightarrow[\text{2. } CH_3CH_2CH_2CH_2Br]{\text{1. } NaNH_2, NH_3} CH_3CH_2CH_2CH_2C\equiv CH$$

$$\downarrow \begin{array}{l}\text{1. BH}_3\text{, THF}\\ \text{2. H}_2\text{O}_2\text{, }^-\text{OH}\end{array}$$

$$CH_3CH_2CH_2CH_2CH_2CHO$$

9.43

(a)

$$CH_3CH_2C\equiv CCH_2CH_3 \xrightarrow[\text{Lindlar catalyst}]{D_2}$$

(b)

$$CH_3CH_2C\equiv CCH_2CH_3 \xrightarrow{\text{Li in ND}_3}$$

(c)

$$CH_3CH_2CH_2C\equiv CH \xrightarrow[\text{NH}_3]{\text{NaNH}_2} [CH_3CH_2CH_2C\equiv C:^-Na^+] \xrightarrow{D_3O^+} CH_3CH_2CH_2C\equiv CD$$

(d)

9.44

$$HC\equiv CH \xrightarrow[\text{2. } BrCH_2(CH_2)_6CH_2Br]{\text{1. 2 } NaNH_2, NH_3}$$

A dihalide is used to form the ring.

9.45 Muscalure is a C₂₃ alkene. The only functional group present is the double bond between C₉ and C₁₀. Since our synthesis begins with acetylene, we can assume that the double bond can be produced by hydrogenation of a triple bond.

(Z)-9-Tricosene

General Problems

9.46 (a) An acyclic alkane with eight carbons has the formula C_8H_{18}. C_8H_{10} has eight fewer hydrogens, or four fewer pairs of hydrogens, than C_8H_{18}. Thus, C_8H_{10} contains four degrees of unsaturation (rings/double bonds/triple bonds).

(b) Because only one equivalent of H_2 is absorbed over the Lindlar catalyst, *one* triple bond is present.

(c) Three equivalents of H_2 are absorbed when reduction is done over a palladium catalyst; two of them hydrogenate the triple bond already found to be present. Therefore, one *double* bond must also be present.

(d) C_8H_{10} must therefore contain one ring.

(e) Many structures are possible.

9.47

9.48

H$_2$C—C≡C—C≡C—CH$_2$
| |
H$_2$C—C≡C—C≡C—CH$_2$
            A

8 H$_2$ / Pd/C →

1. O$_3$
2. Zn, H$_3$O$^+$ →

2 HO$_2$CCH$_2$CH$_2$CO$_2$H
+
2 HO$_2$C—CO$_2$H

9.49

(a)

trans-5-Decene

Br$_2$ / CH$_2$Cl$_2$ →

H   Br
| |
         | |
H   Br

2 NaNH$_2$, NH$_3$ ↓

H$_2$ / Lindlar catalyst ←

cis-5-Decene

(b)

cis-5-Decene

Br$_2$ / CH$_2$Cl$_2$ →

H   Br
| |
         | |
H   Br

2 NaNH$_2$, NH$_3$ ↓

Li in NH$_3$ ←

trans-5-Decene

9.50

$$CH_3\overset{O}{\overset{\|}{C}}CH_3 \xrightarrow[\text{2. H}_3\text{O}^+]{\text{1. HC≡C:}^-\text{Na}^+} CH_3\underset{CH_3}{\overset{OH}{\underset{|}{\overset{|}{C}}}}C≡CH \xrightarrow[\text{H}_2\text{SO}_4]{\text{H}_2\text{O}} H_2C=\underset{}{\overset{CH_3}{\overset{|}{C}}}C≡CH$$

H$_2$ / Lindlar catalyst ↓

H$_2$C=$\overset{CH_3}{\overset{|}{C}}$CH=CH$_2$

2-Methyl-1,3-butadiene

9.51

9.52

1-Octen-3-ol

The addition of acetylide occurs by the same route as shown in Problem 9.41.

9.53 (1) Erythrogenic acid contains six degrees of unsaturation (see Sec. 7.2 for the method of calculating unsaturation equivalents for compounds containing elements other than C and H).

(2) One of these double bonds is contained in the carboxylic acid functional group $-CO_2H$; thus, five other degrees of unsaturation are present.

(3) Because five equivalents of H_2 are absorbed on catalytic hydrogenation, erythrogenic acid contains no rings.

(4) The presence of both aldehyde and carboxylic acid products of ozonolysis indicates that both double and triple bonds are present in erythrogenic acid.

(5) Only two ozonolysis products contain aldehyde functional groups; these fragments must have been double-bonded to each other in erythrogenic acid.
$H_2C=CH(CH_2)_4C\equiv$

(6) The other ozonolysis products result from cleavage of triple bonds. However, not enough information is available to tell in which order the fragments were attached. The two possible structures are:

  A $H_2C=CH(CH_2)_4C\equiv C-C\equiv C(CH_2)_7CO_2H$

  B $H_2C=CH(CH_2)_4C\equiv C(CH_2)_7C\equiv CCO_2H$

One method of distinguishing between the two possible structures is to treat erythrogenic acid with two equivalents of H_2, using Lindlar catalyst. The resulting trialkene can then be ozonized. The fragment that originally contained the carboxylic acid can then be identified. (**A** is the structure of erythrogenic acid.)

9.54

9.55

This simplest cumulene is pictured above. The carbons at the end of the cumulated double bonds are sp^2-hybridized and form one π bond to the "interior" carbons. The interior carbons are sp-hybridized; each carbon forms two π bonds – one to an "exterior" carbon and one to the other interior carbon. If you build a model of this cumulene, you can see that the substituents all lie in the same plane. This cumulene can thus exhibit cis–trans isomerism, just as simple alkenes can.

In general, the substituents of any compound with an odd number of adjacent double bonds lie in a plane; these compounds can exhibit cis–trans isomerism.

9.56 In order for a base to strong enough to deprotonate an acid, the cognate acid of the base must a weaker acid (have a higher pK_a) than the acid being deprotonated. The pK_a of 1-butyne is 25.

(a)   The pK_a of H_2O (the conjugate acid of hydroxide) is 15. Because water is a stronger acid than 1-butyne, hydroxide *will not deprotonate* 1-butyne.

KOH

(b)   The pK_a of dimethylsulfoxide (the conjugate acid of dimyslate ion) is 35. Because dimethylsulfoxide is a weaker acid than 1-butyne, dimsylate *will deprotonate* 1-butyne.

(c)   The pK_a of butane (the conjugate acid of n-butyl lithium) is approximately 60. Because butane is a weaker acid than 1-butyne, n-butyl lithium *will deprotonate* 1-butyne.

$$CH_3CH_2CH_2CH_2Li$$

(d)   The pK_a of acetone (the conjugate acid of acetone enolate) is 19. Because acetone is a stronger acid than 1-butyne, acetone enolate *will not deprotonate* 1-butyne.

9.57 The overall trend for carbocation stability is shown below. Recall that resonance adds additional stability to carbocations.

(a)

(b)

(c)

Chapter 10 – Organohalides

Chapter Outline

I.   Names and properties of alkyl halides (Section 10.1).
   A. Naming alkyl halides.
      1. Rules for naming alkyl halides:
         a. Find the longest chain and name it as the parent.
            i.   If a double or triple bond is present, the parent chain must contain it.
         b. Number the carbon atoms of the parent chain, beginning at the end nearer the first substituent, whether alkyl or halo.
         c. Number each substituent.
            i.   If more than one of the same kind of substituent is present, number each, and use the prefixes *di-, tri-, tetra-* and so on.
            ii. If different halogens are present, number all and list them in alphabetical order.
         d. If the parent chain can be numbered from either end, start at the end nearer the substituent that has alphabetical priority.
      2. Some alkyl halides are named by first citing the name of the alkyl group and then citing the halogen.
   B. Structure of alkyl halides.
      1. Alkyl halides have approximately tetrahedral geometry.
      2. Bond lengths increase with increasing size of the halogen bonded to carbon.
      3. Bond strengths decrease with increasing size of the halogen bonded to carbon.
      4. Carbon–halogen bonds are polar, and many halomethanes have dipole moments.
      5. Alkyl halides behave as electrophiles in polar reactions.
II. Preparation of alkyl halides (Sections 10.2–10.5).
   A. Radical halogenation of alkanes (Section 10.2).
      1. The sequence of steps: initiation, propagation, termination.
      2. Complications of radical halogenation.
         a. The reaction continues on to produce di- and polysubstituted products.
         b. If more than one type of hydrogen is present, more than one type of monosubstituted product is formed.
         c. The reactivity order of different types of hydrogen towards chlorination is: primary < secondary < tertiary.
            i.   This reactivity order is due to the bond dissociation energies for formation of the alkyl radicals.
            ii. The stability order of alkyl radicals: primary < secondary < tertiary.
   B. Allylic bromination of alkenes (Sections 10.3–10.4).
      1. Reaction of an alkene with NBS causes bromination at the position allylic to the double bond (Section 10.3).
      2. This reaction occurs by a radical chain mechanism.
         a. Br· abstracts an allylic hydrogen.
         b. The allylic radical reacts with Br_2 to form an allylic bromide, plus Br·.

3. Reaction occurs at the allylic position because an allylic C–H bond is weaker than most other C–H bonds, and an allylic radical is more stable.

4. Reasons for stability of an allylic radical (Section 10.4).

   a. The carbon with the unpaired electron is sp^2-hybridized, and its p orbital can overlap with the p orbitals of the double-bond carbons.

   b. The radical intermediate is thus stabilized by resonance.

      i. This stability is due to delocalization (spreading out) of the unpaired electron over an extended π network.

   c. Reaction of the allylic radical with Br₂ can occur at either end of the π orbital system.

      i. A mixture of products may be formed if the alkene is unsymmetrical.

      ii. These products aren't usually formed in equal quantities: reaction to form the more substituted double bond is favored.

   d. Products of allylic bromination can be dehydrohalogenated to form dienes.

C. Alkyl halides from alcohols (Section 10.5).

   1. Tertiary alkyl chlorides, bromides or iodides can be prepared by the reaction of a tertiary alcohol with HCl, HBr or HI.

      a. Reaction of secondary or primary alcohols occurs under more drastic conditions, which may destroy other acid-sensitive functional groups.

   2. Primary and secondary alkyl chlorides and bromides can be formed by treatment of the corresponding alcohols with SOCl₂ or PBr₃, respectively.

      a. Reaction conditions are mild, less acidic, and are less likely to cause acidcatalyzed rearrangements.

   3. Alkyl fluorides can be prepared using either (CH₃CH₂)₂NSF₃ or HF in pyridine.

III. Reactions of alkyl halides (Sections 10.6–10.7).

A. Grignard reagents (Section 10.6).

   1. Organohalides react with Mg to produce organomagnesium halides, RMgX.

      a. These compounds are known as Grignard reagents.

   2. Grignard reagents can be formed from alkyl, alkenyl and aryl halides.

      a. Steric hindrance is no barrier to formation of Grignard reagents.

   3. The carbon bonded to Mg is negatively polarized and is nucleophilic.

   4. Grignard reagents react with weak acids to form hydrocarbons.

B. Organometallic coupling reagents (Section 10.7).

   1. Alkyl halides can react with Li to form alkyllithiums.

   2. These alkyllithiums can combine with CuI to form lithium diorganocopper compounds (R₂CuLi), which are known as Gilman reagents.

   3. R₂CuLi compounds can react with alkyl halides (except for fluorides) to form hydrocarbon products.

   4. Organometallic coupling reactions are useful for forming large molecules from small pieces.

      a. The reaction can be carried out on alkyl, vinyl and aryl halides.

      b. The mechanism is not a typical polar nucleophilic substitution.

5. A related reaction is the Suzuki-Miyaura reaction – a palladium-catalyzed coupling of aryl or vinyl organotin reagents with organohalides.

IV. Oxidation and reduction in organic chemistry (Section 10.8).

   A. In organic chemistry, an oxidation is a reaction that results in a loss in electron density by carbon.

      1. This loss may be due to two kinds of reactions:

         a. Bond formation between carbon and a more electronegative atom (usually O, N or halogen).

         b. Bond breaking between carbon and a less electronegative atom (usually H).

      2. Examples include chlorination of alkanes and reaction of alkenes with Br_2.

   B. A reduction is a reaction that results in a gain of electron density by carbon.

      1. This gain may be due to two kinds of reactions:

         a. Bond formation between carbon and a less electronegative atom.

         b. Bond breaking between carbon and a more electronegative atom.

      2. Examples include conversion of a Grignard reagent to an alkane, and reduction of an alkene with H_2.

   C. Alkanes are at the lowest oxidation level, and CO_2 is at the highest level.

   D. A reaction that converts a compound from a lower oxidation level to a higher oxidation level is an oxidation.

   E. A reaction that converts a compound from a higher oxidation level to a lower oxidation level is an reduction.

Solutions to Problems

10.1   The rules that were given for naming alkanes in Section 3.4 are used for alkyl halides. A halogen is treated the same as an alkyl substituent but is named as a halo group.

(a) $CH_3CH_2CH_2CH_2I$

1-Iodobutane

(b)
$$\underset{CH_3}{CH_3CHCH_2CH_2Cl}$$
1-Chloro-3-methylbutane

(c)
$$\underset{CH_3}{\overset{CH_3}{BrCH_2CH_2CH_2CCH_2Br}}$$
1,5-Dibromo-2,2-dimethylpentane

(d)
$$\underset{Cl}{\overset{CH_3}{CH_3CCH_2CH_2Cl}}$$
1,3-Dichloro-3-methylbutane

(e)
$$\overset{I\ \ \ CH_2CH_2Cl}{CH_3CHCHCH_2CH_3}$$
1-Chloro-3-ethyl-4-iodopentane

(f)
$$\overset{Br\qquad\ Cl}{CH_3CHCH_2CH_2CHCH_3}$$
2-Bromo-5-chlorohexane

10.2

(a)
$$\overset{H_3C\quad Cl}{\underset{CH_3}{CH_3CH_2CH_2C-CHCH_3}}$$
2-Chloro-3,3-dimethylhexane

(b)
$$\overset{Cl\quad CH_3}{\underset{Cl}{CH_3CH_2CH_2C-CHCH_3}}$$
3,3-Dichloro-2-methylhexane

(c)
$$\overset{CH_2CH_3}{\underset{Br}{CH_3CH_2CCH_2CH_3}}$$
3-Bromo-3-ethylpentane

(d)

1,1-Dibromo-4-isopropylcyclohexane

(e)
$$\overset{Cl}{CH_3CH_2CH_2CH_2CH_2CCH_2CHCH_3}$$
$$CH_3CH_2CHCH_3$$
4-*sec*-Butyl-2-chlorononane

(f)

1,1-Dibromo-4-*tert*-butylcyclohexane

10.3

	Product	Site of chlorination
	CH₃ | CH₃CH₂CH₂CHCH₂Cl 1-Chloro-2-methylpentane	a
	Cl | CH₃CH₂CH₂C(CH₃)₂ 2-Chloro-2-methylpentane	b
	Cl | CH₃CH₂CHCH(CH₃)₂ 3-Chloro-2-methylpentane	c
	Cl | CH₃CHCH₂CH(CH₃)₂ 2-Chloro-4-methylpentane	d
	ClCH₂CH₂CH₂CH(CH₃)₂ 1-Chloro-4-methylpentane	e

e d c b a
CH₃CH₂CH₂CH(CH₃)₂   —Cl₂/hv→
2-Methylpentane

Chlorination at sites b and e yields achiral products. The products of chlorination at sites a, c and d are chiral; each product is formed as a racemic mixture of enantiomers.

10.4

a
CH₃
|
CH₃–CH₂–C—CH₃
d   c | a
H
b
2-Methylbutane

Type of –H	a	b	c	d
Number of –H of each type	6	1	2	3
Relative reactivity	1.0	5.0	3.5	1.0
Number times reactivity	6.0	5.0	7.0	3.0
Percent chlorination	29%	24%	33%	14%

CH₃
|
CH₃CH₂CHCH₃   —Cl₂/hv→

CH₃
|
CH₃CH₂CHCH₂Cl   +   CH₃CH₂CCH₃
29%   Cl 24%

CH₃
|
+   CH₃CHCHCH₃   +   ClCH₂CH₂CHCH₃
Cl 33%   14%

10.5

10.6

Abstraction of hydrogen by a bromine radical yields an allylic radical.

The allylic radical reacts with Br₂ to produce **A** and **B**

A   **B**

Product **B** is favored because reaction at the primary end of the allylic radical yields a product with a trisubstituted double bond.

10.7   (a)

5-Methylcycloheptene   3-Bromo-5-methylcycloheptene   3-Bromo-6-methyl-cycloheptene

(b)

Two different allylic radicals can form, and four different bromohexenes can be produced.

10.8   Remember that halogen acids are used for converting tertiary alcohols to alkyl halides. PBr₃ and SOCl₂ are used for converting secondary and primary alcohols to alkyl halides. (CH₃CH₂)₂NSF₃ and HF in pyridine can be used to form alkyl fluorides.

(a)

(b)

(c)

(d)

10.9   Table 9.1 shows that the pK_a of CH_3−H is 60. Since CH_4 is a very weak acid, ⁻:CH_3 is a very strong base. Alkyl Grignard reagents are similar in base strength to ⁻:CH_3, but alkynyl Grignard reagents are somewhat weaker bases. Both reactions (a) and (b) occur as written.

(a)

$$CH_3MgBr \ + \ H-C \equiv C-H \ \longrightarrow \ CH_4 \ + \ H-C \equiv C-MgBr$$

  stronger   stronger   weaker   weaker
   base   acid   acid   base

(b)

$$CH_3MgBr \ + \ NH_3 \ \longrightarrow \ CH_4 \ + \ H_2N-MgBr$$

  stronger   stronger   weaker   weaker
   base   acid   acid   base

10.10   Just as Grignard reagents react with *proton* donors to convert R−MgX into R−H, they also react with *deuterium* donors to convert R−MgX into R−D. In this case:

10.11   (a)   The methyl group has an allylic relationship to the double bond. Thus, an organometallic coupling reaction between 3-bromocyclohexene and lithium dimethylcopper gives the desired product. 3-Bromocyclohexene can be formed by allylic bromination of cyclohexene with NBS.

3-Methylcyclohexene

(b)   We are asked to synthesize an eight–carbon product from a four–carbon starting material. Thus, an organometallic coupling reaction between 1-bromobutane and lithium dibutylcopper gives octane as the product. The Gilman reagent is formed from 1-bromobutane.

$$2 \ CH_3CH_2CH_2CH_2Br \ \xrightarrow[\text{pentane}]{4 \ Li} \ 2 \ CH_3CH_2CH_2CH_2Li \ \xrightarrow[\text{ether}]{CuI} \ (CH_3CH_2CH_2CH_2)_2Cu^-Li^+$$
$$+ \ 2 \ LiBr \qquad\qquad + \ LiI$$

$$CH_3CH_2CH_2CH_2Br \ + \ (CH_3CH_2CH_2CH_2)_2Cu^-Li^+ \ \xrightarrow{\text{ether}} \ CH_3(CH_2)_6CH_3 \ + \ LiBr$$
Octane

$$+ \ CH_3CH_2CH_2CH_2Cu$$

(c) The synthesis in (b) suggests a route to the product. Decane can be synthesized from 1-bromopentane and lithium dipentylcopper. 1-Bromopentane is formed by hydroboration of 1-pentene, followed by treatment of the resulting alcohol with PBr₃.

$$CH_3CH_2CH_2CH=CH_2 \xrightarrow[\text{2. H}_2\text{O}_2,\ ^-\text{OH}]{\text{1. BH}_3,\ \text{THF}} CH_3CH_2CH_2CH_2CH_2OH$$

$$\downarrow \text{PBr}_3,\ \text{ether}$$

$$CH_3CH_2CH_2CH_2CH_2Br$$

$$2\ CH_3CH_2CH_2CH_2CH_2Br \xrightarrow[\text{pentane}]{\text{4 Li}} 2\ CH_3CH_2CH_2CH_2CH_2Li\ +\ 2\ LiBr$$

$$\downarrow \text{CuI, ether}$$

$$(CH_3CH_2CH_2CH_2CH_2)_2Cu^-Li^+\ +\ LiI$$

$$CH_3CH_2CH_2CH_2CH_2Br\ +\ (CH_3CH_2CH_2CH_2CH_2)_2Cu^-Li^+ \xrightarrow{\text{ether}} CH_3(CH_2)_8CH_3$$
$$\text{Decane}$$

$$+\ LiBr\ +\ CH_3CH_2CH_2CH_2CH_2Cu$$

10.12 (a) As described in Worked Example 10.2, the oxidation level of a compound can be found by adding the number of C−O, C−N, and C−X bonds and subtracting the number of C−H bonds. Cyclohexane, the first compound shown, has 12 C−H bonds, and has an oxidation level of −12. Cyclohexanone has 2 C−O bonds (from the double bond) and 10 C−H bonds, for an oxidation level of −8. 1-Chlorocyclohexene has one C−Cl bond and 9 C−H bonds, and also has an oxidation level of −8. Benzene has 6 C−H bonds, for an oxidation level of −6.

In order of increasing oxidation level:

(b)

$$CH_3CH_2NH_2 \quad < \quad H_2NCH_2CH_2NH_2 \quad < \quad CH_3CN$$
$$\text{oxidation level} = -4 \qquad \text{oxidation level} = -2 \qquad \text{oxidation level} = 0$$

10.13 (a) The aldehyde carbon of the reactant has an oxidation level of 1 (2 C−O bonds minus 1 C−H bond). The alcohol carbon of the product has an oxidation level of −1 (1 C−O bond minus 2 C−H bonds). The reaction is a reduction because the oxidation level of the product is lower than the oxidation level of the reactant.

(b)   The oxidation level of the upper carbon of the double bond in the reactant changes from 0 to +1 in the product; the oxidation level of the lower carbon of the double bond changes from 0 to −1. The total oxidation level, however, is the same for both product and reactant, and the reaction is neither an oxidation nor a reduction.

Visualizing Chemistry

10.14

(a)   *cis*-1-Chloro-3-methylcyclohexane

(b)   4-Chloro-2-methyl-2-heptene

10.15 (a)

(b)

10.16 The name of the compound is (*R*)-2-bromopentane. Reaction of (*S*)-2-pentanol with PBr₃ to form (*R*)-2-bromopentane occurs with a change in stereochemistry because the configuration at the chirality center changes from *S* to *R*.

Mechanism Problems

10.17 (a)

Mechanism:

Initiation

Propagation

Propagation

Termination

Although there are several possible termination
steps, only this one provides the product.

(b)

Mechanism:

Initiation

Propagation

Propagation

Termination

Although there are several possible termination
steps, only this one provides the product.

(c)

Mechanism:

Initiation

Propagation

Propagation

Termination

Although there are several possible termination
steps, only this one provides the product.

10.18 (a)

Mechanism:

(b)

Mechanism:

(c)

Mechanism:

10.19

iminol
intermediate

succinimide

Mechanism:

10.20

10.21

$$(C_4H_9)_3SnH \xrightarrow{h\nu} (C_4H_9)_3Sn\cdot + H\cdot$$

$$(C_4H_9)_3Sn\cdot + RX \longrightarrow (C_4H_9)_3SnX + R\cdot$$

$$(C_4H_9)_3SnH + R\cdot \longrightarrow (C_4H_9)_3Sn\cdot + RH$$

Additional Problems

Naming Alkyl Halides

10.22

(a)

H₃C Br Br   CH₃

CH₃CHCHCHCH₂CHCH₃

3,4-Dibromo-2,6-dimethylheptane

(b)

I

CH₃CH=CHCH₂CHCH₃

5-Iodo-2-hexene

(c)

Br   Cl CH₃

CH₃CCH₂CHCHCH₃

CH₃

2-Bromo-4-chloro-2,5-dimethylhexane

(d)

CH₂Br

CH₃CH₂CHCH₂CH₂CH₃

3-(Bromomethyl)hexane

(e)

ClCH₂CH₂CH₂C≡CCH₂Br

1-Bromo-6-chloro-2-hexyne

10.23

(a)

CH₃ Cl

CH₃CH₂CHCHCHCH₃

Cl

2,3-Dichloro-4-methylhexane

(b)

Br   CH₃

CH₃CH₂CCH₂CHCH₃

CH₂CH₃

4-Bromo-4-ethyl-2-methylhexane

(c)

CH₃ I CH₃

CH₃C—CHCCH₃

CH₃   CH₃

3-Iodo-2,2,4,4-tetramethylpentane

(d)

Br   CH₂CH₃

cis-1-Bromo-2-ethylcyclopentane

10.24

(a)   2-methylbutane

There are seven different products. All of the compounds whose names include the absolute configuration (R or S) in their name are chiral and optically active.

2-methylbutane →(Cl₂, Light)→ (R)-1-chloro-2-methylbutane   (S)-1-chloro-2-methylbutane   (R)-2-chloro-2-methylbutane   (S)-2-chloro-2-methylbutane

(R)-2-chloro-3-methylbutane   (S)-2-chloro-3-methylbutane   1-chloro-3-methylbutane

(b) methylcyclopropane

There are six different products. All of the compounds whose names include the absolute configuration (R and/or S) in their name are chiral and optically active.

methylcyclopropane

1-chloro-1-methylcyclopropane

(1R,2R)-1-chloro-2-methylcyclopropane

(1S,2R)-1-chloro-2-methylcyclopropane

1-chloro-1-methylcyclopropane

(1R,2S)-1-chloro-2-methylcyclopropane

(1S,2S)-1-chloro-2-methylcyclopropane

(c) 2,2-dimethylpentane

There are six different products. All of the compounds whose names include the absolute configuration (R or S) in their name are chiral and optically active.

2,2-dimethylpentane

1-chloro-2,2-dimethylpentane

(R)-3-chloro-2,2-dimethylpentane

(S)-3-chloro-2,2-dimethylpentane

(S)-2-chloro-4,4-dimethylpentane

(R)-2-chloro-4,4-dimethylpentane

1-chloro-4,4-dimethylpentane

Synthesizing Alkyl Halides

10.25

(a)

Chlorocyclopentane

(b)

Methylcyclopentane

(c)

3-Bromocyclopentene

(d)

$$\text{1. Hg(OAc)}_2, \text{H}_2\text{O} \quad \text{2. NaBH}_4$$

Cyclopentanol

Hydroboration/oxidation can also be used to form cyclopentanol.

(e)

HBr   →   Br $\xrightarrow[\text{pentane}]{\text{2 Li}}$ Li

$$2 \quad \text{—Li} \quad \xrightarrow[\text{ether}]{\text{CuI}} \quad \left[\text{—CuLi} \right]_2 \quad \xrightarrow[\text{ether}]{\text{from (b)}}$$

Cyclopentylcyclopentane

(f)

Br

$$\xrightarrow[\text{CH}_3\text{CH}_2\text{OH}]{\text{KOH}}$$

from (c)   1,3-Cyclopentadiene

10.26

(a)

H₃C OH   H₃C Br

$$\xrightarrow[\text{Ether}]{\text{HBr}}$$ + H₂O

(b) $\text{CH}_3\text{CH}_2\text{CH}_2\text{CH}_2\text{OH} \xrightarrow{\text{SOCl}_2} \text{CH}_3\text{CH}_2\text{CH}_2\text{CH}_2\text{Cl}$

(c)

$$\xrightarrow[\substack{\text{CCl}_4 \\ h\nu}]{\text{NBS}}$$

Br   +   Br

The major product contains a tetrasubstituted double bond, and the minor product contains a trisubstituted double bond.

(d)

OH   Br

$$\xrightarrow[\text{Ether}]{\text{PBr}_3}$$

(e)

$$CH_3CH_2\overset{\overset{\displaystyle Br}{\displaystyle |}}{C}HCH_3 \xrightarrow[\text{Ether}]{\text{Mg}} CH_3CH_2\overset{\overset{\displaystyle MgBr}{\displaystyle |}}{C}HCH_3 \xrightarrow{H_2O} CH_3CH_2CH_2CH_3$$
                                                              A   **B**

(f)

$$2\ CH_3CH_2CH_2CH_2Br \xrightarrow[\text{Pentane}]{\text{4 Li}} 2\ CH_3CH_2CH_2CH_2Li \xrightarrow{\text{CuI}} (CH_3CH_2CH_2CH_2)_2CuLi$$
                                                                    A   **B**

(g) $CH_3CH_2CH_2CH_2Br\ +\ (CH_3)_2CuLi \xrightarrow{\text{Ether}} CH_3CH_2CH_2CH_2CH_3 + CH_3Cu + LiBr$

10.27 Abstraction of hydrogen by Br• can produce either of two allylic radicals. The first radical, resulting from abstraction of a secondary hydrogen, is more likely to be formed.

$$CH_3\overset{\bullet}{C}HCH=CHCH_3 \longleftrightarrow CH_3CH=CH\overset{\bullet}{C}HCH_3 \quad \text{(identical resonance forms)}$$

and

$$CH_3CH_2CH=CH\overset{\bullet}{C}H_2 \longleftrightarrow CH_3CH_2\overset{\bullet}{C}HCH=CH_2$$

Reaction of the radical intermediates with a bromine source leads to a mixture of products:

$$CH_3\overset{\overset{\displaystyle Br}{\displaystyle |}}{C}HCH=CHCH_3 \qquad\qquad \textit{cis-} \text{ and } \textit{trans-}\text{4-Bromo-2-pentene}$$

and

$$CH_3CH_2CH=CHCH_2Br \qquad\qquad \textit{cis-} \text{ and } \textit{trans-}\text{1-Bromo-2-pentene}$$

$$CH_3CH_2\overset{\overset{\displaystyle Br}{\displaystyle |}}{C}HCH=CH_2 \qquad\qquad \text{3-Bromo-1-pentene}$$

The major product is 4-bromo-2-pentene, instead of the desired product, 1-bromo-2-pentene.

10.28 Three different allylic radical intermediates can be formed. Bromination of these intermediates can yield as many as five bromoalkenes. This is definitely not a good reaction to use in a synthesis even if the products could be separated.

(allylic – secondary hydrogen abstracted)   3-Bromo-2-methylcyclohexene

(allylic – secondary hydrogen abstracted)

3-Bromo-1-methyl-
cyclohexene

3-Bromo-3-methyl-
cyclohexene

(allylic – primary hydrogen abstracted)   1-(Bromomethyl)-
cyclohexene

2-Bromomethylene-
cyclohexane

10.29

$$CH_3CH=CHCH_2CH=CH_2$$

with NBS, hv, CCl_4 forming two resonance-stabilized radical intermediates.

Left intermediate:

$$\overset{\cdot}{C}H_2CH=CHCH_2CH=CH_2$$
$$\updownarrow$$
$$H_2C=CH\overset{\cdot}{C}HCH_2CH=CH_2$$

$\downarrow Br_2$

$$BrCH_2CH=CHCH_2CH=CH_2$$
6-Bromo-1,4-hexadiene

$+$

$$\underset{\underset{Br}{|}}{H_2C=CHCHCH_2CH=CH_2}$$
3-Bromo-1,5-hexadiene

Right intermediate:

$$CH_3CH=CH\overset{\cdot}{C}HCH=CH_2$$
$$\updownarrow$$
$$CH_3CH=CHCH=CH\overset{\cdot}{C}H_2$$
$$\updownarrow$$
$$CH_3\overset{\cdot}{C}HCH=CHCH=CH_2$$

$\downarrow Br_2$

$$CH_3CH=CHCH=CHCH_2Br$$
1-Bromo-2,4-hexadiene

$+$

$$\underset{\underset{Br}{|}}{CH_3CH=CHCHCH=CH_2}$$
3-Bromo-1,4-hexadiene

$+$

$$\underset{\overset{Br}{|}}{CH_3CHCH=CHCH=CH_2}$$
5-Bromo-1,3-hexadiene

The intermediate on the right is more stable because the unpaired electron is delocalized over more atoms than in the intermediate on the left, and the resulting products should predominate.

10.30 Two allylic radicals can form:

The second radical is much more likely to form because it is both allylic and benzylic, and it yields the following products:

and

Oxidation and Reduction

10.31 Remember that the oxidation level is found by subtracting the number of C−H bonds from the number of C−O, C−N, and C−X bonds. The oxidation levels are shown beneath the structures. In order of increasing oxidation level:

(a)

$$CH_3CH{=}CHCH_3, \quad CH_3CH_2CH{=}CH_2 < \quad CH_3CH_2CH_2\overset{O}{\overset{\|}{C}}H \quad < \quad CH_3CH_2CH_2\overset{O}{\overset{\|}{C}}OH$$

$$-8 \qquad\qquad -8 \qquad\qquad\quad -6 \qquad\qquad\quad -4$$

(b)

$$CH_3CH_2CH_2NH_2, \quad CH_3CH_2CH_2Br \quad < \quad BrCH_2CH_2CH_2Cl \quad < \quad CH_3\overset{O}{\overset{\|}{C}}CH_2Cl$$

$$-6 \qquad\qquad\quad -6 \qquad\qquad\qquad -4 \qquad\qquad\quad -2$$

10.32

$$-6 \qquad\qquad -6 \qquad\qquad -4 \qquad\qquad -6 \qquad\qquad -6$$

All of the compounds except **3** (which is more oxidized) have the same oxidation level.

10.33 (a)   This reaction is an oxidation.

(b)   The reaction is neither an oxidation nor a reduction because the oxidation level of the reactant is the same as the oxidation level of the product.

(c)   This reaction is a reduction.

General Problems

10.34 The general trend for radical stability is shown below.

Hint:

(a)

secondary > primary > vinylic

(b)

allylic > secondary > vinylic

(c)

tertiary > secondary > primary

10.35 Table 6.3 shows that the bond dissociation energy for $C_6H_5CH_2-H$ is 375 kJ/mol. This value is comparable in size to the bond dissociation energy for a bond between carbon and an allylic hydrogen, and thus it is relatively easy to form the $C_6H_5CH_2\bullet$ radical. The high bond dissociation energy for formation of $C_6H_5\bullet$, 472 kJ/mol, indicates the bromination on the benzene ring will not occur. The only product of reaction with NBS is $C_6H_5CH_2Br$.

10.36

10.37

(a)

$$CH_3CH=CHCH=CHCH=CHCH_2^+$$

$$\updownarrow$$

$$CH_3CH=CHCH=CHCHCH=CH_2^+$$

$$\updownarrow$$

$$CH_3CH=CHCHCH=CHCH=CH_2^+$$

$$\updownarrow$$

$$CH_3CHCH=CHCH=CHCH=CH_2^+$$

(b)

(c)

$$CH_3C\equiv\overset{+}{N}-\overset{..}{\underset{..}{O}}:^- \longleftrightarrow CH_3\overset{\overset{..}{}}{C}=\overset{+}{N}=\overset{..}{O} \longleftrightarrow CH_3\overset{+}{C}=\overset{..}{N}-\overset{..}{\underset{..}{O}}:^-$$

10.38

Abstraction of a hydrogen atom from the chirality center of (*S*)-3-methylhexane produces an achiral radical intermediate, which reacts with bromine to form a 1:1 mixture of *R* and *S* enantiomeric, chiral bromoalkanes. The product mixture is optically inactive.

10.39

+ other products

Abstraction of a hydrogen atom from carbon 4 yields a chiral radical intermediate. Reaction of this intermediate with chlorine does not occur with equal probability from each side, and the two diastereomeric products are not formed in 1:1 ratio. The first product is optically active, and the second product is a meso compound.

10.40 All these reactions involve addition of a dialkylcopper reagent [$(CH_3CH_2CH_2CH_2)_2CuLi$] to the same alkyl halide. The dialkylcopper is prepared by treating 1-bromobutane with lithium, followed by addition of CuI:

$$2\ CH_3CH_2CH_2CH_2Br \xrightarrow[\text{Pentane}]{4\ Li} 2\ CH_3CH_2CH_2CH_2Li \xrightarrow[\text{Ether}]{CuI} (CH_3CH_2CH_2CH_2)_2CuLi$$

Butylcyclohexane

10.41 (a)   Fluoroalkanes don't usually form Grignard reagents.
(b)   Two allylic radicals can be produced.

Instead of a single product, as many as four bromoalkene products may result.
(c)   Dialkylcopper reagents don't react with fluoroalkanes.

10.42 A Grignard reagent can't be prepared from a compound containing an acidic functional group because the Grignard reagent is immediately quenched by the proton source. For example, the $-CO_2H$, $-OH$, $-NH_2$, and $RC\equiv CH$ functional groups are too acidic to be used for preparation of a Grignard reagent. $BrCH_2CH_2CH_2NH_2$ is another compound that doesn't form a Grignard reagent.

10.43

Reaction of the ether with HBr can occur by either path **A** or path **B**. Path **A** is favored because its cation intermediate can be stabilized by resonance.

10.44

10.45 As we saw in Chapter 7, tertiary carbocations (R_3C^+) are more stable than either secondary or primary carbocations, due to the ability of the three alkyl groups to stabilize positive charge. If the substrate is also allylic, as in the case of $H_2C=CHC(CH_3)_2Br$, positive charge can be further delocalized. Thus, $H_2C=CHC(CH_3)_2Br$ should form a carbocation faster than $(CH_3)_3CBr$ because the resulting carbocation is more stable.

10.46

alkoxide ion   carboxylate ion

Carboxylic acids are more acidic than alcohols because the negative charge of a carboxylate ion is stabilized by resonance. This resonance stabilization favors formation of carboxylate anions over alkoxide anions, and increases K_a for carboxylic acids. Alkoxide anions are not resonance stabilized.

10.47 (a)

(b)

(c)

10.48   Hint: When solving these problems it is easiest to create a table. First, all of the unique products must be identified. Each product will have a different name. Then one needs first determine the number of hydrogens that will lead to the each different product. As a double check, the sum of the number of hydrogens leading to each product must be equal to the total number of hydrogens in the starting material. Once identified, the type of each hydrogen (1°, 2°, 3°) must be assigned. The rate for each type of hydrogen (1:82:1640 for 1°:2°:3° hydrogens, respectively) should be multiplied by number of hydrogens to give the contribution of each product to the whole. That fraction is then calculated by dividing each individual contribution by the total contribution and that fraction converted into a percentage.

   (a)   methylcyclobutane

Product / Name	#H's giving product	type of H	rate	contrib	fraction	%
1-bromomethylcyclobutane	3	1°	1	3	3/2135	0.1
1-bromo-1-methylcyclobutane	1	3°	1640	1640	1640/2135	76.8
(1R,2R)-1-bromo-2-methylcyclobutane	1	2°	82	82	82/2135	3.8
(1S,2R)-1-bromo-2-methylcyclobutane	1	2°	82	82	82/2135	3.8
(1R,2S)-1-bromo-2-methylcyclobutane	1	2°	82	82	82/2135	3.8

 (1S,2S)-1-bromo-2-methylcyclobutane	1	2°	82	82	82/2135	3.8
 cis-1-bromo-3-methylcyclobutane	1	2°	82	82	82/2135	3.8
 trans-1-bromo-3-methylcyclobutane	1	2°	82	82	82/2135	3.8
Totals	10			2135		100

(b)   3,3-dimethylpentane

Product / Name	#H's giving product	type of H	rate	contrib	fraction	%
 1-chloro-3,3-dimethylpentane	6	1	1	6	6/340	1.8
 (R)-2-chloro-3,3-dimethylpentane	2	2	82	164	164/340	48.2
 (S)-2-chloro-3,3-dimethylpentane	2	2	82	164	164/340	48.2

	#H's giving product	type of H	rate	contrib	fraction	%
3-chloromethyl-3-methylpentane	6	1	1	6	6/340	1.8
Totals	16			340		100

(c) 3-methylpentane

Product / Name	#H's giving product	type of H	rate	contrib	fraction	%
(R)-1-chloro-3-methylpentane	3	1	1	3	3/1974	0.2
(S)-1-chloro-3-methylpentane	3	1	1	3	3/1974	0.2
(2R,3R)-2-chloro-3-methylpentane	1	2	82	82	82/1974	4.2
(2S,3R)-2-chloro-3-methylpentane	1	2	82	82	82/1974	4.2
(2R,3S)-2-chloro-3-methylpentane	1	2	82	82	82/1974	4.2

(2S,3S)-2-chloro-3-methylpentane	1	2	82	82	82/1974	4.2
3-chloro-3--methylpentane	1	3	1640	1640	1640/1974	83.1
3-chloro-3--methylpentane	3	1	1	3	3/1974	0.2
Totals	14			1974		100

10.49 (a) The primary alcohol would go exclusively through an S$_N$2 mechanism. This secondary alcohol will undergo a mixture of S$_N$1 and S$_N$2 substitution. However, formation a secondary carbocation is relatively slow, as is the S$_N$2 reaction compared to the primary alcohol. Thus, the primary alcohol will be the faster reaction.

(b) Tertiary alcohols undergo the fastest of S$_N$1 reactions through a tertiary carbocation, while a secondary alcohol is a relatively slow substrates for S$_N$2 because to the sterics of. Thus, the tertiary alcohol would be expected to react faster.

(c) Tertiary alcohols undergo the fastest of S$_N$1 reactions through a tertiary carbocation, while the primary alcohol of neopentyl alcohol is one for the slowest S$_N$2 substrates for substitution because to the sterics of the tert-butyl group. Thus, the tertiary alcohol would be expected to react faster.

10.50

(a)

Catalytic cycle:

(b)

Catalytic cycle:

Chapter 11 – Reactions of Alkyl Halides: Nucleophilic Substitutions and Eliminations

Chapter Outline

I. Substitution Reactions (Sections 11.1–11.6).
  A. S_N2 reactions (Sections 11.1–11.3).
    1. The discovery of S_N2 reactions (Section 11.1).
      a. Walden discovered that (+) malic acid and (−) malic acid could be interconverted (Section 11.1).
      b. This discovery meant that one or more reactions must have occurred with inversion of configuration at the chirality center.
      c. Nucleophilic substitution of tosylate ion by acetate ion occurs with inversion of configuration.
        i. Nucleophilic substitution reactions of primary and secondary alkyl halides always proceed with inversion of configuration.
    2. The S_N2 reaction (Section 11.2).
      a. Kinetics.
        i. The kinetics of a reaction measure the relationship between reactant concentrations and product concentrations and the rate of reaction.
        ii. In an S_N2 reaction, reaction rate depends on the concentration of both alkyl halide and nucleophile (bimolecular reaction).
          (a). This type of reaction is a second-order reaction.
        iii. In a second-order reaction, rate = k x [RX] x [Nu].
          (a). The constant, k, is the rate constant.
      b. Mechanism.
        i. The reaction takes place in a single step, without intermediates.
        ii. The nucleophile attacks the substrate from a direction directly opposite to the leaving group.
          (a). This type of attack accounts for inversion of configuration.
        iii. In the transition state, the new bond forms at the same time as the old bond breaks.
        iv. Negative charge is shared between the attacking nucleophile and the leaving group.
        v. In the transition state, the three remaining bonds to carbon are in a planar arrangement.
        vi. Both substrate and nucleophile are involved in the step whose rate is measured.
    3. Characteristics of the S_N2 reaction (Section 11.3).
      a. Changes in the energy levels of reactants or of the transition state affect the reaction rate.
      b. Changes in the substrate.
        i. Reaction rate is decreased if the substrate is bulky.
        ii. Substrates, in order of increasing reactivity: tertiary, neopentyl, secondary, primary, methyl.

iii. S_N2 reactions can occur only at relatively unhindered sites.

iv. Vinylic and aryl halides are unreactive to S_N2 substitutions.

c.   Changes in the nucleophile.

i.   Any species can act as a nucleophile if it has an unshared electron pair.

(a). If the nucleophile has a negative charge, the product is neutral.

(a). If the nucleophile is neutral, the product is positively charged.

ii. The reactivity of a nucleophile is dependent on reaction conditions.

iii. In general, nucleophilicity parallels basicity.

iv. Nucleophilicity increases going down a column of the periodic table.

v.   Negatively charged nucleophiles are usually more reactive than neutral nucleophiles.

d.   Changes in the leaving group.

i.   In general, the best leaving groups are those that best stabilize negative charge.

ii. Usually, the best leaving groups are the weakest bases.

iii. Good leaving groups lower the energy of the transition state.

iv. Poor leaving groups include F^-, HO^-. RO^-, and H_2N^-.

(a). Poor leaving groups can be converted to better leaving groups.

e.   Changes in the solvent.

i.   Polar, protic solvents slow S_N2 reactions by lowering the reactivity of the nucleophile.

ii. Polar, aprotic solvents raise the ground-state energy of the nucleophile and make it more reactive.

f.   A summary:

i.   Steric hindrance in the substrate raises the energy of the transition state, increasing $\Delta G^{\ddagger}$, and decreasing the reaction rate.

ii. More reactive nucleophiles have a higher ground-state energy, decreasing $\Delta G^{\ddagger}$, and increasing the reaction rate.

iii. Good leaving groups decrease the energy of the transition state, decreasing $\Delta G^{\ddagger}$, and increasing the reaction rate.

iv. Polar protic solvents solvate the nucleophile, lowering the ground-state energy, increasing $\Delta G^{\ddagger}$, and decreasing the reaction rate. Polar aprotic solvents don't solvate the nucleophile, raising the ground-state energy, decreasing $\Delta G^{\ddagger}$, and increasing the reaction rate.

B. S_N1 Reactions (Sections 11.4–11.5).

1.   The S_N1 reaction (Section 11.4).

a.   Under certain reaction conditions, tertiary halides are much more reactive than primary and methyl halides.

i.   These reactions must be occurring by a mechanism other than S_N2.

b. Kinetics of the S_N1 reaction.

i.   The rate of reaction of a tertiary alkyl halide with water depends only on the concentration of the alkyl halide (unimolecular reaction).

   ii. The reaction is a first order process, with reaction rate = k x [RX].

   iii. The rate expression shows that only RX is involved in the slowest, or ratelimiting, step, and the nucleophile is involved in a different, faster step.

   iv. The rate expression also shows that there must be at least two steps in the reaction.

   v. In an S$_N$1 reaction, slow dissociation of the substrate is followed by rapid reaction with the nucleophile.

c. Stereochemistry of S$_N$1 reactions.

   i. An S$_N$1 reaction of an enantiomer produces racemic product because an S$_N$1 reaction proceeds through a planar, achiral intermediate.

   ii. Few S$_N$1 reactions proceed with complete racemization.

   iii. The ion pair formed by the leaving group and the carbocation sometimes shields one side of the carbocation from attack before the leaving group can diffuse away.

2. Characteristics of the S$_N$1 reaction (Section 11.5).

   a. As in S$_N$2 reactions, factors that lower $\Delta G^{\ddagger}$ favor faster reactions.

   b. Changes in the substrate.

      i. The more stable the carbocation intermediate, the faster the S$_N^{1}$ reaction.

      ii. Substrates, in order of increasing reactivity: methyl, primary, secondary and allyl and benzyl, tertiary.

      iii. Allylic and benzylic substrates are also reactive in S$_N$2 reactions.

   c. Changes in the leaving group.

      i. The best leaving groups are the conjugate bases of strong acids.

      ii. In S$_N$1 reactions, water can act as a leaving group.

   d. Changes in the nucleophile have no effect on S$_N$1 reactions.

   e. Changes in the solvent.

      i. Polar solvents (high dielectric constant) increase the rates of S$_N$1 reactions.

      ii. Polar solvents stabilize the carbocation intermediate more than the reactants and lower $\Delta G^{\ddagger}$.

      iii. Polar solvents stabilize by orienting themselves around the carbocation, with electron-rich ends facing the positive charge.

   f. A summary:

      i. The best substrates are those that form stable carbocations.

      ii. Good leaving groups lower the energy of the transition state leading to carbocation formation and increase the reaction rate.

      iii. The nucleophile doesn't affect the reaction rate, but it must be nonbasic.

      iv. Polar solvents stabilize the carbocation intermediate and increase the reaction rate.

C. Biological substitution reactions (Section 11.6).

   1. Both S$_N$1 and S$_N$2 reactions occur often in biochemical pathways.

   2. In S$_N$1 reactions, the leaving group is often an organodiphosphate.

   3. S$_N$2 reactions are involved in biological methylations.

II. Elimination reactions (Sections 11.7–11.11).

    A. Introduction (Section 11.7).

        1. In addition to bringing about substitution, a basic nucleophile can also cause elimination of HX from an alkyl halide to form a carbon–carbon double bond.

        2. A mixture of double-bond products is usually formed, but the product with the more substituted double bond is the major product.

           a. This observation is the basis of Zaitsev's rule.

        3. Double-bond formation can occur by several mechanistic routes, but at this point, we will study only three mechanisms.

    B. The E2 reaction (Sections 11.8–11.9).

        1. General features (Section 11.8).

           a. An E2 reaction occurs when an alkyl halide is treated with strong base.

           b. The reaction occurs in one step, without intermediates.

           c. E2 reactions follow second-order kinetics.

           d. E2 reactions show the deuterium isotope effect .

              i. In a reaction in which a C−H bond is cleaved in the rate-limiting step, substitution of −D for −H results in a decrease in rate.

              ii. Because this effect is observed in E2 reactions, these reactions must involve C−H bond breaking in the rate-limiting step.

           e. E2 reactions always occur with periplanar geometry.

              i. Periplanar geometry is required because of the need for overlap of the sp^3 orbitals of the reactant as they become π orbitals in the product.

              ii. Anti periplanar geometry is preferred because it allows the substituents of the two carbons to assume a staggered relationship.

              iii. Syn periplanar geometry occurs only when anti periplanar geometry isn't possible.

           f. The preference for anti periplanar geometry results in the formation of double bonds with specific *E, Z* configurations.

        2. Elimination reactions and cyclohexane conformations (Section 11.9).

           a. The chemistry of substituted cyclohexanes is controlled by their conformations.

           b. The preference for anti periplanar geometry for E2 reactions can be met only if the atoms to be eliminated have a trans-diaxial relationship.

           c. Neomenthyl chloride reacts 200x faster than menthyl chloride because the groups to be eliminated are trans diaxial in the most favorable conformation, and the Zaitsev product is formed.

           d. For menthyl chloride, reaction must proceed through a higher energy conformation, and non-Zaitsev product is formed.

    C. The E1 and E1cB reactions (Sections 11.10–11.11).

        1. An E1 reaction occurs when the intermediate carbocation of an S_N1 loses H^+ to form a C=C bond.

        2. E1 reactions usually occur in competition with S_N1 reactions.

         3. E1 reactions show first-order kinetics.

    4.   There is no geometric requirement for the groups to be eliminated, and the most stable (Zaitsev) product is formed.

  D. The E1cB reaction (cB = conjugate base).

    1.   The E1cB reaction takes place through a carbanion intermediate.

    2.   The rate-limiting step involves base-induced abstraction of a proton.

    3.   Often the leaving group is poor.

    4.   A carbonyl group stabilizes the anion.

    5.   The E1cB is fairly common in biochemical pathways (Section 11.11).

III. Summary of reactivity (Section 11.12).

  A. Primary halides.

    1. S_N2 reaction is usually observed.

    2.   E1 reaction occurs if a strong, bulky base is used.

    3.   E1cB reaction occurs if the leaving group is two carbons away from a carbonyl group.

  B. Secondary halides.

    1. S_N2 and E2 reactions occur in competition.

    2.   Strong bases promote E2 elimination.

    3.   Secondary halides (especially allylic and benzylic halides) can react by S_N1 and E1 routes if weakly basic nucleophiles and protic solvents are used.

    4.   E1cB reaction occurs if the leaving group is two carbons away from a carbonyl group.

  C. Tertiary halides.

    1.   Under basic conditions, E2 elimination is favored.

    2. S_N1 and E1 products are formed under nonbasic conditions.

    3.   E1cB reaction occurs if the leaving group is two carbons away from a carbonyl group.

Solutions to Problems

11.1 As described in Worked Example 11.1, identify the leaving group and the chirality center. Draw the product carbon skeleton, inverting the configuration at the chirality center, and replace the leaving group (bromide) with the nucleophilic reactant (acetate).

11.2 Use the suggestions in the previous problem to draw the correct product.

11.3

11.4 All of the nucleophiles in this problem are relatively reactive. See Table 11.1.

(a) $CH_3CH_2CH_2CH_2Br + NaI \longrightarrow CH_3CH_2CH_2CH_2I + NaBr$

(b) $CH_3CH_2CH_2CH_2Br + KOH \longrightarrow CH_3CH_2CH_2CH_2OH + KBr$

(c) $CH_3CH_2CH_2CH_2Br + HC{\equiv}C^-\ Li^+ \longrightarrow CH_3CH_2CH_2CH_2C{\equiv}CH + LiBr$

(d) $CH_3CH_2CH_2CH_2Br + NH_3 \longrightarrow CH_3CH_2CH_2CH_2NH_3{}^+\ Br^-$

11.5 (a) $(CH_3)_2N^-$ is more nucleophilic because it is more basic than $(CH_3)_2NH$ and because a negatively charged nucleophile is more nucleophilic than a neutral nucleophile.

(b) $(CH_3)_3N$ is more nucleophilic than $(CH_3)_3B$. $(CH_3)_3B$ is non-nucleophilic because it has no lone electron pair.

(c) H_2S is more nucleophilic than H_2O because nucleophilicity increases in going down a column of the periodic table.

11.6 In this problem, we are comparing two effects – the effect of the substrate and the effect of the leaving group. Tertiary substrates are less reactive than secondary substrates, which are less reactive than primary substrates.

Least reactive ⟶ Most reactive

$(CH_3)_3CCl < (CH_3)_2CHCl < CH_3Br < CH_3OTos$

tertiary	secondary	good	excellent
carbon	carbon	leaving	leaving
		group	group

11.7

Polar protic solvents (curve 1) stabilize the charged transition state by solvation and also stabilize the nucleophile by hydrogen bonding.

Polar aprotic solvents (curve 2) stabilize the charged transition state by solvation, but do not hydrogen-bond to the nucleophile. Since the energy level of the nucleophile is higher, $\Delta G^\ddagger$ is smaller and the reaction is faster in polar aprotic solvents than in polar protic solvents.

Nonpolar solvents (curve 3) stabilize neither the nucleophile nor the transition state. $\Delta G^\ddagger$ is therefore higher in nonpolar solvents than in polar solvents, and the reaction rate is slower. Benzene, ether, and chloroform are in this category.

11.8

In this S_N1 reaction, attack by acetate can occur on either side of the planar, achiral carbocation intermediate, resulting in a mixture of both the R and S enantiomeric acetates. The ratio of enantiomers is probably close to 50:50.

11.9 If reaction had proceeded with complete inversion, the product would have had a specific rotation of +53.6°. If complete racemization had occurred, $[\alpha]_D$ would have been zero. The observed rotation was +5.3°. Since $\dfrac{+5.3°}{+53.6°} = 0.099$, 9.9% of the original tosylate was inverted. The remaining 90.1% of the product must have been racemized.

11.10 The *S* substrate reacts with water to form a mixture of *R* and *S* alcohols. The ratio of enantiomers is close to 50:50.

11.11 S_N1 reactivity is related to carbocation stability. Thus, substrates that form the most stable carbocations are the most reactive in S_N1 reactions.

Least reactive $\longrightarrow$ *Most reactive*

$$H_2C{=}CHBr \ll CH_3CH_2Br < CH_3\overset{\overset{\displaystyle Br}{|}}{C}HCH_3 < H_2C{=}CH\overset{\overset{\displaystyle Br}{|}}{C}HCH_3$$

      vinylic   primary   secondary   allylic

11.12

The two bromobutenes form the same allylic carbocation in the rate-limiting step.

11.13 Both substrates have allylic groups and might react either by an S_N1 or an S_N2 route. The reaction mechanism is determined by the leaving group, the solvent, or the nucleophile.

(a) This reaction probably occurs by an S_N1 mechanism. HCl converts the poor –OH leaving group into an excellent $-OH_2{}^+$ leaving group, and the polar solvent stabilizes the carbocation intermediate.

(b) This reaction takes place with a negatively charged nucleophile in a polar, aprotic solvent. It is very likely that the reaction occurs by an S_N2 mechanism.

11.14 Redraw linalyl diphosphate so that has the same orientation as limonene.

After dissociation of PPi, the cation cyclizes by attack of the double bond π electrons. Removal of an −H by base yields limonene.

Linalyl diphosphate

Limonene

: Base

11.15 Form the double bond by removing HX from the alkyl halide reactant in as many ways as possible. The major elimination product in each case has the most substituted double bond (Zaitsev's rule).

(a)

$$CH_3CH_2CHCHCH_3 \xrightarrow{Base} CH_3CH_2CH{=}CCH_3 + CH_3CH{=}CHCHCH_3$$

major
(trisubstituted double bond)

minor
(disubstituted double bond)

(b)

$$CH_3CHCH_2C-CHCH_3 \xrightarrow{Base} CH_3CHCH_2C{=}CCH_3 + CH_3CHCH{=}CCHCH_3$$

major
(tetrasubstituted double bond)

minor
(trisubstituted double bond)

$$+ CH_3CHCH_2CCHCH_3$$

minor
(disubstituted double bond)

(c)

$$\xrightarrow{Base}$$

major
(trisubstituted double bond)

minor
(monosubstituted double bond)

11.16 For maximum yield, the alkyl halide reactant should not give a mixture of products on elimination.

(a)

$$CH_3CHCH_2CH_2CHCH_2CH_2Br \xrightarrow[\text{CH}_3\text{CH}_2\text{OH}]{\text{KOH}} CH_3CHCH_2CH_2CHCH_2CH=CH_2$$

(with CH₃ substituents on the indicated carbons)

The 2-bromo isomer yields a mixture of alkene products.

(b)

$$\xrightarrow[\text{CH}_3\text{CH}_2\text{OH}]{\text{KOH}}$$

11.17 Draw the reactant with correct stereochemistry.

(1R,2R)-1,2-Dibromo-1,2-diphenylethane

Convert this drawing into a Newman projection, and draw the conformation having anti periplanar geometry (staggered) for −H and −Br.

The alkene resulting from E2 elimination is (Z)-1-bromo-1,2-diphenylethylene.

11.18 As in the previous problem, draw the structure, convert it to a Newman projection, and rotate the groups so that the −H and −Br to be eliminated have an anti periplanar (staggered) relationship.

The major product is (Z)-3-methyl-2-pentene. A small amount of 3-methyl-1-pentene is also formed.

11.19

| trans | cis |

The more stable conformations of each of the two isomers are pictured above; the larger *tert*–butyl group is always equatorial in the more stable conformation. The cis isomer reacts faster under E2 conditions because −Br and −H are in the anti periplanar arrangement that favors E2 elimination.

11.20

(a)

$$CH_3CH_2CH_2CH_2Br \quad + \quad NaN_3 \quad \longrightarrow \quad CH_3CH_2CH_2CH_2N_3$$
    primary   substitution product

The reaction occurs by an S_N2 mechanism because the substrate is primary, the nucleophile is nonbasic, and the product is a substitution product.

(b)

$$\underset{\text{secondary}}{CH_3CH_2\overset{\overset{\displaystyle Cl}{|}}{C}HCH_2CH_3} \quad + \quad \underset{\text{strong base}}{KOH} \quad \longrightarrow \quad \underset{\text{elimination product}}{CH_3CH_2CH=CHCH_3}$$

This is an E2 reaction since a secondary halide reacts with a strong base to yield an elimination product.

(c)

tertiary   substitution product

This is an S$_N$1 reaction. Tertiary substrates form substitution products only by the S$_N$1 route.

(d)

This is an E1cB reaction because the leaving group is two carbons away from a carbonyl group.

Visualizing Chemistry

11.21 (a) (i) $CH_3CH_2Cl + Na^+ \ ^-SCH_3 \longrightarrow CH_3CH_2SCH_3 + NaCl$

(ii) $CH_3CH_2Cl + Na^+ \ ^-OH \longrightarrow CH_3CH_2OH + NaCl$

Both reactions yield S$_N$2 substitution products because the substrate is primary and both nucleophiles are strong.

(b)

(i)

(ii)

The substrate is tertiary, and the nucleophiles are basic. Two elimination products are expected; the major product has the more substituted double bond, in accordance with Zaitsev's rule.

(c)

(i)

In (i), the secondary substrate reacts with the good, but weakly basic, nucleophile to yield substitution product. In (ii), NaOH is a poorer nucleophile but a stronger base, and both substitution and elimination product are formed.

11.22

Reaction of the secondary bromide with the weakly basic acetate nucleophile occurs by an S_N2 route, with inversion of configuration, to produce the R acetate.

11.23

The S substrate has a secondary allylic chloride group and a primary hydroxyl group. S_N2 reaction occurs at the secondary carbon to give the R cyano product because hydroxide is a poor leaving group.

11.24

Rotate the left side of the molecule so that the groups to be eliminated have an anti periplanar relationship. The double bond in the product has the E configuration.

Mechanism Problems

11.25   The mechanism for each problem is S$_N$2.

   (a)   The electrophile is a 1° alkyl halide reacting with a good nucleophile.

CH$_3$CH$_2$CH$_2$OTs   +   NaBr $\xrightarrow{\text{DMF}}$ CH$_3$CH$_2$CH$_2$Br   +   Na$^+$ + $^-$OTs

Mechanism: Na$^+$ $^-$Br   CH$_3$CH$_2$CH$_2$—OTs

   (b)   The electrophile is a 1° alkyl halide reacting with a good nucleophile.

   (c)   The electrophile is a 1° alkyl halide reacting with a good nucleophile.

   (d)   The electrophile is a 2° alkyl halide reacting with a good nucleophile.

11.26   Each mechanism goes through an S$_N$1 mechanism. In each case there is opportunity to form a carbocation prior to substitution.

     (a)   Formation of a tertiary carbocation

Mechanism:

     (b)   Formation a resonance stabilized carbocation

Mechanism:

     (c)   Use of a poor nucleophilic polar, protic solvent (note that the product is wrong in the for text file!)

Mechanism:

11.27   In each case the mechanism is E1 because a carbocation can be formed before the proton is removed.

     (a)

                                 major   minor

Mechanism:

(b)

Mechanism:

(c)

major   minor   trace

Mechanism:

major   minor   trace

11.28 Each mechanism is an E2 elimination.

(a) E2 involving a strong base and a secondary alkyl halide. The E2 mechanism requires the leaving group and H to be removed to be antiperiplanar to one another. In the chair conformation that means that the leaving group and the proton must be axial. Only the less substituted product is formed because only the bold proton is antiperiplanar to the Cl.

Mechanism:

(b) E2 involving a strong, bulky base and a primary alkyl halide

Mechanism:

(c)   E2 involving a strong base and a secondary alkyl halide

Mechanism:

11.29 Each mechanism is E1CB. In each case there is an OH β to (two carbons from) the carbonyl in the presence of a strong base.

(a)

Mechanism:

(b)

Mechanism:

(c)

Mechanism:

11.30 (a) In this reaction a β-hydroxyketone react with a strong base. These are the conditions necessary for an E1cB mechanism.

(b) In this reaction a primary alkyl halide reacts with a good nucleophile. These are the conditions necessary for an SN2 mechanism.

(c) In this reaction a tertiary alkyl halide reacts with a poor nucleophile which is a polar protic solvent. These are the conditions that favor for an S_N1 mechanism.

(d) In this reaction a tertiary alkyl halide reacts with a strong base. These are the conditions that favor for an E2 mechanism. Even though a tertiary alkyl halide and a polar protic solvent are present, the strong base results elimination must faster than the formation of a carbocation.

11.31

Rotate around the C–C bond so that –OH and –Br are 180° apart.

This reaction is an intramolecular S_N2 displacement.

11.32 The first step in an S_N1 displacement is dissociation of the substrate to form a planar, sp^2-hybridized carbocation and a leaving group. The carbocation that would form from dissociation of this alkyl halide can't become planar because of the rigid structure of the ring skeleton. Because it's not possible to form the necessary carbocation, an S_N1 reaction can't occur. In addition, approach by a nucleophile from the back side of the alkyl halide is blocked by the rigid ring system, and S_N2 displacement can't take place either.

11.33 Step 1: NAD^+ oxidizes an alcohol to a ketone.

: Base   *S*-Adenosylhomocysteine

Step 2:   Base brings about an E1cB elimination reaction that has homocysteine as the leaving group.

Homocysteine

11.34

Formal charge = −1

Formal charge = +1

S_N2 attack by the lone pair electrons associated with carbon gives the nitrile product. Attack by the lone pair electrons associated with nitrogen yields the isonitrile product.

11.35

Base removes a proton, yielding an E1cB carb-anion intermediate.

Arginine, the protonated leaving group, departs, forming fumarate.

Arginine   Fumarate

11.36 The two pieces of evidence indicate that the reaction proceeds by an S$_N$2 mechanism: S$_N$2 reactions proceed much faster in polar aprotic solvents such as DMF, and methyl esters react faster than ethyl esters. This reaction is an S$_N$2 displacement on a methyl ester by iodide ion.

Other experiments can provide additional evidence for an S$_N$2 mechanism. We can determine if the reaction is second-order by varying the concentration of LiI. We can also vary the type of nucleophile to distinguish an S$_N$2 mechanism from an S$_N$1 mechanism, which does not depend on the identity of the nucleophile.

11.37

Two inversions of configuration equal a net retention of configuration.

11.38

Step 1: Protonation.
Step 2: Elimination of leaving group.
Step 3: Removal of proton by base.

This reaction proceeds by an E1 mechanism.

Additional Problems

Nucleophilic Substitution Reactions

11.39

Most reactive ⟶ *Least reactive*

$$CH_3CH_2CH_2CH_2Br \;>\; \underset{\substack{|\\CH_3}}{CH_3CHCH_2Br} \;>\; \underset{\substack{|\\Br}}{CH_3CH_2CHCH_3} \;>\; \underset{\substack{|\\CH_3}}{\overset{\substack{Br\\|}}{CH_3CCH_3}}$$

1-Bromobutane   1-Bromo-2-   2-Bromobutane   2-Bromo-2-
                  methylpropane   methylpropane

11.40 An alcohol is converted to an ether by two different routes in this series of reactions. The two resulting ethers have identical structural formulas but differ in the sign of specific rotation. Therefore, at some step or steps in these reaction sequences, inversion of configuration at the chiral carbon must have occurred. Let's study each step of the series to find where inversion is occurring.

In step 1, the alcohol reacts with potassium metal to produce a potassium alkoxide. Since the bond between carbon and oxygen has not been broken, no inversion occurs in this step.

The potassium alkoxide acts as a nucleophile in the S_N2 displacement on CH_3CH_2Br in step 2. It is the C–Br bond of bromoethane, however, not the C–O bond of the alkoxide that is broken. No inversion at the carbon chirality center occurs in step 2.

The starting alcohol reacts with tosyl chloride in step 3. Again, because the O–H bond, rather than the C–O bond, of the alcohol is broken, no inversion occurs at this step.

Inversion must therefore occur at step 4 when the ⁻OTos group is displaced by CH_3CH_2OH. The C–O bond of the tosylate (–OTos) is broken, and a new C–O bond is formed.

Notice the specific rotations of the two enantiomeric products. The product of steps 1 and 2 should be enantiomerically pure because neither reaction has affected the C–O bond. Reaction 4 proceeds with some racemization at the chirality center to give a smaller absolute value of $[\alpha]_D$.

11.41 (a) CH_3I reacts faster than CH_3Br because I^- is a better leaving group than Br^-.

   (b) CH_3CH_2I reacts faster with OH^- in dimethylsulfoxide (DMSO) than in ethanol. Ethanol, a protic solvent, hydrogen-bonds with hydroxide ion and decreases its reactivity.

   (c)   Under the S_N2 conditions of this reaction, CH_3Cl reacts faster than $(CH_3)_3CCl$. Approach of the nucleophile to the bulky $(CH_3)_3CCl$ molecule is hindered.

   (d) $H_2C=CHCH_2Br$ reacts faster because vinylic halides such as $H_2C=CHBr$ are unreactive to substitution reactions.

11.42 To predict nucleophilicity, remember these guidelines:

   (1)   In comparing nucleophiles that have the same attacking atom, nucleophilicity parallels basicity. In other words, a more basic nucleophile is a more effective nucleophile.

   (2)   Nucleophilicity increases in going down a column of the periodic table.

   (3)   A negatively charged nucleophile is usually more reactive than a neutral nucleophile.

	More Nucleophilic	*Less Nucleophilic*	*Reason*
(a)	⁻NH_2	NH_3	Rule 1 or 3
(b)	$CH_3CO_2^-$	H_2O	Rule 1 or 3
(c)	F^-	BF_3	BF_3 is not a nucleophile
(d)	$(CH_3)_3P$	$(CH_3)_3N$	Rule 2
(e)	I^-	Cl^-	Rule 2
(f)	⁻$C\equiv N$	⁻OCH_3	Reactivity chart, Section 11.3

11.43

CH₃CH₂CHCH₂I + ⁻CN ⟶ CH₃CH₂CHCH₂CN + I⁻

primary halide

This is an S_N2 reaction, whose rate depends on the concentration of both alkyl halide and nucleophile. Rate = k x [RX] x [Nu:⁻]

(a) Halving the concentration of cyanide ion and doubling the concentration of alkyl halide doesn't change the reaction rate. The two effects cancel.

(b) Tripling the concentrations of both cyanide ion and alkyl halide causes a ninefold increase in reaction rate.

11.44

tertiary halide

This is an S_N1 reaction, whose rate depends only on the concentration of 2-iodo-2-methylbutane. Rate = k x [RX].

(a) Tripling the concentration of alkyl halide triples the rate of reaction.

(b) Halving the concentration of ethanol by dilution with diethyl ether reduces the polarity of the solvent and decreases the rate.

11.45 (a)   CH₃Br + Na⁺ ⁻C≡CCH(CH₃)₂ ⟶ CH₃C≡CCH(CH₃)₂ + NaBr

Not CH₃C≡C⁻ Na⁺ + BrCH(CH₃)₂. The strong base CH₃C≡C⁻ brings about elimination, producing CH₃C≡CH and H₂C=CHCH₃.

(b)

(c)   CH₃CH₂CH₂CH₂Br + NaCN ⟶ CH₃CH₂CH₂CH₂CN + NaBr

(d)   CH₃CH₂CH₂Br + excess NH₃ ⟶ CH₃CH₂CH₂NH₂ + NH₄⁺Br⁻

major

11.46 (a)   The difference in this pair of reactions is in the *leaving group*. Since ⁻OTos is a better leaving group than ⁻Cl (see Section 11.5), S_N2 displacement by iodide on CH₃–OTos proceeds faster.

(b)   The *substrates* in these two reactions are different. Bromoethane is a primary bromoalkane, and bromocyclohexane is a secondary bromoalkane. Since

S_N2reactions proceed faster at primary than at secondary carbon atoms, S_N2 displacement on bromoethane is a faster reaction.

(c)   Ethoxide ion and cyanide ion are different *nucleophiles*. Since CN^- is more reactive than $CH_3CH_2O^-$ in S_N2 reactions, S_N2 displacement on 2-bromopropane by CN^- proceeds at a faster rate.

(d)   The *solvent* in each reaction is different. The S_N2 reaction on bromoethane in polar, aprotic acetonitrile proceeds faster than the reaction in nonpolar benzene.

11.47 (R)-2-Bromooctane is a secondary bromoalkane, which undergoes S_N2 substitution. Since S_N2 reactions proceed with inversion of configuration, the configuration at the carbon chirality center is inverted. (This does not necessarily mean that all *R* isomers become *S* isomers after an S_N2 reaction. The *R,S* designation refers to the priorities of groups, which may change when the nucleophile is varied.)

11.48 After 50% of the starting material has reacted, the reaction mixture consists of 50% (R)-2-bromooctane and 50% (S)-2-bromooctane. At this point, the R starting material is completely racemized.

Elimination Reactions

11.49 (a)

(b) $H_2C=CHBr$, like other vinylic organohalides, does not undergo nucleophilic substitutions.

(c)

This alkyl halide gives the less substituted cycloalkene (non-Zaitsev product). Elimination to form the Zaitsev product does not occur because the $-Cl$ and $-H$ involved cannot assume the anti periplanar geometry preferred for E2 elimination.

(d) $(CH_3)_3C-OH$ + HCl $\xrightarrow{0°}$ $(CH_3)_3C-Cl$ + H_2O

11.50 Because 1-bromopropane is a primary haloalkane, the reaction proceeds by either a S_N2 or E2 mechanism, depending on the basicity and the amount of steric hindrance in the nucleophile.

(a) $CH_3CH_2CH_2Br + NaNH_2 \longrightarrow CH_3CH_2CH_2NH_2 + NaBr$

Propene (elimination product) is also formed because $NaNH_2$ is a strong base.

(b) $CH_3CH_2CH_2Br + K^+ {}^-OC(CH_3)_3 \longrightarrow CH_3CH=CH_2 + HOC(CH_3)_3 + KBr$
$+ CH_3CH_2CH_2OC(CH_3)_3$

$K^+ {}^-OC(CH_3)_3$ is a strong, bulky base that brings about elimination as well as some substitution.

(c) $CH_3CH_2CH_2Br + NaI \longrightarrow CH_3CH_2CH_2I + NaBr$

(d) $CH_3CH_2CH_2Br + NaCN \longrightarrow CH_3CH_2CH_2CN + NaBr$

(e) $CH_3CH_2CH_2Br + Na^+ {}^-C{\equiv}CH \longrightarrow CH_3CH_2CH_2C{\equiv}CH + NaBr$

(f) $CH_3CH_2CH_2Br \xrightarrow{Mg} CH_3CH_2CH_2MgBr \xrightarrow{H_2O} CH_3CH_2CH_3$

11.51

A   **B**

Both Newman projections place −H and −Cl in the correct anti periplanar geometry for E2 elimination.

$A^{\ddagger}$ $B^{\ddagger}$

trans-1,2-Diphenylethylene

Either transition state $A^{\ddagger}$ or $B^{\ddagger}$ can form when 1-chloro-1,2-diphenylethane undergoes E2 elimination. Crowding of the two large phenyl groups in $A^{\ddagger}$ makes this transition state (and the product resulting from it) of higher energy than transition state $B^{\ddagger}$. Formation of the product from $B^{\ddagger}$ is therefore favored, and *trans*-1,2-diphenylethylene is the major product.

11.52

The alkene shown above has the most highly substituted double bond, and, according to Zaitsev's rule, is the major product. The following minor products may also form.

11.53

Diastereomer *8* reacts much more slowly than other isomers in an E2 reaction because no pair of hydrogen and chlorine atoms can adopt the anti periplanar orientation preferred for E2 elimination.

General Problems

11.54 (a)   Substitution does not take place with secondary alkyl halides when a strong, bulky base is used. Elimination occurs instead and produces $H_2C=CHCH_2CH_3$ and $CH_3CH=CHCH_3$.

   (b)   Fluoroalkanes don't undergo S_N2 reactions because F^- is a poor leaving group.

   (c) $SOCl_2$ in pyridine converts primary and secondary alcohols to chlorides by an S_N2 mechanism. 1-Methyl-1-cyclohexanol is a tertiary alcohol and does not undergo S_N2 substitution. Instead, E2 elimination occurs to give 1-methylcyclohexene.

11.55 (a)

primary   secondary   secondary
                                                              with resonance

(b)

primary   <   secondary   <   tertiary

(c)

primary   <   secondary   <   tertiary
with resonance

11.56 S$_N$1 reactivity:

Least reactive ⟶ Most reactive

(a)

most stable carbocation

(b)   (CH$_3$)$_3$COH   <   (CH$_3$)$_3$CCl   <   (CH$_3$)$_3$CBr

best leaving group

(c)

most stable carbocation

11.57 S_N2 reactivity:

Least reactive ⟶ *Most reactive*

(a)

$$H_3C-\underset{\underset{CH_3}{|}}{\overset{\overset{CH_3}{|}}{C}}-Cl \quad < \quad CH_3CH_2\underset{\underset{Cl}{|}}{C}HCH_3 \quad < \quad CH_3CH_2CH_2Cl$$
primary substrate

(b)

$$H_3C-\underset{\underset{CH_3}{|}}{\overset{\overset{CH_3}{|}}{C}}-CH_2Br \quad < \quad CH_3\underset{\underset{Br}{|}}{\overset{\overset{CH_3}{|}}{C}}HCHCH_3 \quad < \quad CH_3\overset{\overset{CH_3}{|}}{C}HCH_2Br$$
least sterically
hindered substrate

(c) $CH_3CH_2CH_2OCH_3 \quad < \quad CH_3CH_2CH_2Br \quad < \quad CH_3CH_2CH_2OTos$
best leaving group

11.58 (a) The product of this S_N2 reaction (2° alkyl halide, good nucleophile) is achiral (no chiral centers). Therefore, it will not rotate plane polarized light.

(b) This is an E2 reaction (3° alkyl halide, strong base) and the product achiral (no chiral centers). Therefore, it will not rotate plane polarized light.

(c) This is an S_N1 reaction (3° alkyl halide, polar protic solvent/nucleophile) and results in a pair of diastereomers. Therefore, the product mixture *will rotate plane polarized light.*

11.59 According to Cahn–Ingold–Prelog rules (Section 5.5), the nucleophile (–CN) has a lower priority than the leaving group (–OTos). Thus, even though the reaction proceeds with inversion of configuration, the priorities of the substituents also change, and the configuration remains *S*.

11.60

This is an excellent method of ether preparation because iodomethane is very reactive in S_N2 displacements.

Reaction of a secondary haloalkane with a basic nucleophile yields both substitution and elimination products. This is a less satisfactory method of ether preparation.

11.61

11.62 Draw 4-bromo-1-butanol in an orientation that resembles the product.

Tetrahydrofuran

Step 1: Methoxide removes a proton from the hydroxyl group of 4-bromo-1-butanol.

Step 2: S$_N$2 displacement of Br$^-$ by O$^-$ yields the cyclic ether tetrahydrofuran (THF).

11.63 In a molecule containing a double bond, all atoms bonded to the sp^2 carbons must lie in a common plane. For this compound, planar geometry at the "bridgehead" of the ring system is not possible because the rigid ring framework won't allow it. Thus, E2 elimination does not take place because the product containing a bridgehead double bond can't form.

11.64

Draw a Newman projection of the tosylate of (2R,3S)-3-phenyl-2-butanol, and rotate the projection until the −OTos and the −H on the adjoining carbon atom are anti periplanar. Even though this conformation has several gauche interactions, it is the only conformation in which −OTos and −H are 180° apart.

(Z)-2-Phenyl-2-butene

Elimination yields the Z isomer of 2-phenyl-2-butene.

11.65 Using the same argument from the previous problem, you can show that elimination from the tosylate of (2R,3R)-3-phenyl-2-butanol gives the E-alkene.

(E)-2-Phenyl-2-butene

The (2S,3S) isomer also forms the E alkene; the (2S,3R) isomer yields the Z alkene.

11.66

E2 reactions require that the two atoms to be eliminated have a periplanar relationship. Since it's impossible for bromine at C1 and the hydrogen at C2 to be periplanar, elimination occurs in the non-Zaitsev direction to yield 3-methylcyclohexene.

11.67

This tertiary bromoalkane reacts by S$_N$1 and E1 routes to yield alcohol and alkene products. The alcohol products are diastereomers.

11.68

(Z)-2-Chloro-2-butene-1,4-dioic acid   (E)-2-Chloro-2-butene-1,4-dioic acid

Hydrogen and chlorine are anti to each other in the *Z* isomer and are syn in the *E* isomer Since the *Z* isomer reacts fifty times faster than the *E* isomer, elimination must proceed more favorably when the substituents to be eliminated are anti to one another. This is the same stereochemical result that occurs in E2 eliminations of alkyl halides.

11.69 (a)

Mechanism:

(b)

Mechanism:

(c)

Mechanism:

11.70 Since 2-butanol is a secondary alcohol, substitution can occur by either an S_N1 or S_N2 route, depending on reaction conditions. Two factors favor an S_N1 mechanism in this case. (1) The reaction is run under acidic conditions in a polar, protic solvent (water). (2) Dilute acid converts a poor leaving group (^-OH) into a good leaving group (OH_2), which dissociates easily.

Protonation of the
hydroxyl oxygen..

is followed by loss of
water to form a planar
carbocation.

Attack of water from
either side of the planar
carbocation yields
racemic product.

11.71 The chiral tertiary alcohol (*R*)-3-methyl-3-hexanol reacts with HBr by an S$_N$1 pathway. HBr protonates the hydroxyl group, which dissociates to yield a planar, achiral carbocation. Reaction with the nucleophilic bromide anion can occur from either side of the carbocation to produce (±)3-bromo-3-methylhexane.

11.72 Since carbon-deuterium bonds are slightly stronger than carbon-hydrogen bonds, more energy is required to break a C−D bond than to break a C−H bond. In a reaction where either a carbon-deuterium or a carbon–hydrogen bond can be broken in the rate-limiting step, a higher percentage of C−H bond-breaking occurs because the energy of activation for C−H breakage is lower.

Transition state $\mathbf{A}^{\ddagger}$ is of higher energy than transition state $\mathbf{B}^{\ddagger}$ because more energy is required to break the C–D bond. The product that results from transition state $\mathbf{B}^{\ddagger}$ is thus formed in greater abundance.

11.73

(2R,3S)-2-Bromo-3-methyl-2-phenylpentane

(E)-3-Methyl-2-phenyl-2-pentene

The (2S,3R) isomer also yields E product.

11.74 One of the steric requirements of E2 elimination is the need for periplanar geometry, which optimizes orbital overlap in the transition state leading to alkene product. Two types of periplanar arrangements of substituents are possible — syn and anti.

A model of the deuterated bromo compound shows that the deuterium, bromine, and the two carbon atoms that will constitute the double bond all lie in a plane. This arrangement of atoms leads to syn elimination. Even though anti elimination is usually preferred, it doesn't occur for this compound because the bromine, hydrogen, and two carbons can't achieve the necessary geometry.

11.75

We concluded in Problem 11.62 that E2 elimination in compounds of this bicyclic structure occurs with syn periplanar geometry. In compound **A**, −H and −Cl can be eliminated via the syn-periplanar route. Since neither syn nor anti periplanar elimination is possible for **B**, elimination occurs by a slower, E1 route.

11.76 Because Cl⁻ is a relatively poor leaving group and acetate is a relatively poor nucleophile, a substitution reaction involving these two groups proceeds at a very slow rate. I⁻, however, is both a good nucleophile and a good leaving group. 1-Chlorooctane thus reacts preferentially with iodide to form 1-iodooctane. Only a small amount of 1-iodooctane is formed (because of the low concentration of iodide ion), but 1-iodooctane is more reactive than 1-chlorooctane toward substitution by acetate. Reaction with acetate produces 1-octyl acetate and regenerates iodide ion. The whole process can now be repeated with another molecule of 1-chlorooctane. The net result is production of 1-octyl acetate, and no iodide is consumed.

11.77 Two optically inactive compounds are possible structures for compound **X**.

11.78

At lower temperatures, a tosylate is formed from the reaction of *p*-toluenesulfonyl chloride and an alcohol. The new bond is formed between the toluenesulfonyl group and the oxygen of the alcohol. At higher temperatures, the chloride anion can displace the – Otos group, which is an excellent leaving group, to form an organochloride.

11.79 Notice that the chiral methyl group has the (*R*) configuration in both *N*-methyltetrahydrofolate and in methionine. This fact suggests that methylation proceeds with two inversions of configuration which, in fact, has been shown to be the case.

11.80

The intermediate is a charged quaternary ammonium compound that results from S$_N$2 substitutions on three CH$_3$I molecules by the amine nitrogen. E2 elimination occurs because the neutral N(CH$_3$)$_3$ molecule is a good leaving group.

11.81 (a)

(b)

(c)

F

(d)   The stereoisomers are *E,Z* double bond isomers and are diastereomers.

Review Unit 4: Alkynes; Alkyl Halides; Substitutions and Eliminations

Major Topics Covered (with vocabulary):

Alkynes:
alkyne enyne vicinal tautomer Lindlar catalyst acetylide anion alkylation
Organic Synthesis.

Alkyl halides:
allylic position delocalization Grignard reagent Gilman reagent Suzuki –Miyaura reaction

Oxidation and reduction in organic chemistry:

Substitution reactions:
nucleophilic substitution reaction Walden inversion reaction rate kinetics second-order reaction rate constant S_N2 reaction bimolecular nucleophilicity leaving group solvation S_N1 reaction first-order reaction rate-limiting step ion pair dielectric polarization

Elimination reactions:
Zaitsev's rule E2 reaction syn periplanar geometry anti periplanar geometry deuterium isotope effect E1 reaction E1cB reaction

Types of Problems:

After studying these chapters, you should be able to:

– Predict the products of reactions involving alkynes.
– Choose the correct alkyne starting material to yield a given product.
– Deduce the structure of an alkyne from its molecular formula and products of cleavage.
– Carry out syntheses involving alkynes.
– Draw, name, and synthesize alkyl halides.
– Understand the mechanism of radical halogenation and the stability order of radicals.
– Prepare Grignard reagents and dialkylcopper reagents and use them in synthesis.
– Predict the oxidation level of a compound.
– Formulate the mechanisms of S_N2, S_N1 and elimination reactions.
– Predict the effect of substrate, nucleophile, leaving group and solvent on substitution and elimination reactions.
– Predict the products of substitution and elimination reactions.
– Classify substitution and elimination reactions by type.

Points to Remember:

* Although it is very important to work backwards when planning an organic synthesis, don't forget to pay attention to the starting material, also. Planning a synthesis is like solving a maze from the middle outward: keeping your eye on the starting material can keep you from running into a dead end.

* The reagent Li/NH_3 is used to reduce an alkyne to a trans alkene; the reagent $NaNH_2/NH_3$ is used to form an acetylide anion. It is easy to confuse the two reagents.

*   In naming alkyl halides by the IUPAC system, remember that a halogen is named as a substituent on an alkane. When numbering the alkyl halide, the halogens are numbered in the same way as alkyl groups and are cited alphabetically.

*   The definition of oxidation and reduction given in Chapter 10 expands the concept to reactions that you might not have considered to be oxidations or reductions. As you learn new reactions, try to classify them as oxidations, reductions or neither.

*   Predicting the outcome of substitutions and eliminations is only straightforward in certain cases. For primary halides, S_N2 and E2 reactions are predicted. For tertiary halides, S_N1, E2 and E1(to a certain extent) are the choices. The possibilities for secondary halides are more complicated. In addition, many reactions yield both substitution and elimination products, and both inversion and retention of configuration may occur in the same reaction.

Self-Test:

A Terbinafine (an antifungal)

B

What is the configuration of the double bond in the side chain of **A**? What products result from treatment of **A** with $KMnO_4$, H_3O^+ (neither the aromatic ring nor the amine are affected)? How might the triple bond have been introduced?

Provide a name for **B**. Predict the products of reaction of **B** with (a) 1 equiv HBr (b) H_2, Pd/C (c) BH_3, THF, then H_2O_2, HO^- (d) O_3, then Zn, H_3O^+.

C

D
Ethchlorvynol
(a sedative)

E

Name **C**. Draw all stereoisomers of **C**, label them, and describe their relationship. Predict the products of reaction of **C** with:(a) NaOH; (b) Mg, then H_2O; (c) product of (b) + Br_2, hv (show the major product); (d) $(CH_3CH_2)_2CuLi$.

Draw the *R* enantiomer of **D**. Predict the products of reaction of **D** with: (a) HBr; (b) product of (a) + aqueous ethanol. Describe the reactivity of the $^-$Cl atom in substitution and elimination reactions.

How might **E** be synthesized from the appropriate alkylbenzene? From the appropriate alcohol? Predict the reactivity of **E** in substitution and elimination reactions.

Multiple Choice:

1.   An enol is a tautomer of an:
        (a) alcohol   (b) alkyne   (c) alkene   (d) ketone

2.   Which reaction proceeds through a vinylic radical?
        (a) Hg-catalyzed hydration of an alkyne   (b) Li/NH$_3$ reduction of an alkyne
        (c) catalytic hydrogenation of an alkyne   (d) treatment of an alkyne with a strong base

3.   Which of the following reagents is not used in a Suzuki–Miyaura reaction?
        (a) aromatic boronic acid   (b) lithium   (c) Pd catalyst   (d) potassium carbonate

4.   Monochlorination of 2,3-dimethylbutane yields what percent of 2-chloro-2,
        3-dimethylbutane?
        (a) 16%   (b) 35%   (c) 45%   (d) 55%

5.   How many monobromination products can be formed by NBS bromination of 2-ethyl-
        1-pentene? Include double-bond isomers.
        (a) 3   (b) 4   (c) 5   (d) 6

6.   Which of the following reactions is an oxidation?
        (a) hydroxylation   (b) hydration   (c) hydrogenation   (d) addition of HBr

7.   All of the following are true of S$_N$2 reactions except:
        (a)   The rate varies with the concentration of nucleophile
        (b)   The rate varies with the type of nucleophile
        (c)   The nucleophile is involved in the rate-determining step
        (d)   The rate of the S$_N$2 reaction of a substrate and a nucleophile is the same as the rate of
        the E2 reaction of the same two compounds.

8.   Which of the following is true of S$_N$1 reactions?
        (a)   The rate varies with the concentration of nucleophile
        (b)   The rate varies with the type of nucleophile
        (c)   he rate is increased by use of a polar solvent.
        (d)   he nucleophile is involved in the rate-determining step.

9.   Which base is best for converting 1-bromohexane to 1-hexene?
        (a) (CH$_3$)$_3$CO$^-$ (b) $^-$CN   (c) $^-$OH   (d) $^-$C≡CH

10.   Which of the following is both a good nucleophile and a good leaving group?
        (a) $^-$OH   (b) $^-$CN   (c) $^-$Cl   (d) $^-$I

11.   In the reaction of (2R,3S)-3-methyl-2-pentanol with tosyl chloride, what is the
        configuration of the product?
        (a)   a mixture of all four possible stereoisomers
        (b)   (2R,3S) and (2S,3S)
        (c)   (2R,3S)
        (d)   (2S,3S)

Chapter 12 – Structure Determination: Mass Spectroscopy and Infrared Spectroscopy

Chapter Outline

I. Mass Spectrometry (Sections 12.1–12.4).
    A. General features of mass spectrometry (Section 12.1).
        1. Purpose of mass spectrometry.
            a. Mass spectrometry is used to measure the molecular weight of a compound.
            b. Mass spectrometry can also provide information on the structure of an unknown compound.
        2. Technique of mass spectrometry.
            a. A small amount of sample is vaporized into the *ionization source* and is bombarded by a stream of high-energy electrons.
            b. An electron is dislodged from a molecule, producing a cation radical.
            c. Most of the cation radicals fragment; the fragments may be positively charged or neutral.
            d. In the *deflector*, a strong magnetic field deflects the positively charged fragments, which are separated by m/z ratio.
            e. A *detector* records the fragments as peaks on a graphic display.
        3. Important terms.
            a. The mass spectrum is presented as a bar graph, with masses (m/z) on the x axis and intensity (relative abundance) on the y axis.
            b. The base peak is the tallest peak and is assigned an intensity of 100%.
            c. The parent peak, or molecular ion (M^+), corresponds to the unfragmented cation radical.
            d. In large molecules, the base peak is often not the molecular ion.
    B. Interpreting mass spectra (Sections 12.2–12.4).
        1. Molecular weight (Section 12.2).
            a. Mass spectra can frequently provide the molecular weight of a sample.
                i.   Double-focusing mass spectrometers can provide mass measurements accurate to 0.0005 amu.
                ii. Some samples fragment so easily that M^+ is not seen.
            b. If you know the molecular weight of the sample, you can often deduce its molecular formula.
            c. There is often a peak at M+1 that is due to contributions from ^{13}C and ^2H.
        2. Fragmentation patterns of hydrocarbons.
            a. Fragmentation patterns can be used to identify a known compound, because a given compound has a unique fragmentation "fingerprint".
            b. Fragmentation patterns can also provide structural information.
                i.   Most hydrocarbons fragment into carbocations and radicals.
                ii. The positive charge remains with the fragment most able to stabilize it.
                iii. It is often difficult to assign structures to fragments.
                iv. For hexane, major fragments correspond to the loss of methyl, ethyl, propyl, and butyl radicals.

3. Fragmentation patterns of common functional groups (Section 12.3).
   a. Alcohols.
      i. Alcohols can fragment by alpha cleavage, in which a C–C bond next to the –OH group is broken.
         (a). The products are a cation and a radical.
      ii. Alcohols can also dehydrate, leaving an alkene cation radical with a mass 18 units less than M$^+$.
   b. Amines also undergo alpha cleavage, forming a cation and a radical.
   c. Carbonyl compounds.
      i. Aldehydes and ketones with a hydrogen 3 carbons from the carbonyl group can undergo the McLafferty rearrangement.
         (a). The products are a cation radical and a neutral alkene.
      ii. Aldehydes and ketones also undergo alpha cleavage, which breaks a bond between the carbonyl group and a neighboring carbon.
         (a). The products are a cation and a radical.
C. Mass spectrometry in biological systems: TOF instruments (Section 12.4).
   1. Time-of-flight (TOF) instruments are used to produce charged molecules with little fragmentation.
   2. The ionizer can be either ESI or MALDI.
      a. In an ESI source, the sample is dissolved in a polar solvent and sprayed through a steel capillary tube.
         i. As the sample exits, it is subjected to a high voltage, which protonates the sample.
         ii. The solvent is evaporated, yielding protonated sample molecules.
      b. In a MALDI source, the sample is absorbed onto a matrix compound.
         i. The matrix compound is ionized by a burst of laser light.
         ii. The matrix compound transfers energy to the sample, protonating it.
   3. The samples are focused into a small packet and given a burst of energy.
      a. Each molecule moves at a velocity that depends on the square root of its mass.
   4. The analyzer is an electrically grounded tube that detects the charged molecules by velocity.
II. Spectroscopy and the electromagnetic spectrum (Section 12.5).
   A. The nature of radiant energy.
      1. Different types of electromagnetic radiation make up the electromagnetic spectrum.
      2. Electromagnetic radiation behaves both as a particle and as a wave.
      3. Electromagnetic radiation can be characterized by three variables.
         a. The wavelength (λ) measures the distance from one maximum to the next.
         b. The frequency (ν) measures the number of wave maxima that pass a fixed point per unit time.
         c. The amplitude is the height measured from the midpoint to the maximum.
      4. Wavelength times frequency equals the speed of light.
      5. Electromagnetic energy is transmitted in discrete energy bundles called quanta.

a. $\varepsilon = h\nu$, where ε is energy per photon.

b.   Energy varies directly with frequency but inversely with wavelength.

c. $E = 1.20 \times 10^{-2}$ kJ/mol $\div \lambda$ (cm) for a "mole" of photons.

B.   Electromagnetic radiation and organic molecules.

　　1.   When an organic compound is struck by a beam of electromagnetic radiation, it absorbs radiation of certain wavelengths, and transmits radiation of other wavelengths.

　　2.   If we determine which wavelengths are absorbed and which are transmitted, we can obtain an absorption spectrum of the compound.

　　　　a.   For an infrared spectrum:

　　　　　　i.   The horizontal axis records wavelength.

　　　　　　ii.   The vertical axis records percent transmittance.

　　　　　　iii.   The baseline runs across the top of the spectrum.

　　　　　　iv.   Energy absorption is a downward spike (low percent transmittance).

　　3.   The energy a molecule absorbs is distributed over the molecule.

　　4.   There are many types of spectroscopies that differ in the region of the electromagnetic spectrum that is being used.

III.   Infrared Spectroscopy (Sections 12.6–12.8).

A.   Infrared radiation (Section 12.6).

　　1.   The infrared (IR) region of the electromagnetic spectrum extends from 7.8×10^{-7} m to 10^{-4} m.

　　　　a.   Organic chemists use the region from 2.5×10^{-6} m to 2.5×10^{-5} m.

　　　　b.   Wavelengths are usually given in μm, and frequencies are expressed in wavenumbers, which are the reciprocal of wavelength.

　　　　c.   The useful range of IR radiation is 4000 cm^{-1} to 400 cm^{-1}; this corresponds to energies of 48.0 kJ/mol to 4.80 kJ/mol.

　　2.   IR radiation causes bonds to stretch and bend and causes other molecular vibrations.

　　3.   Energy is absorbed at a specific frequency that corresponds to the frequency of the vibrational motion of a bond.

　　4.   If we measure the frequencies at which IR energy is absorbed, we can find out the kinds of bonds a compound contains and identify functional groups.

B.   Interpreting IR spectra (Sections 12.7–12.8).

　　1.   General principles (Section 12.7).

　　　　a.   Most molecules have very complex IR spectra.

　　　　　　i.   This complexity means that each molecule has a unique fingerprint that allows it to be identified by IR spectroscopy.

　　　　　　ii.   Complexity also means that not all absorptions can be identified.

　　　　b.   Most functional groups have characteristic IR absorption bands that change very little from one compound to another.

　　　　c.   The significant regions of IR absorptions:

      i. $4000 \text{ cm}^{-1} - 2500 \text{ cm}^{-1}$ corresponds to absorptions by C–H, O–H, and N–H bonds.

      ii. $2500 \text{ cm}^{-1} - 2000 \text{ cm}^{-1}$ corresponds to triple-bond stretches.

      iii. $2000 \text{ cm}^{-1} - 1500 \text{ cm}^{-1}$ corresponds to double bond stretches.

      iv. The region below 1500 cm^{-1} is the fingerprint region, where many complex bond vibrations occur that are unique to a molecule.

  d. The frequency of absorption of different bonds depends on two factors:

      i.   The strength of the bond.

      ii. The difference in mass between the two atoms in the bond.

2. IR spectra of some common functional groups (Section 12.8).

  a. Alkanes.

      i.   C–C absorbs at 800–1300 cm^{-1}.

      ii   C–H absorbs at 2850–2960 cm^{-1}.

  b. Alkenes.

      i.   =C–H absorbs at 3020–3100 cm^{-1}.

      ii. C=C absorbs at 1650–1670 cm^{-1}.

      iii. $RCH{=}CH_2$ absorbs at 910 and 990 cm^{-1}.

      iv. $R_2C{=}CH_2$ absorbs at 890 cm^{-1}.

  c. Alkynes.

      i.   –C≡C– absorbs at 2100–2260 cm^{-1}.

      ii. ≡C–H absorbs at 3300 cm^{-1}.

  d. Aromatic compounds.

      i.   =C–H absorbs at 3030 cm^{-1}.

      ii. Ring absorptions occur at 1660–2000 cm^{-1} and at 1450–1600 cm^{-1}.

  e. The alcohol O–H bond absorbs at 3400–3650 cm^{-1}.

  f. The N–H bond of amines absorbs at 3300–3500 cm^{-1}.

  g. Carbonyl compounds.

      i.   Saturated aldehydes absorb at 1730 cm^{-1}; unsaturated aldehydes absorb at 1705 cm^{-1}.

      ii. Saturated ketones absorb at 1715 cm^{-1}; unsaturated ketones absorb at 1690 cm^{-1}.

      iii. Saturated esters absorb at 1735 cm^{-1}; unsaturated esters absorb at 1715 cm^{-1}.

Solutions to Problems

12.1   If the isotopic masses of the atoms C, H, and O had integral values of 12 amu, 1 amu and 16 amu, many molecular formulas would correspond to a molecular weight of 288 amu. Because isotopic masses are not integral, however, only one molecular formula is associated with a molecular ion at 288.2089 amu.

To reduce the number of possible formulas, assume that the difference in molecular weight between 288 and 288.2089 is due mainly to hydrogen. Divide 0.2089 by 0.00783, the amount by which the atomic weight of one ^{1}H atom differs from 1. The answer, 26.67, gives a "ballpark" estimate of the number of hydrogens in testosterone. Then, divide 288 by 12, to determine the maximum number of carbons. Since $288 \div 12 = 24$, we know that testosterone can have no more than 22 carbons if it also includes hydrogen and oxygen. Make a list of *reasonable* molecular formulas containing C, H and O whose mass is 288 and which contain 20–30 hydrogens. Tabulate these, and calculate their exact masses using the exact atomic mass values in the text. The only possible formula for testosterone is $C_{19}H_{28}O_2$.

Isotopic mass

Molecular formula	Mass of carbons	Mass of hydrogens	Mass of oxygens	Mass of molecular ion
$C_{20}H_{32}O$	240.0000 amu	32.2504 amu	15.9949 amu	288.2453 amu
$C_{19}H_{28}O_2$	228.0000	28.2191	31.9898	288.2089
$C_{18}H_{24}O_3$	216.0000	24.1879	47.9847	288.1726

12.2

$$CH_3CH_2CH=C\overset{\displaystyle CH_3}{\underset{\displaystyle CH_3}{}}$$

2-Methyl-2-pentene

$$CH_3CH_2CH_2CH=CHCH_3$$

2-Hexene

$$^+CH_2CH=C\overset{\displaystyle CH_3}{\underset{\displaystyle CH_3}{}}$$

$m/z = 69$

$$^+CH_2CH=CHCH_3$$

$m/z = 55$

Fragmentation occurs to a greater extent at the weakest carbon–carbon bonds, producing a relatively stable cation. Spectrum (a), which has a dominant peak at $m/z = 69$, corresponds to 2-methyl-2-pentene, and spectrum (b), which has $m/z = 55$ as its base peak, corresponds to 2-hexene.

12.3   In a mass spectrum, the molecular ion is both a cation and a radical. When it fragments, two kinds of cleavage can occur. (1) Cleavage can form a radical and a cation (the species observed in the mass spectrum). Alpha cleavage shows this type of pattern. (2) Cleavage can form a neutral molecule and a different radical cation (the species observed in the mass spectrum). Alcohol dehydration and the McLafferty rearrangement show this cleavage pattern.

For each compound, calculate the mass of the molecular ion and identify the functional groups present. Draw the fragmentation products and calculate their masses.

(a)

In theory, alpha cleavage can take place on either side of the carbonyl group to produce cations with $m/z = 43$ and $m/z = 71$. In practice, cleavage occurs on the more substituted side of the carbonyl group, and the first cation, with $m/z = 43$, is observed.

(b)

Dehydration of cyclohexanol produces a cation radical with $m/z = 82$.

(c)

The cation radical fragment resulting from McLafferty rearrangement has $m/z = 58$.

(d)

Alpha cleavage of triethylamine yields a cation with $m/z = 86$.

12.4 Identify the functional groups present in the molecule and recall the kinds of fragmentations those functional groups produce. 2-Methyl-2-pentanol produces fragments that result from both dehydration and from alpha cleavage. Two different alpha cleavage products are possible.

Peaks might appear at $M^+ = 102$ (molecular ion), 87, 84, 59.

12.5 We know that: (1) energy increases as wavelength decreases, and (2) the wavelength of X-radiation is smaller than the wavelength of infrared radiation. Thus, we estimate that an X ray is of higher energy than an infrared ray.

$\varepsilon = h\nu = hc/\lambda$; $h = 6.62 \times 10^{-34}$ J·s; $c = 3.00 \times 10^8$ m/s

for $\lambda = 1 \times 10^{-6}$ m (infrared radiation):

$$\varepsilon = \frac{\left(6.62 \times 10^{-34}\,\text{J·s}\right)\left(3.00 \times 10^8\,\text{m/s}\right)}{1.0 \times 10^{-6}\,\text{m}} = 2.0 \times 10^{-19}\,\text{J}$$

for $\lambda = 3.0 \times 10^{-9}$ m (X radiation):

$$\varepsilon = \frac{\left(6.62 \times 10^{-34}\,\text{J·s}\right)\left(3.00 \times 10^8\,\text{m/s}\right)}{3.0 \times 10^{-9}\,\text{m}} = 6.6 \times 10^{-17}\,\text{J}$$

Confirming our estimate, the calculation shows that an X ray is of higher energy than infrared radiation.

Convert radiation in m to radiation in Hz by the equation:

$$\nu = \frac{c}{\lambda} = \frac{3.00 \times 10^8\,\text{m/s}}{9.0 \times 10^{-6}\,\text{m}} = 3.3 \times 10^{13}\,\text{Hz}$$

The equation $\varepsilon = h\nu$ shows that the greater the value of ν, the greater the energy. Thus, radiation with $\nu = 3.3 \times 10^{13}$ Hz ($\lambda = 9.0 \times 10^{-6}$ m) is higher in energy than radiation with $\nu = 4.0 \times 10^9$ Hz.

12.6

(a) $E = \dfrac{1.20 \times 10^{-4}\,\text{kJ/mol}}{\lambda(\text{in m})} = \dfrac{1.20 \times 10^{-4}\,\text{kJ/mol}}{5.0 \times 10^{-11}}$

   $= 2.4 \times 10^{6}\,\text{kJ/mol for a gamma ray}$

(b) $E = 4.0 \times 10^{4}$ kJ/mol for an X ray.

(c) $v = \dfrac{c}{\lambda};\ \lambda = \dfrac{c}{v} = \dfrac{3.0 \times 10^{8}\,\text{m/s}}{6.0 \times 10^{15}\,\text{Hz}} = 5.0 \times 10^{-8}\,\text{m}$

   $E = \dfrac{1.20 \times 10^{-4}\,\text{kJ/mol}}{5.0 \times 10^{-8}} = 2.4 \times 10^{3}\,\text{kJ/mol for ultraviolet light}$

(d) $E = 2.8 \times 10^{2}$ kJ/mol for visible light.

(e) $E = 6.0$ kJ/mol for infrared radiation

(f) $E = 4.0 \times 10^{-2}$ kJ/mol for microwave radiation.

12.7 (a) A compound with a strong absorption at 1710 cm^{-1} contains a carbonyl group and is either a ketone or aldehyde.

(b) A compound with a nitro group has a strong absorption at 1540 cm^{-1}.

(c) A compound showing both carbonyl (1720 cm^{-1}) and $-$OH (2500–3000 cm^{-1} broad) absorptions is a carboxylic acid.

12.8 To use IR spectroscopy to distinguish between isomers, find a strong IR absorption that is present in one isomer but absent in the other.

(a) CH$_3$CH$_2$OH
Strong hydroxyl band
at 3400–3640 cm$^-$

CH$_3$OCH$_3$
No band in the region
3400–3640 cm^{-1}

(b) CH$_3$CH$_2$CH$_2$CH$_2$CH=CH$_2$
Alkene bands at 3020–3100
cm^{-1} and
at 1640–1680^{-1}

No bands in alkene region

(c) CH$_3$CH$_2$CO$_2$H
Strong, broad band
at 2500–3100 cm^{-1}

HOCH$_2$CH$_2$CHO
Strong band at 3400–3640
cm^{-1}

12.9   Based on what we know at this point, we can identify four absorptions in this spectrum.

   (a)   Absorptions in the region 1450 cm^{-1}–1600 cm^{-1} are due to aromatic ring –C=C– motions.

   (b)   The absorption at 2100 cm^{-1} is due to a –C≡C– stretch.

   (c)   Absorptions in the range 3000 cm^{-1} – 3100 cm^{-1} are due to aromatic ring =C–H stretches.

   (d)   The absorption at 3300 cm^{-1} is due to a ≡C–H stretch.

12.10 (a)   An ester next to a double bond absorbs at 1715 cm^{-1}. The alkene double bond absorbs at 1640–1680 cm^{-1}.

   (b)   The aldehyde carbonyl group absorbs at 1730 cm^{-1}. The alkyne C≡C bond absorbs at 2100–2260 cm^{-1}, and the alkyne H–C≡ bond absorbs at 3300 cm^{-1}.

   (c)   The most important absorptions for this compound are due to the alcohol group (a broad, intense band at 3400–3650 cm^{-1}) and to the carboxylic acid group, which has a C=O absorption in the range 1710–1760 cm^{-1} and a broad O–H absorption in the range 2500–3100 cm^{-1}. Absorptions due to the aromatic ring [3030 cm^{-1} (w) and 1450–1600 cm^{-1} (m)] may also be seen.

12.11

The compound contains nitrile and ketone groups, as well as a carbon–carbon double bond. The nitrile absorption occurs at 2210–2260 cm^{-1}. The ketone shows an absorption at 1690 cm^{-1}, a lower value than usual because the ketone is next to the double bond. The double bond absorption occurs at 1640–1680 cm^{-1}.

Visualizing Chemistry

12.12

Compound	Significant IR Absorption	Due to:
(a)	1540 cm^{-1} 1730 cm^{-1} 3030 cm^{-1} 1450–1600 cm^{-1}	nitro group (1) aldehyde (2) aromatic ring C–H(3) aromatic ring C=C(3)

(b)

1735 cm^{-1}	ester (1)
3020–3100 cm^{-1}	vinylic stretch C–H(2)
910 cm^{-1}, 990 cm^{-1}	C=CH$_2$ bend(3)
1640–1680 cm^{-1}	alkene C=C

(c)

1715 cm^{-1}	ketone (1)
3400–3650 cm^{-1}	alcohol (2)

12.13 (a)   The mass spectrum of this ketone shows fragments resulting from both McLafferty rearrangement and alpha cleavage.

McLafferty rearrangement:

Alpha cleavage:

(b)   Two different fragments can arise from alpha cleavage of this amine:

The second product results from cleavage of a bond in the five-membered ring. Due to the symmetry of the amine, only one peak is observed.

Additional Problems

Mass Spectrometry

12.14

M^+	Molecular Formula	Degree of Unsaturation	Possible Structure
(a) 132	$C_{10}H_{12}$	5	
(b) 166	$C_{13}H_{10}$	9	
	$C_{12}H_{22}$	2	
(c) 84	C_6H_{12}	1	

12.15 (a)   The compound contains no more than 7 carbons. As in Problem 12.1, divide the mass to the right of the decimal point by 0.00783 to arrive at an approximate value for the number of hydrogens (10.8). Since the compound has an even mass (and an even number of hydrogens), it contains an even number of nitrogens, or no nitrogens. The two most likely formulas are $C_6H_{10}O$ ($M^+ = 98.0732$) and $C_5H_{10}N_2$ ($M^+ = 98.0844$). The latter formula agrees precisely with the given molecular ion.

(b) The compound contains no more than 9 carbons, and approximately 4.1 hydrogens. The number of hydrogens must be odd, since M^+ is odd. Assume the molecule has 5 hydrogens, and adjust the numbers of nitrogens and oxygens until you arrive at the correct value for M^+. The formula is $C_6H_5NO_2$.

12.16 Reasonable molecular formulas for camphor are $C_{10}H_{16}O$, $C_9H_{12}O_2$, $C_8H_8O_3$ (see Problem 12.1). The actual formula, $C_{10}H_{16}O$ ($M^+ = 152.1201$), corresponds to three degrees of unsaturation. The ketone functional group accounts for one of these. Since camphor is a saturated compound, the other two degrees of unsaturation are due to two rings.

Camphor

12.17 Carbon is tetravalent, and nitrogen is trivalent. If a C–H unit (formula weight 13) is replaced by an N atom (formula weight 14), the molecular weight of the resulting compound increases by one. Since all neutral hydrocarbons have even-numbered molecular weights (C_nH_{2n+2}, C_nH_{2n}, and so forth) the resulting nitrogen-containing compounds have odd-numbered molecular weights. If two C–H units are replaced by two N atoms, the molecular weight of the resulting compound increases by two and remains an even number.

12.18 Because M^+ is an odd number, pyridine contains an odd number of nitrogen atoms. If pyridine contained one nitrogen atom (atomic weight 14) the remaining atoms would have a formula weight of 65, corresponding to $-C_5H_5$. C_5H_5N is, in fact, the molecular formula of pyridine.

12.19 Subtract the isotopic mass of the two nitrogens from the value of M^+, and divide the quantity to the right of the decimal point by 0.00783 to find the approximate number of hydrogens in nicotine. The molecular formula of nicotine is $C_{10}H_{14}N_2$. To find the equivalent hydrocarbon formula, subtract the number of nitrogens from the number of hydrogens. The equivalent hydrocarbon formula of nicotine, $C_{10}H_{12}$, indicates five degrees of unsaturation — two of them due to the two rings and the other three due to three double bonds.

Nicotine

12.20 Use the technique described in Problem 12.1 to find the molecular formula of cortisone. Cortisone contains approximately 25 hydrogens. Make a table of possible molecular formulas for cortisone that have around 25 hydrogens and calculate the exact molecular weights corresponding to these formulas.

Molecular formula	Mass of carbons	Isotopic mass Mass of hydrogens	Mass of oxygens	Mass of molecular ion
$C_{27}H_{20}O$	324.0000 amu	20.1565 amu	15.9949 amu	360.1514 amu
$C_{25}H_{28}O_2$	300.0000	28.2191	31.9898	360.2089
$C_{24}H_{24}O_3$	288.0000	24.1878	47.9847	360.1725
$C_{21}H_{28}O_5$	252.0000	28.2191	79.9746	360.1937

The molecular weight of $C_{21}H_{28}O_5$ corresponds to the observed molecular weight of cortisone. (Note that only the last formula has the correct degree of unsaturation, 8).

12.21 In order to simplify this problem, neglect the ^{13}C and ^{2}H isotopes in determining the molecular ions of these compounds.

(a) The formula weight of $-CH_3$ is 15, and the atomic masses of the two bromine isotopes are 79 and 81. The two molecular ions of bromoethane occur at $M^+ = 94$ (50.7%) and $M^+ = 96$ (49.3%).

(b) The formula weight of $-C_6H_{13}$ is 85, and the atomic masses of the two chlorine isotopes are 35 and 37. The two molecular ions of 1-chlorohexane occur at $M^+ = 120$ (75.8%) and $M^+ = 122$ (24.2%).

12.22 Each carbon atom has a 1.10% probability of being ^{13}C and a 98.90% probability of being ^{12}C. The ratio of the height of the ^{13}C peak to the height of the ^{12}C peak for a one-carbon compound is $(1.10/98.9) \times 100\% = 1.11\%$. For a six-carbon compound, the contribution to $(M+1)^+$ from ^{13}C is $6 \times (1.10/98.9) \times 100\% = 6.66\%$. For benzene, the relative height of $(M+1)^+$ is 6.66% of the height of M^+.

A similar line of reasoning can be used to calculate the contribution to $(M+1)^+$ from ^{2}H. The natural abundance of ^{2}H is 0.015%, so the ratio of a ^{2}H peak to a ^{1}H peak for a one-hydrogen compound is 0.015%. For a six-hydrogen compound, the contribution to $(M+1)^+$ from ^{2}H is $6 \times 0.015\% = 0.09\%$.

For benzene, $(M+1)^+$ is 6.75% of M^+. Notice that ^{2}H contributes very little to the size of $(M+1)^+$.

12.23 (a) The molecular formula of the ketone is $C_5H_{10}O$, and the fragments correspond to the products of alpha cleavage (McLafferty rearrangement fragments have even-numbered values of m/z). Draw all possible ketone structures, show the charged products of alpha cleavage, and note which fragments correspond to those listed.

$$\left[\begin{array}{c} \overset{\displaystyle O}{\underset{\displaystyle \|}{}} \\ H_3C \text{—} C \text{—} CH_2CH_2CH_3 \end{array}\right]^{+\cdot} \xrightarrow[\text{cleavage}]{\text{Alpha}} \left[H_3C \text{—} C \equiv O\right]^{+} + \left[O \equiv C \text{—} CH_2CH_2CH_3\right]^{+}$$

$$M^+ = 86 \qquad\qquad\qquad m/z = 43 \qquad\qquad m/z = 71$$

$$\left[\begin{array}{c} \overset{\displaystyle O}{\underset{\displaystyle \|}{}} \quad CH_3 \\ H_3C \text{—} C \text{—} CHCH_3 \end{array}\right]^{+\cdot} \xrightarrow[\text{cleavage}]{\text{Alpha}} \left[H_3C \text{—} C \equiv O\right]^{+} + \left[\begin{array}{c} CH_3 \\ O \equiv C \text{—} CHCH_2CH_3 \end{array}\right]^{+}$$

$$M^+ = 86 \qquad\qquad\qquad m/z = 43 \qquad\qquad m/z = 71$$

$$\left[\begin{array}{c} \overset{\displaystyle O}{\underset{\displaystyle \|}{}} \\ CH_3CH_2 \text{—} C \text{—} CH_2CH_3 \end{array}\right]^{+\cdot} \xrightarrow[\text{cleavage}]{\text{Alpha}} \left[CH_3CH_2 \text{-} C \equiv O\right]^{+} + \left[O \equiv C \text{—} CH_2CH_3\right]^{+}$$

$$M^+ = 86 \qquad\qquad\qquad m/z = 57 \qquad\qquad m/z = 57$$

Either of the first two compounds shows the observed fragments in its mass spectrum.

(b) $C_5H_{12}O$ is the formula of an alcohol with $M^+ = 88$. The fragment at $m/z = 70$ is due to the product of dehydration of M^+. The other two fragments are a result of alpha cleavage. Draw the possible C_5 alcohol isomers, and draw their products of alpha cleavage. The tertiary alcohol shown fits the data.

$$\left[\begin{array}{c} OH \\ CH_3CH_2 \text{—} C \text{—} CH_3 \\ CH_3 \end{array}\right]^{+\cdot} \xrightarrow[\text{cleavage}]{\text{Alpha}} \left[\begin{array}{c} OH \\ \| \\ CH_3CH_2 \diagdown C \diagup CH_3 \end{array}\right]^{+} + \left[\begin{array}{c} OH \\ \| \\ H_3C \diagdown C \diagup CH_3 \end{array}\right]^{+}$$

$$M^+ = 88 \qquad\qquad\qquad m/z = 73 \qquad\qquad m/z = 59$$

$$\left[\begin{array}{c} OH \\ CH_3CH_2 \text{—} C \text{—} CH_3 \\ CH_3 \end{array}\right]^{+\cdot} \xrightarrow{\text{Dehydration}} \left[\begin{array}{c} H_3C \diagdown \quad \diagup CH_3 \\ C = C \\ H \diagup \quad \diagdown CH_3 \end{array}\right]^{+\cdot}$$

$$M^+ = 88 \qquad\qquad\qquad m/z = 70$$

12.24

$$CH_3$$
$$|$$
$$CH_3CH_2CH_2CHCH_3$$
2-Methylpentane

The molecular ion, at $m/z = 86$, is present in very low abundance. The base peak, at $m/z = 43$, represents a stable secondary carbocation.

12.25 Before doing the hydrogenation, familiarize yourself with the mass spectra of cyclohexene and cyclohexane. Note that M^+ is different for each compound. After the reaction is underway, inject a sample from the reaction mixture into the mass spectrometer. If the reaction is finished, the mass spectrum of the reaction mixture should be superimposable with the mass spectrum of cyclohexane.

12.26 (a)   This ketone shows mass spectrum fragments that are due to alpha cleavage and to the McLafferty rearrangement. The molecular ion occurs at $M^+ = 148$, and major fragments have $m/z = 120, 105$, and 71. (Note that only charged species are shown.)

(b)   The fragments in the mass spectrum of this alcohol ($C_8H_{16}O$) result from dehydration and alpha cleavage. Major fragments have m/z values of 128 (the same value as the molecular ion), 110, and 99.

(c)   Amines fragment by alpha cleavage. In this problem, cleavage occurs in the ring, producing a fragment with the same value of m/z as the molecular ion (99).

Infrared Spectroscopy

12.27 $CH_3CH_2C\equiv CH$ shows absorptions at 2100–2260 cm^{-1} ($C\equiv C$) and at 3300 cm^{-1} ($C\equiv C-H$) that are due to the terminal alkyne bond.

$H_2C=CHCH=CH_2$ has absorptions in the regions 1640–1680 cm^{-1} and 3020–3100 that are due to the double bonds. It also shows absorptions at 910 cm^{-1} and 990 cm^{-1} that are due to monosubstituted alkene bonds. No absorptions occur in the alkyne region.

$CH_3C\equiv CCH_3$. For reasons we won't discuss, symmetrically substituted alkynes such as 2-butyne do not show a $C\equiv C$ bond absorption in the IR. This alkyne is distinguished from the other isomers in that it shows no absorptions in either the alkyne or alkene regions.

12.28 Two enantiomers have identical physical properties (other than the sign of specific rotation). Thus, their IR spectra are also identical.

12.29 Since diastereomers have different physical properties and chemical behavior, their IR spectra are also different.

12.30 (a) Absorptions at 3300 cm^{-1} and 2150 cm^{-1} are due to a terminal triple bond. Possible structures:

CH$_3$CH$_2$CH$_2$C≡CH   (CH$_3$)$_2$CHC≡CH

(b) An IR absorption at 3400 cm^{-1} is due to a hydroxyl group. Since no double bond absorption is present, the compound must be a cyclic alcohol.

(c) An absorption at 1715 cm^{-1} is due to a ketone. The only possible structure is CH$_3$CH$_2$COCH$_3$.

(d) Absorptions at 1600 cm^{-1} and 1500 cm^{-1} are due to an aromatic ring. Possible structures:

12.31

(a) HC≡CCH$_2$NH$_2$ CH$_3$CH$_2$C≡N
     Alkyne absorptions at   Nitrile absorption at
     3300 cm^{-1}, 2100–2260 cm^{-1} 2210–2260 cm^{-1}
     Amine absorption at
     3300–3500 cm^{-1}

(b) CH$_3$COCH$_3$ CH$_3$CH$_2$CHO
     Strong ketone absorption   Strong aldehyde absorption
     at 1715 cm^{-1} at 1730 cm^{-1}

12.32 Spectrum (b) differs from spectrum (a) in several respects. Note in particular the absorptions at 715 cm^{-1} (strong), 1140 cm^{-1} (strong), 1650 cm^{-1} (medium), and 3000 cm^{-1} (medium) in spectrum (b). The absorptions at 1650 cm^{-1} (C=C stretch) and 3000 cm^{-1} (=C–H stretch) can be found in Table 12.1. They allow us to assign spectrum (b) to cyclohexene and spectrum (a) to cyclohexane.

12.33 Only absorptions with medium to strong intensity are listed.

(a) aromatic ring C=C
1450–1600 cm^{-1}
aromatic ring C–H
3030 cm^{-1}
carboxylic acid C=O
1710–1760 cm^{-1}
carboxylic acid O–H
2500–3100 cm^{-1}

(b) aromatic ring C=C
1450–1600 cm^{-1}
aromatic ring C–H
3030 cm^{-1}
aromatic ester
1715 cm^{-1}

(c) aromatic ring C=C
1450–1600 cm^{-1}
aromatic ring C–H
3030 cm^{-1}
alcohol O–H
3400–3650 cm^{-1}
nitrile C≡N
2210–2260 cm^{-1}

(d) alkene C=C
1640–1680 cm^{-1}
alkene =C–H
3020–3100 cm^{-1}
ketone
1715 cm^{-1}

(e) $CH_3CCH_2CH_2COCH_3$ ester 1735 cm^{-1} ketone 1715 cm^{-1}

12.34 (a) $CH_3C≡CCH_3$ exhibits no terminal ≡C–H stretching vibration at 3300 cm^{-1}, as $CH_3CH_2C≡CH$ does.

(b) $CH_3COCH=CHCH_3$, a ketone next to a double bond, shows a strong ketone absorption at 1690 cm^{-1}; $CH_3COCH_2CH=CH_2$ shows a ketone absorption at 1715 cm^{-1} and monosubstituted alkene absorptions at 910 cm^{-1} and 990 cm^{-1}.

(c) CH_3CH_2CHO exhibits an aldehyde band at 1730 cm^{-1}; $H_2C=CHOCH_3$ shows characteristic monosubstituted alkene absorptions at 910 cm^{-1} and 990 cm^{-1}.

12.35

Compound	Distinguishing Absorption	Due to:
(a) $CH_3CH_2CCH_3$	1715 cm^{-1}	C=O (ketone)
(b) $(CH_3)_2CHCH_2C≡CH$	2100–2260 cm^{-1} 3300 cm^{-1}	C≡C C≡C–H
(c) $(CH_3)_2CHCH_2CH=CH_2$	910 cm^{-1}, 990 cm^{-1} 1640–1680 cm^{-1} 3020–3100 cm^{-1}	RCH=CH$_2$ C=C =C–H

(d) $\overset{\overset{\displaystyle O}{\parallel}}{CH_3CH_2CH_2COCH_3}$ 1735 cm^{-1} C=O (ester)

(e)   1690 cm^{-1} ketone next to
                    aromatic ring
    1450–1600 cm^{-1} aromatic ring
    3030 cm^{-1} aromatic ring

(f)   1710 cm^{-1} aldehyde next to
                    aromatic ring
    3400–3650 cm^{-1} alcohol
    1450–1600 cm^{-1} aromatic ring
    3030 cm^{-1} aromatic ring

12.36

1-Methylcyclohexanol   1-Methylcyclohexene

The infrared spectrum of the starting alcohol shows a broad absorption at 3400–3640 cm$^-$1 due to an O–H stretch. The alkene product exhibits medium intensity absorbances at 1645–1670 cm^{-1} and at 3000–3100 cm^{-1}. Monitoring the disappearance of the alcohol absorption makes it possible to decide when reaction is complete. It is also possible to monitor the appearance of the alkene absorptions.

12.37

3-Bromo-3-methylpentane   3-Methyl-2-pentene   2-Ethyl-1-butene

The IR spectra of both products show the characteristic absorptions of alkenes in the regions 3020–3100 cm^{-1} and 1650 cm^{-1}. However, in the region 700–1000 cm^{-1}, 2-ethyl-1-butene shows a strong absorption at 890 cm^{-1} that is typical of 2,2-disubstituted $R_2C=CH_2$ alkenes. The presence or absence of this peak should help to identify the product. (3-Methyl-2-pentene is the major product of the dehydrobromination reaction.)

General Problems

12.38 The following expressions are needed:

$\varepsilon = hv = hc/\lambda = hc\tilde{v}$, where $\tilde{v}$ is the wavenumber. The last expression shows that, as $\tilde{v}$ increases, the energy needed to cause IR absorption increases, indicating greater bond strength. Thus an ester C=O bond $\left(\tilde{v} = 1735 \text{ cm}^{-1}\right)$ is stronger than a ketone C=O bond $\left(\tilde{v} = 1715 \text{ cm}^{-1}\right)$.

12.39 Possible molecular formulas containing carbon, hydrogen, and oxygen and having $M^+ = 150$ are $C_{10}H_{14}O$, $C_9H_{10}O_2$, and $C_8H_6O_3$. The first formula has four degrees of unsaturation, the second has five degrees of unsaturation, and the third has six degrees of unsaturation. Since carvone has three double bonds (including the ketone) and one ring, or four degrees of unsaturation, $C_{10}H_{14}O$ is the correct molecular formula for carvone.

Carvone

12.40 The intense absorption at 1690 cm^{-1} is due to a ketone next to a double bond.

12.41 The peak of maximum intensity (base peak) in the mass spectrum occurs at $m/z = 67$. This peak does *not* represent the molecular ion, however, because M$^+$ of a hydrocarbon must be an even number. Careful inspection reveals the molecular ion peak at $m/z = 68$. M$^+$ = 68 corresponds to a hydrocarbon of molecular formula C_5H_8 with a degree of unsaturation of two.

Fairly intense peaks in the mass spectrum occur at $m/z = 67, 53, 40, 39,$ and 27. The peak at $m/z = 67$ corresponds to loss of one hydrogen atom, and the peak at $m/z = 53$ represents loss of a methyl group. The unknown hydrocarbon thus contains a methyl group.

Significant IR absorptions occur at 2130 cm^{-1} (–C≡C– stretch) and at 3320 cm^{-1} (≡C–H stretch). These bands indicate that the unknown hydrocarbon is a terminal alkyne. Possible structures for C_5H_8 are CH$_3$CH$_2$CH$_2$C≡CH and (CH$_3$)$_2$CHC≡CH. [1-Pentyne is correct.]

12.42 The molecular ion, M$^+$ = 70, corresponds to the molecular formula C_5H_{10}. This compound has one double bond or one ring.

The base peak in the mass spectrum occurs at $m/z = 55$. This peak represents loss of a methyl group from the molecular ion and indicates the presence of a methyl group in the unknown hydrocarbon. All other peaks occur with low intensity.

In the IR spectrum, it is possible to distinguish absorptions at 1660 cm^{-1} and at 3000 cm^{-1} due to a double bond. (The 2960 cm^{-1} absorption is rather hard to detect because it occurs as a shoulder on the alkane C–H stretch at 2850–2960 cm^{-1}.)Since no absorptions occur in the region 890 cm^{-1}–990 cm^{-1}, we can exclude terminal alkenes as possible structures. The remaining possibilities for C$_5$H$_{10}$ are CH$_3$CH$_2$CH=CHCH$_3$ and (CH$_3$)$_2$C=CHCH$_3$. [2-Methyl-2-butene is correct.]

12.43

(a)

$$CH_3CH_2\overset{\overset{\displaystyle O}{\|}}{C}\underset{\underset{\displaystyle CH_3}{|}}{H}CH$$

(b)

$$CH_3CH_2CH_2CH_2C\equiv N \qquad CH_3\underset{\underset{\displaystyle CH_3}{|}}{\overset{\overset{\displaystyle CH_3}{|}}{C}}C\equiv N \qquad CH_3\underset{\underset{\displaystyle CH_3}{|}}{C}HCH_2C\equiv N$$

12.44 The simplest way to distinguish between the two isomers is by taking their IR spectra. The aldehyde carbonyl group absorbs at 1730 cm^{-1}, and the ketone carbonyl group absorbs at 1715 cm^{-1}.

The mass spectra of the two isomers also differ. Like ketones, aldehydes also undergo alpha cleavage and McLafferty rearrangements.

McLafferty rearrangement:

The fragments from the McLafferty rearrangements differ in values of *m/z*.

Alpha cleavage:

The fragments resulting from alpha cleavage also differ in values of *m/z*.

12.45

The absorption at 3400 cm^{-1} is due to a hydroxyl group.

12.46 The absorption at 3400^{-1} is due to an alcohol.

12.47

The absorption at 1710 cm^{-1} is due to the carbonyl group of a carboxylic acid, and the absorption at 2500–3100 cm^{-1} is due to the −OH group of the carboxylic acid.

12.48 Ethyl octanoate

12.49 2-Methyl-1-pentene

12.50 Chlorobenzene

Chapter 13 – Structure Determination:
Nuclear Magnetic Resonance Spectroscopy

Chapter Outline

I. Principles of Nuclear Magnetic Resonance Spectroscopy (Sections 13.1–13.3).

  A. Theory of NMR Spectroscopy (Section 13.1).

    1. Many nuclei behave as if they were spinning about an axis.

      a. The positively charged nuclei produce a magnetic field that can interact with an externally applied magnetic field.

      b. The ^{13}C nucleus and the ^{1}H nucleus behave in this manner.

      c. In the absence of an external magnetic field the spins of magnetic nuclei are randomly oriented.

    2. When a sample containing these nuclei is placed between the poles of a strong magnet, the nuclei align themselves either with (parallel to) the applied field or against (antiparallel to) the applied field, measured in Tesla (T).

      a. The parallel orientation is slightly lower in energy and is slightly favored.

    3. If the sample is irradiated with radiofrequency energy of the correct frequency, the nuclei of lower energy absorb energy and "spin-flip" to the higher energy state.

      a. The magnetic nuclei are in resonance with the applied radiation.

      b. The frequency of the rf radiation needed for resonance depends on the applied magnetic field strength and on the identity of the magnetic nuclei.

        i. In a strong magnetic field, higher frequency rf energy is needed.

        ii. At a magnetic field strength of 4.7 T, rf energy of 200 MHz is needed to bring a ^{1}H nucleus into resonance, and energy of 50 MHz for ^{13}C.

    4. Nuclei with an odd number of protons and nuclei with an odd number of neutrons show magnetic properties.

  B. The nature of NMR absorptions (Section 13.2).

    1. Not all ^{13}C nuclei and not all ^{1}H nuclei absorb at the same frequency.

      a. Each magnetic nucleus is surrounded by electrons that set up their own magnetic fields.

      b. These small fields oppose the applied field and shield the magnetic nuclei.

        i. $\textbf{\textit{B}}_{effective} = \textbf{\textit{B}}_{applied} - \textbf{\textit{B}}_{local}$.

        ii. This expression shows that the magnetic field felt by a nucleus is less than the applied field.

      c. These shielded nuclei absorb at slightly different values of magnetic field strength.

      d. A sensitive NMR spectrometer can detect these small differences.

      e. Thus, NMR spectra can be used to map the carbon–hydrogen framework of a molecule.

    2. NMR spectra.

      a. The horizontal axis shows effective field strength, and the vertical axis shows intensity of absorption.

      b. Each peak corresponds to a chemically distinct nucleus.

      c. Zero absorption is at the bottom.

        d. Absorptions due to both ^{13}C and ^{1}H can't both be observed at the same time and are displayed on separate spectra.

   3. Operation of an NMR spectrometer

        a. A solution of a sample is placed in a thin glass tube between the poles of a magnet.

        b. The strong magnetic field causes the nuclei to align in either of the two possible orientations.

        c. The strength of the applied magnetic field is varied, holding the rf frequency constant.

        d. Chemically distinct nuclei come into resonance at slightly different values of $\boldsymbol{B}$.

        e. A detector monitors the absorption of rf energy.

        f. The signal is amplified and recorded as a peak.

   4. Time scale of NMR absorptions.

        a. The time scale (10^{-3} s) of NMR spectra is much slower than that of most other spectra.

        b. If a process occurs faster than the time scale of NMR, absorptions are observed as "time-averaged" processes.

           i. NMR records only a single spectrum of the time-averaged process.

        c. NMR can be used to measure rates and activation energies of fast processes.

           i. Because cyclohexane ring-flips are very fast at room temperature, only a single peak is observed for equatorial and axial hydrogens at room temperature.

          ii. At $-90°$C, both axial and equatorial hydrogens can be identified.

C. Chemical Shifts (Section 13.3).

   1. On NMR spectra, field strength increases from left (downfield) to right (upfield).

        a. Nuclei that absorb downfield require a lower field strength for resonance and are deshielded.

        b. Nuclei that absorb upfield require a higher field strength and are shielded.

   2. TMS is used as a reference point in both ^{13}C NMR and ^{1}H NMR.

        a. The TMS (tetramethylsilane) absorption occurs upfield of most other absorptions, and is set as the zero point.

   3. The chemical shift is the position along the x-axis where a nucleus absorbs energy.

   4. NMR charts are calibrated by using an arbitrary scale – the delta (δ) scale.

        a. One δ equals 1 ppm of the spectrometer operating frequency.

        b. By using this system, all chemical shifts occur at the same value of δ, regardless of the spectrometer operating frequency.

   5. NMR absorptions occur over a narrow range.

        a. ^{1}H absorptions occur 0–10 δ downfield from TMS.

        b. ^{13}C absorptions occur 1–220 δ downfield from TMS.

        c. The chances of accidental overlap can be reduced by using an instrument with a higher field strength.

II. ^{13}C NMR spectroscopy (Sections 13.4–13.7).

    A. Signal averaging and FT (fourier-transform)-NMR (Section 13.4).

        1. The low natural abundance of ^{13}C (1.1%) makes it difficult to observe ^{13}C peaks because of background noise.

        2. If hundreds of individual runs are averaged, the background noise cancels.

            a. This technique takes a long time.

        3. In FT-NMR, all signals are recorded simultaneously.

            a. The sample is irradiated with a pulse of rf energy that covers all useful frequencies.

            b. The resulting complex signal must be mathematically manipulated before display.

            c. FT-NMR takes only a few seconds per spectrum.

        4. FT-NMR and signal averaging provide increased speed and sensitivity.

            a. Only a few mg of sample are needed for ^{13}C NMR spectra.

            b. Only a few μg of sample are needed for ^{1}H NMR spectra.

    B. Characteristics of ^{13}C NMR spectroscopy (Section 13.5).

        1. Each distinct carbon shows a single line.

        2. The chemical shift depends on the electronic environment within a molecule.

            a. Carbons bonded to electronegative atoms absorb downfield.

            b. Carbons with sp^3 hybridization absorb in the range 0–90 δ.

            c. Carbons with sp^2 hybridization absorb in the range 110–220 δ.

                i. Carbonyl carbons absorb in the range 160–220 δ.

        3. Molecular symmetry reduces the number of absorptions.

        4. Peaks aren't uniform in size and are not integrated.

    C. DEPT ^{13}C NMR spectra (Section 13.6).

        1. With DEPT experiments, the number of hydrogens bonded to each carbon can be determined.

        2. DEPT experiments are run in three stages.

            a. A broadband decoupled spectrum gives the chemical shifts of all carbons.

            b. A DEPT-90 spectrum shows signals due only to CH carbons.

            c. A DEPT-135 spectrum shows CH_3 and CH resonances as positive signals, and CH_2 resonances as negative signals.

        3. Interpretation of DEPT spectra.

            a. Subtract all peaks in the DEPT-135 spectrum from the broadband-decoupled spectrum to find C.

            b. Use DEPT-90 spectrum to identify CH.

            c. Use negative DEPT-135 peaks to identify CH_2.

            d. Subtract DEPT-90 peaks from positive DEPT–135 peaks to identify CH_3.

    D. Uses of ^{13}C NMR spectroscopy (Section 13.7).

        a. ^{13}C NMR spectroscopy can show the number of nonequivalent carbons in a molecule and can identify symmetry in a molecule.

III. ^{1}H NMR Spectroscopy (Sections 13.8–13.13).

    A. Proton equivalence (Section 13.8).

        1. ^{1}H NMR can be used to determine the number of nonequivalent protons in a molecule.

        2. If it is not possible to decide quickly if two protons are equivalent, replace each proton by $-$X.

            a. If the protons are unrelated, the products formed by replacement are constitutional isomers.

            b. If the protons are chemically identical, the same product will form, regardless which proton is replaced, and the protons are homotopic.

            c. If the replacement products are enantiomers, the protons are enantiotopic.

            d. If the molecule contains a chirality center, the replacement products are diastereomers, and the protons are diastereotopic.

    B. Chemical shifts in ^{1}H NMR spectroscopy (Section 13.9).

        1. Chemical shifts are determined by the local magnetic fields surrounding magnetic nuclei.

            a. More strongly shielded nuclei absorb upfield.

            b. Less shielded nuclei absorb downfield.

        2. Most ^{1}H NMR chemical shifts are in the range 0–10 δ.

            a. Protons that are sp^3-hybridized absorb at higher field strength.

            b. Protons that are sp^2-hybridized absorb at lower field strength.

            c. Protons on carbons that are bonded to electronegative atoms absorb at lower field strength.

        3. The ^{1}H NMR spectrum can be divided into 5 regions:

            a. Saturated (0–1.5 δ).

            b. Allylic (1.5–2.5 δ).

            c. H bonded to C next to an electronegative atom (2.5–4.5 δ).

            d. Vinylic (4.5–6.5 δ).

            e. Aromatic (6.5–8.0 δ).

            f. Aldehyde and carboxylic acid protons absorb even farther downfield.

    C. Integration of ^{1}H NMR signals: proton counting (Section 13.10).

        1. The area of a peak is proportional to the number of protons causing the peak.

        2. Modern NMR instruments provide a digital readout of relative peak areas, although older instruments showed a stair-step line.

    D. Spin–spin splitting (Section 13.11).

        1. The tiny magnetic field produced by one nucleus can affect the magnetic field felt by a neighboring nucleus.

        2. Protons that have n equivalent neighboring protons show a peak in their ^{1}H NMR spectrum that is split into $n + 1$ smaller peaks (a multiplet).

        3. This splitting is caused by the coupling of spins of neighboring nuclei.

        4. The distance between peaks in a multiplet is called the coupling constant (J).

            a. The value of J is usually 0–18 Hz.

b. The value of J is determined by the geometry of the molecule and is independent of the spectrometer operating frequency.

c. The value of J is shared between both groups of hydrogens whose spins are coupled.

d. By comparing values of J, it is possible to know the atoms whose spins are coupled.

5. Three rules for spin–spin splitting in ^{1}H NMR:

a. Chemically identical protons don't show spin–spin splitting.

b. The signal of a proton with n equivalent neighboring protons is split into a multiplet of $n + 1$ peaks with coupling constant J.

c. Two groups of coupled protons have the same value of J.

6. Spin–spin splitting isn't seen in ^{13}C NMR.

a. Although spin–spin splitting can occur between carbon and other magnetic nuclei, the spectrometer operating conditions suppress it.

b. Coupling between the spins of two ^{13}C nuclei isn't seen because of the low probability that two ^{13}C nuclei might be adjacent.

E. Complex spin–spin splitting (Section 13.12).

1. At times the signals in a ^{1}H NMR absorption overlap accidentally.

2. Also, signals may be split by two or more nonequivalent kinds of protons.

a. To understand the effect of multiple coupling, it helps to draw a tree diagram.

b. In this type of multiplet, the peaks on one side of the multiplet may be larger than those on the other side.

i. The larger peaks are on the side nearer to the coupled partner.

ii. This helps identify the nuclei whose spins are coupled.

F. ^{1}H NMR can be used to identify the products of reactions. (Section 13.13).

Solutions to Problems

13.1

$$E = \frac{1.20 \times 10^{-4} \, \text{kJ/mol}}{\lambda(\text{in m})}$$

$$\lambda = \frac{c}{v} = \frac{3.0 \times 10^8 \, \text{m/s}}{v}; \quad v = 187 \, \text{MHz} = 1.87 \times 10^8 \, \text{Hz}$$

$$\lambda = \frac{3.0 \times 10^8 \, \text{m/s}}{1.87 \times 10^8 \, \text{Hz}} = 1.60 \, \text{m}$$

$$E = \frac{1.20 \times 10^{-4} \, \text{kJ/mol}}{1.60} = 7.5 \times 10^{-5} \, \text{kJ/mol}$$

Compare this value with $E = 8.0 \times 10^{-5}$ kJ/mol for ^{1}H (given in the text). It takes slightly less energy to spin-flip a ^{19}F nucleus than to spin-flip a ^{1}H nucleus.

13.2

$$\lambda = \frac{c}{v} = \frac{3.0 \times 10^8 \, \text{m/s}}{v}; \quad v = 300 \, \text{MHz} = 3.0 \times 10^8 \, \text{Hz}$$

$$\lambda = \frac{3.0 \times 10^8 \, \text{m/s}}{3.0 \times 10^8 \, \text{Hz}} = 1.0 \, \text{m}$$

$$E = \frac{1.20 \times 10^{-4} \, \text{kJ/mol}}{1.0} = 1.20 \times 10^{-4} \, \text{kJ/mol}$$

Increasing the spectrometer frequency from 200 MHz to 300 MHz increases the amount of energy needed for resonance.

13.3

2-Chloropropene has three kinds of protons. Protons b and c differ because one is cis to the chlorine and the other is trans.

13.4

$$\delta = \frac{\text{Observed chemical shift (in Hz)}}{200 \, \text{MHz}}$$

(a) $\quad \delta = \dfrac{1454 \, \text{Hz}}{200 \, \text{MHz}} = 7.27 \, \delta$ for $CHCl_3$

(b) $\quad \delta = \dfrac{610 \, \text{Hz}}{200 \, \text{MHz}} = 3.05 \, \delta$ for CH_3Cl

(c) $\quad \delta = \dfrac{693 \, \text{Hz}}{200 \, \text{MHz}} = 3.46 \, \delta$ for CH_3OH

(d) $\quad \delta = \dfrac{1060 \, \text{Hz}}{200 \, \text{MHz}} = 5.30 \, \delta$ for CH_2Cl_2

13.5   **(a)** $\delta = \dfrac{\text{Observed chemical shift (\#Hz away from TMS)}}{\text{Spectrometer frequency in MHz}}$

Units of δ are parts per million. In this problem, $\delta = 2.1$ ppm

$2.1 \text{ ppm} = \dfrac{\text{Observed chemical shift}}{200 \text{ (MHz)}}$

$420 \text{ Hz} = \text{Observed chemical shift}$

(b)   If the ^{1}H NMR spectrum of acetone were recorded at 500 MHz, the position of absorption would still be 2.1 δ because measurements given in ppm or δ units are independent of the operating frequency of the NMR spectrometer.

(c) $2.1 \delta = \dfrac{\text{Observed chemical shift}}{500 \text{ (MHz)}}$;   Observed chemical shift $= 1050$ Hz

13.6

(a)

Methylcyclopentane

Four resonance lines are observed because of symmetry.

(b)

1-Methylcyclohexene

Seven lines are seen because no two carbons are equivalent

(c)

1,2-Dimethylbenzene

Four resonance lines are seen.

(d)

2-Methyl-2-butene

Five resonance lines are observed. Carbons 1 and 2 are nonequivalent because of the double bond stereo-chemistry.

(e)

Five resonance lines are seen.

(f)

Seven resonance lines are seen.

13.7   Many other structures can be drawn.

(a)

(b)

$$\begin{array}{l} \text{H}_3\text{C } 5 \\ \quad \backslash \\ \quad 4 \text{ CHCH}_2\text{CH}_2\text{CH}_3 \\ \quad / \qquad 3 \quad 2 \quad 1 \\ \text{H}_3\text{C } 5 \end{array}$$

Two of the 6 carbons are equivalent.

(c)

$$\begin{array}{l} \quad 3 \\ \text{H}_3\text{C} \\ \quad \backslash \\ \quad 2 \text{ CHCH}_2\text{Cl} \\ \quad / \quad 1 \\ \text{H}_3\text{C} \\ \quad 3 \end{array}$$

Two of the 4 carbons are equivalent.

13.8   Methyl propanoate has 4 unique carbons, and each one absorbs in a specific region of the ^{13}C spectrum. The absorption (4) has the lowest value of δ and occurs in the $-\text{CH}_3$ region of the ^{13}C spectrum. Absorption (3) occurs in the $-\text{CH}_2-$ region. The methyl group (1) is next to an electronegative atom and absorbs downfield from the other two absorptions. The carbonyl carbon (2) absorbs the farthest downfield.

	δ (ppm)	Assignment
	9.3	4
$$\begin{array}{c} \text{O} \\ \| \\ \text{CH}_3\text{CH}_2\text{COCH}_3 \\ \text{4} \quad \text{3} \quad \text{2} \quad \text{1} \end{array}$$	27.6	3
	51.4	1
	174.6	2

13.9   The top spectrum shows all eight ^{13}C NMR peaks. The middle spectrum (DEPT-90) shows only peaks due to CH carbons. From the DEPT-90 spectrum, the absorption at 124 δ can be assigned to the vinyl carbon (5), and the absorption at 68 δ can be assigned to the $-\text{OH}$ carbon (2).

The DEPT-135 spectrum shows all but the quaternary carbon (6), which appears in the top spectrum at 132 δ. The top half of the DEPT-135 spectrum shows absorptions due to CH_3 carbons and CH carbons (which we have already identified). The 3 remaining peaks on the top of the DEPT-135 spectrum are due to methyl groups. Although we haven't learned enough to distinguish among these peaks, the peak at 23 δ is due to carbon (1). The other two peaks arise from carbons (7) and (8) (18 δ, 26 δ).

The bottom half of the DEPT-135 shows the two CH_2 carbons. Carbon (3) absorbs at 39 δ (negative), and carbon (4) absorbs at 24 δ (negative).

	Carbon	Chemical Shift (δ)
	1	23
	2	68
	3	39 (negative)
	4	24 (negative)
	5	124
	6	132
	7, 8	18, 26

6-Methyl-5-hepten-2-ol

13.10 Identify the carbons as CH_3, CH_2, CH or quaternary, and use Figure 13.7 to find approximate values for chemical shifts. (When an actual spectrum is given, it is easier to assign the carbons to the chemical shifts.) Remember: DEPT-90 spectra identify CH carbons, and DEPT-135 spectra identify CH_3 carbons (positive peaks), CH carbons (positive peaks already identified), and CH_2 carbons (negative peaks). Quaternary carbons are identified in the broadband-decoupled spectrum, in which all peaks appear.

Carbon	Chemical Shift (δ)	DEPT-90?	DEPT-135?
1	10–30	no	yes (positive)
2	30–50	no	yes (negative)
3	160–220	no	no
4	110–150	yes	yes (positive)
5	110–150	no	no
6	10–30	no	yes (positive)
7	50–90	no	yes (positive)

13.11 Always start this type of problem by calculating the degree of unsaturation of the unknown compound. $C_{11}H_{16}$ has 4 degrees of unsaturation. Since the unknown hydrocarbon is aromatic, a benzene ring accounts for all four degrees of unsaturation.

Next, look for elements of symmetry. Although the molecular formula indicates 11 carbons, only 7 peaks appear in the ^{13}C NMR spectrum, indicating a plane of symmetry. Four of the 7 peaks are due to aromatic carbons, indicating a benzene ring that is probably monosubstituted. (Prove to yourself that a monosubstituted benzene ring has 4 different kinds of carbons).

The DEPT-90 spectrum shows that 3 of the kinds of carbons in the aromatic ring are CH carbons. The positive peaks in the DEPT-135 spectrum include these three peaks, along with the peak at 29.5 δ, which is due to a CH_3 carbon. The negative peak in the DEPT-135 spectrum is due to a CH_2 carbon.

Two peaks remain unidentified and are thus quaternary carbons; one of them is aromatic.

At this point, the unknown structure is a monosubstituted benzene ring with a substituent that contains CH_2, C, and CH_3 carbons. A structure for the unknown compound that satisfies all data:

13.12

The two possible products are easy to distinguish by using ^{13}C NMR. 2-Bromo-1-hexene, the actual product formed, shows no peaks in its DEPT-90 ^{13}C NMR spectrum because it has no CH carbons. The other possible product, 1-bromo-1-hexene, shows 2 peaks in its DEPT-90 spectrum.

13.13 First, check for protons that are unrelated (none appear in this problem). Next, look for molecules that already have chirality centers. Replacement of a $-CH_2-$ proton by X in (d) and (e) produces a second chirality center, and the two possible replacement products are diastereomers. Thus, the indicated protons in (d) and (e) are diastereotopic.

Finally, for the other molecules, mentally replace each of the two hydrogens in the indicated set with X, a different group. In (a), the resulting products are enantiomers, and the protons are enantiotopic. Replacement of the protons in (b) produces two chirality centers (the carbon bearing the hydroxyl group is now chiral) and the indicated protons are diastereotopic. Replacement of one of the methyl protons in each of the groups in (c) produces a pair of double-bond isomers that are diastereomers; these protons are diastereotopic. The protons in (f) are homotopic, producing only one signal.

(a)   enantiotopic   (b)   diastereotopic   (c)   diastereotopic

(d)   Diastereotopic   (e)   Diastereotopic   (f)   Homotopic

13.14

	Compound	Kinds of non-equivalent protons		Compound	Kinds of non-equivalent protons
(a)	CH₃CH₂Br (1, 2)	2	(b)	CH₃OCH₂CHCH₃ with CH₃ (1, 2, 3, 4)	4
(c)	CH₃CH₂CH₂NO₂ (1, 2, 3)	3	(d)	(aromatic ring, positions 1, 2, 3, 4)	4
(e)	(alkene)	5	(f)	(alkene)	3

(a) CH_3CH_2Br (1, 2)   —   2

(b) $CH_3OCH_2CHCH_3$ with CH_3 branch   (1, 2, 3, 4)   —   4

(c) $CH_3CH_2CH_2NO_2$ (1, 2, 3)   —   3

(d)   aromatic ring substituted with CH_3 —   4

(e)
$$H_3C \backslash\ C=C\ /H$$
CH_3CH_2 (1), H_3C (3), H (4), H (5)   —   5

The two vinylic protons are nonequivalent.

(f)
CH_3CH_2 (1, 2) — $C=C$ — CH_2CH_3 (2, 1), H (3), H (3)   —   3

plane of symmetry

13.15

diasterotopic

(S)-Malate

Because (S)-malate already has a chirality center (starred), the two protons next to it are diastereotopic and absorb at different values. The ¹H NMR spectrum of (S)-malate has four absorptions.

13.16

	Compound	δ	Kind of proton
(a)	C_6H_{12}	1.43	secondary alkyl
(b)	CH_3COCH_3	2.17	methyl ketone
(c)	C_6H_6	7.37	aromatic
(d)	CH_2Cl_2	5.30	protons adjacent to two halogens
(e)	OHCCHO	9.70	aldehyde
(f)	$(CH_3)_3N$	2.12	methyl protons adjacent to nitrogen

13.17

Proton	δ	Kind of proton
1	1.0	primary alkyl
2	1.8	allylic
3	6.1	vinylic
4	6.3	vinylic (different from proton 3)
5	7.2	aromatic
6	6.8	aromatic
7	3.8	ether

This compound has seven different kinds of protons. Notice that the two protons labeled 5 are equivalent, as are the two protons labeled 6 because of rotation around the bond joining the aromatic ring and the alkenyl side chain.

13.18

p-Xylene

There are two absorptions in the ^1H NMR spectrum of *p*–xylene. The four ring protons absorb at 7.05 δ, and the six methyl-group protons absorb at 2.23 δ. The peak ratio of methyl protons:ring protons is 3:2.

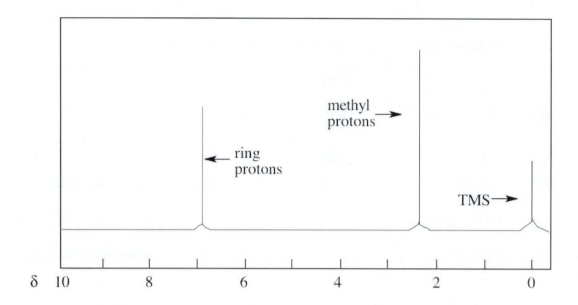

13.19

	Compound	Protons	Number of Adjacent Protons	Splitting
(a)	$\overset{1}{C}HBr_2\overset{2}{C}H_3$	1	3	quartet
		2	1	doublet
(b)	$\overset{1}{C}H_3O\overset{2}{C}H_2\overset{3}{C}H_2Br$	1	0	singlet
		2	2	triplet
		3	2	triplet
(c)	$\overset{1}{C}lCH_2\overset{2}{C}H_2\overset{1}{C}H_2Cl$	1	2	triplet
		2	4	quintet
(d)	$\overset{1}{H_3C}\ \ \overset{}{O}$ $\overset{1}{C}H_3\overset{2}{C}H\overset{\parallel}{C}O\overset{3}{C}H_2\overset{4}{C}H_3$	1	1	doublet
		2	6	septet
		3	3	quartet
		4	2	triplet
(e)	$\overset{4}{\ \ }$ $O\ \ CH_3$ $\overset{1}{C}H_3\overset{2}{C}H_2\overset{\parallel}{C}O\overset{3}{C}H\overset{4}{C}H_3$	1	2	triplet
		2	3	quartet
		3	6	septet
		4	1	doublet

	Compound	Protons	Number of Adjacent Protons	Splitting
(f)		1	2	triplet
		2	1	doublet
		3	1	multiplet
		4	1	multiplet

The splitting patterns for protons 3 and 4 are complex and are not explained in the text.

13.20 Calculate the degree of unsaturation, and note the number of peaks to see if symmetry is present.

(a) This compound has no degrees of saturation and only one kind of hydrogen. The only possible structure is CH_3OCH_3.

(b) Again, this compound has no degrees of unsaturation and has two kinds of hydrogens. The compound is 2-chloropropane.

(c) This compound, with no degrees of unsaturation, has two different kinds of hydrogen, each of which has two neighboring hydrogens.

(d) $C_4H_8O_2$; one degree of unsaturation and 3 different kinds of hydrogen.

(a) CH_3OCH_3 (b)

$$CH_3\overset{\overset{\displaystyle Cl}{|}}{C}HCH_3$$

(c) $ClCH_2CH_2OCH_2CH_2Cl$ (d)

$$CH_3CH_2\overset{\overset{\displaystyle O}{||}}{C}CH_3$$

$$or$$

$$CH_3\overset{\overset{\displaystyle O}{||}}{C}CH_2CH_3$$

13.21 The molecular formula ($C_4H_{10}O$) indicates that the compound has no multiple bonds or rings. The ^1H NMR spectrum shows two signals, corresponding to two types of hydrogens in the ratio 1.50:1.00, or 3:2. Since the unknown contains 10 hydrogens, four protons are of one type and six are of the other type.

The upfield signal at 1.22 δ is due to saturated primary protons. The downfield signal at 3.49 δ is due to protons on carbon adjacent to an electronegative atom – in this case, oxygen.

The signal at 1.23 δ is a triplet, indicating two neighboring protons. The signal at 3.49 δ is a quartet, indicating three neighboring protons. This splitting pattern is characteristic of an ethyl group. The compound is diethyl ether, $CH_3CH_2OCH_2CH_3$.

13.22

(*E*)-3-Bromo-1-phenyl-1-propene

Coupling of the C2 proton to the C1 vinylic proton occurs with $J = 16$ Hz and causes the signal of the C2 proton to be split into a doublet. The C2 proton is also coupled to the two C3 protons with $J = 8$ Hz. This splitting causes each leg of the C2 proton doublet to be split into a triplet, producing six lines in all. Because of the size of the coupling constants, two of the lines coincide, and a five-line multiplet is observed.

$J_{1-2} = 16$ Hz

$J_{2-3} = 8$ Hz

13.23

Focus on the ^{1}H NMR methyl group absorption. In the left product, the methyl group signal is unsplit; in the right product, it appears as a doublet. In addition, the right product shows a downfield absorption in the 2.5 δ – 4.0 δ region due to the proton bonded to a carbon that is also bonded to an electronegative atom. If you were to take the ^{1}H NMR spectrum of the reaction product, you would find an unsplit methyl group, and you could conclude that the product was 1-chloro-1-methylcyclohexane.

Visualizing Chemistry

13.24

(a)

1. doublet
2. septet
3. singlet

(b)

1. singlet
2. doublet
3. doublet
4. doublet
5. triplet

13.25 None of the hydrogens or carbons are equivalent.

¹H NMR: 5 signals   **¹³C NMR**: 7 signals

13.26 The compound has 5 different types of carbons and 4 different types of hydrogens.

¹³C

¹H

13.27 If you assign *R, S* configurations to the two carbons bonded to the methyl group, it is apparent that *cis*-1, 2-dimethylcyclohexane is a meso compound. When the cyclohexane ring undergoes a ring-flip, the ring passes through an intermediate that has a plane of symmetry. Both the ^{13}C NMR spectrum and the ^{1}H NMR spectrum show 4 peaks.

13.28 (a) Because cysteine has a chirality center, the indicated protons are diastereotopic.

(b) Imagine replacing first one, then the other, of the indicated protons with a substituent X. The two resulting compounds would be enantiomers. The protons are thus enantiotopic.

Additional Problems

Chemical Shifts and NMR Spectroscopy

13.29 $\delta = \dfrac{\text{Observed chemical shift (in Hz)}}{200 \text{ MHz}}$

(a) 2.18 δ   (b) 4.78 δ   (c) 7.52 δ

13.30 δ × 300 MHz = Observed chemical shift (in Hz)

(a) 30 Hz   (b) 1035 Hz   (c) 1890 Hz   (d) 2310 Hz

13.31 (a) Since the symbol "δ" indicates ppm downfield from TMS, chloroform absorbs at 7.3 ppm.

(b) $\delta = \dfrac{\text{Observed chemical shift (in Hz)}}{\text{Spectrometer frequency in MHz}}$

$7.3 \text{ ppm} = \dfrac{\text{chemical shift}}{360 \text{ MHz}}$; 7.3 ppm × 360 MHz = chemical shift

2600 Hz = chemical shift

(c) The value of δ is still 7.3 because the chemical shift measured in δ is independent of the operating frequency of the spectrometer.

13.32 ^{13}C NMR absorptions occur over a range of 250 ppm, while ^{1}H NMR absorptions generally occur over a range of only 10 ppm. The spread of peaks in ^{13}C NMR is therefore much greater, so accidental overlap is less likely. In addition, normal ^{13}C NMR spectra are uncomplicated by spin–spin splitting, and the total number of lines is smaller.

13.33 A nucleus that absorbs at 6.50 δ is less shielded than a nucleus that absorbs at 3.20 δ and thus requires a weaker applied field to come into resonance. A shielded nucleus feels a smaller effective field, and a stronger applied field is needed to bring it into resonance.

¹NMR Spectroscopy

13.34

	Compound	Kinds of non-equivalent protons		Compound	Kinds of non-equivalent protons
(a)		4	(b)	CH₃CH₂CH₂OCH₃	4
(c)		2	(d)		6
(e)		5			

13.35

Lowest Chemical Shift ⟶ Highest Chemical Shift

CH₄	<	Cyclohexane	<	CH₃COCH₃	<	CH₂, H₂C=CH₂	<	Benzene
0.23		1.43		2.17		5.30 5.33		7.37

13.36

(a)

¹³C: 2 absorptions
¹H: 1 absorption

(b)

¹³C: 5 absorptions
(at room temperature)

¹H: 3 absorptions
(at room temperature)

(c)

¹³C: 2 absorptions
¹H: 1 absorption

(d)

¹³C: 4 absorptions

¹H: 2 absorptions

(e)

H$_3$C—⟨benzene ring⟩—CH$_3$

^{13}C: 3 absorptions
^{1}H: 2 absorptions

(f)

H$_3$C   CH$_3$
   ⟨cyclopropane⟩

^{13}C: 3 absorptions
^{1}H: 2 absorptions

13.37

(a) (CH$_3$)$_4$C   (b) ⟨cyclopentane⟩   (c) ⟨1,4-dioxane ring with two O⟩

13.38

Compound	Number of peaks	Peak Assignment	Splitting Pattern	
(a) $\overset{1}{(CH_3)_3}\overset{2}{CH}$	2	1	doublet	(9H)
		2	multiplet (dectet)	(1H)
(b) $\overset{1}{CH_3}\overset{2}{CH_2}\overset{O}{\overset{\|}{C}}\overset{3}{OCH_3}$	3	1	triplet	(3H)
		2	quartet	(2H)
		3	singlet	(3H)
(e) $\begin{matrix} \overset{2}{H} & \overset{1}{CH_3} \\ \backslash & / \\ C=C \\ / & \backslash \overset{2}{} \\ \overset{1}{H_3C} & H \end{matrix}$	2	1	doublet	(6H)
		2	quartet	(2H)

13.39

	Peak Assignment	Splitting Pattern	
$\overset{1}{CH_3}\overset{2}{CH_2}\overset{O}{\overset{\|}{C}}\overset{3}{O}\overset{}{CH}(\overset{4}{CH_3})_2$	1	triplet	(3H)
	2	quartet	(2H)
	3	septet	(1H)
	4	doublet	(6H)

13.40 (a) enantiotopic   (b) diastereotopic   (c) diastereotopic

Refer to Problem 13.13 for help. The protons in (c) are diastereotopic because the molecules that result from replacement of the indicated hydrogens are diastereomers (prove it to yourself with models).

13.41 (a)   homotopic   (b)   enantiotopic   (c)   diastereotopic

13.42 Use of ^{13}C NMR to distinguish between the two isomers has been described in the text in Section 13.7. ^{1}H NMR can also be useful.

Isomer **A** has only four kinds of protons because of symmetry. Its vinylic proton absorption (4.5–6.5 δ) represents two hydrogens. Isomer **B** contains six different kinds of protons. Its ^{1}H NMR shows an unsplit methyl group signal and one vinylic proton signal of relative area 1. These differences make it possible to distinguish between **A** and **B**.

13.43 First, check each isomer for structural differences that are recognizable in the ^{1}H NMR spectrum. If it's not possible to pick out distinguishing features immediately, it may be necessary to sketch an approximate spectrum of each isomer for comparison.

(a)   CH₃CH=CHCH₂CH₃ has two vinylic protons with chemical shifts at 5.4–5.5 δ. Because ethylcyclopropane shows no signal in this region, it should be easy to distinguish one isomer from the other.

(b)   CH₃CH₂OCH₂CH₃ has two kinds of protons, and its ^{1}H NMR spectrum consists of two peaks — a triplet and a quartet. CH₃OCH₂CH₂CH₃ has four different types of protons, and its spectrum is more complex. In particular, the methyl group bonded to oxygen shows an unsplit singlet absorption.

(c)   Each compound shows three peaks in its ^{1}H NMR spectrum. The ester, however, shows a downfield absorption due to the −CH₂− hydrogens next to oxygen. No comparable peak shows in the spectrum of the ketone.

(d)   Each isomer contains four different kinds of protons — two kinds of methyl protons and two kinds of vinylic protons. For the first isomer, the methyl peaks are both singlets, whereas for the second isomer, one peak is a singlet and one is a doublet.

13.44

(a)

$$(CH_3)_2CHCCH_3$$

with labels: 1, 3 (on the $C=O$), 2

    1 = 0.95 δ (isopropyl group)
    2 = 2.10 δ (methyl ketone)
    3 = 2.43 δ (isopropyl group)

(b)

$$\begin{array}{c} {}^{1}H_3C \\ Br \end{array} C=C \begin{array}{c} H \\ H \end{array} \Bigg\} 2.3$$

    1 = 2.32 δ (methyl group attached to double bond)
    2,3 = 5.35 δ, 5.54 δ (vinylic H)

13.45

(a)

$$CH_3CHCH_2Br$$
with CH₃ on the central carbon

(b)

$$CH_3CHCH_2CH_2Cl$$
with Cl on the central carbon

¹³C NMR Spectroscopy

13.46

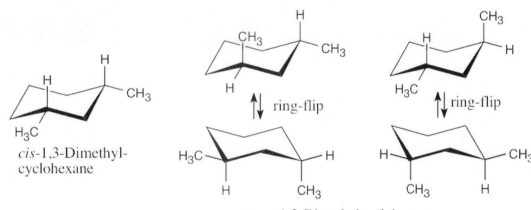

cis-1,3-Dimethyl-cyclohexane

trans-1,3-Dimethylcyclohexane

cis-1,3-Dimethylcyclohexane is a meso compound. Because of symmetry, it shows 5absorptions in its ^{13}C NMR spectrum. trans-1,3-Dimethylcyclohexane exists as a pair of enantiomers, which, at room temperature, undergo ring-flips that average the absorptions due to nonequivalent carbons. Like the cis isomer, the racemic mixture of trans enantiomers shows 5 absorptions in its ^{13}C spectrum.

13.47 – 13.48

Compound	Number of ^{13}C Absorptions	Carbons Showing Peaks in DEPT-135 ^{13}C NMR Spectrum		
		Positive Peaks	Negative Peaks	No Peaks
(a)	5	carbon 1	carbons 3,4,5	carbon 2
(b) $\overset{1}{C}H_3\overset{2}{C}H_2\overset{3}{O}CH_3$	3	carbons 1,3	carbon 2	
(c)	6	carbons 1,3	carbons 4,5,6	carbon 2
(d)	6	carbons 1,3,4,6	carbon 5	carbon 2
(e)	4	carbons 1,2	carbons 3,4	

	Number of ^{13}C	Carbons Showing Peaks in DEPT-135 ^{13}C NMR Spectrum		
Compound	Absorptions	Positive Peaks	Negative Peaks	No Peaks
(f)	4		carbons 2,3,4	carbons 1

13.49 Either ^1H NMR or ^{13}C NMR can be used to distinguish among these isomers. In either case, it is first necessary to find the number of different kinds of protons or carbon atoms.

Compound	Kinds of Protons	Kinds of Carbon atoms	Number of ^1H NMR peaks	Number of ^{13}C NMR peaks
H_2C-CH_2 / H_2C-CH_2	1	1	1	1
$H_2C=CHCH_2CH_3$	5	4	5	4
$CH_3CH=CHCH_3$	2	2	2	2
$(CH_3)_2C=CH_2$	2	3	2	3

^{13}C NMR is the simplest method for identifying these compounds because each isomer differs in the number of absorptions in its ^{13}C NMR spectrum. ^1H NMR can also be used to distinguish among the isomers because the two isomers that show two ^1H NMR peaksdiffer in their splitting patterns.

13.50

	Number of Peaks		Distinguishing Absorptions
	^{13}C	7	Two vinylic peaks
	^1H	5	Unsplit vinylic peak, relative area 1
	^{13}C	5	One vinylic peak
	^1H	4	Split vinylic peak, relative area 2

The two isomers have different numbers of peaks in both ^1H NMR and ^{13}C NMR. In addition, the distinguishing absorptions in the vinylic region of both the ^1H and ^{13}C spectra make it possible to identify each isomer by its NMR spectrum.

The ketone IR absorption of 3-methyl-2-cyclohexenone occurs near 1690 cm^{-1} because the double bond is one bond away from the ketone group. The ketone IR absorption of 3-cyclopentenyl methyl ketone occurs near 1715 cm^{-1}, the usual position for ketone absorption.

13.51

Carbon	δ (ppm)
1	14
2	61
3	166
4	
5	
6	127-133 (4 peaks)
7	

Ethyl benzoate

General Problems

13.52 (a),(b)   C$_3$H$_6$O contains one double bond or ring. Possible structures for C$_3$H$_6$O include:

H$_2$C—CH$_2$ / H$_2$C—O
Cyclic ether

H$_2$C—CHCH$_3$ (epoxide with O)
Cyclic ether

H$_2$C=CHOCH$_3$
Ether, double bond

H$_2$C=CHCH$_2$OH
Alcohol, double bond

H$_2$C / H$_2$C >C< OH / H
Cyclic alcohol

CH$_3$CCH$_3$ (C=O)
Ketone (acetone)

CH$_3$CH$_2$CH (C=O)
Aldehyde

(c)   Saturated ketones absorb at 1715 cm^{-1} in the infrared. Only the last two compounds above show an infrared absorption in this region.

(d)   Because the aldehyde from part (b) has three different kinds of protons, its ^{1}H NMR spectrum shows three peaks. The ketone, however, shows only one peak. Since the unknown compound of this problem shows only one ^{1}H NMR absorption (in the methyl ketone region), it must be acetone.

13.53   The unknown compound has no degrees of unsaturation and has two different kinds of hydrogens. The unknown compound is BrCH$_2$CH$_2$CH$_2$Br.

13.54 Possible structures for C$_4$H$_7$ClO$_2$ (one degree of unsaturation) are CH$_3$CH$_2$CO$_2$CH$_2$Cl and ClCH$_2$CO$_2$CH$_2$CH$_3$. Chemical shift data can distinguish between them.

O
‖
CH$_3$CH$_2$COCH$_2$Cl
A

O
‖
ClCH$_2$COCH$_2$CH$_3$
B

In **A**, the protons attached to the carbon bonded to both oxygen and chlorine (−OCH$_2$Cl) are expected to absorb far downfield (5.0–6.0 δ). Because no signal is present in this region of the ^{1}H NMR spectrum given, the unknown must be **B**. In addition, the quartet absorbing at 4.26 δ is typical of a CH$_2$ group next to an electronegative atom and coupled with a methyl group.

13.55

(a)

1 = 2.18 δ (allylic)
2 = 4.16 δ (H,Cl bonded to same C)
3 = 5.71 δ (vinylic)

(b)

1 = 1.30 δ (saturated)
2 = 7.30 δ (aromatic)

(c)

O
3   2 ‖   1
BrCH$_2$CH$_2$CCH$_3$

1 = 2.11 δ (next to C=O)
2 = 3.52 δ (next to C=O, Br)
3 = 4.40 δ (H,Br bonded to same C)

(d)

2   1   3
CH$_2$CH$_2$CH$_2$Br

1 = 2.15 δ
2 = 2.75 δ (benzylic)
3 = 3.38 δ (H,Br bonded to same C)
4 = 7.22 δ (aromatic)

The *E* isomer is also a satisfactory answer

In (b) and (d), the aromatic ring hydrogens coincidentally have the same chemical shift.

13.56

CH₃O, C—H_c
H_a—C≡C—C
H_b

| Proton *a* | Proton *b* | Proton *c* |
| 3.08 δ | 4.52 δ | 6.35 δ |

$J_{a\text{-}b} = 3$ Hz
$J_{a\text{-}c} = 1$ Hz

$J_{b\text{-}c} = 7$ Hz
$J_{a\text{-}b} = 3$ Hz

$J_{b\text{-}c} = 7$ Hz
$J_{a\text{-}c} = 1$ Hz

13.57 Compound **A** (4 multiple bonds and/or rings) must be symmetrical because it exhibits only six peaks in its ^{13}C NMR spectrum. Saturated carbons account for two of these peaks (δ = 15, 28 ppm), and unsaturated carbons account for the other four (δ = 119, 129, 131, 143 ppm).

^{1}H NMR shows a triplet (3 H at 1.20 δ), and a quartet (2 H at 2.58 δ), indicating the presence of an ethyl group. The other signals (4 H at 7.07 δ , 7.39 δ are due to aromaticprotons.

Br —⟨ ⟩— CH₂CH₃   **A**

13.58

(a)   O
      ‖
CH₃CH₂CH₂CCH₃

(b)   ⟨ ⟩—CH₃ with Br

(c)   ⟨ ⟩—CH₂CH₂Br

13.59 The peak in the mass spectrum at *m/z* = 84 is probably the molecular ion of the unknown compound and corresponds to a formula of C_6H_{12} − one double bond or ring. The base peak, at *m/z* = 55, corresponds to the loss of an ethyl group.

^{13}C NMR shows three different kinds of carbons and indicates a symmetrical hydrocarbon. The absorption at 132 δ is due to a vinylic carbon atom. A reasonable structure for the unknown is 3-hexene. The data do not distinguish between cis and trans isomers.

CH₃CH₂CH=CHCH₂CH₃   3-Hexene

13.60 Compound **A**, a hydrocarbon having $M^+ = 96$, has the formula C_7H_{12}, indicating two degrees of unsaturation. Because it reacts with BH_3, Compound **A** contains a double bond. From the broadband decoupled ^{13}C NMR spectrum, we can see that C_7H_{12} is symmetrical, since it shows only five peaks.

The DEPT-135 spectrum of Compound **A** indicates three different CH_2 carbons, one $=CH_2$ carbon and one $-C=$ carbon; the last two carbons are shown to be sp^2-hybridized by their chemical shifts. In the DEPT-135 spectrum of Compound **B**, the absorptions due to double bond carbons have been replaced by a CH carbon and a CH_2 carbon bonded to anelectronegative group.

Compound A		Compound B	
1	106.9 δ	1	68.2 δ
2	149.7 δ	2	40.5 δ
3	35.7 δ	3,4,5	29.9 δ, 26.1 δ, 26.9 δ
4, 5	26.8 δ, 28.7 δ		

13.61 The IR absorption indicates that **C** is an alcohol. From M^+, we can arrive at a molecular formula of $C_5H_{10}O$, which indicates one degree of unsaturation. The broadband-decoupled spectrum shows five peaks; two are due to a double bond, and one is due to a carbon bonded to an electronegative atom (O).

The DEPT spectra show that **C** contains 4 CH_2 carbons and one CH carbon, and that **C** has a monosubstituted double bond. $HOCH_2CH_2CH_2CH=CH_2$ is a likely structure for **C**.

13.62 Compound **D** is very similar to Compound **C**. The DEPT spectra make it possible to distinguish between the isomers. **D** has 2 CH carbons, one CH_3 carbon, and 2 CH_2 carbons, and, like **C**, has a monosubstituted double bond. The peak at 74.4 δ is due to asecondary alcohol.

$$\overset{\displaystyle OH}{\underset{\displaystyle |}{CH_3CH_2CHCH}} = CH_2 \qquad \text{Compound D}$$

13.63 Compound **E**, $C_7H_{12}O_2$, has two degrees of unsaturation and has two equivalent carbons because its broadband-decoupled spectrum shows only 6 peaks. Two carbons absorb in the vinylic region of the spectrum; because one is a CH carbon and the other is a CH_2 carbon, **E** contains a monosubstituted double bond. The peak at 165.8 δ (not seen in the DEPT spectra) is due to a carbonyl group.

Carbon	δ (ppm)
1	19.1
2	28.0
3	70.5
4	165.8
5	129.8
6	129.0

Compound E

13.64

Compound F

Carbon	δ (ppm)
1	132.4
2	32.2
3,4	{ 29.3 27.6

Compound G

Carbon	δ (ppm)
1	56.0
2	39.9
3,4	{ 27.7 25.1

13.65 Make a model of one enantiomer of 3-methyl-2-butanol and orient it as a staggered Newman projection along the C2-C3 bond. The *S* enantiomer is pictured.

Because of the chirality center at C2, the two methyl groups at the front of the projection are diastereotopic. Since the methyl groups aren't equivalent, their carbons show slightly different signals in the ^{13}C NMR.

13.66 Commercial 2,4-pentanediol is a mixture of three stereoisomers: (*R,R*), (*S,S*), and (*R,S*). The meso isomer shows three signals in its ^{13}C NMR spectrum. Its diastereomers, the *R,R* and *S,S* enantiomeric pair, also show three signals, but two of these signals occur at different δ values from the meso isomer. This is expected, because diastereomers differ in physical and chemical properties. One resonance from the meso compound accidentally overlaps with one signal from the enantiomeric pair.

13.67 The product (M⁺= 88) has the formula $C_4H_8O_2$. The IR absorption indicates that the product is an ester. The ^1H NMR shows an ethyl group and an $-OCH_3$ group.

$$CH_3CH_2\overset{\overset{\displaystyle O}{\|}}{C}OH \xrightarrow[\text{H}^+ \text{ catalyst}]{CH_3OH} CH_3CH_2\overset{\overset{\displaystyle O}{\|}}{C}OCH_3$$

13.68 The product is a methyl ketone.

$$\underset{\underset{\displaystyle CH_3}{|}}{CH_3CH}C\equiv N \xrightarrow[\text{2. H}_3\text{O}^+]{\text{1. CH}_3\text{MgBr}} \underset{\underset{\displaystyle CH_3}{|}}{CH_3CH}\overset{\overset{\displaystyle O}{\|}}{C}CH_3$$

13.69 Ethyl 2-nitropropanoate

13.70 3-Methyl-3-buten-1-ol

13.71 Prenyl acetate (3-methyl-2-butenyl ethanoate)

Review Unit 5: Spectroscopy

Major Topics Covered (with vocabulary):

Mass Spectrometry:

cation radical mass spectrum base peak double-focusing mass spectrometer molecular ion
alpha cleavage McLafferty rearrangement dehydration MALDI ESI TOF mass analyzer

The Electromagnetic Spectrum:

electromagnetic radiation wavelength frequency hertz amplitude quanta absorption
spectrum

Infrared Spectroscopy:

wavenumber fingerprint region

Nuclear Magnetic Resonance Spectroscopy:

nuclear magnetic resonance rf energy effective magnetic field shielding downfield upfield
chemical shift delta scale FT-NMR DEPT ^{13}C NMR homotopic enantiotopic
diastereotopic integration multiplet spin-spin splitting coupling $n + 1$ rule coupling
constant tree diagram

Types of Problems:

After studying these chapters, you should be able to:

– Write molecular formulas corresponding to a given molecular ion.
– Use mass spectra to determine molecular weights and base peaks, to distinguish between hydrocarbons, and to identify selected functional groups by their fragmentation patterns.
– Calculate the energy of electromagnetic radiation, and convert from wavelength to wavenumber and *vice versa*.
– Identify functional groups by their infrared absorptions.
– Use IR and MS to monitor reaction progress.
– Calculate the relationship between delta value, chemical shift, and spectrometer operating frequency.
– Identify nonequivalent carbons and hydrogens, and predict the number of signals appearing in the ^{1}H NMR and ^{13}C NMR spectra of compounds.
– Assign resonances to specific carbons or hydrogens of a given structure.
– Propose structures for compounds, given their NMR spectra.
– Predict splitting patterns, using tree diagrams if necessary.
– Use NMR to distinguish between isomers and to identify reaction products.

Points to Remember:

*   In mass spectrometry, the molecular ion is a cation radical. Further fragmentations of the molecular ion can be of two types – those that produce a cation plus a radical, and those that produce a different cation radical plus a neutral atom. In all cases, the fragment bearing the charge – whether cation or cation radical – is the one that is detected.

*   Although mass spectrometry has many uses in research, we are interested in it for only a limited amount of data. The most important piece of information it provides for us is the molecular weight of an unknown. A mass spectrum can also show if an unknown is branched or straight-chain (branched hydrocarbons have more complex spectra than their straight-chain isomers). Finally, if we know if certain groups are present, we can obtain structural information about an unknown compound. For example if we know that a ketone is present, we can look for peaks that correspond to alpha cleavage and/or McLafferty rearrangement fragments.

*   The position of an IR absorption is related to both the strength of the bond and to the nature of the two atoms that form the bond. For example, a carbon-carbon triple bond absorbs a higher frequency than a carbon-carbon double bond, which absorbs at a higher frequency than a carbon-carbon single bond. Bonds between two atoms of significantly different mass absorb at higher frequencies than bonds between two atoms of similar mass.

*   Not all IR absorptions are due to bond stretches. Many of the absorptions in the fingerprint region of an IR spectrum are due to bending and out-of-plane motions.

*   It is confusing, but true, that larger δ values in an NMR spectrum are associated with nuclei that are less shielded, and that these nuclei require a lower field strength for resonance. Nuclei with small values of δ are more shielded and require a higher field strength for resonance.

*   Both ^{13}C NMR and ^{1}H NMR are indispensable for establishing the structure of an organic compound. ^{13}C NMR indicates if a molecule is symmetrical and shows the types of carbons in a molecule (by DEPT NMR). ^{1}H NMR shows how the carbons are connected (by spin-spin splitting) and how many protons are in the molecule (by integration). Both types of spectra show (by chemical shift) the electronic environment of the magnetic nuclei.

Self-Test:

Compound **A** is a hydrocarbon with $M^{+} = 78$. What is its molecular formula? What is its degree of unsaturation? Draw three possible formulas for **A**. The ^{13}C NMR spectrum of **A** shows 3 peaks – at 18.5 δ, 69.4 δ and 82.4 δ, and the ^{1}H NMR spectrum shows two peaks. What is the structure of **A**? What significant absorptions would you see in the IR spectrum of **A**?

Compound **B** has the molecular formula C_8H_{14}, and shows 3 peaks in its ^{1}H NMR spectrum – at 1.7 δ (6H), 2.1 δ (4H) and 4.7 δ (4H). All 3 peaks are singlets. **B** also shows an IR absorption at 890 cm^{-1}. What is a possible structure for **B**? If you're still not sure, the following peaks were observed in the ^{13}C NMR spectrum of **B**: 22 δ, 36 δ, 110 δ, 146 δ. The peaks at 36 δ and 110 δ were negative signals in the DEPT-135 spectrum, and the peak at 22 δ was a positive signal.

Compound **C** is a hydrocarbon with M$^+$ = 112. What are possible molecular formulas for **C**? The five peaks in the ^{1}H NMR spectrum of **C** are all singlets and occur at the following δ values: 0.9 δ (9 H), 1.8 δ (3 H), 1.9 δ (2 H), 4.6 δ (1 H) and 4.8 δ (1 H). An IR absorption at 890 cm^{-1} is also present What is the structure of **C**?

D

Describe the ^{13}C NMR and ^{1}H NMR spectra of **D**. For the ^{1}H NMR spectrum, include the spin-spin splitting patterns, peak areas, and positions of the chemical shifts. Give two significant absorptions that you might see in the IR spectrum. Would you expect to see products of McLafferty rearrangement in the mass spectrum of **D**? Of alpha cleavage?

Multiple choice:

1.   Which of the following formulas could not arise from a compound with $M^+ = 142$ that contains C, H, and possibly O?
(a) $C_{11}H_{10}$ (b) $C_{10}H_8O$ (c) $C_9H_{18}O$ (d) $C_8H_{14}O_2$

2.   Which of the following mass spectrum fragments is a cation, rather than a cation radical?
(a) molecular ion
(b) product of alpha cleavage
(c) product of McLafferty rearrangement
(d) product of dehydration of an alcohol

3.   Which element contributes significantly to $(M+1)^+$?
(a) N   (b) H   (c) C   (d) O

4.   In which type of spectroscopy is the wavelength of absorption the longest?
(a) NMR spectroscopy   (b) infrared spectroscopy   (c) ultraviolet spectroscopy
(d) X-ray spectroscopy

5.   Which functional group is hard to detect in an IR spectrum?
(a) aldehyde   (b) $-C\equiv CH$ (c) alcohol   (d) ether

6.   IR spectroscopy is especially useful for:
(a) determining if an alkyne triple bond is at the end of a carbon chain or is in the middle
(b) predicting the type of carbonyl group that is present in a compound
(c) deciding if a double bond is monosubstituted or disubstituted
(d) all of these situations

7.   If a nucleus is strongly shielded:
(a) The effective field is smaller than the applied field, and the absorption is shifted downfield.
(b) The effective field is larger than the applied field, and the absorption is shifted upfield.
(c) The effective field is smaller than the applied field, and the absorption is shifted upfield.
(d) The effective field is larger than the applied field, and the absorption is shifted downfield.

8.   When the operating frequency of an ^{1}H NMR spectrometer is changed:
(a) The value of chemical shift in δ and of the coupling constant remain the same.
(b) The values of chemical shift in Hz and of the coupling constant change.
(c) The value of chemical shift in Hz remains the same, but the coupling constant changes.
(d) The values of chemical shift in δ and of the coupling constant change.

9. ^{13}C NMR can provide all of the following data except:
      (a) he presence or absence of symmetry in a molecule
      (b) the connectivity of the carbons in a molecule
      (c) the chemical environment of a carbon
      (d) the number of hydrogens bonded to a carbon

10.   Which kind of carbon is detected in DEPT-90 ^{13}C NMR spectroscopy?
      (a) primary carbon   (b) secondary carbon   (c) tertiary carbon   (d) quaternary carbon

11.   The protons on carbon 3 of (*R*)-2-bromobutane are:
      (a) homotopic   (b) enantiotopic   (c) diastereotopic   (d) unrelated

Chapter 14 – Conjugated Compounds and Ultraviolet Spectroscopy

Chapter Outline

I. Conjugated Dienes (Sections 14.1–14.6).
    A. Preparation and stability of conjugated dienes (Section 14.1).
        1. Base-induced elimination of allylic halides is the most common method.
        2. The C2–C3 bond length of 1,3-butadiene is 6 pm shorter than a C−C single bond.
        3. Stability of conjugated dienes.
            a. Heats of hydrogenation show that conjugated dienes are somewhat more stable than nonconjugated dienes.
            b. Because conjugated dienes are more stable and contain less energy, they release less heat on hydrogenation.
        4. Molecular orbital description of 1,3-butadiene.
            a. The stability of 1,3-butadiene may be due to the greater amount of s character of the C−C single bond between the double bonds.
            b. Molecular orbital theory offers another explanation.
                i.   If we combine 4 adjacent p orbitals, we generate a set of 4 molecular orbitals.
                ii. Bonding electrons go into the lower two MOs.
                iii. The lowest MO has a bonding interaction between C2 and C3 that gives that bond partial double-bond character.
                iv. The π electrons of butadiene are delocalized over this entire π framework.
    B. Reactions of conjugated dienes (Sections 14.2–14.6).
        1. Electrophilic addition to conjugated dienes (Sections 14.2–14.3).
            a. Conjugated dienes react in electrophilic addition reactions to give products of both 1,2-addition and 1,4-addition (Section 14.2).
                i.   Addition of an electrophile gives an allylic carbocation intermediate that is resonance-stabilized.
                ii. Addition of the nucleophile in the second step of the reaction can occur at either end of the allylic carbocation to yield two products.
            b. The ratio of products can vary if the reaction is carried out under conditions of kinetic control or of thermodynamic control (Section 14.3).
                i.   Under conditions of kinetic control (lower temperature), the product whose formation has the lower energy of activation forms in greater amounts.
                ii. Under conditions of thermodynamic control (high temperature), the more stable product (the product whose formation has a larger negative value of $\Delta G°$) forms in greater amounts.
                iii. In electrophilic addition reactions of conjugated dienes, the 1,2 (kinetic) adduct forms preferentially at low T, and the 1,4 (thermodynamic) adduct forms preferentially at high temperature.
        2. The Diels–Alder cycloaddition reaction (Sections 14.4–14.5).
            a. How the reaction occurs (Section 14.4).
                i.   A diene can react with certain alkenes to form a cyclic product.
                ii. This reaction, the Diels–Alder reaction, forms two new C−C bonds in a single step.

   iii. The reaction occurs by a pericyclic mechanism, which takes place in a step by a cyclic redistribution of electrons.

   iv. In the reaction, σ overlap occurs between the two alkene *p* orbitals and the two *p* orbitals on carbons 1 and 4 of the diene.

   v. The two alkene carbons and C1 and C4 of the diene rehybridize from sp^2 to sp^3, and C2 and C3 of the diene remain sp^2 hybridized.

  b. The dienophile (Section 14.5).

   i. The dienophile must have an electron-withdrawing group and maycontain a triple bond.

   ii. The stereochemistry of the dienophile is maintained during the reaction.

   iii. Only endo product is formed because orbital overlap is greater in the transition state than for exo product.

    (a) A substituent in a bicyclic ring system is endo if it is syn to thel arger of the other two bridges.

  c. The diene.

   i. A diene must adopt an *s*-cis ("cis-like") conformation in order to undergo the Diels–Alder reaction.

   ii. Some dienes can rotate to achieve an *s*-cis conformation; those that are rigid can't react.

   iii. Dienes that have fixed *s*-cis geometry are very reactive.

 3. Diene polymers (Section 14.6).

  a. Like simple alkenes, conjugated dienes can polymerize.

   i. Because double bonds remain in the polymer, cis–trans isomerism is possible.

   ii. Polymerization can be initiated by either a radical or by acid.

   iii. Polymerization occurs by 1,4-addition.

  b. Natural rubber is a polymer of isoprene with *Z* double-bond stereochemistry, and *gutta-percha* is a polymer of isoprene with *E* double-bond stereochemistry.

  c. Synthetic rubber and neoprene (a polymer of chloroprene) are also diene polymers.

  d. Rubber needs to be hardened by vulcanization.

   i. Heating rubber with sulfur forms cross-links that lock the chains together.

  e. Rubber's ability to stretch and contract is due to the irregular shapes of the polymer chains.

II. Ultraviolet spectroscopy (Sections 14.7–14.9).

 A. Principles of ultraviolet spectroscopy (Section 14.7).

  1. The ultraviolet region of interest is between the wavelengths 200 nm and 400 nm.

  2. The energy absorbed is used to promote a π electron in a conjugated system from a lower-energy orbital to a higher energy orbital.

 B. Ultraviolet spectrum of 1,3-butadiene.

  1. When 1,3-butadiene is irradiated with ultraviolet light, a π electron is promoted from the highest occupied molecular orbital (HOMO) to the lowest unoccupied molecular orbital (LUMO).

     2. UV radiation of 217 nm is necessary to promote this transition.

     3. This transition is known as a $\pi \rightarrow \pi^*$ transition.

C. The ultraviolet spectrum.

     1. A UV spectrum is a plot of absorbance (A) vs. wavelength in nanometers.

         a. The absorbance is $A = \log [I_0/I]$.

         b. $I_0 =$ intensity of incident light.

         c. $I =$ intensity of transmitted light.

         d. The baseline is zero absorbance.

     2. For a specific substance, A is related to the molar absorptivity (ε).

         a. Molar absorptivity, characteristic of a specific compound, is the absorbance of a sample whose concentration is 1 mol/L with a path length of 1 cm.

         b. $A = \varepsilon \times c \times l$.

         c. The range of ε is 10,000 – 25,000 L/mol·cm.

     3. UV spectra usually consist of a single broad peak, whose maximum is λ_{max}.

D. Interpreting UV spectra (Section 14.8).

     1. The wavelength necessary for a $\pi \rightarrow \pi^*$ transition depends on the energy difference between HOMO and LUMO.

     2. By measuring this difference, it is possible to learn about the extent of conjugation in a molecule.

     3. As the extent of conjugation increases, λ_{max} increases.

     4. Different types of conjugated systems have characteristic values of λ_{max}.

E. Conjugation, color, and the chemistry of vision (Section 14.9).

     1. Compounds with extensive systems of conjugated bonds absorb in the visible range of the electromagnetic spectrum (400 – 800 nm).

     2. When "white light" strikes a conjugated molecule, the wavelength needed for excitation is absorbed, and all other light is transmitted.

Solutions to Problems

14.1   We would expect $\Delta H_{\text{hydrog}} = -126 + (-126) = -252$ kJ/mol for allene if the heat of hydrogenation for each double bond were the same as that for an isolated double bond. The measured ΔH_{hydrog}, -298 kJ/mol, is 46 kJ/mol more negative than the expected value. Thus, allene is higher in energy (less stable) than a nonconjugated diene, which in turn is less stable than a conjugated diene.

14.2

$$CH_3CH = CHCH = CH_2 \qquad 1,3\text{-Pentadiene}$$

Product	*Name*	*Results from:*
CH₃CH=CHCHCH₃ (Cl on carbon)	4-Chloro-2-pentene	1,2 addition 1,4 addition
CH₃CH₂CHCH=CH₂ (Cl)	3-Chloro-1-pentene	1,2 addition
CH₃CH₂CH=CHCH₂Cl	1-Chloro-2-pentene	1,4 addition

14.3

A and **D**, which are resonance-stabilized, are formed in preference to **B** and **C**, which are not. The positive charge of allylic carbocation **A** is delocalized over two secondary carbons, while the positive charge of carbocation **D** is delocalized over one secondary and one primary carbon. We therefore predict that carbocation **A** is the major intermediate formed, and that 4-chloro-2-pentene predominates. Note that this product results from both 1,2 and 1,4 addition.

14.4

14.5

3-Bromo-1-butene
(1,2 adduct)

+ Br⁻

1-Bromo-2-butene
(1,4 adduct)

Allylic halides can undergo slow dissociation to form stabilized carbocations (S$_N$¹ reaction). Both 3-bromo-1-butene and 1-bromo-2-butene form the same allylic carbocation, pictured above, on dissociation. Addition of bromide ion to the allylic carbocation then occurs to form a mixture of bromobutenes. Since the reaction is run under equilibrium conditions, the thermodynamically more stable 1-bromo-2-butene predominates.

14.6

1,4 adducts are more stable than 1,2 adducts because disubstituted double bonds are more stable than monosubstituted double bonds (see Chapter 7).

14.7 Draw the reactants in an orientation that shows where the new bonds will form. Form the new bonds by connecting the two reactants, removing two double bonds, and relocating the remaining double bond so that it lies between carbon 2 and carbon 3 of the diene. The substituents on the dienophile retain their trans relationship in the product. The product is a racemic mixture.

14.8   Good dienophiles have an electron-withdrawing group conjugated with a double bond.

Good dienophiles: (a) $H_2C{=}CHCCl$ with $\overset{O}{\overset{\|}{}}$ (d)

Poor dienophiles: (b) $H_2C{=}CHCH_2CH_2COCH_3$ with $\overset{O}{\overset{\|}{}}$ (c)   (e)

Compound (a) and (d) are good dienophiles because they have electron-withdrawing groups conjugated with a carbon–carbon double bond. Alkene (c) is a poor dienophile because it has no electron-withdrawing functional group. Compounds (b) and (e) are poor dienophiles because their electron-withdrawing groups are not conjugated with the double bond.

14.9   (a)   This diene has an *s*-cis conformation and should undergo Diels–Alder cycloaddition.

   (b)   This diene has an *s*-trans conformation. Because the double bonds are in a fused ring system, it is not possible for them to rotate to an *s*-cis conformation.

   (c)   Rotation can occur about the single bond of this *s*-trans diene. The resulting *s*-cis, however, has an unfavorable steric interaction of a methyl group with a hydrogen at carbon 1. Rotation to the *s*-cis conformation is possible but not favored.

s-trans
(more stable)

s-cis
(less stable)

14.10 Rotation of the diene to the *s*-cis conformation must occur in order for reaction to take place.

s-trans   *s*-cis

14.11 The initiator may be either a radical or a cation. Diene polymerization is a 1,4 addition process that forms a polymer whose monomer units have a 4 carbon chain that contains a double bond every 4 bonds.

14.12

etc.

14.13.

$$200 \text{ nm} = 200 \times 10^{-9}\text{m} = 2 \times 10^{-7}\text{m}$$

$$400 \text{ nm} = 400 \times 10^{-9}\text{m} = 4 \times 10^{-7}\text{m}$$

for $\lambda = 2 \times 10^{-7}$m:

$$E = \frac{1.20 \times 10^{-4} \text{ kJ/mol}}{\lambda \text{ (in m)}} = \frac{1.20 \times 10^{-4} \text{ kJ/mol}}{2.0 \times 10^{-7}} = 6.0 \times 10^{2} \text{ kJ/mol}$$

for $\lambda = 4 \times 10^{-7}$m:

$$E = \frac{1.20 \times 10^{-4} \text{ kJ/mol}}{\lambda \text{ (in m)}} = \frac{1.20 \times 10^{-4} \text{ kJ/mol}}{4.0 \times 10^{-7}} = 3.0 \times 10^{2} \text{ kJ/mol}$$

The energy of electromagnetic radiation in the region of the spectrum from 200 nm to 400 nm is 300–600 kJ/mol.

	UV	IR	¹H NMR (at 200 MHz)
Energy (in kJ/mol)	300–600	4.8–48	8.0×10^{-5}

The energy required for UV transitions is greater than the energy required for IR or ¹H NMR transitions.

14.14 $\varepsilon = \dfrac{A}{c \times l}$ Where ε = molar absorptivity (in L/mol·cm)

In this problem: A = absorbance

$\varepsilon = 50{,}100 = 5.01 \times 10^4$ L/mol·cm l = sample pathlength (in cm)

$l = 1.00$ cm c = concentration (in mol/L)

$A = 0.735$

$$c = \frac{A}{\varepsilon \times l} = \frac{0.735}{5.01 \times 10^4 \, \text{L/mol·cm} \times 1.00 \, \text{cm}} = 1.47 \times 10^{-5} \, \text{M}$$

14.15 All compounds having alternating single and multiple bonds should show ultraviolet absorption in the range 200–400 nm. Only compound (a) is not UV-active. All of the compounds pictured below show UV absorptions.

Visualizing Chemistry

14.16

14.17

14.18 In order to undergo Diels–Alder reaction, this *s*-trans diene would have to rotate to an *s*-cis arrangement. In an *s*-cis conformation, however, the two circled methyl groups experience steric strain by being too close to each other, preventing the molecule from adopting this conform ation. Thus, Diels–Alder reaction doesn't occur.

14.19

3-Methyl-2,4-hexadiene   4-Methyl-1,3-hexadiene   2-Ethyl-1,3-pentadiene

2-Bromo-4-methyl-3-hexene   4-Bromo-4-methyl-2-hexene

Mechanism Problems

14.20   Note: The major product of the reaction depends on the reaction temperature. At higher temperatures (> ? 40 °C) the more substituted akene (the thermodynamic product) is favored, while at lower temperatures (0 °C) the less substituted akene (the kinetic product) is favored.

(a)

Mechanism:

(b)

Mechanism:

(c)

Mechanism:

14.48 Diels–Alder reactions are reversible when the products are much more stable (of lower energy) than the reactants. In this case, the reactant is a nonconjugated diene, and the products are benzene (a stable, conjugated molecule) and ethylene.

14.49 (a) A Diels–Alder reaction between α-pyrone (diene) and the alkyne dienophile yields the following product.

The double bonds in this product are not conjugated, and a more stable product can be formed by loss of CO_2.

This process can occur in a manner similar to the reverse Diels–Alder reaction of the previous problem.

(b)

Mechanism:

14.23

Luminol dianion

Mechanism:

14.24

Additional Problems

Conjugated Dienes

14.25 All of these compounds can exhibit *E/Z* isomerism.

(a)

3-Methyl-2,4-hexadiene

(b) $H_2C\!=\!CHCH\!=\!CHCH\!=\!CHCH_3$

1,3,5-Heptatriene

(c) $CH_3CH\!=\!C\!=\!CHCH\!=\!CHCH_3$

2,3,5-Heptatriene

(d)

3-Propyl-1,3-pentadiene

14.26 Excluding double-bond isomers:

Conjugated dienes:

$CH_3CH\!=\!CHCH\!=\!CH_2$

1,3-Pentadiene

2-Methyl-1,3-butadiene

Cumulated dienes:

$CH_3CH_2CH\!=\!C\!=\!CH_2$

1,2-Pentadiene

$CH_3CH\!=\!C\!=\!CHCH_3$

2,3-Pentadiene

$H_2C\!=\!C\!=\!C(CH_3)_2$

3-Methyl-1,2-butadiene

Nonconjugated diene:

$H_2C\!=\!CHCH_2CH\!=\!CH_2$

1,4-Pentadiene

14.27

(a)

(b)

(c)

(d)

(e)

(f)

meso

enantiomers

14.28

A
tertiary/primary
allylic carbocation

addition
to carbon 1

addition
to carbon 4

B
secondary/primary
allylic carbocation

Tertiary/primary allylic carbocation **A** is more stable than secondary/primary allylic carbocation **B**. Since the products formed from the more stable intermediate predominate, 3,4-dibromo-3-methyl-1-butene is the major product of 1,2 addition of bromine to isoprene. In both cases, the product with the more substituted double bond (1,4 addition product) predominates.

14.29 Any unsubstituted cyclic 1,3-diene cyclic diene gives the same product from 1,2- and 1,4 addition. For example:

14.30

1-Phenyl-1,3-butadiene

Protonation of carbon 1:

Protonation of carbon 2:

Protonation of carbon 3:

Protonation of carbon 4:

Carbocation **D** is most stable because it can use the π systems of both the benzene ring and the side chain to delocalize positive charge. 3-Chloro-1-phenyl-1-butene is the major product because it results from cation **D** and because its double bond can be conjugated with the benzene ring to provide extra stability.

Diels–Alder Reaction

14.31 (a)

(b)

If two equivalents of cyclohexadiene are present for each equivalent of dienophile, you can also obtain a second product:

14.32 This conformation of 2,3-di-*tert*-butyl-1,3-butadiene, in which the *tert*-butyl groups have a cis relationship, suffers from steric strain due to the bulky substituents. Instead, the molecule adopts the *s*-trans conformation, which relieves the strain but does not allow Diels–Alder reaction to take place.

14.33 The diene rotates to the *s*-cis conformation. The trans relationship of the two ester groups in the dienophile is preserved in the product.

14.34

cis-1,3-Pentadiene   *trans*-1,3-Pentadiene

Both pentadienes are more stable in *s*-trans conformations. To undergo Diels–Alder reactions, however, they must rotate about the single bond between the double bonds to assume *s*-cis conformations.

cis-1,3-Pentadiene   *trans*-1,3-Pentadiene

When *cis*-1,3-pentadiene rotates to the *s*-cis conformation, steric interaction occurs between the methyl-group protons and a hydrogen on C1. Since it's more difficult for *cis*-1,3-pentadiene to assume the *s*-cis conformation, it is less reactive in the Diels–Alder reaction.

14.35 HC≡CC≡CH can't be used as a Diels–Alder diene because it is linear. The end carbons are too far apart to be able to react with a dienophile in a cyclic transition state. Furthermore, the product of Diels–Alder addition would be impossibly strained, with two *sp*-hybridized carbons in a six-membered ring.

14.36

(chemical reaction schemes)

Two different orientations of the dienophile ester group are possible, and two different products can form.

14.37 The most reactive dienophiles contain electron-withdrawing groups.

Most reactive ⟶ *Least reactive*

(NC)₂C=C(CN)₂>   H₂C=CHCHO   >   H₂C=CHCH₃> (CH₃)₂C=C(CH₃)₂
Four electron-   One electron-   One electron-   Four electron-
withdrawing groups   withdrawing group   donating group   donating groups

The methyl groups of 2,3-dimethyl-2-butene also decrease reactivity for steric reasons.

14.38 The difference in reactivity of the three cyclic dienes is due to steric factors. As the non-diene part of the molecule becomes larger, the carbon atoms at the end of the diene portion of the ring are forced farther apart. Overlap with the π system of the dienophile in the cyclic transition state is poorer, and reaction is slower.

14.39 Although an electron-withdrawing group increases the reactivity of a dienophile, it decreases the reactivity of a diene.

14.40 First, find the cyclohexene ring formed by the Diels–Alder reaction. After you locate the new bonds, you should then be able to identify the diene and the dienophile.

(a)

bonds formed   diene   dienophile

(b)

diene   +   dienophile

(c)

diene   +   dienophile   +   diene

(d)

diene   +   dienophile

14.41

(a)

(b)

(c)

Diene Polymers

14.42 A vinyl branch in a diene polymer is the result of an occasional 1,2 double bond addition to the polymer chain, rather than the usual 1,4 addition. Branching can also occur in cationic polymerization for the same reason.

$$\text{CH}_2\text{CH}=\text{CHCH}_2 \quad \text{CH}_2=\text{CH} \longrightarrow \text{CH}_2\text{CH}=\text{CHCH}_2\text{CH}_2\text{CH} \xrightarrow{\text{etc.}}$$

14.43

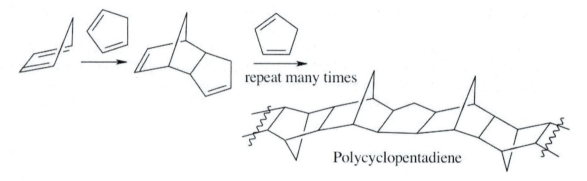

Ozone causes oxidative cleavage of the double bonds in rubber and breaks the polymer chain.

14.44 Polycyclopentadiene is the product of successive Diels–Alder additions of cyclopentadiene to a growing polymer chain. Strong heat causes depolymerization of the chain and reformation of cyclopentadiene monomer units.

repeat many times

Polycyclopentadiene

UV Spectroscopy

14.45 **Note:** In general the more conjugated double that a molecule has the longer the wavelength at which it will absorb UV light.

(a)

no conjugated double bonds   <   two conjugated double bonds   <   three conjugated double bonds

(b)

two conjugated
double bonds

<

three conjugated
double bonds

<

four conjugated
double bonds

(c)

two conjugated double bonds
and a benzene ring

<

a double bond conjugated
to a benzene ring

<

two double bonds conjugated
to a benzene ring

14.46 Only compounds having alternating multiple bonds show $\pi \rightarrow \pi^*$ ultraviolet absorptions in the 200–400 nm range. Of the compounds shown, only pyridine (b) absorbs in this range.

14.47 To absorb in the 200–400 nm range, an alkene must be conjugated. Since the double bonds of allene aren't conjugated, allene doesn't absorb light in the UV region.

14.48 The value of λ_{max} in the ultraviolet spectrum of dienes becomes larger with increasing alkyl substitution. Since energy is inversely related to λ_{max}, the energy needed to produce ultraviolet absorption decreases with increasing substitution.

Diene	# of –CH₃ groups	λ_{max}(nm)	$\lambda_{max} - \lambda_{max}$ (butadiene)
	0	217	0
	1	220	3
	1	223	6
	2	226	9
	2	227	10
	3	232	15
	4	240	23

Each alkyl substituent causes an increase in λ_{max} of 3–6 nm.

14.49

1,3,5-Hexatriene
$\lambda_{max} = 258$ nm

2,3-Dimethyl-1,3,5-hexatriene
$\lambda_{max} \approx 268$ nm

In Problem 14.41, we concluded that one alkyl group increases λ_{max} of a conjugated diene by approximately 5 nm. Since 2,3-dimethyl-1,3,5-hexatriene has two methyl substituents, its UV λ_{max} should be about 10 nm longer than the λ_{max} of 1,3,5-hexatriene.

14.50 (a) ß-Ocimene, $C_{10}H_{16}$, has three degrees of unsaturation. Catalytic hydrogenation yields a hydrocarbon of formula $C_{10}H_{22}$. ß-Ocimene thus contains three double bonds and no rings.

(b) The ultraviolet absorption at 232 nm indicates that ß-ocimene is conjugated.

(c) The carbon skeleton, as determined from hydrogenation, is:

$$\underset{\text{2,6-Dimethyloctane}}{CH_3CH_2\overset{\overset{\displaystyle CH_3}{|}}{C}HCH_2CH_2CH_2\overset{\overset{\displaystyle CH_3}{|}}{C}HCH_3}$$

Ozonolysis data are used to determine the location of the double bonds. The acetone fragment, which comes from carbon atoms 1 and 2 of 2,6-dimethyloctane, fixes the position of one double bond. Formaldehyde results from ozonolysis of a double bond at the other end of ß-ocimene. Placement of the other fragments to conform to the carbon skeleton yields the following structural formula for ß-ocimene.

$$\underset{\beta\text{-Ocimene}}{H_2C=CH\overset{\overset{\displaystyle CH_3}{|}}{C}=CHCH_2CH=\overset{\overset{\displaystyle CH_3}{|}}{C}CH_3}$$

(d)

General Problems

14.51 (a)

(b)

(c)

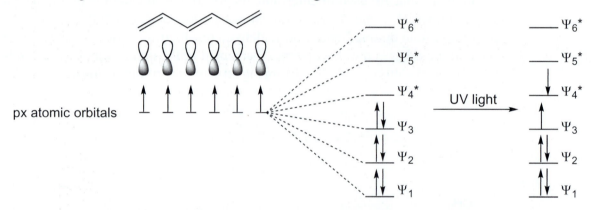

3° carbocation
more stable

2° carbocation
less stable

14.52 All of the questions can be answered with the figure below.

px atomic orbitals

Ψ_6^*

Ψ_5^*

Ψ_4^*

UV light

Ψ_6^*

Ψ_5^*

Ψ_4^*

Ψ_3

Ψ_2

Ψ_1

Ψ_3

Ψ_2

Ψ_1

(a)   The three pi-bonds that are conjugated each have two p-atomic orbitals. Thus, there are six p-atomic orbitals in the conjugated system.

(b)   The total number of molecular orbitals must equal the number of atomic orbitals used to create them. Therefore there are six molecular orbitals. (Ψ_1, Ψ_2, Ψ_3, Ψ_4, Ψ_5, Ψ_6)

(c)   The bonding molecular orbitals are lower in energy than the p-atomic orbital energies. Thus, there are three bonding orbitals (Ψ_1, Ψ_2, Ψ_3).

(d)   The bonding molecular orbitals are higher in energy than the p-atomic orbital energies. Thus, there are three antibonding orbitals (Ψ_4^*, Ψ_5^*, Ψ_6^*; antibonding orbitals are denoted with an *).

(e)   Because orbitals fill from lowest energy to highest energy and there are six electrons Ψ_1, Ψ_2 and Ψ_3 are filled.

(f)   The electron would move from the Ψ_3 (the highest occupied molecular orbital; HOMO) to Ψ_4^* (the lowest unoccupied molecular orbital; LUMO).

14.53

	CH₃CH₂C≡CCH₂H₃ 3-Hexyne	CH₃CH=CHCH=CHCH₃ 2,4-Hexadiene	CH₃CH₂CH=C=CHCH₃ 2,3-Hexadiene
¹H NMR:	2 peaks (triplet, quartet) below 2.0 δ	3 peaks, two in region 4.5–6.5 δ	5 peaks, two in region 4.5–6.5 δ
¹³C NMR:	3 peaks, 8–55 δ (2) 65–85 δ (1)	3 peaks, 8–30 δ (1) 100–150 δ (2)	6 peaks, 8–55 δ (3) 100–150 δ (2) ~200 δ (1)(*sp* carbon)
UV absorp- tion?	no	yes	no

The compounds shown are:

$CH_3CH_2C{\equiv}CCH_2H_3$ 3-Hexyne

$CH_3CH{=}CHCH{=}CHCH_3$ 2,4-Hexadiene

$CH_3CH_2CH{=}C{=}CHCH_3$ 2,3-Hexadiene

2,4-Hexadiene can easily be distinguished from the other two isomers because it is the only isomer that absorbs in the UV region. The other two isomers show significant differences in their ¹H and ¹³C NMR spectra and can be identified by either technique.

14.54

Conjugation with the oxygen lone pair electrons makes the double bond more nucleophilic

Reaction with HCl yields a cation intermediate that can be stabilized by the oxygen electrons.

Addition of Cl⁻ leads to the observed product.

There are two reasons why the other regioisomer is not formed: (1) Carbon 1 is less nucleophilic than carbon 2; (2) The cation intermediate that would result from protonation at carbon 1 can't be stabilized by the oxygen electrons and does not form.

14.55

Aldrin

14.56

Diels–Alder reaction   E2 elimination

14.57 The first equivalent of maleic anhydride adds to the *s*-cis bond of the triene.

The new double bond has an *s*-cis relationship to the remaining double bond of the triene starting material. A second equivalent of maleic anhydride adds to the diene to form the product shown.

14.58 Much of what was proven for ß-ocimene (Problem 14.43) is also true for myrcene, since both hydrocarbons have the same carbon skeleton and contain conjugated double bonds. The difference between the two isomers is in the placement of the double bonds.

The ozonolysis fragments from myrcene are 2-oxopentanedial (five carbon atoms), acetone (three carbon atoms), and two equivalents of formaldehyde (one carbon atom each). Putting these fragments together in a manner consistent with the data gives the following structural formula for myrcene:

14.59 (a) Hydrocarbon **A** must have two double bonds and two rings, since the sole ozonolysis product contains all the carbons and a diketone–dialdehyde is formed.

(b) Rotation about the central single bond of **II** allows the double bond to assume the *s-cis* conformation necessary for a Diels–Alder reaction. Rotation is not possible for **I**.

(c)

14.60

14.61 $c = \dfrac{A}{\varepsilon \times l} = \dfrac{0.065}{11{,}900 \ \text{L/mol}\cdot\text{cm} \times 1.00 \ \text{cm}} = \dfrac{6.5 \times 10^{-2}}{1.19 \times 10^{4} \ \text{L/mol}} = 5.5 \times 10^{-6} \ \text{M}$

14.62

14.63

The stereochemistry of the product resulting from Diels–Alder reaction of the (2E,4Z) diene differs at the starred carbon from that of the (2E,4E) diene. Not only is the stereochemistry of the dienophile maintained during the Diels–Alder reaction, the stereochemistry of the diene is also maintained.

14.64 Although it is usually best to work backwards in a synthesis problem, it sometimes helps to work *both* forwards and backwards. In this problem, we know that the starting materials are a diene and a dienophile. This suggests that the synthesis involves a Diels–Alder reaction. The product is a dialdehyde in which the two aldehyde groups have a cis relationship, indicating that they are the products of ozonolysis of a double bond that is part of a ring. These two pieces of information allow us to propose the following synthesis:

The −CHO groups are cis to the ester in the product.

14.65 The lone pair electrons from nitrogen can overlap with the double bond π electrons in a manner similar to the overlap of the π electrons of two conjugated double bonds. This electron contribution from nitrogen makes an enamine double bond electron-rich.

The orbital picture of an enamine shows a 4 p-electron system that resembles the system of a conjugated diene.

14.66 Double bonds can be conjugated not only with other multiple bonds but also with the lone-pair electrons of atoms such as oxygen and nitrogen. p-Toluidine has the same number of double bonds as benzene, yet its λ_{max} is 31 nm greater. The electron pair of the nitrogen atom can conjugate with the π electrons of the three double bonds of the ring, extending the π system and increasing λ_{max}.

Chapter 15 – Benzene and Aromaticity

Chapter Outline

I. Introduction to aromatic compounds (Sections 15.1–15.2).

    A. Sources of aromatic hydrocarbons (Section 15.1).

        1. Some aromatic hydrocarbons are obtained from distillation of coal tar.

        2. Other aromatic hydrocarbons are formed when petroleum is passed over a catalyst during refining.

    B. Naming aromatic compounds.

        1. Many aromatic compounds have nonsystematic names.

        2. Monosubstituted benzenes are named in the same way as other hydrocarbons, with -benzene as the parent name.

            a. Alkyl-substituted benzenes are named in two ways:

                i. If the alkyl substituent has six or fewer carbons, the hydrocarbon is named as an alkyl-substituted benzene.

                ii. If the alkyl substituent has more than six carbons, the compound is named as a phenyl-substituted alkane.

            b. The $C_6H_5CH_2-$ group is a benzyl group, and the C_6H_5- group is a phenyl group.

        3. Disubstituted benzenes are named by the ortho(*o*), meta(*m*), para(*p*) system.

            a. A benzene ring with two substituents in a 1,2 relationship is *o*-disubstituted.

            b. A benzene ring with two substituents in a 1,3 relationship is *m*-disubstituted.

            c. A benzene ring with two substituents in a 1,4 relationship is *p*-disubstituted.

            d. The *o, m, p*–system of nomenclature is also used in describing reactions.

        4. Benzenes with more than two substituents are named by numbering the position of each substituent.

            a. Number so that the lowest possible combination of numbers is used.

            b. Substituents are listed alphabetically.

        5. Any of the nonsystematic names in Table 15.1 can be used as a parent name.

    C. Structure and stability of benzene (Section 15.2).

        1. Stability of benzene.

            a. Benzene doesn't undergo typical alkene reactions.

            b. Benzene reacts slowly with Br_2 to give substitution, not addition, products.

            c. $\Delta H°_{hydrog}$ of benzene is 150 kJ/mol less than that predicted for $3 \times \Delta H°_{hydrog}$ of cyclohexene, indicating that benzene has extra stability.

        2. Structure of benzene.

            a. All carbon–carbon bonds of benzene have the same length.

            b. The electron density in all bonds is identical.

            c. Benzene is planar, with all bond angles 120°.

            d. All carbons are sp^2-hybridized and identical, and each carbon has one electron in a *p* orbital perpendicular to the plane of the ring.

            e. Resonance theory explains that benzene is a resonance hybrid of two forms.

            f. Benzene is represented in this book as one line-bond structure, rather than as a hexagon with a circle to represent the double bonds.

3. Molecular orbital picture of benzene.
   a. It is impossible to define 3 localized π bonds; the electrons are delocalized over the ring.
   b. Six molecular orbitals (MOs) can be constructed for benzene.
      i. The 3 lower-energy MOs are bonding MOs.
      ii. The 3 higher energy MOs are antibonding.
      iii. One pair of bonding orbitals is degenerate, as is one pair of antibonding orbitals.
      iv. The 6 bonding electrons of benzene occupy the 3 bonding orbitals and are delocalized over the ring.

II. Aromaticity (Sections 15.3–15.6).
   A. The Hückel $4n + 2$ rule (Section 15.3).
      1. For a compound to be aromatic, it must possess the qualities we have already mentioned and, in addition, must fulfill Hückel's Rule.
      2. Hückel's Rule: A molecule is aromatic only if it has a planar, monocyclic system of conjugation with a total of $4n + 2$ π electrons (where n is an integer).
      3. Molecules with (4, 8, 12 ...) π electrons are antiaromatic.
      4. Examples:
         a. Cyclobutadiene ($n = 4$) is antiaromatic.
         b. Benzene ($n = 6$) is aromatic.
         c. Planar cyclooctatetraene ($n = 8$) is antiaromatic.
            i. Cyclooctatetraene is stable, but its chemical behavior is like an alkene, rather than an aromatic compound.
            ii. Cyclooctatetraene is tub-shaped, and its bonds have two different lengths.
      5. Why $4n + 2$?
         a. For aromatic compounds, there is a single lowest-energy MO that can accept two π electrons.
         b. The next highest levels occur in degenerate pairs that can accept 4 π electrons.
         c. For all aromatic compounds and ions, a stable species occurs only when $(4n + 2)$ π electrons are available to completely fill the bonding MOs.
   B. Aromatic ions (Section 15.4).
      1. Any cyclic conjugated molecule with $4n + 2$ electrons can be aromatic, even if it is an ion.
      2. The cyclopentadienyl anion.
         a. Although cyclopentadiene isn't aromatic, removal of H^+ produces a six-π-electron cyclic anion that is aromatic.
         b. Cyclopentadiene has a $pK_a = 16$, indicating that a stable anion is formed on removal of H^+.
         c. Both the cyclopentadienyl cation (4 π electrons) and the cyclopentadienyl radical (5 π electrons) are unstable.
      3. The cycloheptatrienyl cation.

    a. Removal of H⁻ from cycloheptatriene produces the cycloheptatrienyl cation, which has 6 π electrons and is stable.

    b. The cycloheptatrienyl radical and anion are unstable.

C. Aromatic heterocycles (Section 15.5).

  1. A heterocycle (a cyclic compound containing one or more elements in addition to carbon in the ring) can also be aromatic.

  2. Pyridine.

    a. The nitrogen atom of pyridine contributes one π electron to the π system of the ring, making pyridine aromatic.

    b. The nitrogen lone pair is not involved with the ring π system.

  3. Pyrrole.

    a. The nitrogen of pyrrole contributes both lone-pair electrons to the ring π system, making pyrrole aromatic.

    b. The nitrogen atom makes a different contribution to the π ring system in and in pyridine.

  4. Pyrimidine and imidazole rings are important in biological chemistry.

D. Polycyclic aromatic compounds (Section 15.6).

  1. Although Hückel's Rule strictly applies only to monocyclic compounds, some polycyclic compounds show aromatic behavior.

  2. Naphthalene has a Hückel number of π electrons and shows chemical and physical properties common to aromatic compounds.

  3. There are many heterocyclic analogs of naphthalene.

    i. Tryptophan, adenine, and guanine are biologically important polycyclic aromatic compounds.

III. Spectroscopy of aromatic compounds (Section 15.7).

A. IR spectroscopy.

  1. A C–H stretch occurs at 3030 cm⁻¹.

  2. As many as 4 absorptions occur in the region 1450–1600 cm⁻¹.

  3. Weak absorptions are visible in the range 1660–2000 cm⁻¹.

  4. Strong absorptions in the region 690–900 cm⁻¹, due to C–H out-of-plane bending, can be used to determine the substitution pattern of an aromatic ring.

B. UV spectroscopy.

  1. The conjugated π system of an aromatic ring gives rise to an intense absorption at 205 nm and weaker absorptions in the range 255–275 nm.

C. NMR spectroscopy.

  1. ¹H NMR.

    a. Hydrogens directly bonded to an aromatic ring absorb in the region 6.5–8.0 δ.

      i. Spin–spin coupling can give information about the substitution pattern.

      ii. Aromatic protons are deshielded because the applied magnetic field sets up a ring-current, which produces a small magnetic field that reinforces the applied field outside of the ring and deshields the aromatic protons.

      iii. If protons reside on the inside of an aromatic ring system, they are strongly shielded and absorb far upfield.

      iv. The presence of a ring-current, evidenced by chemical shift, is a test of aromaticity.

  b. Benzylic protons absorb at 2.3–3.0 δ.

2. ^{13}C NMR.

  a. Aromatic carbons absorb in the range 110–140 δ.

  b. Since alkene carbons also absorb in this region, ^{13}C NMR is not uniquely useful in identifying an aromatic ring.

Solutions to Problems

15.1   An ortho disubstituted benzene has two substituents in a 1,2 relationship. A meta disubstituted benzene has two substituents in a 1,3 relationship. A para disubstituted benzene has two substituents in a 1,4 relationship.

(a) meta disubstituted   (b) para disubstituted   (c) ortho disubstituted

15.2   Remember to give the lowest possible numbers to substituents on trisubstituted rings.

(a)

m-Bromochlorobenzene

(b)

(3-Methylbutyl)benzene

(c)

p-Bromoaniline

(d)

2,5-Dichlorotoluene

(e)

1-Ethyl-2,4-dinitrobenzene

(f)

1,2,3,5-Tetramethylbenzene

15.3

(a)

p-Bromochlorobenzene

(b)

p-Bromotoluene

(c)

m-Chloroaniline

(d)

1-Chloro-3,5-dimethylbenzene

15.4

Pyridine

The electronic descriptions of pyridine and benzene are very similar. The pyridine ring is formed by the σ overlap of carbon and nitrogen sp^2 orbitals. In addition, six p orbitals, perpendicular to the plane of the ring, hold six electrons. These six p orbitals form six π molecular orbitals that allow electrons to be delocalized over the π system of the pyridine ring. The lone pair of nitrogen electrons occupies an sp^2 orbital that lies in the plane of the ring.

15.5

Cyclodecapentaene has $4n + 2$ π electrons (n = 2), but it is not flat. If cyclodecapentaene were flat, the starred hydrogen atoms would crowd each other across the ring. To avoid this interaction, the ring system is distorted from planarity.

15.6

A compound that can be described by several resonance forms has a structure that can be represented by no single form. The structure of the cyclopentadienyl anion is a hybrid of all of the above structures and contains only one kind of carbon atom and one kind of hydrogen atom. All carbon–carbon bond lengths are equivalent, as are all carbon–hydrogen bonds lengths. Both the ^{1}H NMR and ^{13}C NMR spectra show only one absorption.

15.7 When cyclooctatetraene accepts two electrons, it becomes a $(4n + 2)$ π electron aromatic ion. Cyclooctatetraenyl dianion is planar with a carbon–carbon bond angle of 135°, that of a regular octagon.

15.8   This diagram resembles Figure 15.5 but has one less antibonding orbital.

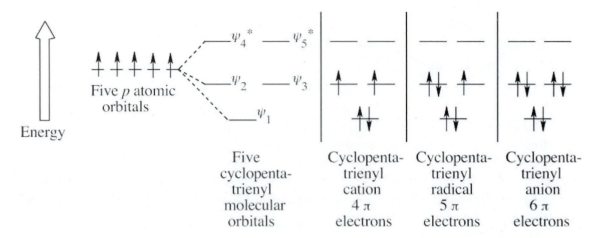

15.9   Furan is the oxygen analog of pyrrole. Furan is aromatic because it has 6 π electrons in a cyclic, conjugated system. Oxygen contributes two lone-pair electrons from a p orbital perpendicular to the plane of the ring.

Furan

15.10

The heterocyclic thiazolium ring contains six π electrons. Each carbon contributes one electron, nitrogen contributes one electron, and sulfur contributes two electrons to the ring π system. The thiazolium ion is aromatic because it has 6 π electrons in a cyclic, planar, conjugated system.

15.11

Azulene

Azulene is aromatic because it has a conjugated cyclic π electron system containing ten π electrons (a Hückel number).

15.12

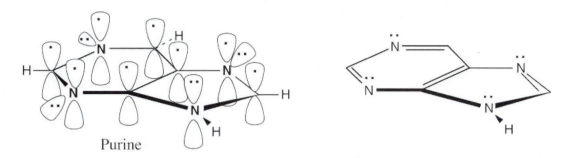

Purine

Purine is a ten-π-electron aromatic molecule. The N−H nitrogen atom in the five-membered ring donates both electrons of its lone pair to the π electron system, and each of the other three nitrogens donates one electron to the π electron system.

Visualizing Chemistry

15.13

(a)

(b)

m-Isopropylphenol

o-Nitrobenzoic acid

15.14 The all-cis cyclodecapentaene shown here is not aromatic because it is not planar. All hydrogens, however, are equivalent and show one absorption in the vinylic region of the molecule's ^{1}H NMR spectrum. If the molecule were planar and therefore aromatic, the absorption would appear between 6.5–8.0 δ.

15.15 1,6-Methanonaphthalene has ten π electrons and is sufficiently planar to behave as an aromatic molecule. The perimeter hydrogens absorb in the aromatic region of the ^{1}H NMR spectrum (6.9–7.3 δ). Interaction of the applied magnetic field with the perimeter π electrons sets up a ring current (see Section 15.7) that strongly shields the CH$_2$ protons and causes them to absorb far upfield (−0.5 δ).

15.16 Three resonance forms for the carbocation of the formula C$_{13}$H$_9$ are shown below, and more can be drawn. These forms show that the positive charge of the carbocation can be stabilized in the same way as an allylic or benzylic carbocation is stabilized – by overlap with the neighboring π electrons of the ring system.

15.17

Molecules with dipole moments are polar because electron density is drawn from one part of the molecule to another. In azulene, electron density is drawn from the seven-membered ring to the five-membered ring, satisfying Hückel's rule for both rings and producing a dipole moment. The five-membered ring resembles the cyclopentadienyl anion in having six π electrons, while the seven-membered ring resembles the cycloheptatrienyl cation. The electrostatic potential map shows that the five-membered ring is more electron-rich (red) than the seven-membered ring.

Additional Problems

Naming Aromatic Compounds

15.18

(a) 2-Methyl-5-phenylhexane

(b) *m*-Bromobenzoic acid

(c) 1-Bromo-3,5-dimethylbenzene

(d) *o*-Bromopropylbenzene

(e) 1-Fluoro-2,4-dinitrobenzene

(f) *p*-Chloroaniline

15.19

(a)

3-Methyl-1,2-benzenediamine

(b)

1,3,5-Benzenetriol

(c)

$CH_3CH_2CH_2CHCHCH_3$ with CH_3

3-Methyl-2-phenylhexane

(d)

o-Aminobenzoic acid

(e)

m-Bromophenol

(f)

2,4,6-Trinitrophenol

15.20 (a)

o-Dinitrobenzene

m-Dinitrobenzene

p-Dinitrobenzene

(b)

1-Bromo-2,3-dimethylbenzene

2-Bromo-1,3-dimethylbenzene

2-Bromo-1,4-dimethylbenzene

1-Bromo-2,4-dimethylbenzene

4-Bromo-1,2-dimethylbenzene

1-Bromo-3,5-dimethylbenzene

(c)

2,3,4-Trinitrophenol

2,3,5-Trinitrophenol

2,3,6-Trinitrophenol

2,4,5-Trinitrophenol

2,4,6-Trinitrophenol

3,4,5-Trinitrophenol

15.21 All aromatic compounds of formula C₇H₇Cl have one ring and three double bonds.

o-Chlorotoluene   *m*-Chlorotoluene   *p*-Chlorotoluene   Benzyl chloride
                                                                          (Chloromethyl)-
                                                                          benzene

15.22 Six of these compounds are illustrated and named in Problem 15.20 (b). The other eight are:

o-(Bromomethyl)toluene   m-(Bromomethyl)toluene   p-(Bromomethyl)toluene

(1-Bromoethyl)benzene   (2-Bromoethyl)benzene

o-Bromoethylbenzene   m-Bromoethylbenzene   p-Bromoethylbenzene

Structures of Organic Compounds

15.23 All compounds in this problem have four double bonds and/or rings and must be substituted benzenes, if they are to be aromatic. They may be substituted by methyl, ethyl, propyl, or butyl groups.

(a)

(b)

(c)

(d)

15.24

The bond between carbons 1 and 2 is represented as a double bond in two of the three resonance structures, but the bond between carbons 2 and 3 is represented as a double bond in only one resonance structure. The C1–C2 bond thus has more double-bond character in the resonance hybrid, and it is shorter than the C2–C3 bond. The C3–C4, C5–C6, and C7–C8 bonds also have more double-bond character than the remaining bonds.

15.25

15.26, 15.27

The circled bond is represented as a double bond in four of the five resonance forms of phenanthrene. This bond has more double-bond character and thus is shorter than the other carbon–carbon bonds of phenanthrene.

15.28

If *o*-xylene exists only as structure **A**, ozonolysis would cause cleavage at the bonds indicated and would yield two equivalents of pyruvaldehyde and one equivalent of glyoxal for each equivalent of **A** consumed. If *o*-xylene exists only as structure **B**, ozonolysis would yield one equivalent of 2,3-butanedione and two equivalents of glyoxal. If *o*–xylene exists as a resonance hybrid of **A** and **B**, the ratio of ozonolysis products would be glyoxal : pyruvaldehyde : 2,3-butanedione = 3:2:1. Since this ratio is identical to the experimentally determined ratio, we know that **A** and **B** contribute equally to the structure of *o*-xylene. Note that these data don't distinguish between the resonance hybrid structure and the alternate possibility, equilibrium between two isomeric *o*-xylenes.

Aromaticity and Hückel's Rule

15.29

The product of the reaction of 3-chlorocyclopropene with $AgBF_4$ is the cyclopropenyl cation $C_3H_3^+$. The resonance structures of the cation indicate that all hydrogen atoms are equivalent, and the ^1H NMR spectrum, which shows only one type of hydrogen atom, confirms this equivalence. The cyclopropenyl cation contains two π electrons and is aromatic according to Hückel's rule. (Here, $n = 0$.)

15.30

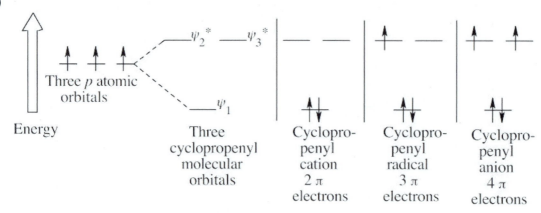

The cyclopropenyl cation is aromatic, according to Hückel's rule.

15.31

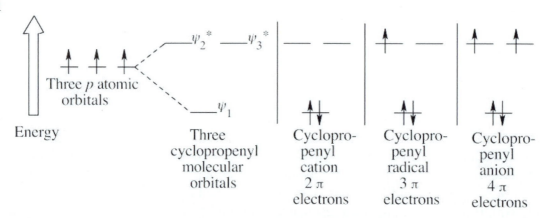

In resonance structure **A**, methylcyclopropenone is a cyclic conjugated compound with three π electrons in its ring. Because the electronegative oxygen attracts the π electrons of the carbon–oxygen π bond, however, a second resonance structure **B** can be drawn in which both carbonyl π electrons are located on oxygen, leaving only two π electrons in the ring. Since 2 is a Hückel number, the methylcyclopropenone ring is aromatic and is expected to be stable.

15.32

As in the previous problem, we can draw resonance forms in which both carbonyl π electrons are located on oxygen. The cycloheptatrienone ring in **B** contains six π electrons and is aromatic according to Hückel's rule. The cyclopentadienone ring in **D** contains four π electrons and is antiaromatic.

15.33 Check the number of electrons in the π system of each compound. The species with a Hückel (4n + 2) number of π electrons is the most stable.

cation
8 π electrons

radical
9 π electrons

anion
10 π electrons

The 10 π electron anion is the most stable.

15.34 Treat 1,3,5,7-cyclononatetraene with a strong base to remove a proton.

15.35 As with azulene, redistribution of the π electrons of calicene produces a resonance form in which both rings are aromatic and which has a dipole moment (see Problem 15.17).

15.36 Pentalene has eight π electrons and is antiaromatic. Pentalene dianion, however, has ten π electrons and is a stable, aromatic ion.

15.37

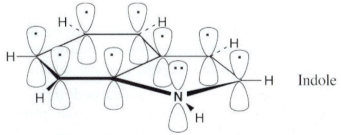   Indole

Indole, like naphthalene, has ten π electrons in two rings and is aromatic. Two π electrons come from the nitrogen atom.

15.38 The 1,2,4-triazole ring is aromatic because it has 6 π electrons in a cyclic, conjugated system.

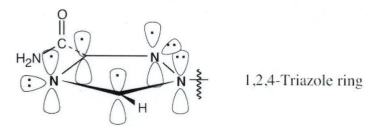

1,2,4-Triazole ring

Spectroscopy

15.39 Compound **A** has four multiple bonds and/or rings. Possible structures that yield three monobromo substitution products are:

Only structure I shows a six-proton singlet at 2.30 δ, because it contains two identical benzylic methyl groups unsplit by other protons. The presence of four protons in the aromatic region of the ^{1}H NMR spectrum confirms that I is the correct structure.

15.40 The molecular weight of the hydrocarbon (120) corresponds to the molecular formula C_9H_{12}, which indicates four double bonds and/or rings. The ^{1}H NMR singlet at 7.25 δ indicates five aromatic ring protons. The septet at 2.90 δ is due to a benzylic proton that has six neighboring protons.

CH(CH₃)₂   Isopropylbenzene

15.41 Based on degree of unsaturation, both hydrocarbons are disubstituted benzenes. The compound in (a) has two ethyl groups, and the compound in (b) has an isopropyl group and a methyl group. The IR data must be used to find the pattern of substitution.

Data in Section 15.7 show that an IR absorption of 745 cm^{-1} corresponds to an *o*-disubstituted benzene, and an absorption of 825 cm^{-1} corresponds to a *p*-disubstituted benzene.

(a) *o*-Diethylbenzene

(b) *p*-Isopropyltoluene

General Problems

15.42

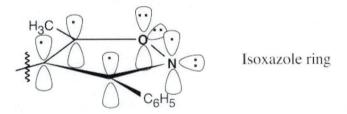

Protonation of 4-pyrone gives structure **A**, which has resonance forms **B**, **C**, **D**, **E** and **F**. In **E** and **F**, a lone pair of electrons of the ring oxygen is delocalized into the ring to produce a six π electron system, which should be aromatic according to Hückel's rule.

15.43 The isoxazole ring is aromatic for the same reasons as the 1,2,4-triazole ring is aromatic (Problem 15.38).

Isoxazole ring

15.44 The second resonance form of *N*-phenylsydnone shows the aromaticity of the five-membered ring more clearly. In this form, the ring oxygen contributes two electrons to the ring π system, each nitrogen contributes one electron and each carbon contributes one electron (the carbonyl oxygen bears a formal negative charge). This 6 π electron, cyclic conjugated system obeys Hückel's rule.

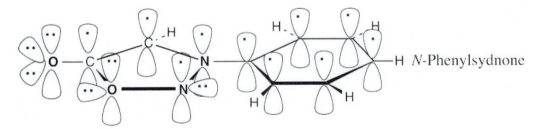

N-Phenylsydnone

15.45

The cycloheptatrienyl cation has six π electrons (a Hückel number) and is aromatic.

15.46

1-Phenyl-2-butene   1-Phenyl-1-butene

The alkene double bond is protonated to yield an intermediate carbocation, which loses a proton to give a product in which the double bond is conjugated with the aromatic ring, as shown by the increased value of λ_{max}.

15.47 All of these compounds have 4 degrees of unsaturation and are substituted benzenes. The benzylic absorptions (2.3–3.0 δ) identify the hydrogens next to the aromatic ring. Remember that the data from the IR spectrum can be used to assign the substitution pattern of the ring.

(a)

(b)

(c)

p-Bromoethylbenzene   *o*-Ethyltoluene   *p*-tert-Butyltoluene

15.48 The compound has nine degrees of unsaturation. The ^{1}H NMR spectrum shows that the compound is symmetrical and that the only absorptions occur in the vinylic and aromatic regions of the spectrum. The IR spectrum shows peaks due to a monosubstituted benzene ring and to $R_2C=CH_2$ (890 cm^{-1}).

15.49 4-Methoxyphenylacetone

15.50 2,6-Dichloroaniline

15.51

15.52

The ortho and para products predominate because the intermediate carbocation is more stabilized. The third resonance form drawn for ortho-para attack places the positive charge at the methyl-substituted carbon, which is a more stable tertiary carbocation.

15.53 **Note:** In the cyclic compounds, the charge can be placed on every atom in the ring, while in the linear compounds the charge can only reside on every other atom.

(a)

(b)

15.54 After the two triple bonds are connected a new 14-membered ring is formed. This new ring is aromatic. Because of the ring current created by the new aromatic ring the protons are shifted downfield.

15.55 The tautomer has an aromatic ring making it the more stable of the two tautomers.

15.56 (a)

Methyl Orange

(b)

C.I. Acid Red 74

Chapter 16 – Chemistry of Benzene: Electrophilic Aromatic Substitution

Chapter Outline

I. Electrophilic aromatic substitution reactions (Sections 16.1–16.3).
   A. Bromination of aromatic rings (Section 16.1).
      1. Characteristics of electrophilic aromatic substitution reactions.
         a. The accessibility of the π electrons of an aromatic ring make it a nucleophile.
         b. Aromatic rings are less reactive to electrophiles than are alkenes.
            i. A catalyst is needed to make the reacting molecule more electrophilic.
      2. Mechanism of bromination.
         a. Br_2 complexes with $FeBr_3$ to produce a positively polarized bromine.
         b. The polarized electrophile is attacked by the π electrons of the ring in a slow, rate-limiting step.
         c. The cation intermediate is doubly allylic but is much less stable than the starting aromatic compound.
         d. The carbocation intermediate loses H^+ from the bromine-bearing carbon in a fast step to regenerate an aromatic ring.
   B. Other aromatic substitution reactions (Section 16.2).
      1. Fluorination, chlorination and iodination.
         a. Fluorination is achieved by using the reagent F-TEDA-BF_4.
         b. Chlorine reacts in the presence of $FeCl_3$ to yield chlorinated rings.
         c. Iodination occurs only in the presence of an oxidizing agent.
      2. Nitration.
         a. A mixture of HNO_3 and H_2SO_4 is used for nitration.
         b. The reactive electrophile is NO_2^+.
         c. Products of nitration can be reduced with Fe or $SnCl_2$ to yield an arylamine.
      3. Sulfonation.
         a. Rings can be sulfonated by a mixture of SO_3 and H_2SO_4 to yield sulfonic acids.
         b. The reactive electrophile is either SO_3 or HSO_3^+.
         c. Sulfonation is reversible.
      4. Hydroxylation.
         a. Direct hydroxylation of an arylamine is rarely done in the laboratory.
         b. In enzyme-catalyzed biological hydroxylations, the reactive species is an "OH^+" equivalent.
   C. Alkylation of aromatic rings (Section 16.3).
      1. The Friedel–Crafts alkylation introduces an alkyl group onto an aromatic ring.
      2. An alkyl chloride, plus an $AlCl_3$ catalyst, produces an electrophilic carbocation.
      3. There are several limitations to using the Friedel–Crafts reaction.
         a. Only alkyl halides – not aryl or vinylic halides – can be used.
         b. Friedel–Crafts reactions don't succeed on rings that have amino substituents or deactivating groups.
         c. Polyalkylation is often seen.
         d. Rearrangements of the alkyl carbocation often occur.

i.   Rearrangements may occur by hydride shifts or by alkyl shifts.
D. Acylation of aromatic rings.
1. Friedel–Crafts acylation occurs when an aromatic ring reacts with a carboxylic acid chloride (ROCl).
2. The reactive electrophile is an acyl cation, which doesn't rearrange.
3. Polyacylation never occurs in acylation reactions.
II. Substituent effects in substituted aromatic rings (Sections 16.4–16.6).
A. Types of substituent effects (Section 16.4).
1. Substituents affect the reactivity of an aromatic ring.
2. Substituents affect the orientation of further substitution.
3. Substituents can be classified into three groups:
a. Ortho- and para-directing activators.
b. Ortho- and para-directing deactivators.
c. Meta-directing deactivators.
B. Explanation of substituent effects (Section 16.5).
1. All activating groups donate electrons to an aromatic ring.
2. All deactivating groups withdraw electrons from a ring.
3. Two kinds of effects are responsible for reactivity and orientation.
a. Inductive effects are due to differences in bond polarity.
b. Resonance effects are due to overlap of a p orbital of a substituent with a p orbital on an aromatic ring.
i.   Carbonyl, cyano and nitro substituents withdraw electrons.
(a) These substituents have the structure $-Y=Z$.
ii. Halogen, hydroxyl, alkoxyl and amino substituents donate electrons.
(a) These substituents have the structure $-Y:$.
iii. Resonance effects are greatest at the ortho and para positions.
c. Resonance and inductive effects don't always act in the same direction.
4. Alkyl groups – ortho- and para-directing activators.
a. Alkyl groups inductively donate electrons to a ring.
b. Alkyl groups are o,p-directors because the carbocation intermediates are best stabilized when attack occurs at the ortho and para positions.
5. $-OH$, $-NH_2$ groups – ortho- and para-directing activators.
a. $-OH$, $-NH_2$ donate electrons by resonance involving the ring and the group.
b. The intermediates of ortho- and para-attack are more stabilized by resonance than are intermediates of meta attack.
6. Halogens – ortho- and para-directing deactivators.
a. The electron-withdrawing inductive effect of halogen outweighs its electron-donating resonance effect.
b. The resonance effect orients substitution to the o,p positions.
c. The inductive effect deactivates the ring.
7. Meta-directing deactivators.

  a. Meta-directing deactivators act through both inductive and resonance effects.

  b. Because resonance effects destabilize ortho and para positions the most, substitution ion occurs at the meta position.

 C. Trisubstituted benzenes: additivity of effects (Section 16.6).

  1. If the effects of both groups are additive, the product of substitution is easy to predict.

  2. If the directing effects of the groups are opposed, the more powerful activating group determines the product, although mixtures sometimes result.

  3. For steric reasons, substitution rarely occurs between two groups that are meta to each other.

III. Other reactions of aromatic rings (Sections 16.7–16.10).

 A. Nucleophilic aromatic substitution (Section 16.7).

  1. An aryl halide with electron-withdrawing groups can undergo nucleophilic aromatic substitution.

  2. This reaction occurs through an addition/elimination mechanism.

  3. Addition of the nucleophile proceeds through an intermediate Meisenheimer complex that is stabilized by *o, p* electron-withdrawing substituents on the ring.

  4. The halide is eliminated to yield product.

 B. Benzyne (Section 16.8).

  1. At high temperatures and with strong base, halobenzenes without electron-withdrawing substituents can be converted to phenols.

  2. This reaction occurs by an elimination/addition reaction that involves a benzyne intermediate.

   a. Strong base causes elimination of HX from the aryl halide to generate benzyne.

   b. A nucleophile adds to benzyne to give the product.

  3. The benzyne intermediate can be trapped in a Diels–Alder reaction.

  4. Benzyne has the electronic structure of a distorted alkyne and has one very weak π bond.

 C. Oxidation of aromatic compounds (Section 16.9).

  1. Oxidation of alkylbenzene side chains.

   a. Strong oxidizing agents cause the oxidation of alkyl side chains with benzylic hydrogens.

   b. The products of side-chain oxidation are benzoic acids.

   c. Reaction proceeds by a complex radical mechanism.

  2. Bromination of alkylbenzene side chains.

   a. NBS brominates alkylbenzene side chains at the benzylic position.

   b. Bromination occurs by the mechanism described for allylic bromination and requires a radical initiator.

   c. The intermediate benzylic radical is stabilized by resonance.

 D. Reduction of aromatic compounds (Section 16.10).

  1. Catalytic hydrogenation of aromatic rings.

   a. It is possible to selectively reduce alkene bonds in the presence of aromatic rings because rings are relatively inert to catalytic hydrogenation.

  b. With a stronger catalyst, aromatic rings can be reduced to cyclohexanes.

 2. Reduction of aryl alkyl ketones.

  a. Aryl alkyl ketones can undergo catalytic hydrogenation to form alkylbenzenes.

  b. Acylation plus reduction is a route to alkyl substitution without rearrangement.

  c. This reaction only occurs with aryl alkyl ketones and also reduces nitro groups to amino groups.

IV. Synthesis of polysubstituted benzenes (Section 16.11).

 A. To synthesize substituted benzenes, it is important to introduce groups so that they have the proper orienting effects.

 B. It is best to use retrosynthetic analysis (work backward from the product) to plan a synthesis.

Solutions to Problems

16.1

o-Bromotoluene   *m*-Bromotoluene   *p*-Bromotoluene

16.2

The π electrons of benzene attack the fluorine of F-TEDA-BF$_4$, and the nonaromatic intermediate loses –H to give the fluorinated product.

16.3

o-Xylene

Chlorination at either position "a" of *o*-xylene yields product **A**, and chlorination at either position "b" yields product **B**.

m-Xylene

Three products might be expected to form on chlorination of *m*–xylene.

Only one product results from chlorination of p–xylene because all sites are equivalent.

16.4

Benzene can be protonated by strong acids. The resulting intermediate can lose either deuterium or hydrogen. If −H is lost, deuterated benzene is produced. Attack by deuterium can occur at all positions of the ring and leads to eventual replacement of all hydrogens by deuterium. Only the first step is shown.

16.5   Carbocation rearrangements of alkyl halides occur (1) if the initial carbocation is primary or secondary, and (2) if it is possible for the initial carbocation to rearrange to a more stable secondary or tertiary cation.

(a)   Although $CH_3 \overset{+}{C} H_2$ CH_3CH_2 is a primary carbocation, it can't rearrange to a more stable cation.

(b) $CH_3CH_2CH(Cl)CH_3$ forms a secondary carbocation that doesn't rearrange.

(c) $CH_3CH_2 \overset{+}{C} H_2$ rearranges to the more stable $CH_3 \overset{+}{C} HCH_3$ CH_3CHCH_3.

(d) $(CH_3)_3 C \overset{+}{C} H_2$ (primary) undergoes an alkyl shift to yield $(CH_3)_2 \overset{+}{C} CH_2CH_3$ (tertiary).

(e)   The cyclohexyl carbocation doesn't rearrange.

In summary:

No rearrangement: (a) CH₃CH₂Cl, (b) CH₃CH₂CH(Cl)CH₃, (e) chlorocyclohexane

Rearrangement: (c) CH₃CH₂CH₂Cl, (d) (CH₃)₃CCH₂Cl

16.6

Isobutyl carbocation
(primary)

tert-Butyl carbocation
(tertiary)

tert-Butylbenzene

The isobutyl carbocation, initially formed when 1-chloro-2-methylpropane and AlCl₃ react, rearranges via a hydride shift to give the more stable *tert*–butyl carbocation, which can then alkylate benzene to form *tert*–butylbenzene.

16.7   To identify the carboxylic acid chloride used in the Friedel–Crafts acylation of benzene, break the bond between benzene and the ketone carbon and replace it with a −Cl.

(a)

(b)

16.8   Use Figure 16.11 to find the activating and deactivating effects of groups.

Most Reactive ⟶ *Least Reactive*

(a)   Phenol > toluene > benzene > nitrobenzene

(b)   Phenol > benzene > chlorobenzene > benzoic acid

(c)   Aniline > benzene > bromobenzene > benzaldehyde

16.9   Refer to Figure 16.11 in the text for the directing effects of substituents. You should memorize the effects of the most important groups. As in Worked Example 16.2, identify the directing effect of the substituent, and draw the product.

(a)

Even though bromine is a deactivator, it is an ortho-para director.

(b)

The $-NO_2$ group is a meta-director.

(c)

(d)

No catalyst is necessary because aniline is highly activating.

16.10 An acyl substituent is deactivating. Once an aromatic ring has been acylated, it is much less reactive to further substitution. An alkyl substituent is activating, however, so an alkyl-substituted ring is more reactive than an unsubstituted ring, and polysubstitution occurs readily.

16.11 (Trifluoromethyl)benzene is less reactive toward electrophilic substitution than toluene. The electronegativity of the three fluorine atoms causes the trifluoromethyl group to be electron-withdrawing and deactivating toward electrophilic substitution. The electrostatic potential map shows that the aromatic ring of (trifluoromethyl)benzene is more electronpoor, and thus less reactive, than the ring of toluene (red).

16.12

less favored

more favored

For acetanilide, resonance delocalization of the nitrogen lone pair electrons to the aromatic ring is *less favored* because the positive charge on nitrogen is next to the positively polarized carbonyl group. Resonance delocalization to the carbonyl oxygen *is* favored because of the electronegativity of oxygen. Since the nitrogen lone pair electrons are less available to the ring than in aniline, the reactivity of the ring toward electrophilic substitution is decreased, and acetanilide is less reactive than aniline toward electrophilic substitution.

16.13

Ortho attack:

Meta attack:

Para attack:

The circled resonance forms are unfavorable, because they place two positive charges adjacent to each other. The intermediate from meta attack is thus favored.

16.14 (a)

Both groups are ortho, para directors and direct substitution to the same positions. Attack doesn't occur between the two groups for steric reasons.

(b)

Both groups are ortho, para directors, but direct to different positions. Because
−NH₂ group is a more powerful activator, substitution occurs ortho and para to it.

(c)

Both groups are deactivating, but they orient substitution toward the same positions.

16.15 (a)

Although both groups are ortho, para directors, the methyl group directs the
orientation of the substituents because it is a stronger activating group than
bromine.

(b)

The methoxyl group directs substitution to the positions ortho and para to it.

16.16   Hydroxide is used to form the nucleophilic phenoxide anion.

Meisenheimer complex

Oxyfluorfen

Step 1: Addition of the nucleophile.
Step 2: Elimination of fluoride ion.

The nitro group makes the ring electron-poor and vulnerable to attack by the nucleophilic RO⁻ group. It also stabilizes the negatively charged Meisenheimer complex.

16.17

p-Bromotoluene

p-Methylphenol   *m*-Methylphenol

m-Bromotoluene

o-Methylphenol   *m*-Methylphenol

p-Methylphenol

Treatment of *m*-bromotoluene with NaOH leads to two possible benzyne intermediates, which react with water to yield three methylphenol products.

16.18 Oxidation takes place at the benzylic position.

(a)

m-Nitrobenzoic acid

(b)

p-tert-Butylbenzoic acid

Treatment with KMnO₄ oxidizes the methyl group but leaves the *tert*-butyl group untouched.

16.19

Bond	CH₃CH₂—H	⟨benzyl⟩ CH₂—H	H₂C=CHCH₂—H
Bond dissociation energy	421 kJ/mol	375 kJ/mol	369 kJ/mol

Bond dissociation energies measure the amount of energy that must be supplied to cleave a bond into two radical fragments. A radical is thus higher in energy and less stable than the compound from which it came. Since the C—H bond dissociation energy is 421 kJ/mol for ethane and 375 kJ/mol for a methyl group C—H bond of toluene, less energy is required to form a benzyl radical than to form an ethyl radical. A benzyl radical is thus more stable than a primary alkyl radical by 46 kJ/mol. The bond dissociation energy of an allyl C—H bond is 369 kJ/mol, indicating that a benzyl radical is nearly as stable as an allyl radical.

16.20

Styrene

16.21

16.22 **(a)** In order to synthesize the product with the correct orientation of substituents, benzene must be nitrated before it is chlorinated.

m-Chloronitrobenzene

(b) Chlorine can be introduced into the correct position if benzene is first acylated. The chlorination product can then be reduced.

m-Chloroethylbenzene

(c) Friedel–Crafts acylation, followed by chlorination, reduction, and nitration, is the only route that gives a product in which the alkyl group and chlorine have a meta relationship.

4-Chloro-1-nitro-2-propylbenzene

(d) In planning this pathway, remember that the ring must be sulfonated after Friedel–Crafts alkylation because a sulfonated ring is too deactivated for alkylation to occur. Performing the reactions in this order allows the first two groups to direct bromine to the same position.

3-Bromo-2-methyl-
benzenesulfonic acid

16.23 (a)   Friedel–Crafts acylation, like Friedel–Crafts alkylation, does not occur at an aromatic ring carrying a strongly electron-withdrawing group.

(b)   There are two problems with this synthesis as it is written:

1.   Rearrangement often occurs during Friedel–Crafts alkylations using primary halides.

2.   Even if *p*-chloropropylbenzene could be synthesized, introduction of the second −Cl group would occur ortho, not meta, to the alkyl group.

A possible route to this compound:

Visualizing Chemistry

16.24 (a)   The methoxyl group is an ortho–para director.

(1)

p-Bromomethoxybenzene   *o*-Bromomethoxybenzene

(2)

p-Methoxyacetophenone   *o*-Methoxyacetophenone

(b) Both functional groups direct substituents to the same position.

(1)

3-Bromo-4-methylbenzaldehyde

(2)

3-Acetyl-4-methylbenzaldehyde

16.25 In the lowest-energy conformation of this biphenyl, the aromatic rings are tilted. If the rings had a planar relationship, steric strain between the methyl groups and the ring hydrogens on the second ring would occur. Complete rotation around the single bond doesn't take place because the repulsive interaction between the methyl groups causes a barrier to rotation.

16.26

16.27 Imagine two routes for synthesis of *m*-nitrotoluene:

(1) Alkylation of benzene, followed by nitration, doesn't succeed because an alkyl group is an *o,p*-director.

(2) Nitration of benzene, followed by alkylation, doesn't succeed because nitrobenzene is unreactive to Friedel–Crafts alkylation.

Thus, it isn't possible to synthesize *m*-nitrotoluene by any route that we have studied in this chapter.

Mechanism Problems

Mechanisms of Electrophilic Substitutions

16.28

ICl can be represented as $\overset{\delta^+}{I}-\overset{\delta^-}{Cl}$ because chlorine is a more electronegative element than iodine. Iodine can act as an electrophile in electrophilic aromatic substitution reactions.

16.29

This mechanism is the reverse of the sulfonation mechanism illustrated in the text. H^+ is the electrophile in this reaction.

16.30

Phosphoric acid protonates 2-methylpropene, forming a *tert*-butyl carbocation. This carbocation acts as an electrophile in a Friedel–Crafts reaction to yield *tert*-butylbenzene.

16.31   When an electrophile reacts with an aromatic ring bearing a $(CH_3)_3N^+-$ group:

Ortho attack:

This is a destabilizing resonance form because two positive charges are next to each other.

Meta attack:

Para attack:

This form is destabilizing.

The N,N,N-trimethylammonium group has no electron-withdrawing resonance effect because it has no vacant p orbitals to overlap with the π orbital system of the aromatic ring. The $(CH_3)_3N^+-$ group is inductively deactivating, however, because it is positively charged. It is meta-directing because the cationic intermediate resulting from meta attack is somewhat more stable than those resulting from ortho or para attack.

16.32   The aromatic ring is deactivated toward electrophilic aromatic substitution by the combined electron-withdrawing inductive effect of electronegative nitrogen and oxygen. The lone pair of electrons of nitrogen can, however, stabilize by resonance the ortho and para substituted intermediates but not the meta intermediate.

Ortho attack:

Meta attack:

Para attack:

16.33

$$CHCl_3 + AlCl_3 \rightleftharpoons {}^+CHCl_2 \ AlCl_4^-$$

(Dichloromethyl)benzene

(Dichloromethyl)benzene can react with two additional equivalents of benzene by the same mechanism to produce triphenylmethane.

Triphenylmethane

16.34

Resonance structures show that bromination occurs in the ortho and para positions of the rings. The positively charged intermediate formed from ortho or para attack can be stabilized by resonance contributions from the second ring of biphenyl, but this stabilization is not possible for meta attack.

16.35

HO—OH $\xrightarrow[\text{catalyst}]{\text{acid}}$ HO—$\overset{+}{\text{OH}_2}$ reactive electrophile

Attack of
π electrons
on reactive
electrophile

Loss of proton

The reactive electrophile (protonated H_2O_2) is equivalent to ^+OH.

Additional Mechanism Practice

16.36

Protonation of the double bond at carbon 2 of 1-phenylpropene leads to an intermediate
that can be stabilized by resonance involving the benzene ring.

16.37

(1)

Formaldehyde is protonated to form a carbocation.

(2)

The formaldehyde cation acts as the electrophile in a substitution reaction at the "6" position of 2,4,5-trichlorophenol.

(3)

The product from step 2 is protonated by strong acid to produce a cation.

(4)

Hexachlorophene

This cation is attacked by a second molecule of 2,4,5-trichlorophenol to produce hexachlorophene.

16.38

Benzyne
+ CO₂ + N₂

A Diels–Alder product

16.39

+ Cl⁻

Step 1: Attack of the nucleophile diethylamine.

Step 2: Loss of proton.

Step 3: Loss of Cl⁻.

This reaction is an example of nucleophilic aromatic substitution. Dimethylamine is a nucleophile, and the pyridine nitrogen acts as an electron-withdrawing group that can stabilize the negatively-charged intermediate.

16.40

Step 1:   Formation of primary carbocation.

Step 2:   Rearrangement to a secondary carbocation.

Step 3:   Attack of ring π electrons on the carbocation.

Step 4:   Loss of H^+.

This reaction takes place despite the fact that an electron-withdrawing group is attached to the ring. Apparently, the cyclization reaction is strongly favored.

16.41

(1)

Carbon monoxide is protonated to form an acyl cation.

(2)

The acyl cation reacts with benzene by a Friedel–Crafts acylation mechanism.

16.42

Protonation of   Loss of
aromatic ring   *tert*-butyl carbocation

16.43

Step 1:   S$_N$2 displacement takes place when the negatively charged oxygen of dimethyl sulfoxide attacks the benzylic carbon of benzyl bromide, displacing Br$^-$.

Step 2:   Base removes a benzylic proton, and dimethyl sulfide is eliminated in an E2 reaction.

16.44   The Smiles Rearrangement is an intramolecular Nucleophilic Aromatic Substitution.

16.45

(a)

Methyl Orange

Mechanism:

Methyl Orange

(b)

Allura Red

Mechanism:

Allura Red

(c)

Lithol Rubine BK

Mechanism:

Lithol Rubine BK

Additional Problems

Reactivity and Orientation of Electrophilic Substitutions

16.46

Group:	Identification:	Reason:
(a) $\overset{..}{N}(CH_3)_2$	*o,p*-activator	Reaction intermediates are stabilized by electron donation by the amine nitrogen.
(b) cyclopentyl	*o,p*-activator	Reaction intermediates are stabilized by the electron donating inductive effect of the alkyl group.
(c) $\overset{..}{\underset{..}{O}}CH_2CH_3$	*o,p*-activator	Reaction intermediates are stabilized by electron donation by the ether oxygen.
(d) :O: with cyclohexyl	*m*-deactivator	Reaction intermediates are destabilized by electron withdrawal by the carbonyl oxygen.

16.47

(a)

$$\xrightarrow[\text{H}_2\text{SO}_4]{\text{HNO}_3}$$

o-Bromonitrobenzene   +   *p*-Bromonitrobenzene

(b)

$$\xrightarrow[\text{H}_2\text{SO}_4]{\text{HNO}_3}$$

m-Nitrobenzonitrile

(c)

$$\xrightarrow[\text{H}_2\text{SO}_4]{\text{HNO}_3}$$

m-Nitrobenzoic acid

(d)

$$\xrightarrow[\text{H}_2\text{SO}_4]{\text{HNO}_3}$$

m-Dinitrobenzene

(e)

m-Nitrobenzenesulfonic acid

(f)

o-Methoxynitrobenzene   *p*-Methoxynitrobenzene

Only methoxybenzene reacts faster than benzene (See Figure 16.11).

16.48 *Most reactive* ⟶ *Least reactive*

(a)   Benzene > Chlorobenzene > *o*-Dichlorobenzene

(b)   Phenol > Nitrobenzene > *p*-Bromonitrobenzene

(c)   *o*-Xylene > Fluorobenzene > Benzaldehyde

(d)   *p*-Methoxybenzonitrile > *p*-Methylbenzonitrile > Benzonitrile

16.49

(a)

o-Bromotoluene   *p*-Bromotoluene

(b)

5-Bromo-2-methylphenol   3-Bromo-4-methylphenol

Both groups direct substitution to the same position.

(c)

No reaction. AlCl₃ combines with –N̈H₂ to form a complex that deactivates the ring toward Friedel–Crafts alkylation.

(d)

No reaction. The ring is deactivated.

(e)

2,4-Dichloro-6-methylphenol

(f)

No reaction.The ring is deactivated.

(g)

No reaction.The ring is deactivated.

(h)

1,4-Dibromo-2,5-dimethylbenzene

Alkylation occurs in the indicated position because the methyl group is more activating than bromine, and because substitution rarely takes place between two groups.

16.50 (a)

4-Chloro-3-nitrophenol   2-Chloro-5-nitrophenol

The —OH group directs the orientation of substitution.

(b)

4-Chloro-1,2-
dimethylbenzene

1-Chloro-2,3-
dimethylbenzene

(c)

3-Chloro-4-nitro-
benzoic acid

+

2-Chloro-4-nitro-
benzoic acid

Both groups are deactivating to a similar extent, and both possible products form.

(d)

4-Bromo-3-chlorobenzenesulfonic acid

16.51

(a)

+

p-Fluorobenzene-
sulfonic acid

o-Fluorobenzene-
sulfonic acid

(b)

+

2-Bromo-4-hydroxy-
benzenesulfonic acid

4-Bromo-2-hydroxy-
benzenesulfonic acid

(c)

2,4-Dichlorobenzene-
sulfonic acid

(d)

3,5-Dibromo-2-hydroxy-
benzenesulfonic acid

16.52 *Most reactive* ⟶ *Least reactive*

Phenol > Toluene > *p*-Bromotoluene > Bromobenzene

Aniline and nitrobenzene don't undergo Friedel–Crafts alkylations.

16.53 (a)

o-Ethylaniline

Catalytic hydrogenation reduces both the aromatic ketone and the nitro group.

(b)

3,4-Dibromoaniline   2,3-Dibromoaniline

Nitration, followed by reduction with Fe, produces substituted anilines.

(c)

o-Benzenedicarboxylic acid
(Phthalic acid)

Aqueous KMnO₄ oxidizes alkyl side chains to benzoic acids.

(d)

4-Chloro-2-isopropyl-   5-Chloro-2-methoxy-
methoxybenzene   propylbenzene

The methoxyl group directs substitution because it is a more powerful activating group. Rearranged and unrearranged side chains are present in the products.

16.54

(a)

(b)

(c)

(d)

Organic Synthesis

16.55

(a)

o-Bromobenzoic acid

(b)

p-Methoxytoluene

The reactions in (b) can be performed in either order.

(c)

2,4,6-Trinitrobenzoic acid

(d)

m-Bromoaniline

16.56 When synthesizing substituted aromatic rings, it is necessary to introduce substituents in the proper order. A group that is introduced out of order will not have the proper directing effect. Remember that in many of these reactions a mixture of ortho and para isomers may be formed.

(a)

p-Chloroacetophenone

(b)

m-Bromonitrobenzene

(c)

o-Bromobenzenesulfonic acid

(d)

m-Chlorobenzenesulfonic acid

16.57

(a)

2-Bromo-4-nitrotoluene

(b)

1,3,5-Trinitrobenzene

(c)

2,4,6-Tribromoaniline

No catalyst is needed for bromination because aniline is very activated toward substitution.

(d)

m-Fluorobenzoic acid

16.58 (a) Chlorination of toluene occurs at the ortho and para positions. To synthesize the given product, first oxidize toluene to benzoic acid and then chlorinate.

(b) *p*-Chloronitrobenzene is inert to Friedel–Crafts alkylation because the ring is deactivated.

(c)   The first two steps in the sequence are correct, but H₂/Pd reduces the nitro group as well as the ketone.

General Problems

16.59   Attack occurs on the unsubstituted ring because bromine is a deactivating group. Attack occurs at the ortho and para positions of the ring because the positively charged intermediate can be stabilized by resonance contributions from bromine and from the second ring (Problem 16.43).

16.60   When directly bonded to a ring, the –CN group is a meta-directing deactivator for both inductive and resonance reasons. In 3-phenylpropanenitrile, however, the saturated side chain does not allow resonance interactions of –CN with the aromatic ring, and the –CN group is too far from the ring for its inductive effect to be strongly felt. The side chain acts as an alkyl substituent, and ortho–para substitution is observed.

In 3-phenylpropenenitrile, the –CN group interacts with the ring through the π electrons of the side chain. Resonance forms show that –CN deactivates the ring toward electrophilic substitution, and substitution occurs at the meta position.

16.61   (a)

Activated   Activated
by –O–   by –O–
    and –CH₃

Substitution occurs in the more activated ring. The position of substitution is determined by the more powerful activating group – in this case, the ether oxygen.

(b)

The left ring is more activated than the right ring. −NHR is an ortho-para director.

(c)

Substitution occurs at the ortho and para positions of the more activated ring. Substitution doesn't occur between −C₆H₅ and −CH₃ for steric reasons.

(d)

Substitution occurs at the meta positions of the ring on the left because it is less deactivated.

16.62

Deactivated   Activated
by C=O   by –N̈–

Attack occurs in the activated ring and yields ortho and para bromination products. The intermediate is resonance-stabilized by overlap of the nitrogen lone pair electrons with the π electrons of the substituted ring.

Similar drawings can be made of the resonance forms of the intermediate resulting from ortho attack. Even though the nitrogen lone-pair electrons are less available for delocalization than the lone-pair electrons of aniline (Problem 16.12), the −NH− group is nevertheless more activating than the C=O group.

16.63 Reaction of (R)-2-chlorobutane with AlCl$_3$ produces an ion pair [CH$_3$$^+$CHCH$_2CH_3$ $^-$AlCl$_4$]. The planar, sp^2-hybridized carbocation is achiral, and its reaction with benzene can occur on either side of the carbocation to yield racemic product.

16.64 All of these syntheses involve NBS bromination of the benzylic position of a side chain.

(a)

(b)

(c)

16.65 The product is a substituted phenol, whose −OH group directs the orientation of the −C(CH₃)₃ groups. The precursor to MON-0585 is synthesized by a Friedel–Crafts alkylation of phenol by the appropriate hydrocarbon halide. This compound is synthesized by NBS bromination of the product of alkylation of benzene with 2-chloropropane.

16.66

The trivalent boron atom in phenylboronic acid has only six outer-shell electrons and is a Lewis acid. It is possible to write resonance forms for phenylboronic acid in which an electron pair from the phenyl ring is delocalized onto boron. In these resonance forms, the ortho and para positions of phenylboronic acid are the most electron-deficient, and substitutions occur primarily at the meta position.

16.67 Resonance forms for the intermediate from attack at C1:

Resonance forms for the intermediate from attack at C2:

There are seven resonance forms for attack at C1 and six for attack at C2. Look carefully at the forms, however. In the first four resonance structures for C1 attack, the second ring is still fully aromatic. In the other three forms, however, the positive charge has been delocalized into the second ring, destroying the ring's aromaticity. For C2 attack, only the first two resonance structures have a fully aromatic second ring. Since stabilization is lost when aromaticity is disrupted, the intermediate from C2 attack is less stable than the intermediate from C1 attack, and C1 attack is favored.

16.68

Meisenheimer complex

Step 1:   Addition of the nucleophile −OCH₃.

Step 2:   Elimination of −Cl⁻.

The carbonyl oxygens make the chlorine-containing ring electron-poor and open to attack by the nucleophile ⁻OCH₃. They also stabilize the negatively charged Meisenheimer complex.

16.69

Step 1:   Abstraction of proton and elimination of Br⁻.

Step 1:   Addition of NH₃ to the benzyne intermediate to form two aniline products.

The reaction of an aryl halide with potassium amide proceeds through a benzyne intermediate. Ammonia can then add to either end of the triple bond to produce the two methylanilines observed.

16.70 (a)

Protonation of the cyclic ether creates a carbocation intermediate that can react in a Friedel–Crafts alkylation.

(b)

The intermediate alkylates benzene, forming an alcohol product.

(c)

Protonation of the alcohol, followed by loss of water, generates a second carbocation.

(d)

This carbocation undergoes internal alkylation to yield the observed product.

16.71 Both of these syntheses test your ability to carry out steps in the correct order.

(a)

+ ortho isomer

(b)

16.72 Problem 16.51 shows the mechanism of the addition of HBr to 1-phenylpropene and shows how the aromatic ring stabilizes the carbocation intermediate. For the methoxyl-substituted styrene, an additional resonance form can be drawn in which the cation is stabilized by the electron-donating resonance effect of the oxygen atom. For the nitro-substituted styrene, the cation is destabilized by the electron-withdrawing effect of the nitro group.

favorable   :O:⁻ unfavorable

Thus, the intermediate resulting from addition of HBr to the methoxyl-substituted styrene is more stable, and reaction of *p*-methoxystyrene is faster.

16.73

$\mu = 1.53$ D $\mu = 1.52$ D $\mu = 2.91$ D

–Br has a strong electron-withdrawing inductive effect.

–NH₂ has a strong electron-donating resonance effect.

The polarities of the two groups add to produce a net dipole moment almost equal to the sum of the individual moments.

16.74

(a)   CH₃CH₂COCl, AlCl₃;

(b)   H₂, Pd/C;

(c)   Br₂, FeBr₃;

(d)   NBS, (PhCO₂)₂; (e) KOH, ethanol

16.75

An electron-withdrawing substituent *destabilizes* a positively charged intermediate (as in electrophilic aromatic substitution) but *stabilizes* a negatively charged intermediate. For the dissociation of a phenol, an −NO₂ group stabilizes the phenoxide anion by resonance, thus lowering $\Delta G°$ and pK_a. In the starred resonance form for *p*-nitrophenol, the negative charge has been delocalized onto the oxygens of the nitro group.

16.76   For the same reason described in the previous problem, a methyl group destabilizes the negatively charged intermediate, thus raising $\Delta G°$ and pK_a, making this phenol less acidic.

16.77

(a)

HNO_3
H_2SO_4

NO_2

H NO₂ — use LaTeX: H NO_2

H NO_2 H NO_2 H NO_2

O_2N

Addition of the electrophile to the meta-position
avoids the unstable resonance form where the
carbocation is adjacent to the electron
withdrawing group. This type of resonance form
is seen when the electrophile adds either ortho-
or para to an electron withdrawing group

(b)

$(CH_3)_2CHCl$, $AlCl_3$

$CH(CH_3)_2$

CH_3

$(CH_3)_2HC$

CH_3

+

H
$CH(CH_3)_2$
CH_3

H
$CH(CH_3)_2$
CH_3

H
$CH(CH_3)_2$
CH_3

resonance form stabilized
by tertiar carbocation

$(CH_3)_2HC$
H
CH_3

$(CH_3)_2HC$
H
CH_3

$(CH_3)_2HC$
H
CH_3

resonance form stabilized
by tertiar carbocation

(c)

Addition of the electrophile to the *meta*-position avoids the unstable resonance form where the carbocation is adjacent to the electron withdrawing group. This type of resonance form is seen when the electrophile adds either *ortho*- or *para*- to an electron withdrawing group

(d)

resonance form stabilized by extra bond formed with eletron donating oxygen

resonance form stabilized by extra bond formed with eletron donating oxygen

16.78 **Note:** The reactivity is explained by recognizing that nitrogen is a powerful electron donor. Electron donors deactivate arenes towards toward nulceophilic aromatic substitution. The reaction becomes increasingly more difficult as each molecule of ammonia is added.

s-trichlorotriazine   melamine

Review Unit 6: Conjugation and Aromaticity

Major Topics Covered (with vocabulary):

Conjugated dienes:
delocalization   1,4-addition   allylic position   thermodynamic control   kinetic control
vulcanization   Diels–Alder cycloaddition   dienophile   *endo* product   *exo* product
s-cis conformation

Ultraviolet spectroscopy:
highest occupied molecular orbital (HOMO)   lowest unoccupied molecular orbital (LUMO)
molar absorptivity

Aromaticity:
aromatic   arene   phenyl group   benzyl group   ortho, meta, para substitution   degenerate
Hückel $4n + 2$ rule   antiaromatic   heterocycle   polycyclic aromatic compound   ring current

Chemistry of aromatic compounds:
electrophilic aromatic substitution   sulfonation   F-TEDA-BF$_4$ Friedel–Crafts alkylation
polyalkylation   Friedel–Crafts acylation   ortho- and para-directing activator   ortho- and
para-directing deactivator   meta-directing deactivator   inductive effect   resonance effect
nucleophilic   aromatic substitution   Meisenheimer complex   benzyne   benzylic position

Types of Problems:

After studying these chapters, you should be able to:

- Predict the products of electrophilic addition to conjugated molecules.
- Understand the concept of kinetic *vs.* thermodynamic control of reactions.
- Recognize diene polymers, and draw a representative segment of a diene polymer.
- Predict the products of Diels–Alder reactions, and identify compounds that are good dienophiles and good dienes.
- Calculate the energy required for UV absorption, and use molar absorptivity to calculate concentration.
- Predict if and where a compound absorbs in the ultraviolet region.
- Name and draw substituted benzenes.
- Draw resonance structures and molecular orbital diagrams for benzene and other cyclic conjugated molecules.
- Use Hückel's rule to predict aromaticity.
- Draw orbital pictures of cyclic conjugated molecules.
- Use NMR, IR and UV data to deduce the structures of aromatic compounds.
- Predict the products of electrophilic aromatic substitution reactions.
- Formulate the mechanisms of electrophilic aromatic substitution reactions.
- Understand the activating and directing effects of substituents on aromatic rings, and use inductive and resonance arguments to predict orientation and reactivity.
- Predict the products of other reactions of aromatic compounds.
- Synthesize substituted benzenes.

Points to Remember:

*   It's not always easy to recognize Diels–Alder products, especially if the carbon-carbon double bond of the initial product has been hydrogenated. If no hydrogenation has taken place, look for a double bond in a six-membered ring and at least one electron-withdrawing group across the ring from the double bond. When a bicyclic product has been formed, it has probably resulted from a Diels-Alder reaction in which the diene is cyclic.

*   To be aromatic, a molecule must be planar, cyclic, conjugated, and it must have $4n + 2$ electrons in its π system.

*   The carbocation intermediate of electrophilic aromatic substitution loses a proton to yield the aromatic product. In all cases, a base is involved with proton removal, but the nature of the base varies with the type of substitution reaction. Although this book shows the loss of the proton, it often doesn't show the base responsible for proton removal. This doesn't imply that the proton flies off, unassisted; it just means that the base involved has not been identified in the problem.

*   Nucleophilic aromatic substitution reactions and substitution reactions proceeding through benzyne intermediates take place by different routes. In the first reaction, the substitution takes place by an addition, followed by an elimination. In the second case, the substitution involves an elimination, followed by an addition. Virtually all substitutions are equivalent to an addition and an elimination (in either order).

*   Activating groups achieve their effects by making an aromatic ring more electron-rich and reactive toward electrophiles. Ortho and para directing groups achieve their effects by stabilizing the positive charge that results from ortho or para addition of an electrophile to the aromatic ring. The intermediate resulting from addition to a ring with an ortho or para director usually has one resonance form that is especially stable. The intermediate resulting from addition to a ring with a meta director usually has a resonance form that is especially unfavorable when addition occurs ortho or para to the functional group. Meta substitution results because it is less unfavorable than ortho or para substitution.

Self-test:

A α-Farnesene

α-Farnesene (**A**), an important biological intermediate in the synthesis of many natural products, has double bonds that are both conjugated and unconjugated. Show the products you would expect from conjugate addition of HBr; of Br_2. What products would you expect from ozonolysis of **A**? Give one or more distinctive absorptions that you might see in the IR spectrum of **A** and distinguishing features of the ^{1}H NMR of **A**. Would you expect **A** to be UV-active?

B

C
Paroxypropione

D

Describe the π orbitals in the ring of **B**. Might this ring be described as aromatic?

Paroxypropione (**C**) is a hormone inhibitor. Predict the products of reaction of **C** with: (a) Br_2, $FeBr_3$; (b) CH_3Cl, $AlCl_3$; (c) $KMnO_4$, H_3O^+; (d) H_2, Pd/C. If the product of (d) is treated with the reagents in (a) or (b), does the orientation of substitution change? What significant information can you obtain from the IR spectrum of **C**?

Name **D**. Plan a synthesis of **D** from benzene. Describe the ^{1}H NMR of **D** (include spin-spin splitting). Where might **D** show an absorption in a UV spectrum?

Multiple choice:

1. What are the hybridizations of the carbons in 1,2-butadiene, starting with C1?
   (a) sp^2, sp^2, sp^2, sp^2 (b) sp^2, sp^2, sp^2, sp^3 (c) sp^2, sp, sp^2, sp^3 (d) sp, sp, sp^2, sp^3

2. In a reaction in which the less stable (ls) product is formed at lower temperature, and the more stable product (ms) is formed at higher temperature:
   (a) $\Delta G_{ms}° > \Delta G_{ls}°$ and $\Delta G_{ms}^{\ddagger} > \Delta G_{ls}^{\ddagger}$
   (b) $\Delta G_{ms}° > \Delta G_{ls}°$ and $\Delta G_{ls}^{\ddagger} > \Delta G_{ms}^{\ddagger}$
   (c) $\Delta G_{ms}° < \Delta G_{ls}°$ and $\Delta G_{ms}^{\ddagger} > \Delta G_{ls}^{\ddagger}$
   (d) $\Delta G_{ms}° < \Delta G_{ls}°$ and $\Delta G_{ls}^{\ddagger} > \Delta G_{ms}^{\ddagger}$
   Note: In this problem, a large value for $\Delta G°$ means a large <u>negative</u> value.

3. Which of the following combinations is most likely to undergo a successful Diels–Alder reaction?

4. Which of the following groups, when bonded to the terminal carbon of a conjugated π system, probably affects the value of λ_{max} the least?
   (a) $-NH_2$ (b) $-Cl$ (c) $-OH$ (d) $-CH_3$

5. If the value of λ_{max} for an unsubstituted diene is approximately 220 nm, and each additional double bond increases the value of λ_{max} by 30 nm, what is the minimum number of double bonds present in a compound that absorbs in the visible range of the electromagnetic spectrum?
   (a) 6   (b) 7   (c) 8   (d) 9

6. Which of the following compounds is aromatic?
   (a)   (b)   (c)   (d)

7. How many benzene isomers of $C_7H_6Br_2$ can be drawn?
   (a) 10   (b) 11   (c) 12   (d) 14

8. Which of the following functional groups isn't a meta-directing deactivator?
   (a) $-NO_2$ (b) $-CONHCH_3$ (c) $-N(CH_3)_3{}^+$ (d) $-NHCOCH_3$

9.	Which of the following compounds can't be synthesized by an electrophilic aromatic substitution reaction that we have studied?
	(a) *m*-Cresol (b) *p*-Chloroaniline (c) 2,4-Toluenedisulfonic acid (d) *m*-Bromotoluene

10.	In only one of the following compounds can you reduce the aromatic ring without also reducing the side chain. Which compound is it?
	(a)	*p*-Bromoanisole
	(b)	Acetophenone (methyl phenyl ketone)
	(c)	Styrene
	(d)	Phenylacetylene

Chapter 17 – Alcohols and Phenols

Chapter Outline

I. Naming alcohols and phenols (Section 17.1).
  A. Alcohols are classified as primary, secondary or tertiary, depending on the number of organic groups bonded to the −OH carbon.
  B. Rules for naming simple alcohols.
    1. The longest chain containing the −OH group is the parent chain, and the parent name replaces -e with -ol.
    2. Numbering begins at the end of the chain nearer the −OH group.
    3. The substituents are numbered according to their position on the chain and cited in alphabetical order.
  C. Phenols are named according to rules discussed in Section 15.1 for aromatic compounds.

II. Properties of alcohols and phenols (Section 17.2).
  A. Hydrogen-bonding of alcohols and phenols.
    1. Alcohols have sp^3 hybridization and a nearly tetrahedral bond angle.
    2. Alcohols and phenols have elevated boiling points, relative to hydrocarbons, due to hydrogen-bonding.
      a. In hydrogen-bonding, an −OH hydrogen is attracted to a lone pair of electrons on another molecule, resulting in a weak electrostatic force that holds the molecules together.
      b. These weak forces must be overcome in boiling.
  B. Acidity and basicity of alcohols and phenols.
    1. Alcohols and phenols are weakly acidic as well as weakly basic.
    2. Alcohols and phenols can be reversibly protonated to form oxonium ions.
    3. Alcohols and phenols dissociate to a slight extent to form alkoxide ions and phenoxide ions.
    4. Acidity of alcohols.
      a. Alcohols are similar in acidity to water.
      b. Alkyl substituents decrease acidity by preventing solvation of the alkoxide ion.
      c. Electron-withdrawing substituents increase acidity by delocalizing negative charge.
      d. Alcohols don't react with weak bases, but they do react with alkali metals and strong bases.
    5. Acidity of phenols.
      a. Phenols are a million times more acidic than alcohols and are soluble in dilute NaOH.
      b. Phenol acidity is due to resonance stabilization of the phenoxide anion.
      c. Electron-withdrawing substituents increase phenol acidity, and electron-donating substituents decrease phenol acidity.

III. Alcohols (Sections 17.3–17.8).
  A. Preparation of alcohols (Sections 17.3–17.5).
    1. Familiar methods (Section 17.3).

       a. Hydration of alkenes.

          i. Hydroboration/oxidation yields non-Markovnikov products.

          ii. Oxymercuration/reduction yields Markovnikov products.

       b. 1,2-diols can be prepared by OsO_4 hydroxylation, followed by reduction.

          i. This reaction occurs with syn stereochemistry.

          ii. Ring-opening of epoxides produces 1,2-diols with anti stereochemistry.

  2. Reduction of carbonyl compounds (Section 17.4).

       a. Aldehydes are reduced to primary alcohols.

       b. Ketones are reduced to secondary alcohols.

          1. Either $NaBH_4$(milder) or $LiAlH_4$(more reactive) can be used to reduce aldehydes and ketones.

       c. Carboxylic acids and esters are reduced to primary alcohols with $LiAlH_4$.

          i. These reactions occur by addition of hydride to the positively polarized carbon of a carbonyl group.

          ii. Water adds to the alkoxide intermediate during workup to yield alcohol product.

  3. Reaction of carbonyl compounds with Grignard reagents (Section 17.5).

       a. RMgX adds to carbonyl compounds to give alcohol products.

          i. Reaction of RMgX with formaldehyde yields primary alcohols.

          ii. Reaction of RMgX with aldehydes yields secondary alcohols.

          iii. Reaction of RMgX with ketones yields tertiary alcohols.

          iv. Reaction of RMgX with esters yields tertiary alcohols with at least two identical R groups bonded to the alcohol carbon.

          v. No reaction occurs with carboxylic acids because the acidic hydrogen quenches the Grignard reagent.

       b. Limitations of the Grignard reaction.

          i. Grignard reagents can't be prepared from reagents containing other reactive functional groups.

          ii. Grignard reagents can't be prepared from compounds having acidic hydrogens.

       c. Grignard reagents behave as carbon anions and add to the carbonyl carbon.

          i. A proton from water is added to the alkoxide intermediate to produce the alcohol.

B. Reactions of alcohols (Sections 17.6–17.8).

  1. Conversion to alkyl halides (Section 17.6).

       a. Tertiary alcohols (ROH) are converted to RX by treatment with HX.

          i. The reaction occurs by an S_N1 mechanism.

       b. Primary alcohols are converted by the reagents PBr_3 and $SOCl_2$.

          i. The reaction occurs by an S_N2 mechanism.

  2. Conversion into tosylates.

       a. Reaction with *p*-toluenesulfonyl chloride converts alcohols to tosylates.

       b. Only the O−H bond is broken.

  c. Tosylates behave as halides in substitution reactions.

  d. S$_N$2 reactions involving tosylates proceed with inversion of configuration.

 3. Dehydration to yield alkenes.

  a. Tertiary alcohols can undergo acid-catalyzed dehydration with warm aqueous H$_2$SO$_4$.

   i. Zaitsev products are usually formed.

   ii. The severe conditions needed for dehydration of secondary and primary alcohols restrict this method to tertiary alcohols.

   iii. Tertiary alcohols react fastest because the intermediate carbocation formed in this E1 reaction is more stable.

  b. Secondary alcohols are dehydrated with POCl$_3$ in pyridine.

   i. This reaction occurs by an E2 mechanism.

   ii. Pyridine serves both as a base and as a solvent.

 4. Conversion into esters.

 5. Oxidation of alcohols (Section 17.7).

  a. Primary alcohols can be oxidized to aldehydes or carboxylic acids.

  b. Secondary alcohols can be oxidized to ketones.

  c. Tertiary alcohols aren't oxidized.

  d. Oxidation to ketones and carboxylic acids can be carried out with KMnO$_4$, CrO$_3$, or Na$_2$Cr$_2$O$_7$.

  e. Oxidation of a primary alcohol to an aldehyde is achieved with the Dess–Martin periodinane.

   i. The Dess–Martin periodinane is also used on sensitive alcohols.

  f. Oxidation occurs by a mechanism closely related to an E2 mechanism.

 5. Protection of alcohols (Section 17.8).

  a. It is sometimes necessary to protect an alcohol when it interferes with a reaction involving a functional group in another part of a molecule.

  b. The following reaction sequence may be applied:

   i. Protect the alcohol.

   ii. Carry out the reaction.

   iii. Remove the protecting group.

  c. A trimethylsilyl (TMS) ether can be used for protection.

   i. TMS ether formation occurs by an S$_N$2 route.

   ii. TMS ethers are quite unreactive.

   iii. TMS ethers can be cleaved by aqueous acid or by F⁻ to regenerate the alcohol.

IV. Phenols (Sections 17.9–17.10).

 A. Preparation and uses of phenols (Section 17.9).

  1. Phenols can be prepared by treating chlorobenzene with NaOH.

  2. Phenols can also be prepared from isopropylbenzene (cumene).

   a. Cumene reacts with O$_2$ by a radical mechanism to form cumene hydroperoxide.

   b. Treatment of the hydroperoxide with acid gives phenol and acetone.

   i. The mechanism involves protonation, rearrangement, loss of water, of water to form a hemiacetal, and breakdown to acetone and phenol.
   3. Chlorinated phenols, such as 2,4-D, are formed by chlorinating phenol.
   4. BHT is prepared by Friedel–Crafts alkylation of *p*-cresol with 2-methylpropene.
  B. Reactions of phenols (Section 17.10).
   1. Phenols undergo electrophilic aromatic substitution reactions (Chapter 16).
      a. The −OH group is a *o,p*-director.
   2. Strong oxidizing agents convert phenols to quinones.
      a. Reaction with Fremy's salt to form a quinone occurs by a radical mechanism.
      b. The redox reaction quinone → hydroquinone occurs readily.
      c. Ubiquinones are an important class of biochemical oxidizing agents that function as a quinone/hydroquinone redox system.
 V. Spectroscopy of alcohols and phenols (Section 17.11).
  A. IR spectroscopy.
   1. Both alcohols and phenols show −OH stretches in the region 3300–3600 cm^{-1}.
      a. Unassociated alcohols show a peak at 3600 cm^{-1}.
      b. Associated alcohols show a broader peak at 3300–3400 cm^{-1}.
   2. Alcohols show a C−O stretch near 1050 cm^{-1}.
   3. Phenols show aromatic bands at 1500–1600 cm^{-1}.
   4. Phenol shows monosubstituted aromatic bands at 690 and 760 cm^{-1}.
  B. NMR spectroscopy.
   1. In ^{13}C NMR spectroscopy, carbons bonded to −OH groups absorb in the range 50–80 δ.
   2. ^1H NMR.
      a. Hydrogens on carbons bearing −OH groups absorb in the range 3.5–4.5 δ.
         i. The hydroxyl hydrogen doesn't split these signals.
      b. D_2O exchange can be used to locate the O−H signal.
      c. Spin–spin splitting occurs between protons on the oxygen-bearing carbon and neighboring −H.
      d. Phenols show aromatic ring absorptions, as well as an O−H absorption in the range 3–8 δ.
  C. Mass Spectrometry.
   1. Alcohols undergo alpha cleavage to give a neutral radical and an oxygen-containing cation.
   2. Alcohols also undergo dehydration to give an alkene radical cation.

Solutions to Problems

17.1 The parent chain must contain the hydroxyl group, and the hydroxyl group(s) should receive the lowest possible number.

(a)

$$CH_3CHCH_2CHCHCH_3$$

with OH, OH, and CH_3 substituents

5-Methyl-2,4-hexanediol

(b)

$$CH_2CH_2CCH_3$$

with OH and CH_3

2-Methyl-4-phenyl-2-butanol

(c)

4,4-Dimethylcyclohexanol

(d)

(1S,2S)-2-Bromocyclopentanol

(e)

4-Bromo-3-methylphenol

(f)

2-Cyclopenten-1-ol

17.2

(a)

(Z)-2-Ethyl-2-buten-1-ol

(b)

3-Cyclohexen-1-ol

(c)

trans-3-Chlorocycloheptanol
and enantiomer

(d)

$$CH_3CHCH_2CH_2CH_2OH$$

with OH

1,4-Pentanediol

(e)

2,6-Dimethylphenol

(f)

o-(2-Hydroxyethyl)phenol

17.3   In general, the boiling points of a series of isomers decrease with branching. The more nearly spherical a compound becomes, the less surface area it has relative to a straight chain compound of the same molecular weight and functional group type. A smaller surface area allows fewer van der Waals interactions, the weak forces that cause covalent molecules to be attracted to each other.

In addition, branching in alcohols makes it more difficult for hydroxyl groups to approach each other to form hydrogen bonds. A given volume of 2-methyl-2-propanol therefore contains fewer hydrogen bonds than the same volume of 1-butanol, and less energy is needed to break them in boiling.

17.4   *Least acidic* ⎯⎯⎯⎯⎯⎯⎯⎯⎯⎯⎯⎯⎯⎯→ *Most acidic*

(a) $HC\equiv CH$ < $(CH_3)_2CHOH$ < CH_3OH < $(CF_3)_2CHOH$
        alkyne   hindered   alcohol   alcohol with electron-
                                    alcohol   withdrawing groups

(b)   *p*-Methylphenol <   Phenol   <   *p*-(Trifluoromethyl)phenol
        phenol with electron-   phenol with electron-
        donating groups   withdrawing groups

(c)   Benzyl alcohol <   Phenol   <   *p*-Hydroxybenzoic acid
        alcohol   phenol   carboxylic acid

17.5   We saw in Chapter 16 that a nitro group is electron-withdrawing. Since electron-withdrawing groups stabilize anions, *p*-nitrobenzyl alcohol is more acidic than benzyl alcohol. The methoxyl group, which is electron-donating, destabilizes an alkoxide ion, making *p*-methoxybenzyl alcohol less acidic than benzyl alcohol.

17.6   (a)

2-Methyl-3-pentanol

In a hydroboration/oxidation reaction, the hydroxyl group is bonded to the less substituted carbon.

(b)

2-Methyl-4-phenyl-2-butanol

Markovnikov product results from oxymercuration/reduction.

(c)

meso-5,6-Decanediol

Hydroxylation results in a diol with syn stereochemistry.

17.7   (a)

NaBH₄ reduces aldehydes and ketones without interfering with other functional groups.

(b)

LiAlH₄, a stronger reducing agent, reduces both ketones and esters.

(c)

LiAlH₄ reduces carbonyl functional groups without reducing double bonds.

17.8   (a)

Benzyl alcohol may be the reduction product of an aldehyde, a carboxylic acid, or an ester. NaBH₄ may be used to reduce the aldehyde.

(b)

Reduction of a ketone yields the secondary alcohol. NaBH₄ may also be used here and in (c)

(c)

(d)

$$(CH_3)_2CHCHO \quad or \quad (CH_3)_2CHCO_2H \quad or \quad (CH_3)_2CHCO_2R \xrightarrow[\text{2. H}_3\text{O}^+]{\text{1. LiAlH}_4} (CH_3)_2CHCH_2OH$$

17.9   All of the products have an −OH and a methyl group bonded to what was formerly a ketone carbon.

(a)

(b)

(c)

17.10   First, identify the type of alcohol. If the alcohol is primary, it can only be synthesized from formaldehyde plus the appropriate Grignard reagent. If the alcohol is secondary, it is synthesized from an aldehyde and a Grignard reagent. (Usually, there are two combinations of aldehyde and Grignard reagent). A tertiary alcohol is synthesized from a ketone and a Grignard reagent. If all three groups on the tertiary alcohol are different, there are often three different combinations of ketone and Grignard reagent. If two of the groups on the alcohol carbon are the same, the alcohol may also be synthesized from an ester and two equivalents of Grignard reagent.

(a)   2-Methyl-2-propanol is a tertiary alcohol. To synthesize a tertiary alcohol, start with a ketone.

If two or more alkyl groups bonded to the carbon bearing the −OH group are the same, an alcohol can be synthesized from an ester and a Grignard reagent.

2-Methyl-2-propanol

(b) Since 1-methylcyclohexanol is a tertiary alcohol, start with a ketone.

1-Methylcyclohexanol

(c) 3-Methyl-3-pentanol is a tertiary alcohol. When two of the three groups bonded to the alcohol carbon are the same, either a ketone or an ester can be used as a starting material.

(d) Three possible combinations of ketone plus Grignard reagent can be used to synthesize this tertiary alcohol.

(e)   Formaldehyde must be used to synthesize this primary alcohol.

Benzyl alcohol

(f)   As in (e), use formaldehyde to synthesize a primary alcohol.

$(CH_3)_2CHCH_2CH_2CH_2OH$
4-Methyl-1-pentanol

17.11 First, interpret the structure of the alcohol. This alcohol, 1-ethylcyclohexanol, is a tertiary alcohol that can be synthesized from a ketone. Only one combination of ketone and Grignard reagent is possible.

17.12 Recall from Chapter 11 that $-OH$ is a very poor leaving group in reactions run under S_N2 conditions. A toluenesulfonate, however, is a very good leaving group, and reaction of the toluenesulfonate of the alcohol with ^-CN proceeds readily under S_N2 conditions to give the desired product with inversion of configuration at the chirality center.

17.13 (a)

The major product has the more substituted double bond.

(b)

3-Methylcyclohexene

In E2 elimination, dehydration proceeds most readily when the two groups to be eliminated have an anti periplanar relationship. In this compound, the only hydrogen with the proper stereochemical relationship to the −OH group is at C6. Thus, the non-Zaitsev product 3-methylcyclohexene is formed.

(c)

1-Methylcyclohexene

Here, the hydrogen at C2 is trans to the hydroxyl group, and dehydration yields the Zaitsev product, 1-methylcyclohexene.

(d)

2,3-Dimethyl-2-pentene
major

2-Ethyl-3-methyl-1-butene
minor

(Z)-3,4-Dimethyl-2-pentene
minor

(E)-3,4-Dimethyl-2-pentene
minor

(e)

2-Methyl-2-pentene
major

2-Methyl-1-pentene
minor

17.14 Aldehydes are synthesized from oxidation of primary alcohols, and ketones are synthesized from oxidation of secondary alcohols.

(a)

(b)

(c)

17.15

Starting material	CrO_3, H_3O^+ Product	Periodinane Product
(a) $CH_3CH_2CH_2CH_2CH_2CH_2OH$	$CH_3CH_2CH_2CH_2CH_2CO_2H$	$CH_3CH_2CH_2CH_2CH_2CHO$
(b) $CH_3CH_2CH_2CH_2\overset{\overset{\textstyle OH}{\textstyle \vert}}{C}HCH_3$	$CH_3CH_2CH_2CH_2\overset{\overset{\textstyle O}{\textstyle \Vert}}{C}CH_3$	$CH_3CH_2CH_2CH_2\overset{\overset{\textstyle O}{\textstyle \Vert}}{C}CH_3$
(c) $CH_3CH_2CH_2CH_2CH_2CHO$	$CH_3CH_2CH_2CH_2CH_2CO_2H$	no reaction

17.16

This is an S_N2 reaction in which the nucleophile F^- attacks silicon and displaces an alkoxide ion as leaving group.

17.17

Phosphoric acid protonates 2-methylpropene, forming a *tert*-butyl carbocation.

The *tert*-butyl carbocation acts as an electrophile and alkylates *p*-cresol. Alkylation occurs ortho to the −OH group for both steric and electronic reasons.

A second *tert*-butyl carbocation alkylation forms BHT.

17.18 The infrared spectra of cholesterol and 5-cholestene-3-one each exhibit a unique absorption that makes it easy to distinguish between them. Cholesterol shows an −OH stretch at 3300-3600 cm^{-1}, and 5-cholestene-3-one shows a C=O stretch at 1715 cm^{-1}. In the oxidation of cholesterol to 5-cholestene-3-one, the −OH band disappears and is replaced by a C=O band. When oxidation is complete, no −OH absorption should be visible.

17.19 Under conditions of slow exchange, the −OH signal of a tertiary alcohol (R$_3$COH) is unsplit, the signal of a secondary alcohol (R$_2$CHOH) is split into a doublet, and the signal of a primary alcohol (RCH$_2$OH) is split into a triplet.

(a)   2-Methyl-2-propanol is a tertiary alcohol; its −OH signal is unsplit.

(b)   Cyclohexanol is a secondary alcohol; its −OH absorption is a doublet.

(c)   Ethanol is a primary alcohol; its −OH signal appears as a triplet.

(d)   2-Propanol is a secondary alcohol; its −OH absorption is split into a doublet.

(e)   Cholesterol is a secondary alcohol; its −OH absorption is split into a doublet.

(f)   1-Methylcyclohexanol is a tertiary alcohol; its −OH signal is unsplit.

Visualizing Chemistry

17.20

(a)

(*R*)-5-Methyl-3-hexanol

(b)

cis-3-Methylcyclohexanol

(c)   (d)

(S)-1-Cyclopentylethanol   4-Methyl-3-nitrophenol

17.21 The reduction product is a racemic mixture. Reaction of the (S) enantiomer is shown.

(a)

(S)-5-Methyl-2-hexanol

(b)

(S)-3-Methyl-1-pentanol

17.22

(a)

(b)

major   minor

(c)

(d)

(e)

17.23

(a)

$CH_3CHCH_2CH_2COCH_3$ with CH_3 above and O above the carbonyl

$\xrightarrow[\text{2. }H_3O^+]{\text{1. NaBH}_4}$ no reaction

(b)

$CH_3CHCH_2CH_2COCH_3$ with CH_3 above and O above the carbonyl

$\xrightarrow[\text{2. }H_3O^+]{\text{1. LiAlH}_4}$ $CH_3CHCH_2CH_2CH_2OH$ + $HOCH_3$ with CH_3 above

(c)

$$CH_3CHCH_2CH_2COCH_3 \xrightarrow[\text{2. H}_3\text{O}^+]{\text{1. 2 CH}_3\text{CH}_2\text{MgBr}} CH_3CHCH_2CH_2CCH_2CH_3 + HOCH_3$$

17.24

(3R,4S)-3,4-Di-methyl-3-hexanol

(3S,4S)-3,4-Di-methyl-3-hexanol

The product is a mixture of the (3R,4S) and (3S,4S) diastereomers. The diastereomers are formed in unequal amounts, and the product mixture is optically active. We can't predict which diastereomer will predominate.

Mechanism Problems

17.25

Step 1: Protonation.

Step 2: Loss of H_2O.

Step 3: Alkyl shift to form the tertiary carbocation.

Step 4: Loss of H_3O^+.

17.26 This mechanism consists of the same steps as are seen in Problem 17.25. Two different alkyl shifts result in two different cycloalkenes.

protonation of alcohol

loss of H$_2$O

two different alkyl shifts

loss of H$^+$

Isopropylidenecyclopentane   1,2-Dimethylcyclohexene

17.27

1.   2.

Step 1:   S$_N$2 reaction of Grignard reagent.

Step 2:   Protonation of alkoxide oxygen.

The methyl group and the hydroxyl group have a trans relationship.

17.28

Step 1:   Protonation.

Step 2:   Addition of H₂O.

Step 3:   Loss of H⁺.

17.29

Reaction of 2-butanone with NaBH₄ produces a racemic mixture of (*R*)-2-butanol and (*S*)-2-butanol.

17.30 (a)

Mechanism:

(b)

Mechanism:

(c)

Mechanism:

17.31 (a) Teritary alcohols undergo exclusively S_N1 substitution.

(b) Reactions with PBr3 undergo S_N2 substitution

(c) In this reaction the primary alcohol is first converted to a primary tosylate. Primary tosylates undergo exclusively S_N2 substitution.

17.32 (a)

Mechanism:

(b)

Mechanism:

(c)

Mechanism:

17.33 (a)

Mechanism:

(b)

Mechanism:

(c)

Mechanism:

17.34 (a) *tert*-butyldimethylsilyl chloride (TBS-Cl)

(*R*)-3-bromobutan-1-ol   TBS-Cl

(b) triisopropylsilyl chloride (TIPS-Cl)

TIPS-Cl

(c) triethylsilyl chlorides (TES-Cl)

TES-Cl

17.35

In the POCl₃ reaction the elimination reaction mechanism is E^2. In the elimination catalyzed by sulfuric acid the mechanism is E1 and results in a carbocation that undergoes rearrangement.

Mechanism:

Mechanism:

17.36 (a)

This resonance form destabilized
by having negative charge next to
an electron donating methyl group

(b)

This resonance form is stabilized
by having the negative charge on
the electronegative nitrogen

(c)

Because there is no direct effect through resonance the OCH_3 group does not have a significant effect on the pK_a of the parent phenol

Additional Problems

Naming Alcohols

17.37

(a) (b)

$$CH_3$$
$$HOCH_2CH_2CHCH_2OH$$

2-Methyl-1,4-butanediol

(c)

$$OH$$
$$CH_3CHCHCH_2CH_3$$
$$CH_2CH_2CH_3$$

3-Ethyl-2-hexanol

cis-1,3-Cyclobutanediol

(d)   (e)   (f)

cis-2-Methyl-4-cyclohepten-1-ol   *cis*-3-Phenylcyclopentanol

2-Bromo-4-cyanophenol
or
3-Bromo-4-hydroxy-
benzonitrile

17.38 None of these alcohols has multiple bonds or rings.

CH₃CH₂CH₂CH₂CH₂OH
1-Pentanol

CH₃CH₂CH₂CHCH₃
OH *
2-Pentanol

CH₃CH₂CHCH₂CH₃
OH
3-Pentanol

CH₃CH₂CHCH₂OH
*
CH₃
2-Methyl-1-butanol

CH₃CH₂CCH₃
OH
CH₃
2-Methyl-2-butanol

CH₃CHCHCH₃
OH
* CH₃
3-Methyl-2-butanol

HOCH₂CH₂CHCH₃
CH₃
3-Methyl-1-butanol

CH₃CCH₂OH
CH₃
CH₃
2,2-Dimethyl-1-propanol

2-Pentanol, 2-methyl-1-butanol and 3-methyl-2-butanol have chiral carbons (starred).

17.39 Primary alcohols react with CrO₃ in aqueous acid to form carboxylic acids, secondary alcohols yield ketones, and tertiary alcohols are unreactive to oxidation. Of the eight alcohols in the previous problem, only 2-methyl-2-butanol is unreactive to CrO₃ oxidation.

$$CH_3CH_2CH_2CH_2CH_2OH \xrightarrow[\text{H}_3\text{O}^+]{\text{CrO}_3} CH_3CH_2CH_2CH_2CO_2H$$

$$\underset{\underset{\textstyle OH}{|}}{CH_3CH_2CH_2CHCH_3} \xrightarrow[\text{H}_3\text{O}^+]{\text{CrO}_3} \underset{\overset{\textstyle O}{\parallel}}{CH_3CH_2CH_2CCH_3}$$

$$\underset{\underset{\textstyle OH}{|}}{CH_3CH_2CHCH_2CH_3} \xrightarrow[\text{H}_3\text{O}^+]{\text{CrO}_3} \underset{\overset{\textstyle O}{\parallel}}{CH_3CH_2CCH_2CH_3}$$

$$\underset{\underset{\textstyle CH_3}{|}}{CH_3CH_2CHCH_2OH} \xrightarrow[\text{H}_3\text{O}^+]{\text{CrO}_3} \underset{\underset{\textstyle CH_3}{|}}{CH_3CH_2CHCO_2H}$$

$$\underset{\underset{\textstyle CH_3}{|}}{\underset{\underset{\textstyle OH}{|}}{CH_3CHCHCH_3}} \xrightarrow[\text{H}_3\text{O}^+]{\text{CrO}_3} \underset{\underset{\textstyle CH_3}{|}}{\underset{\overset{\textstyle O}{\parallel}}{CH_3CCHCH_3}}$$

$$\underset{\underset{\textstyle CH_3}{|}}{HOCH_2CH_2CHCH_3} \xrightarrow[\text{H}_3\text{O}^+]{\text{CrO}_3} \underset{\underset{\textstyle CH_3}{|}}{HO_2CCH_2CHCH_3}$$

$$\underset{\underset{\textstyle CH_3}{|}}{\overset{\overset{\textstyle CH_3}{|}}{CH_3CCH_2OH}} \xrightarrow[\text{H}_3\text{O}^+]{\text{CrO}_3} \underset{\underset{\textstyle CH_3}{|}}{\overset{\overset{\textstyle CH_3}{|}}{CH_3CCO_2H}}$$

17.40

Bombykol   (10E,12Z)-10,12-Hexadecadien-1-ol

17.41

Carvacrol   5-Isopropyl-2-methylphenol

Synthesizing Alcohols

17.42 In some of these problems, different combinations of Grignard reagent and carbonyl compound are possible. Remember that aqueous acid is added to the initial Grignard adduct to yield the alcohol.

(a)

$$CH_3CHO + CH_3CH_2MgBr$$
or
$$CH_3CH_2CHO + CH_3MgBr$$

$$\longrightarrow$$

$$\overset{\displaystyle OH}{\underset{\displaystyle}{CH_3CHCH_2CH_3}}$$
2-Butanol

(b)

$$CH_3CH_2CHO \quad + \quad CH_3CH_2MgBr \longrightarrow$$

$$\overset{\displaystyle OH}{\underset{\displaystyle}{CH_3CH_2CHCH_2CH_3}}$$
3-Pentanol

(c)

2-Methyl-2-propen-1-ol

(d)

Triphenylmethanol

(e)

2-Phenyl-2-propanol

(f)

17.43 (a)

Alcohol

$CH_3CH_2CH_2CH_2\overset{\underset{\displaystyle CH_3}{|}}{\overset{\displaystyle CH_3}{|}}CCH_2OH$

Carbonyl precursor(s)

$CH_3CH_2CH_2CH_2\overset{\underset{\displaystyle CH_3}{|}}{\overset{\displaystyle CH_3}{|}}CCHO$

$CH_3CH_2CH_2CH_2\overset{\underset{\displaystyle CH_3}{|}}{\overset{\displaystyle CH_3}{|}}CCO_2H$

$CH_3CH_2CH_2CH_2\overset{\underset{\displaystyle CH_3}{|}}{\overset{\displaystyle CH_3}{|}}CCO_2R$

(b)

$(CH_3)_3CCHCH_3$ (with OH) $(CH_3)_3CCCH_3$ (with O)

(c)

17.44

Grignard Reagent + Carbonyl Compound ⟶ *Product (after dilute acid workup)*

(a)

CH_3MgBr + CH_3CCH_3 (O)

or

$2\ CH_3MgBr$ + CH_3COR (O)

⟶ $(CH_3)_3COH$

(b)

CH_3CH_2MgBr + ⟶

Grignard Reagent + Carbonyl Compound $\longrightarrow$ *Product (after dilute acid workup)*

(c)

$$CH_3CH_2MgBr \ + \ CH_3CH_2\overset{\displaystyle O}{\overset{\displaystyle \|}{C}}\text{—}\bigcirc$$

or

$$2 \ CH_3CH_2MgBr \ + \ RO\overset{\displaystyle O}{\overset{\displaystyle \|}{C}}\text{—}\bigcirc$$

or

$$\bigcirc\text{—}MgBr \ + \ CH_3CH_2\overset{\displaystyle O}{\overset{\displaystyle \|}{C}}CH_2CH_3$$

$$\longrightarrow \qquad CH_3CH_2\overset{\displaystyle OH}{\underset{\displaystyle \bigcirc}{C}}CH_2CH_3$$

(d)

$$CH_3CH_2CH_2MgBr \ + \ CH_3\overset{\displaystyle O}{\overset{\displaystyle \|}{C}}\text{—}\bigcirc$$

or

$$CH_3MgBr \ + \ CH_3CH_2CH_2\overset{\displaystyle O}{\overset{\displaystyle \|}{C}}\text{—}\bigcirc$$

or

$$\bigcirc\text{—}MgBr \ + \ CH_3CH_2CH_2\overset{\displaystyle O}{\overset{\displaystyle \|}{C}}CH_3$$

$$\longrightarrow \qquad CH_3CH_2CH_2\overset{\displaystyle OH}{\underset{\displaystyle \bigcirc}{C}}CH_3$$

(e)

$$H_3C\text{—}\bigcirc\text{—}CH_2MgBr \ + \ H_2C{=}O \ \longrightarrow \ H_3C\text{—}\bigcirc\text{—}CH_2CH_2OH$$

(f)

$$CH_3MgBr \ + \ \bigcirc\text{—}CH_2\overset{\displaystyle O}{\overset{\displaystyle \|}{C}}CH_3$$

or

$$2 \ CH_3MgBr \ + \ \bigcirc\text{—}CH_2\overset{\displaystyle O}{\overset{\displaystyle \|}{C}}OR$$

or

$$\bigcirc\text{—}CH_2MgBr \ + \ CH_3\overset{\displaystyle O}{\overset{\displaystyle \|}{C}}CH_3$$

$$\longrightarrow \qquad \bigcirc\text{—}CH_2\overset{\displaystyle OH}{\underset{\displaystyle CH_3}{C}}CH_3$$

17.45 All of these syntheses involve a Grignard reaction at some step. Both the carbonyl compound and the Grignard reagent must be prepared from alcohols.

(a) $CH_3CH_2OH \xrightarrow{PBr_3} CH_3CH_2Br \xrightarrow[\text{ether}]{Mg} CH_3CH_2MgBr$

cyclohexanol $\xrightarrow[H_3O^+]{CrO_3}$ cyclohexanone $\xrightarrow[\text{2. }H_3O^+]{\text{1. }CH_3CH_2MgBr}$ 1-ethylcyclohexan-1-ol (OH, CH$_2$CH$_3$)

(b) $CH_3CH_2CH_2\overset{\underset{|}{CH_3}}{C}HCH_2OH \xrightarrow{PBr_3} CH_3CH_2CH_2\overset{\underset{|}{CH_3}}{C}HCH_2Br \xrightarrow[\text{ether}]{Mg} CH_3CH_2CH_2\overset{\underset{|}{CH_3}}{C}HCH_2MgBr$

$CH_3OH \xrightarrow[CH_2Cl_2]{\text{Periodinane}} H_2C{=}O$

$CH_3CH_2CH_2\overset{\underset{|}{CH_3}}{C}HCH_2MgBr + H_2C{=}O \xrightarrow[\text{2. }H_3O^+]{\text{1. ether}} CH_3CH_2CH_2\overset{\underset{|}{CH_3}}{C}HCH_2CH_2OH$

(c) benzene $\xrightarrow[FeBr_3]{Br_2}$ bromobenzene (Br) $\xrightarrow[\text{ether}]{Mg}$ phenylmagnesium bromide (MgBr)

$CH_3CH_2CH_2\overset{\underset{|}{OH}}{C}HCH_3 \xrightarrow[H_3O^+]{CrO_3} CH_3CH_2CH_2\overset{\overset{O}{\|}}{C}CH_3$

phenyl-MgBr $+ CH_3CH_2CH_2\overset{\overset{O}{\|}}{C}CH_3 \xrightarrow[\text{2. }H_3O^+]{\text{1. ether}}$ (phenyl)C(HO)(CH$_3$)CH$_2$CH$_2$CH$_3$

(d) $CH_3\overset{\underset{|}{CH_3}}{C}HCH_2CH_2OH \xrightarrow[CH_2Cl_2]{\text{Periodinane}} CH_3\overset{\underset{|}{CH_3}}{C}HCH_2\overset{\overset{O}{\|}}{C}H$

$CH_3\overset{\underset{|}{CH_3}}{C}HCH_2\overset{\overset{O}{\|}}{C}H + CH_3CH_2MgBr \xrightarrow[\text{2. }H_3O^+]{\text{1. ether}} CH_3\overset{\underset{|}{CH_3}}{C}HCH_2\overset{\underset{|}{OH}}{C}HCH_2CH_3$
from (a)

Reactions of Alcohols

17.46

(a) $CH_3CH_2CH_2CH_2CH_2OH \xrightarrow{PBr_3} CH_3CH_2CH_2CH_2CH_2Br$

(b) $CH_3CH_2CH_2CH_2CH_2OH \xrightarrow{SOCl_2} CH_3CH_2CH_2CH_2CH_2Cl$

(c) $CH_3CH_2CH_2CH_2CH_2OH$ $\xrightarrow[H_3O^+]{CrO_3}$ $CH_3CH_2CH_2CH_2CO_2H$

(d) $CH_3CH_2CH_2CH_2CH_2OH$ $\xrightarrow[CH_2Cl_2]{Periodinane}$ $CH_3CH_2CH_2CH_2CHO$

17.47

(a)

CH$_2$CH$_2$OH $\xrightarrow[pyridine]{POCl_3}$ CH=CH$_2$

2-Phenylethanol   Styrene

(b)

CH$_2$CH$_2$OH $\xrightarrow[CH_2Cl_2]{Periodinane}$ CH$_2$CHO

Phenylacetaldehyde

(c)

CH$_2$CH$_2$OH $\xrightarrow[H_3O^+]{CrO_3}$ CH$_2$CO$_2$H

Phenylacetic acid

(d)

CH$_2$CH$_2$OH $\xrightarrow[H_2O]{KMnO_4}$ CO$_2$H

Benzoic acid

(e)

CH=CH$_2$ $\xrightarrow[Pd]{H_2}$ CH$_2$CH$_3$

from (a)   Ethylbenzene

(f)

CH=CH$_2$ $\xrightarrow[\text{2. Zn, }H_3O^+]{\text{1. }O_3}$ CHO

from (a)   Benzaldehyde

(g)

CH=CH$_2$ $\xrightarrow[\text{2. }NaBH_4]{\text{1. }Hg(OAc)_2\text{, }H_2O}$ $\overset{\text{OH}}{\underset{}{\text{CHCH}_3}}$

from (a)   1-Phenylethanol

(h)

CH$_2$CH$_2$OH

PBr$_3$

CH$_2$CH$_2$Br

1-Bromo-2-phenylethane

17.48

(a)

OH
|
CHCH$_3$

1-Phenylethanol

$\xrightarrow[\text{H}_3\text{O}^+]{\text{CrO}_3}$

O
||
CCH$_3$

Acetophenone

(b)

OH
|
CHCH$_3$

$\xrightarrow[\text{H}_2\text{O}]{\text{KMnO}_4}$

CO$_2$H

$\xrightarrow[\text{2. H}_3\text{O}^+]{\text{1. LiAlH}_4}$

CH$_2$OH

Benzyl alcohol

(c)

CO$_2$H

from (b)

$\xrightarrow[\text{FeBr}_3]{\text{Br}_2}$

Br   CO$_2$H

m-Bromobenzoic acid

(d)

O
||
CCH$_3$

from (a)

$\xrightarrow[\text{2. H}_3\text{O}^+]{\text{1. CH}_3\text{MgBr}}$

HO   CH$_3$
  \ /
   C
  / \
     CH$_3$

2-Phenyl-2-propanol

17.49

(a)

OH
H

$\xrightarrow[\text{H}_3\text{O}^+]{\text{CrO}_3}$

O

(b)

OH
H

$\xrightarrow[\text{pyridine}]{\text{POCl}_3}$

(c)

O

from (a)

$\xrightarrow[\text{2. H}_3\text{O}^+]{\text{1. CH}_3\text{MgBr}}$

OH
CH$_3$

(d)

from (c)

Remember that hydroboration proceeds with syn stereochemistry, and the $-H$ and $-OH$ added have a cis relationship.

17.50

(a)

(b)

(c)

(d)

no reaction

Tertiary alcohols aren't oxidized by sodium dichromate.

Spectroscopy

17.51

2.32 δ ⟶ H_3C— —CH_2—OH ← 2.43 δ

7.10 δ, 7.17 δ   4.50 δ

p-Methylbenzyl alcohol

17.52   (a)

OH c

$CH_3CH_2CHCH_2CH_3$
a   b   d   b   a

3-Pentanol

a = 0.93 δ
b = 1.42 δ
c = 1.83 δ
d = 3.41 δ

(b)

OH b

$CHCH_3$
c   a

1-Phenylethanol

a = 1.42 δ
b = 2.43 δ
c = 4.80 δ
d = 7.32 δ

17.53 1. $C_8H_{18}O_2$ has *no* double bonds or rings, based on degree of unsaturation.

2. The IR band at 3350 cm^{-1} shows the presence of a hydroxyl group.

3. The compound is symmetrical (simple NMR).

4. There is no splitting.

$$
\begin{array}{c}
CH_3 \qquad CH_3 \quad \longleftarrow \ 1.24\ \delta \\
| \qquad\quad | \\
HOCCH_2CH_2COH \quad \longleftarrow \ 1.95\ \delta \\
| \qquad\quad | \\
CH_3 \quad CH_3 \\
1.56\ \delta
\end{array}
$$

2,5-Dimethyl-2,5-hexanediol

17.54

e d
H   CH$_3$
  \ /
   C=C
  / \
H   CH$_2$CH$_2$OH
f   c b a

3-Methyl-3-buten-3-ol

The peak absorbing at 1.76 δ (3 H) is due to the d protons. This peak, which occurs in the allylic region of the spectrum, is unsplit.

The peak absorbing at 2.13 δ (1 H) is due to the −OH proton a.

The peak absorbing at 2.30 δ (2 H) is due to protons c. The peak is a triplet because of splitting by the adjacent b protons.

The peak absorbing at 3.72 δ (2 H) is due to the b protons. The adjacent oxygen causes the peak to be downfield, and the adjacent −CH$_2$− group splits the peak into a triplet.

The peaks at 4.79 δ and 4.85 δ (2 H) are due to protons e and f.

17.55 (a) $C_5H_{12}O$, $C_4H_8O_2$, $C_3H_4O_3$

(b) The ^1H NMR data show that the compound has twelve protons.

(c) The IR absorption at 3600 cm^{-1} shows that the compound is an alcohol.

(d) The compound contains five carbons, two of which are identical.

(e) $C_5H_{12}O$ is the molecular formula of the compound.

(f), (g)

b
OH
|
CH$_3$CCH$_2$CH$_3$
 c | d   a
CH$_3$
 c

a = 0.9 δ
b = 1.0 δ
c = 1.2 δ
d = 1.4 δ

2-Methyl-2-butanol

17.56

a = 1.41 δ

b = 2.24 δ

c = 5.00 δ

d = 6.97 δ

General Problems

17.57 In these compounds you want to reduce some, but not all, of the functional groups present. To do this, choose the correct reducing agent.

(a)

H₂ with a palladium catalyst hydrogenates carbon–carbon double bonds without affecting carbonyl double bonds.

(b)

LiAlH₄ reduces carbonyl groups without affecting carbon–carbon double bonds.

(c)

(from b)

17.58

17.59 Remember that electron-withdrawing groups stabilize phenoxide anions and increase acidity. Electron-donating groups decrease phenol acidity.

Least acidic ⟶ *Most acidic*

17.60

Step 1: S$_N$2 substitution.

Step 2: E2 elimination.

17.61

Despite this problem's resemblance to Problem 17.24, the stereochemical outcome is different. Addition of methylmagnesium bromide to the carbonyl group doesn't produce a new chirality center and doesn't affect the chirality center already present. The product is pure (S)-2,3-dimethyl-2-pentanol, which is optically active.

17.62

Step 1:   Protonation.

Step 2:   Loss of H_2O.

Step 3:   Alkyl shift.

Step 4:   Loss of H^+.

This is a carbocation rearrangement involving the shift of an alkyl group. The sequence of steps is the same as those seen in Problems 17.27 and 17.28.

17.63

Testosterone

All of these transformations require the proper sequence of oxidations and reductions. In (d), NaBH₄ can also be used for reduction.

17.64 A phenoxide anion is stabilized by the electron-withdrawing resonance effect of a *p*-nitro group. Methyl groups ortho to the phenol have no effect on acidity, but the methyl groups that flank the nitro group of the 3,5-isomer force the nitro group out of planarity with the ring and reduce orbital overlap with the π orbitals of the ring. The resonance stabilization of the nitro group is reduced, and the pK_a of the phenol becomes higher, indicating lower acidity.

17.65

17.66 (a)   Compound **A** has one double bond or ring.

(b)   The infrared absorption at 3400 cm^{-1} indicates the presence of an alcohol. The weak absorption at 1640 cm^{-1} is due to a C=C stretch.

(c)   (1)   The absorptions at 1.63 δ and 1.70 δ are due to unsplit methyl protons. Because the absorptions are shifted slightly downfield, the protons are adjacent to an unsaturated center.

(2)   The broad singlet at 3.83 δ is due to an alcohol proton.

(3)   The doublet at 4.15 δ is due to two protons bonded to a carbon bearing an electronegative atom (oxygen, in this case).

(4)   The proton absorbing at 5.70 δ is a vinylic proton.

(d)

3-Methyl-2-buten-1-ol

A

17.67

1-Methylcyclopentene   *trans*-2-Methylcyclopentanol   3-Methylcyclopentene

The more stable dehydration product is 1-methylcyclopentene, which can be formed only via syn elimination. The major product of anti elimination is 3-methylcyclopentene. Since this product predominates, the requirement of anti periplanar geometry must be more important than formation of the more stable product.

17.68 The pinacol rearrangement follows a sequence of steps similar to other rearrangements we have studied in this chapter. The second hydroxyl group assists in the alkyl shift.

Pinacol

Pinacolone

Step 1: Protonation.

Step 2: Loss of H_2O.

Step 3: Alkyl shift.

Step 4: Loss of H^+.

17.69 The hydroxyl group is axial in the cis isomer, which is expected to oxidize faster than the trans isomer. (Remember that the bulky *tert*-butyl group is always equatorial in the more stable isomer.)

cis-4-tert-Butylcyclohexanol   faster

trans-4-tert-Butylcyclohexanol   slower

17.70

Bicyclohexylidene

17.71 An alcohol adds to an aldehyde by a mechanism that we will study in a later chapter. The hydroxyl group of the addition intermediate undergoes oxidation (as shown in Section 17.7), and an ester is formed.

17.72

(a)   NaBH$_4$, then H$_3$O$^+$ (b)   PBr$_3$ (c)   Mg, ether, then CH$_2$O

(d)   Dess–Martin periodinane, CH$_2$C$_{12}$ (e)   C$_6$H$_5$CH$_2$MgBr, then H$_3$O$^+$

(f)   POCl$_3$, pyridine

17.73

UDP-Galactose   UDP-Glucose

Step 1: Base deprotonates the C4 hydroxyl group while NAD⁺ oxidizes the alcohol to a ketone.

Step 2: When the ketone is reduced by the NADH formed in Step 1, the configuration at the starred carbon is inverted, and UDP-glucose is formed.

17.74 (a)

OH c

CHCH₂CH₃
d b a

e

1-Phenyl-1-propanol

a = 0.88 δ
b = 1.8 δ
c = 2.32 δ
d = 4.54 δ
e = 7.24 δ

(b)

CH₂OH   a
c   a

CH₃O
b

d,e

p-Methoxybenzyl alcohol

a = 2.60 δ
b = 3.76 δ
c = 4.53 δ
d = 6.85 δ
e = 7.23 δ

17.75 *Structural formula:* $C_8H_{10}O$ contains 4 multiple bonds and/or rings.

Infrared: The broad band at 3500 cm⁻¹ indicates a hydroxyl group. The absorptions at 1500 cm⁻¹ and 1600 cm⁻¹ are due to an aromatic ring. The absorption at 830 cm⁻¹ shows that the ring is *p*-disubstituted. Compound **A** is probably a phenol.

¹H NMR: The triplet at 1.16 δ (3 H) is coupled with the quartet at 2.55 δ (2 H). These two absorptions are due to an ethyl group.

The peaks at 6.74 δ-7.02 δ (4 H) are due to aromatic ring protons. The symmetrical splitting pattern of these peaks indicate that the aromatic ring is *p*-disubstituted.

The singlet absorption at 5.50 δ (1 H) is due to an −OH proton.

Compound **A**

HO⎯⟨ ⟩⎯CH₂CH₃

p-Ethylphenol

17.76

Step 1: The nucleophile ⁻CN adds to the positively polarized carbonyl carbon.

Step 2: The tetrahedral intermediate is protonated to give the addition product.

17.77

The reaction is an S_N^2 displacement of iodide by phenoxide ion.

Chapter 18 – Ethers and Epoxides; Thiols and Sulfides

Chapter Outline

I.   Acyclic ethers (Sections 18.1–18.4).
   A. Naming ethers (Section 18.1).
      1. Ethers with no other functional groups are named by citing the two organic substituents and adding the word "ether".
      2. When other functional groups are present, the ether is an alkoxy substituent.
   B. Properties of ethers.
      1. Ethers have the same geometry as water and alcohols.
      2. Ethers have a small dipole moment that causes a slight boiling point elevation.
      3. Ethers can react slowly with oxygen to give explosive peroxides.
   C. Synthesis of ethers (Section 18.2).
      1. Symmetrical ethers can be synthesized by acid-catalyzed dehydration of alcohols.
         i.   This method is used only with primary alcohols.
      2. Williamson ether synthesis.
         a. Metal alkoxides react with primary alkyl halides and tosylates to form ethers.
         b. The alkoxides are prepared by reacting an alcohol with a strong base, such as NaH.
            i.   Reaction of the free alcohol with the halide can also be achieved with Ag_2O.
         c. The reaction occurs via an S_N^2 mechanism.
            i.   The halide component must be primary.
            ii. In cases where one ether component is hindered, the ether should be synthesized from the alkoxide of the more hindered reagent and the halide of the less hindered reagent.
      3. Alkoxymercuration of alkenes.
         a. Ethers can be formed from the reaction of alcohols with alkenes.
         b. The reaction is carried out in the presence of mercuric trifluoroacetate.
         c. The mechanism is similar to that for hydration of alkenes.
            i. $NaBH_4$ is used for demercuration of the intermediate.
         d. Many different types of ethers can be prepared by this method.
   D. Reactions of ethers (Sections 18.3–18.4).
      1. Ethers are relatively unreactive and often used as solvents.
      2. Acidic cleavage (Section 18.3).
         a. Strong acids can be used to cleave ethers.
         b. Cleavage can occur by S_N^2 or S_N^1 routes.
            i.   Primary and secondary alcohols react by an S_N^2 mechanism, in which the halide attacks the ether at the less hindered site.
               (a) This route selectively produces one halide and one alcohol.
            ii. Tertiary, benzylic, and allylic ethers react by either an S_N1 or an E1 route.
      3. Claisen rearrangement (Section 18.4).
         a. The Claisen rearrangement is specific to allyl aryl ethers or aryl vinyl ethers.
         b. The result of Claisen rearrangement is an *o*-allyl phenol.
         c. The reaction takes place in a single step by a pericyclic mechanism.
            i.   Inversion of the allyl group is evidence for this mechanism.

II. Cyclic ethers (Sections 18.5–18.7).
   A. Epoxides (oxiranes) (Sections 18.5–18.6).
      1. The three-membered ring of epoxides gives them unique chemical reactivity (Section 18.5).
      2. The nonsystematic name *–ene oxide* describes the method of formation.
      3. The systematic prefix *epoxy-* describes the location of the epoxide ring.
      4. Preparation of epoxides.
         a. Epoxides can be prepared by reaction of an alkene with a peroxyacid RCO_3H.
            i. The reaction occurs in one step with syn stereochemistry.
         b. Epoxides are formed when halohydrins are treated with base.
            i. This reaction is an intramolecular Williamson ether synthesis.
      5. Ring-opening reactions of epoxides (Section 18.6).
         a. Acid-catalyzed ring opening.
            i. Acid-catalyzed ring opening produces 1,2 diols.
            ii Ring opening takes place by back-side attack of a nucleophile on the protonated epoxide ring.
               (a) A *trans*-1,2-diol is formed from an epoxycycloalkane.
               (b) If HX is used, the product is a trans halohydrin.
            iii. When both epoxide carbons are primary or secondary, attack occurs primarily at the less hindered site.
            iv. When one epoxide carbon is tertiary, attack occurs at the more highly substituted site.
            v. The mechanism is midway between S_N2 and S_N1 routes.
               (a) The reaction occurs by back-side attack (S_N2), but positive charge is stabilized by a tertiary carbocation-like transition state (S_N1).
         b. Base-catalyzed ring-opening.
            i. Base-catalyzed ring opening occurs because of the reactivity of the strained epoxide ring.
            ii. Ring-opening takes place by an S_N2 mechanism, in which the nucleophile attacks the less hindered epoxide carbon.
            iii. Other nucleophiles can bring about ring opening.
               (a) Epoxides react with Grignard reagents to form a product with two more carbons than the starting alkyl halide.
               (b) Epoxide rings also react with amines in a ring-opening reaction.
   B. Crown ethers (Section 18.7).
      1. Crown ethers are large cyclic ethers.
      2. Crown ethers are named as *x*-crown-*y*, where *x* = the ring size and *y* = # of oxygens.
      3. Crown ethers are able to solvate metal cations.
         a. Different sized crown ethers solvate different cations.
         b. Complexes of crown ethers with ionic salts are soluble in organic solvents.
         c. This solubility allows many reactions to be carried out under aprotic conditions.
         d. The reactivity of many anions in S_N2 reactions is enhanced by crown ethers.

III. Thiols and sulfides (Section 18.8).
    A. Naming thiols and sulfides.
        1. Thiols (sulfur analogs of alcohols) are named by the same system as alcohols, with the suffix *-thiol* replacing *-ol.*
           a. The −SH group is a mercapto- group.
        2. Sulfides (sulfur analogs of ethers) are named by the same system as ethers, with *sulfide* replacing *ether*.
           a. The −SR group is an alkylthio- group.
    B. Thiols.
        1. Thiols stink!
        2. Thiols may be prepared by S_N^2 displacement with a sulfur nucleophile.
           a. The reaction may proceed to form sulfides.
           b. Better yields occur when thiourea is used.
        3. Thiols can be oxidized by Br_2 or I_2 to yield disulfides, RSSR.
           a. The reaction can be reversed by treatment with zinc and acid.
           b. The thiol-disulfide interconversion is an important biochemical interconversion.
    C. Sulfides.
        1. Treatment of a thiol with base yields a thiolate anion, which can react with an alkyl halide to form a sulfide.
        2. Thiolate anions are excellent nucleophiles.
        3. Dialkyl sulfides can react with alkyl halides to form trialkylsulfonium salts, which are also good alkylating agents.
           a. Many biochemical reactions use trialkylsulfonium groups as alkylating agents.
        4. Sulfides are easily oxidized to sulfoxides (R_2SO) and sulfones (R_2SO_2).
           a. Dimethyl sulfoxide is used as a polar aprotic solvent.
IV. Spectroscopy of ethers (Section 18.9).
    A. IR spectroscopy.
        1. Ethers are difficult to identify by IR spectroscopy because many other absorptions occur at 1050–1150 cm^{-1}, where ethers absorb.
    B. NMR spectroscopy.
        1. ^{1}H NMR spectroscopy.
           a. Hydrogens on a carbon next to an ether oxygen absorb downfield (3.4–4.5 δ).
           b. Hydrogens on a carbon next to an epoxide oxygen absorb at a slightly higher field (2.5–3.5 δ).
        2. ^{13}C NMR spectroscopy.
           a. Ether carbons absorb downfield (50–80 δ).

Solutions to Problems

18.1   Ethers can be named either as alkoxy-substituted compounds or by citing the two groups bonded to oxygen, followed by the word "ether".

(a)
$$CH_3CHOCHCH_3$$
with CH_3 groups

Diisopropyl ether

(b)
Propoxycyclopentane

Cyclopentyl propyl ether

(c)
p-Bromoanisole

p-Bromomethoxybenzene

(d)
1-Methoxycyclohexene

(e)
$$CH_3CHCH_2OCH_2CH_3$$
with CH_3 group

Ethyl isobutyl ether

(f) $H_2C=CHCH_2OCH=CH_2$

Allyl vinyl ether

18.2   The first step of the dehydration mechanism is protonation of an alcohol. Water is then displaced by another molecule of alcohol to form an ether. If two different alcohols are present, either one can be protonated and either one can displace water, yielding a mixture of products.

If this procedure were used with ethanol and 1-propanol, the products would be diethyl ether, ethyl propyl ether, and dipropyl ether. If there were equimolar amounts of the alcohols, and if they were of equal reactivity, the product ratio would be diethyl ether: ethyl propyl ether: dipropyl ether = 1:2:1.

18.3   Remember that the halide in the Williamson ether synthesis should be primary or methyl, in order to avoid competing elimination reactions. The alkoxide anions shown are formed by treating the corresponding alcohols with NaH.

(a) $CH_3CH_2CH_2O^-$ + CH_3Br
or
$CH_3CH_2CH_2Br$ + CH_3O^-
$\longrightarrow$ $CH_3CH_2CH_2OCH_3$ + Br^-

Methyl propyl ether

(b)
O^- + CH_3Br $\longrightarrow$ OCH_3 + Br^-

Methyl phenyl ether
(Anisole)

(c)
CH_3CHO^- + CH_2Br $\longrightarrow$ CH_3CHOCH_2 + Br^-
(with CH_3 groups)

Benzyl isopropyl ether

(d)

$$CH_3\underset{\underset{CH_3}{|}}{\overset{\overset{CH_3}{|}}{C}}CH_2O^- \quad + \quad CH_3CH_2Br \quad \longrightarrow \quad CH_3\underset{\underset{CH_3}{|}}{\overset{\overset{CH_3}{|}}{C}}CH_2OCH_2CH_3 \quad + \quad Br^-$$

Ethyl 2,2-dimethylpropyl ether

18.4

1-Ethoxy-1-methylcyclopentane

The reaction mechanism of alkoxymercuration/demercuration of an alkene is similar to other electrophilic additions we have studied. First, the cyclopentene π electrons attack Hg^{2+} with formation of a mercurinium ion. Next, the nucleophilic alcohol displaces mercury. Markovnikov addition occurs because the carbon bearing the methyl group is better able to stabilize the partial positive charge arising from cleavage of the carbon-mercury bond. The ethoxyl and mercuric groups are trans to each other. Finally, removal of mercury by $NaBH_4$ by a mechanism that is not fully understood results in the formation of 1-ethoxy-1-methylcyclopentane.

18.5   Use the Williamson synthesis when one of the ether components can be a primary or benzylic halide. Use alkoxymercuration when one or both components are branched.

(a)   Either method of synthesis is appropriate.

Williamson:

Butyl cyclohexyl ether

Alkoxymercuration:

(b)   Either method is possible, but the Williamson synthesis is simpler.

Benzyl ethyl ether

(c)   Use alkoxymercuration because both parts of the ether are branched.

sec-Butyl tert-butyl ether

(d)   The Williamson synthesis must be used.

Tetrahydrofuran

18.6 The compounds most reactive in the Williamson ether synthesis are also most reactive in any S$_N$2 reaction (review Chapter 11 if necessary).

Most reactive ⟶ *Least reactive*

(a)

CH$_3$CH$_2$Br   >   CH$_3$CHCH$_3$ (with Br)   >>

primary halide   secondary halide   aryl halide (not reactive)

(b)   CH$_3$CH$_2$Br   >   CH$_3$CH$_2$Cl   >>   CH$_3$CH=CHI

better leaving group   poorer leaving group   vinylic (not reactive)

18.7 (a) First, notice the substitution pattern of the ether. Bonded to the ether oxygen are a primary alkyl group and a tertiary alkyl group. When one group is tertiary, cleavage occurs by an S$_N$1 or E1 route to give either an alkene or a tertiary halide and a primary alcohol.

(b) In this problem, the groups are primary and secondary alkyl groups. Br$^-$ attacks at the less hindered primary group, and oxygen remains with the secondary group, to give a secondary alcohol.

18.8

The first step of acid-catalyzed ether cleavage is protonation of the ether oxygen to give an intermediate oxonium ion, which collapses to form an alcohol and a tertiary carbocation. The carbocation then loses a proton to form an alkene, 2-methylpropene. This is an example of E1 elimination. The acid used for cleavage is often trifluoroacetic acid.

18.9

HX first protonates the oxygen atom, and halide then brings about a nucleophilic displacement to form an alcohol and an organic halide. The better the nucleophile, the more effective the displacement. Since I⁻ and Br⁻ are better nucleophiles than Cl⁻, ether cleavage proceeds more smoothly with HI or HBr than with HCl.

18.10 Draw the ether with the groups involved in the rearrangement positioned as they will appear in the product. Six bonds will either be broken or formed in the product; they are shown as dashed lines in the transition state. Redraw the bonds to arrive at the intermediate enone, which rearranges to the more stable phenol.

2-Butenyl   transition state   intermediate   o-(1-Methyl-
phenyl ether   allyl) phenol

18.11 Epoxidation by use of *m*-chloroperoxybenzoic acid (RCO₃H) is a syn addition of oxygen to a double bond. The original bond stereochemistry is retained, and the product is a meso compound.

cis-2-Butene   cis-2,3-Epoxybutane

In the epoxide product, as in the alkene starting material, the methyl groups are cis.

trans-2-Butene   *trans*-2,3-Epoxybutane

Reaction of *trans*-2-butene with *m*-chloroperoxybenzoic acid yields *trans*-2,3-epoxybutane. A mixture of enantiomers is formed because the peroxyacid can attack either the top or bottom of the double bond.

18.12 As discussed in this section, acid-catalyzed epoxide ring opening occurs primarily at the more hindered carbon if one of the epoxide carbons is tertiary. In both parts of this problem, one epoxide carbon is tertiary.

(a)

(b)

18.13 Notice the relationship of the hydroxyl groups in the two diols. In diol (a), the two hydroxyls are cis, and in (b) they are trans. Since ring-opening of epoxides forms *trans*-1,2-diols, only diol (b) can be formed by this route. The *cis*-1,2-diol in (a), results from treatment of 1-methylcyclohexene with OsO$_4$. The enantiomers of the diols are also formed.

18.14 (a)   Attack of the basic nucleophile occurs at the less substituted epoxide carbon.

(b)   Under acidic conditions, ring-opening occurs at the more substituted epoxide carbon when one of the carbons is tertiary.

(c)   Addition of a Grignard reagent takes place at the less substituted epoxide carbon.

18.15

15-Crown-5   12-Crown-4

Bases on ionic radii, the ion-to-oxygen distance in 15-crown-5 is about 40% longer than the ion-to-oxygen distance in 12-crown-4.

18.16   Thiols are named by the same rules as alcohols, with the suffix -ol replaced by the suffix - thiol. Sulfides are named by the same rules as ethers, with "sulfide" replacing "ether".

(a)
$$CH_3CH_2CHSH$$
with CH_3 above

2-Butanethiol

(b)
$$CH_3CCH_2CHCH_2CHCH_3$$
with CH_3, SH, CH_3 above and CH_3 below

2,2,6-Trimethyl-
4-heptanethiol

(c)

2-Cyclopentene-1-thiol

(d)

$$CH_3CHSCH_2CH_3$$
with CH_3 on top

Ethyl isopropyl sulfide

(e)

o-(Dimethylthio)benzene

(f)

3-(Ethylthio)cyclohexanone

18.17 Thiourea is used to prepare thiols from alkyl halides.

$$CH_3CH=CHCOCH_3 \xrightarrow[\text{2. } H_3O^+]{\text{1. LiAlH}_4} CH_3CH=CHCH_2OH \xrightarrow{\text{PBr}_3} CH_3CH=CHCH_2Br$$

Methyl 2-butenoate

$$CH_3CH=CHCH_2Br \downarrow \begin{array}{l} \text{1. } (H_2N)_2C{=}S \\ \text{2. } ^-OH, H_2O \end{array}$$

$$H_2C=CHCH=CH_2 \xrightarrow{\text{HBr}} CH_3CH=CHCH_2Br \xrightarrow[\text{2. } ^-OH, H_2O]{\text{1. } (H_2N)_2C{=}S} CH_3CH=CHCH_2SH$$

1,3-Butadiene   2-Butene-1-thiol

18.18

$$CH_3CH_2\underset{a}{C}\underset{b}{C}-\underset{}{C}\overset{H}{\underset{H}{\diagdown}}_{d,e}$$

1,2-Epoxybutane

a = 1.0 δ
b = 1.5 δ
c = 2.9 δ
d,e = 2.5 δ, 2.7δ

Visualizing Chemistry
18.19

(a)

OCH_2CH_3

H_3C

cis-1-Ethoxy-3-methylcyclohexane

(b)

$$H_3C-\overset{O}{\overset{\diagup \diagdown}{C-C}}\overset{CH_3}{\underset{H}{\diagup}}$$
with Br on phenyl

E-2-(*o*-Bromophenyl)-2,3-epoxybutane

(c)

$$\overset{CH_3}{\underset{H \ SH}{C}}$$ on cyclopentane

(*S*)-1-Cyclopentylethanethiol

18.20 Ring-opening occurs at the tertiary carbon to give carbocation-like stability to the transition state. Bromine approaches 180° from the C—OH bond, as it would in an S_N2 reaction.

18.21 The Grignard reagent attacks the epoxide at the less hindered carbon in an S$_N$2 reaction. The oxygen remains bonded to the tertiary carbon.

18.22 A molecular model shows that approach to the upper face of the double bond is hindered by a methyl group. Reaction with RCO$_3$H occurs at the lower face of the double bond to produce epoxide **A**.

In the reaction of Br$_2$ and H$_2$O, the intermediate bromonium ion also forms at the lower face. Reaction with water yields a bromohydrin which, when treated with base, forms epoxide **B**.

Mechanism Problems

18.23 Note: Each mechanism involves protonation followed by an S$_N$1 displacement of an alcohol.

(a)

Mechanism:

(b)

Mechanism:

(c)

Mechanism:

(d)

Mechanism:

18.24 Note: Each mechanism is S$_N$1, involving protonation of the ether oxygen followed by formation of a tertiary carbocation before addition of the halide.

(a)

Mechanism:

(b)

Mechanism:

(c)

Mechanism:

(d)

CH₃CH₂OH   +

Mechanism:

18.25   **(a)**

1. NaOH

2. CH₃CH₂I

Mechanism:

(b)

HO

1. NaH

2. CH₃CH₂CH₂Br

Mechanism:

(c)

HO

1. NaH

2. CH₃OTs

Mechanism:

(d)

Mechanism:

18.26 (a)

Mechanism:

(b)

Mechanism:

(c)

Mechanism:

18.27 Note: Because none of the reactions contains a starting material with a tertiary epoxide *and* is performed under acidic conditions, each reaction proceeds with S_N2 regiochemistry (the nulcleophile adding to the least hindered carbon) and S_N2 stereochemistry (the nucleophile adding 180° from the epoxide oxygen.)

(a)

Mechanism:

(b)

1. CH_3MgBr

2. H_3O^+

Mechanism:

(c)

HBr

ether

Mechanism:

(d)

18.28 (a)

Mechanism:

(b)

Mechanism:

(c)

Mechanism:

(d)

Mechanism:

18.29 Note: The phenolate ion serves as a nucleophile both to displace chloride ion and to open the expoide. While the mechanism here shows the displacement of chloride first, it is likely that the rate of epoxide opening is competitive. Thus, either epoxide opening or chloride opening may be first. In either case the final product is the same.

epichlorohydrin

Mechanism:

18.30

BBr₃ forms
a Lewis acid
complex with
the ether.

Br⁻ acts as a nucleo-
phile in an S_N2 reaction
to form CH₃Br.

+ CH₃Br

Water cleaves
the Lewis acid
complex.

Step 1: BBr₃ forms a Lewis acid complex with the ether.
Step 2: Br⁻ acts as a nucleophile in an S_N2 reaction to form CH₃Br.
Step 3: Water cleaves the Lewis acid complex.

18.31

Step 1: Protonation.
Step 2: Epoxide opening.
Step 3: Hydride shift.
Step 4: Loss of proton.

Reaction occurs by this route because of the stability of the intermediate tertiary carbocation.

18.32 (a) This is an S_N2 reaction because the rate depends on the concentrations of both reagents.

(b) This S_N2 reaction is a Williamson ether synthesis, in which an alkoxide displaces a halogen. In this reaction, KOH is used to form the phenoxide anion.

Fluoxetine

18.33

Step 1:   Protonation of the tertiary hydroxyl group.
Step 2:   Loss of water to form a tertiary carbocation.
Step 3:   Nucleophilic attack on the carbocation by the second hydroxyl group.

The tertiary hydroxyl group is more likely to be eliminated because the resulting carbocation is more stable.

18.34

This reaction is an S_N2 displacement and can't occur at an aryl carbon. DMF is a polar aprotic solvent that increases the rate of an S_N2 reaction by making anions more nucleophilic.

18.35   The reaction is a nucleophilic aromatic substitution. The intermediate Meisenheimer complex is stabilized by the $-NO_2$ group.

Meisenheimer complex

Acifluorfen

Step 1:   Addition of phenoxide.
Step 2:   Elimination of fluoride.

18.36 (a)

(b)

Step 1:   Protonation.
Step 2:   Loss of water.
Step 3:   Addition of ethanol.
Step 4:   Loss of proton.

18.37   A Claisen rearrangement is followed by a Diels–Alder reaction (Section 14.4). Tautomerization of the Claisen product to a phenol doesn't take place because no hydrogen is available for donation to oxygen.

Additional Problems

Naming Ethers

18.38

(a)

Ethyl 1-ethylpropyl ether

(b)

Di(*p*-chlorophenyl) ether

(c)

3,4-Dimethoxybenzoic acid

(d)

Cyclopentyloxycyclohexane

(e)

4-Allyl-2-methoxyphenol

18.39

(a)

Cyclohexyl isopropyl sulfide

(b)

1,1-(Dimethylthio)-
cyclohexane

(c)

1,2-Epoxycyclopentane

(d)

2-Methyltetrahydrofuran

(e)
$$CH_3CH-O-\triangleright$$
$$\;\;\;\;\;|$$
$$\;\;\;\;CH_3$$

Cyclopropyl isopropyl ether
or
Isopropoxycyclopropane

(f)

o-Nitrobenzenethiol

(g)
$$CH_3CH_2CHCHCHSCHCH_3$$

2-(Isopropylthio)-3,4-
dimethylhexane

(h)
$$CH_3CCH_3$$

2,2-Dimethoxypropane

(i)

o-Dimethoxybenzene

Synthesizing Ethers

18.40

(a)

(b)

$$CH_3CH=CH_2 \xrightarrow[\text{2. NaBH}_4]{\text{1. C}_6\text{H}_5\text{OH, Hg(OCOCF}_3)_2}$$

(c)

$$\xrightarrow{\text{RCO}_3\text{H}}$$

[RCO₃H = *meta*-Chloroperoxybenzoic acid]

(d)

$$\xrightarrow[\text{2. NaBH}_4]{\text{1. Hg(OCOCF}_3)_2, \ (CH_3)_2C=CH_2}$$

(e)

$$\xrightarrow{\text{RCO}_3\text{H}} \xrightarrow[\text{HOCH}_3]{\ ^-\text{OCH}_3} \xrightarrow[\text{2. CH}_3\text{I}]{\text{1. NaH}}$$

+ enantiomer   + enantiomer

(f)

$$\xrightarrow[\text{2. H}_2\text{O}_2, \ ^-\text{OH}]{\text{1. BD}_3} \xrightarrow[\text{2. CH}_3\text{I}]{\text{1. NaH}}$$

+ enantiomer   + enantiomer

18.41

(a)

$$\xrightarrow{\text{NaH}} \quad + \ H_2 \xrightarrow{\text{CH}_3\text{Br}} \quad + \ \text{NaBr}$$

Methyl 1-phenylethyl ether

(b)

$$\xrightarrow[\text{pyridine}]{\text{POCl}_3} \xrightarrow{\text{RCO}_3\text{H}}$$

Styrene   Phenylepoxyethane

(c)

1-Phenylethanethiol

(d)

tert-Butyl 1-phenylethyl ether

18.42

Step 1:   Protonation.
Step 2:   Attack of alcohol oxygen on carbocation.
Step 3:   Loss of proton.

Notice that this reaction is the reverse of acid-catalyzed cleavage of a tertiary ether (Problem 18.8).

18.43   In the trans isomer, the −OH and −Cl are in the trans orientation that allows epoxide formation to occur as described in Section 18.5. Epoxidation can't occur for the cis isomer, however. Instead, the base ⁻OH brings about E2 elimination, producing an enol, which tautomerizes to cyclohexanone.

trans-2-Chlorocyclohexanol

1,2-Epoxycyclohexane

cis-2-Chlorocyclohexanol   enol   Cyclohexanone

Reactions of Ethers and Epoxides

18.44

(a)

(b)

(c)

The enol tautomerizes to an aldehyde.

(d) $(CH_3)_3CCH_2OCH_2CH_3 \xrightarrow[H_2O]{HI} (CH_3)_3CCH_2OH + CH_3CH_2I$

18.45

(a)

$$\xrightarrow[\text{2. NaBH}_4]{\text{1. Hg(OCOCF}_3)_2, \ \text{CH}_3\text{CH}_2\text{OH}}$$

(b)

$$\xrightarrow[\text{H}_2\text{O}]{\text{HI}}$$

$$+ \text{CH}_3\text{I}$$

$$\xrightarrow{\text{PBr}_3}$$

(c)

$$\xrightarrow{\text{RCO}_3\text{H}}$$

+

$$\xrightarrow[\text{H}_2\text{O}]{^{-}\text{OH}}$$

(d)

$$\text{CH}_3\text{CH}_2\text{CH}_2\text{CH}_2\text{C}\equiv\text{CH} \xrightarrow[\substack{\text{Lindlar} \\ \text{catalyst}}]{\text{H}_2} \text{CH}_3\text{CH}_2\text{CH}_2\text{CH}_2\text{CH}=\text{CH}_2$$

$$\left| \begin{array}{l} \text{1. BH}_3, \ \text{THF} \\ \text{2. H}_2\text{O}_2, \ ^{-}\text{OH} \end{array} \right.$$

$$\text{CH}_3\text{CH}_2\text{CH}_2\text{CH}_2\text{CH}_2\text{CH}_2\text{OCH}_3 \xleftarrow[\text{2. CH}_3\text{I}]{\text{1. NaH}} \text{CH}_3\text{CH}_2\text{CH}_2\text{CH}_2\text{CH}_2\text{CH}_2\text{OH}$$

(e)

$$\underset{\text{from (d)}}{\text{CH}_3\text{CH}_2\text{CH}_2\text{CH}_2\text{CH}=\text{CH}_2} \xrightarrow[\text{2. NaBH}_4]{\text{1. Hg(OCOCF}_3)_2, \ \text{CH}_3\text{OH}} \underset{\overset{|}{\text{OCH}_3}}{\text{CH}_3\text{CH}_2\text{CH}_2\text{CH}_2\text{CHCH}_3}$$

18.46

$$\xrightarrow[\text{H}_2\text{O}]{\text{HI}} \text{ICH}_2\text{CH}_2\text{CH}_2\text{CH}_2\text{OH}$$

18.47

cis-5,6-Epoxydecane

protonation of epoxide oxygen

attack of H₂O
attack at carbon a

attack of H₂O
attack at carbon b

loss of proton

+ H_3O^+

The product of acid hydrolysis of *cis*-5,6-epoxydecane is a racemic mixture of *R,R* and *S,S* diols.

18.48

trans-5,6-Epoxydecane

protonation of
epoxide oxygen

attack of H$_2$O
attack at carbon a

attack of H$_2$O
attack at carbon b

loss of proton

The product of acid hydrolysis of *trans*-5,6-epoxydecane is a meso compound that is a diastereomer of the products formed in the previous problem.

18.49

cis-3-tert-Butyl-1,2-epoxycyclohexane

The hydroxyl groups in the product have a trans-diaxial relationship.

18.50 (a) (b)

| (2R,3R)-2,3-Epoxy-3-methylpentane | (2R,3S)-3-Methyl-2,3-pentanediol |

Reaction with aqueous acid causes ring opening to occur at C3 because the positive charge of the transition state is more stabilized at the tertiary carbon. Ring opening produces a diol in which the hydroxyl groups have a trans-diaxial relationship.

(c)   Since ring opening occurs exclusively at C3, the product is the 2R,3S isomer and is chiral. (If ring opening occurred equally at either carbon, the product would be a mixture of chiral enantiomers).

(d)   The product is optically active because only one enantiomer is produced.

18.51

Step 1:   Attack of the hydride nucleophile.
Step 2:   Protonation of the alkoxide anion.

The reaction is an S$_N$2 epoxide cleavage with ":H" as the nucleophile. The exact nature of the attacking nucleophile is not clear.

18.52

Deuterium and −OH have a trans-diaxial relationship in the product.

Spectroscopy

18.53 $M^+ = 116$ corresponds to a sulfide of molecular formula $C_6H_{12}S$, indicating one degree of unsaturation. The IR absorption at 890 cm^{-1} is due to a $R_2C=CH_2$ group.

2-Methyl-4(methylthio)-1-butene

a = 1.74 δ
b = 2.11 δ
c = 2.27 δ
d = 2.57 δ
e = 4.73 δ

18.54

Anethole

Peak	Chemical shift	Multiplicity	Split by:
a	1.84 δ	doublet	c
b	3.76 δ	singlet	
c	6.09 δ	two quartets	a,d
d	6.36 δ	doublet	c
e	6.82 δ.	doublet	f
f	7.23 δ	doublet	e

18.55 (a)

a = 0.99 δ
b = 1.34 δ
c,d = 1.61 δ

(b)

a = 2.31 δ
b = 3.58 δ
c = 4.08 δ
d = 6.90-7.25 δ

(c)

2,2-Dimethoxypropane

General Problems

18.56 (a)

(b)

(c)

(d)

18.57

The anethole ring has two functional groups – an ether and a hydrocarbon side chain with a double bond. The ether is synthesized first – by a Williamson ether synthesis from phenol and CH₃I. The hydrocarbon side chain results from a Friedel–Crafts acylation of the ether. Reduction of the ketone, bromination and dehydrohalogenation are used to introduce the double bond.

18.58

(a)

(b)

(c)

18.59

Step 1:   Attack of the alcohol on the triethyloxonium cation, with loss of diethyl ether.
Step 2:   Loss of proton.

Trialkyloxonium salts are more reactive alkylating agents than alkyl iodides because a neutral ether is an even better leaving group than an iodide ion.

18.60

18.61   The mechanism of Grignard addition to oxetane is the same as the mechanism of Grignard addition to epoxides, described in Section 18.6. The reaction proceeds at a reduced rate because oxetane is less reactive than ethylene oxide. The four-membered ring oxetane is less strained, and therefore more stable, than the three-membered ethylene oxide ring.

18.62

$$\frac{1.06 \text{ g vanillin}}{152 \text{ g/mol}} = 6.97 \times 10^{-3} \text{ mol vanillin}$$

$$\frac{1.60 \text{ g AgI}}{234.8 \text{ g/mol}} = 6.81 \times 10^{-3} \text{ mol AgI}$$

$$6.81 \times 10^{-3} \text{ mol} \rightarrow 6.81 \times 10^{-3} \text{ mol} \rightarrow 6.81 \times 10^{-3} \text{ mol} \rightarrow 6.81 \times 10^{-3} \text{ mol}$$
$$\text{AgI} \qquad\qquad \text{I}^- \qquad\qquad \text{CH}_3\text{I} \qquad\qquad -\text{OCH}_3$$

Thus, 6.97×10^{-3} mol of vanillin contain 6.81×10^{-3} mol of methoxyl groups. Since the ratio of moles vanillin to moles methoxyl is approximately 1:1, each vanillin contains one methoxyl group.

Vanillin

18.63 Disparlure, $C_{19}H_{38}O$, contains one degree of unsaturation, which the ^{1}H NMR absorption at 2.8 δ identifies as an epoxide ring.

6-Methylheptanoic acid   +   Undecanoic acid

18.64

18.65 Use the aldehyde-forming reaction shown in the previous problem.

o-Hydroxyphenylacetaldehyde

18.66

(a) CH₃MgBr, ether;   (b) H₂SO₄, H₂O;   (c) NaH, then CH₃I;
(d) *m*-ClC₆H₄CO₃H;   (e) ⁻OH, H₂O.

18.67

(a)

a = 1.27 δ
b = 3.31 δ
c = 4.57 δ

(b)

a = 3.71 δ
b = 5.17 δ
c = 6.08 δ
d = 7.1-7.6 δ

18.68

Step 1:   Addition of hydride to the ketone.
Step 2:   Displacement of bromide by the alkoxide anion.

The intermediate resulting from addition of H:⁻ is similar to the intermediate in a Williamson ether synthesis. Intramolecular reaction occurs to form the epoxide.

18.69

Chorismate   Prephenate

18.70 Note: For these problems it is best to redraw the molecule so that one can see the six atoms where the electrons will flow as a ring.

(a)   In this case the original Claisen rearrangement product will further tautomerize to a more stable phenol.

(b)

(c)   In this case the original Claisen rearrangement product will further tautomerize to a more stable phenol.

Review Unit 7: Alcohols, Ethers, and Related Compounds

Major Topics Covered (with vocabulary):

The −OH group:
alcohol phenol glycol wood alcohol hydrogen bonding alkoxide ion phenoxide ion acidity constant

Alcohols:
Grignard Reagent   Dess–Martin periodinane tosylate protecting group TMS ether

Phenols:
cumene hydroperoxide quinone hydroquinone ubiquinone

Acyclic ethers:
Williamson ether synthesis   Claisen rearrangement

Cyclic ethers:
Epoxide   oxirane   vicinal glycol   peroxyacid   crown ether   18-crown-6

Thiols and sulfides:
Thiol   sulfide   mercapto group   alkylthio group   disulfide   thiolate ion   trialkylsulfonium salt sulfoxide sulfone

Types of Problems:

After studying these chapters, you should be able to:

– Name and draw structures of alcohols, phenols, ethers, thiols and sulfides.
– Explain the properties and acidity of alcohols and phenols.
– Prepare all of the types of compounds studied.
– Predict the products of reactions involving alcohols, phenols and ethers.
– Formulate mechanisms of reactions involving alcohols, phenols and ethers.
– Identify alcohols, phenols and ethers by spectroscopic techniques.

Points to Remember:

*   The great biochemical importance of hydroxyl groups is due to two factors:(1) Hydroxyl groups make biomolecules more soluble because they can hydrogen-bond with water. (2) Hydroxyl groups can be oxidized to aldehydes, ketones and carboxylic acids. The presence of a hydroxyl group in a biological molecule means that all functional groups derived from alcohols can be easily introduced.

*   Carbon–carbon bond-forming reactions are always more difficult to learn than functional group transformations because it is often difficult to recognize the components that form a carbon skeleton. The product of a Grignard reaction contains a hydroxyl group bonded to at least one alkyl group (usually two or three). When looking at a product that might have been formed by a Grignard reaction, remember that a tertiary alcohol results from the addition of a Grignard reagent to either a ketone or an ester (the alcohol formed from the ester has two identical –R groups), a secondary alcohol results from addition of a Grignard reagent to an aldehyde, and a primary alcohol results from addition of a Grignard reagent to formaldehyde or to ethylene oxide. Remember that any molecule taking part in a Grignard reaction must not contain functional groups that might also react with the Grignard reagent.

* Ethers are quite unreactive, relative to many other functional groups we study, and are often used as solvents for that reason. Concentrated halogen acids can cleave ethers to alcohols and halides. Remember that the halide bonds to the less substituted alkyl group when the ethers are primary or secondary alkyl ethers.

* Epoxide rings can be opened by both acid and base. In basic ring-opening of an unsymmetrical epoxide (and in ring-opening using a Grignard reagent), attack occurs at the less substituted carbon of the epoxide ring. In acidic ring opening, the position of attack depends on the substitution pattern of the epoxide. When one of the epoxide carbons is tertiary, attack occurs at the more substituted carbon, but when the epoxide carbons are both primary or secondary, attack occurs at the less substituted carbon.

* The most useful spectroscopic data for these compounds: (1) A broad IR absorption in the range 3300 cm^{-1}–3600 cm^{-1} shows the presence of the −OH group of an alcohol or a phenol. (2) Hydrogens bonded to the −O−C− carbon of an alcohol or ether absorb in the range 3.5–4.5 δ in an ^1H NMR spectrum or in the range 50–80 δ in a ^{13}C NMR spectrum.

Self- Test:

A	B	
an insecticide	Febuprol (increases bile flow)	Epichlorohydrin

Provide a IUPAC name for **A** and identify chiral carbons. Would you expect **A** to be water-soluble? Label the hydroxyl groups of **A** as primary, secondary or tertiary. What products are formed when **A** reacts with: (a) CrO_3, H_3O^+ ; (b) PBr_3; (c) $(CH_3)_3SiCl$, Et_3N.

Name **B** by IUPAC rules. Show the three components that comprise **B**. The synthesis of **B** involves a ring-opening reaction of the epoxide epichlorohydrin. Use this information to propose a synthesis of **B** from epichlorohydrin and any alcohol or phenol.

C	D
Chlorbenside (larvicide)	Chlorothymol

What type of compound is **C**? Name **C**. Synthesize **C** from benzenethiol and benzene. What products are formed when **C** is treated with: (a) CH_3I; (b) H_2O_2, H_2O; (c) product o (b) + CH_3CO_3H.

Synthesize **D** from *m*-cresol; assume that isomeric product mixtures can be separated. Describe the IR and ^1H NMR spectrum of **D**.

Multiple Choice:

1. Hydrogen bonding affects all of the following except:
   (a) boiling point   (b) solubility   (c) position of $-OH$ absorption in IR spectrum
   (d) chemical shift of $-C-O-$ carbon in ^{13}C NMR.

2. Which of the following alcohols can't be synthesized by a Grignard reaction?
   (a) Benzyl alcohol   (b) Triphenylmethanol   (c) 3-Bromo-1-hexanol   (d) 1-Hexanol

3. Which of the following reactions of a chiral alcohol occurs with inversion of configuration?
   (a) reaction with NaH   (b) reaction with PBr_3 (c) reaction with tosyl chloride
   (d) reaction with $(CH_3)_3SiCl$

4. How many diols of the formula $C_4H_{10}O_2$ are chiral?
   (a) 2 (b) 3 (c) 4 (d) 5

5. Which alcohol is the least acidic?
   (a) 2-Propanol   (b) Methanol   (c) Ethanol(d) 2-Chloroethanol

6. Which of the following compounds can't be reduced to form $C_6H_5CH_2OH$?
   (a) $C_6H_5CO_2H$ (b) C_6H_5CHO (c) $C_6H_5CO_2CH_3$ (d) $C_6H_5OCH_3$

7. The reagent used for dehydration of an alcohol is:
   (a) PCl_3 (b) $POCl_3$ (c) $SOCl_2$ (d) PCC

8. All of the following are products of oxidation of a thiol except:
   (a) a sulfide   (b) a disulfide   (c) a sulfoxide   (d) a sulfone

9. In which of the following epoxide ring-opening reactions does attack of the nucleophile occur at the more substituted carbon of the epoxide ring?

(a)   (b)   (c)   (d)

10. Ethers are stable to all of the following reagents except:
    (a) nucleophiles (b) bases   (c) strong acids (d) dilute acids

Preview of Carbonyl Compounds

Chapter Outline

I. The carbonyl group.
   A. Kinds of carbonyl compounds.
      1. All carbonyl compounds contain an acyl group (R−C=O).
      2. The groups bonded to the acyl group can be of two types:
         a. Groups that can't act as leaving groups.
            i.   Examples: aldehydes and ketones.
         b. Groups that can act as leaving groups.
            ii. Examples: carboxylic acids, esters, amides, acid halides, lactones, acid anhydrides, lactams.
   B. Nature of the carbonyl group.
      1. The carbonyl carbon is sp^2-hybridized.
         a. A π bond is formed between carbon and oxygen.
         b. Carbonyl compounds are planar about the double bond.
      2. The carbon-oxygen bond is polar.
         a. The carbonyl carbon acts as an electrophile.
         b. The carbonyl oxygen acts as a nucleophile.
II. Reactions of carbonyl compounds.
   A. Nucleophilic addition reactions of aldehydes and ketones.
      1. A nucleophile adds to the carbonyl carbon.
      2. The resulting tetrahedral intermediate has two fates:
         a. The negatively charged oxygen can be protonated to form an alcohol.
         b. Loss of water leads to formation of a C=Nu double bond.
   B. Nucleophilic acyl substitution reactions.
      1. A nucleophile adds to the carbonyl carbon.
      2. The resulting tetrahedral intermediate expels a leaving group to form a new carbonyl compound.
      3. This type of reaction takes place with carbonyl compounds other than aldehydes and ketones.
   C. Alpha substitution reactions.
      1. Reaction can occur at the position next to the carbonyl carbon (α position).
         a. This type of reaction is possible because of the acidity of alpha hydrogens.
         b. Reaction with a strong base forms an enolate anion, which behaves as a nucleophile.
      2. All carbonyl compounds can undergo α substitution reactions.
   D. Carbonyl condensation reactions.
      1. Carbonyl condensation reactions occur when two carbonyl compounds react with each other.
      2. The enolate of one carbonyl compound adds to the carbonyl group of a second compound.

Solutions to Problems

1. According to the electrostatic potential maps, the carbonyl carbon of acetyl chloride is more electrophilic and the oxygen of acetone is more nucleophilic. This makes sense, because acetyl chloride has two electron-withdrawing groups that make its carbonyl carbon electron-poor and thus electrophilic. Because acetyl chloride has two electron-withdrawing groups, neither group is as nucleophilic as the carbonyl oxygen of acetone.

2. The reaction of cyanide ion with acetone is a nucleophilic addition reaction.

Cyanide anion adds to the positively polarized carbonyl carbon to form a tetrahedral intermediate. This intermediate is protonated to yield acetone cyanohydrin.

3. (a) This reaction is a *nucleophilic acyl substitution*. Ammonia adds to acetyl chloride, and chloride is eliminated, resulting in formation of an amide.
    (b) In this *nucleophilic addition* reaction, addition of the nucleophile is followed by loss of water.
    (c) Two molecules of cyclopentanone react in this *carbonyl condensation*.

Chapter 19 − Aldehydes and Ketones: Nucleophilic Addition Reactions

Chapter Outline

I. General information about aldehydes and ketones (Sections 19.1-19.3).
   A. Naming aldehydes and ketones (Section 19.1).
      1. Naming aldehydes.
         a. Aldehydes are named by replacing the –e of the corresponding alkane with -al.
         b. The parent chain must contain the –CHO group.
         c. The aldehyde carbon is always carbon 1.
         d. When the –CHO group is attached to a ring, the suffix -carbaldehyde is used.
      2. Naming ketones.
         a. Ketones are named by replacing the –e of the corresponding alkane with -one.
         b. Numbering starts at the end of the carbon chain nearer to the carbonyl carbon.
         c. The word acyl is used when a RCO– group is a substituent.
   B. Preparation of aldehydes and ketones (Section 19.2).
      1. Preparation of aldehydes.
         a. Oxidation of primary alcohols with Dess-Martin periodinane.
         b. Partial reduction of carboxylic acid derivatives.
      2. Preparation of ketones.
         a. Oxidation of secondary alcohols.
         b. Ozonolysis of alkenes with at least one disubstituted unsaturated carbon.
         c. Friedel–Crafts acylation of aromatic compounds.
         d. Preparation from carboxylic acid derivatives.
   C. Oxidation of aldehydes and ketones (Section 19.3).
      1. Aldehydes can be oxidized to carboxylic acids by many reagents.
         a. CrO_3 is used for normal aldehydes.
         b. Oxidation occurs through intermediate 1,1-diols.
      2. Ketones are generally inert to oxidation, but can be oxidized to carboxylic acids with strong oxidizing agents.
II. Nucleophilic addition reactions of aldehydes and ketones (Sections 19.4–19.13).
   A. Characteristics of nucleophilic addition reactions (Section 19.4).
      1. Mechanism of nucleophilic addition reactions.
         a. A nucleophile attacks the electrophilic carbonyl carbon from a direction 105° opposite to the carbonyl oxygen.
         b. The carbonyl group rehybridizes from sp^2 to sp^3, and a tetrahedral alkoxide intermediate is produced.
         c. The attacking nucleophile may be neutral or negatively charged.
            i. Neutral nucleophiles usually have a hydrogen atom that can be eliminated.
         d. The tetrahedral intermediate has two fates:
            i. The intermediate can be protonated to give an alcohol.
            ii. The carbonyl oxygen can be eliminated as –OH to give a product with a C=Nu double bond.

2. Relative reactivity of aldehydes and ketones.
    a. Aldehydes are usually more reactive than ketones in nucleophilic addition reactions for two reasons:
        i.   A nucleophile can approach the carbonyl group of an aldehyde more readily because only one alkyl group is in the way.
        ii. Aldehyde carbonyl groups are more strongly polarized and electrophilic because they are less stabilized by the inductive effect of alkyl groups.
    b. Aromatic aldehydes are less reactive than aliphatic aldehydes because the electron-donating aromatic ring makes the carbonyl carbon less electrophilic.

B. Nucleophilic addition reactions (Section 19.5-19.13).
    1. Hydration (Section 19.5).
        a. Water adds to aldehydes and ketones to give 1,1-diols (often referred to as gem-diols or hydrates).
        b. The reaction is reversible, but generally the equilibrium favors the carbonyl compound.
        c. Reaction is slow in pure water, but is catalyzed by both aqueous acid and base.
            i.   The base-catalyzed reaction is an addition of –OH, followed by protonation of the tetrahedral intermediate by water.
            ii. In the acid-catalyzed reaction, the carbonyl oxygen is protonated, and neutral water adds to the carbonyl carbon.
        d. The catalysts have different effects.
            i.   Base catalysis converts water to a better nucleophile.
            ii. Acid catalysis makes the carbonyl carbon a better electrophile.
        e. Reactions of carbonyl groups with H-Y, where Y is electronegative, are reversible; the equilibrium favors the aldehyde or ketone.
    2. Cyanohydrin formation (Section 19.6).
        a. HCN adds to aldehydes and ketones to give cyanohydrins.
            i.   The reaction is base-catalyzed and proceeds through a tetrahedral intermediate.
            ii. Equilibrium favors the cyanohydrin adduct.
        b. HCN is one of the very few protic acids that add to a carbonyl group
        c. Cyanohydrin formation is useful for the transformations that the –CN group can undergo.
            i.   The –CN group can be reduced, to form an amine.
            ii. The –CN group can be hydrolyzed, to produce a carboxylic acid.
    3. Addition of hydride and Grignard reagents (Section 19.7).
        a. Hydride addition.
            i.   LiAlH$_4$ and NaBH$_4$ act as if they are H:$^-$ donors and add to carbonyl compounds to form tetrahedral alkoxide intermediates.
            ii. In a separate step, water is added to protonate the intermediate, yielding an alcohol.

    b. Addition of Grignard reagents.
        i. Mg^{2+} complexes with oxygen, making the carbonyl group more electrophilic.
        ii. $R:^-$ adds to the carbonyl carbon to form a tetrahedral intermediate.
        iii. Water is added in a separate step to protonate the intermediate, yielding an alcohol.
        iv. Grignard reactions are irreversible because $R:^-$ is not a leaving group.

4. Addition of amines (Section 19.8).
    a. Amines add to aldehydes and ketones to form imines and enamines.
    b. An imine ($R_2C=NR$) is formed when a primary amine adds to an aldehyde or ketone.
        i. The process is acid-catalyzed.
        ii. A proton transfer converts the initial adduct to a carbinolamine.
        iii. Acid-catalyzed elimination of water yields an imine.
        iv. The reaction rate maximum occurs at pH = 4.5. At this pH, $[H^+]$ is high enough to catalyze elimination of water, but low enough so that the amine is nucleophilic.
        v. Some imine derivatives are useful for characterizing aldehydes and ketones.
    c. Enamines ($R_2N=CR-CR_2$) are produced when aldehydes and ketones react with secondary amines.
        i. The mechanism is similar to that of imine formation, except a proton from the α carbon is lost in the dehydration step.

5. Addition of hydrazine: the Wolff-Kishner reaction (Section 19.9).
    a. Hydrazine reacts with aldehydes and ketones in the presence of KOH to form alkanes.
        i. The intermediate hydrazone undergoes base-catalyzed bond migration, loss of N_2 and protonation to form the alkane.
    b. The Wolff-Kishner reduction can also be used to convert an acylbenzene to an alkylbenzene.

6. Addition of alcohols: acetal formation (Section 19.10).
    a. In the presence of an acid catalyst, two equivalents of an alcohol can add to an aldehyde or ketone to produce an acetal.
        i. The initial intermediate, a hemiacetal (hydroxy ether), is formed when the first equivalent of alcohol is added.
        ii. Protonation of –OH, loss of water, with formation of an oxonium ion, and addition of a second molecule of ROH yields the acetal.
    b. Since the reaction is reversible, changing the reaction conditions can drive the reaction in either direction.
    c. Because acetals are inert to many reagents, they can be used as protecting groups in syntheses.
        i. Diols are often used as protecting groups, forming cyclic acetals.

7. The Wittig reaction (Section 19.11).
    a. The Wittig reaction converts an aldehyde or ketone to an alkene.

b. Steps in the Wittig reaction:

   i.   An alkyl halide reacts with triphenylphosphine to form an alkyltriphenylphosphonium salt.

   ii. Butyllithium converts the salt to an ylide (phosphorane).

   iii. The ylide adds to an aldehyde or ketone to form a dipolar betaine.

      (a) In some cases, the addition is a one-step cycloaddition.

   iv. The betaine forms a four-membered ring intermediate (oxaphosphatane), which decomposes to form the alkene and triphenylphosphine oxide.

c. Uses of the Wittig reaction.

   i.   The Wittig reaction can be used to produce mono-, di-, and trisubstituted alkenes, but steric hindrance keeps tetrasubstituted alkenes from forming.

   ii. The Wittig reaction produces pure alkenes of known stereochemistry (excluding E,Z isomers).

8. Biological reductions (Section 19.12).

a. The Cannizzaro reaction is unique in that the tetrahedral intermediate of addition of a nucleophile to an aldehyde can expel a leaving group.

b. Steps in the Cannizzaro reaction.

   i. HO^- adds to an aldehyde with no α hydrogens to form a tetrahedral intermediate.

   ii. H^- is expelled and adds to another molecule of aldehyde.

   iii. The result is a disproportionation reaction, in which one molecule of aldehyde is oxidized and a second molecule is reduced.

c. The Cannizzaro reaction isn't synthetically useful, but it resembles the mode of action of the enzyme cofactor NADH.

9. Conjugate addition to α,β-unsaturated aldehydes / ketones (Section 19.13).

a. Steps in conjugate addition.

   i.   Because the double bond of an α,β-unsaturated aldehyde/ketone is conjugated with the carbonyl group, addition can occur at the β position, which is an electrophilic site.

   ii. Protonation of the α carbon of the enolate intermediate results in a product having a carbonyl group and a nucleophile with a 1,3 relationship.

b. Conjugate addition of amines.

   i.   Primary and secondary amines add to α,β-unsaturated aldehydes and ketones.

   ii. The conjugate addition product is often formed exclusively.

c. Conjugate addition of water.

   i.   Water can add to yield β-hydroxy aldehydes and ketones.

   ii. Conjugate addition of water also occurs in living systems.

d. Conjugate addition of organocopper reagents.

   i.   Conjugate addition of organocopper reagents (R_2CuLi) alkylates the double bond of α,β-unsaturated ketones.

   ii. This type of addition doesn't occur with other organometallic reagents.

        iii. Primary, secondary, tertiary, aryl, and alkenyl groups can be added.

        iv. The mechanism may involve conjugate addition of the diorganocopper anion, followed by transfer of an –R group.

III. Spectroscopy of aldehydes and ketones (Section 19.14).

  A. IR spectroscopy.

    1. The C=O absorption of aldehydes and ketones occurs in the range 1660–1770 cm^{-1}.

      a. The exact position of absorption can be used to distinguish between an aldehyde and a ketone.

      b. The position of absorption also gives information about other structural features, such as unsaturation and angle strain.

      c. The absorption values are constant from one compound to another.

    2. Aldehydes also show absorptions in the range 2720–2820 cm^{-1}.

  B. NMR spectroscopy.

    1. ^{1}H NMR spectroscopy.

      a. Aldehyde protons absorb near 10 δ, and show spin-spin coupling with protons on the adjacent carbon.

      b. Hydrogens on the carbon next to a carbonyl group absorb near 2.0-2.3 δ.

        i. Methyl ketone protons absorb at 2.1 δ.

    2. ^{13}C NMR spectroscopy.

      a. The carbonyl-group carbons absorb in the range 190-215 δ.

      b. These absorptions characterize aldehydes and ketones.

      c. Unsaturation lowers the value of δ.

  C. Mass spectrometry.

    1. Some aliphatic aldehydes and ketones undergo McLafferty rearrangement.

      a. A hydrogen on the γ carbon is transferred to the carbonyl oxygen, the bond between the α carbon and the β carbon is broken, and a neutral alkene fragment is produced.

      b. The remaining cation radical is detected.

    2. Alpha cleavage.

      a. The bond between the carbonyl group and the α carbon is cleaved.

      b. The products are a neutral radical and an acyl cation, which is detected.

Solutions to Problems

19.1   Remember that the principal chain must contain the aldehyde or ketone group and that an aldehyde group occurs only at the end of a chain. The aldehyde carbon is carbon 1 in an acyclic compound, and the suffix *-carbaldehyde* is used when the aldehyde group is attached to a ring.

(a)

$$CH_3CH_2\overset{\overset{\displaystyle O}{\|}}{C}CH(CH_3)_2$$

2-Methyl-3-pentanone

(b)

CH_2CH_2CHO

3-Phenylpropanal

(c)

$$CH_3\overset{\overset{\displaystyle O}{\|}}{C}CH_2CH_2CH_2\overset{\overset{\displaystyle O}{\|}}{C}CH_2CH_3$$

2,6-Octanedione

(d)

trans-2-Methylcyclo-hexanecarbaldehyde

(e) $CH_3CH{=}CHCH_2CH_2CHO$

4-Hexenal

(f)

cis-2,5-Dimethylcyclohexan(

19.2

(a)

$$CH_3\overset{\overset{\displaystyle CH_3}{|}}{C}HCH_2CHO$$

3-Methylbutanal

(b)

$$CH_3\overset{\overset{\displaystyle Cl}{|}}{C}HCH_2\overset{\overset{\displaystyle O}{\|}}{C}CH_3$$

4-Chloro-2-pentanone

(c)

CH_2CHO

Phenylacetaldehyde

(d)

$(CH_3)_3C$ CHO

cis-3-*tert*-Butylcyclo-hexanecarbaldehyde

(e)

$$H_2C{=}\overset{\overset{\displaystyle CH_3}{|}}{C}CH_2CHO$$

3-Methyl-3-butenal

(f)

$$CH_3CH_2\overset{\overset{\displaystyle CH_3}{|}}{C}HCH_2CH_2\overset{\overset{\displaystyle CH_3CHCl}{|}}{C}HCHO$$

2-(1-Chloroethyl)-5-methylheptanal

19.3   We have seen the first two methods of aldehyde preparation in earlier chapters.

(a)

$$CH_3CH_2CH_2CH_2CH_2OH \xrightarrow[\text{CH}_2\text{Cl}_2]{\text{Dess–Martin periodinane}} CH_3CH_2CH_2CH_2CHO$$

1-Pentanol

(b)

$$CH_3CH_2CH_2CH_2CH=CH_2 \xrightarrow[\text{2. Zn, H}_3\text{O}^+]{\text{1. O}_3} CH_3CH_2CH_2CH_2CHO \ + \ CH_2O$$

1-Hexene

(c)

$$CH_3CH_2CH_2CH=CH_2 \xrightarrow[\text{2. H}_2\text{O}_2, \ ^-\text{OH}]{\text{1. BH}_3, \text{ THF}} CH_3CH_2CH_2CH_2CH_2OH$$

1-Pentene   (a)

(d)

$$CH_3CH_2CH_2CH_2CO_2CH_3 \xrightarrow[\text{2. H}_3\text{O}^+]{\text{1. DIBAH}} CH_3CH_2CH_2CH_2CHO$$

19.4   All of these methods are familiar.

(a)

$$CH_3CH_2C\equiv CCH_2CH_3 \xrightarrow[\text{HgSO}_4]{\text{H}_3\text{O}^+} CH_3CH_2CH_2\overset{\displaystyle O}{\overset{\displaystyle \|}{C}}CH_2CH_3$$

(b)

(c)

(d)

19.5

cyanohydrin

Step 1:   Cyanide anion adds to the positively polarized carbonyl carbon to form a tetrahedral intermediate.

Step 2:   This intermediate is protonated to yield the cyanohydrin.

19.6

electron-withdrawing   electron-donating

The electron-withdrawing nitro group makes the aldehyde carbon of *p*-nitrobenzaldehyde more electron-poor (more electrophilic) and more reactive toward nucleophiles than the aldehyde carbon of *p*–methoxybenzaldehyde.

19.7

Chloral hydrate

19.8

The above mechanism is similar to other nucleophilic addition mechanisms we have studied. Since all steps are reversible, we can write the above mechanism in reverse to show how labeled oxygen is incorporated into an aldehyde or ketone.

This exchange is very slow in water but proceeds more rapidly when either acid or base is present.

19.9

2,2,6-Trimethylcyclohexanone

Cyanohydrin formation is an equilibrium process. Because addition of ⁻CN to 2,2,6-trimethylcyclohexanone is sterically hindered by the three methyl groups, the equilibrium lies toward the side of the unreacted ketone.

19.10

Reaction of a ketone or aldehyde with a primary amine yields an imine, in which C=O has been replaced by C=NR. Reaction of a ketone or aldehyde with a secondary amine yields an enamine, in which C=O has been replaced by C–NR$_2$, and the double bond has moved.

19.11

Step 1:   Protonation of nitrogen.

Step 2:   Addition of water.

Step 3:   Loss of proton.

Step 4:   Proton transfer.

Step 5:   Loss of amine.

19.12   The structure is an enamine, which is prepared from a ketone and a secondary amine. Find the amine and ketone components and draw the reaction.

from diethylamine

from cyclopentanone

19.13

19.14 Formation of the hemiacetal is the first step.

(a)

Step 1: Protonation of oxygen.
Step 2: Addition of –OH.
Step 3: Loss of proton.

(b)

Step 1: Protonation.
Step 2: Loss of H₂O.
Step 3: Addition of –OH.
Step 4: Loss of proton.

Protonation of the hemiacetal hydroxyl group is followed by loss of water. Attack by the second hydroxyl group of ethylene glycol forms the cyclic acetal ring.

19.15 Locate the two identical –OR groups to identify the alcohol that was used to form the acetal. (The illustrated acetal was formed from methanol.) Replace these two –OR groups by =O to find the carbonyl compound.

19.16 Locate the double bond that is formed by the Wittig reaction. The simpler or less substituted component comes from the ylide, and the more substituted component comes from the aldehyde or ketone. Triphenylphosphine oxide is a byproduct of all these reactions.

(a)

(b)

(c)

(d)

(e)

The Z isomer is also produced.

(f)

19.17

β-Ionylideneacetaldehyde

β-Carotene

19.18

Step 1: Addition of ⁻OH.

Step 2: Expulsion, addition of ⁻H.

Step 3: Proton transfer.

Step 4: Protonation.

This is an intramolecular Cannizzaro reaction.

19.19 Addition of the *pro-R* hydrogen of NADH takes place at the *Re* face of pyruvate.

19.20 The –OH group adds to the *Re* face at carbon 2, and –H$^+$ adds to the *Re* face at carbon 3, to yield (2R,3S)-isocitrate.

(2R,3S)-Isocitrate

19.21 The product is formed by 1,4 addition of $^-$CN, followed by protonation.

nucleophilic
addition

protonation

19.22 To choose the reactants that form a conjugate addition product, follow these steps:

(1) Give to the aldehyde or ketone carbon the number "1", and count two carbons away from the carbonyl carbon. The double bond in the α,β-unsaturated starting material connected the carbons numbered "2" and "3".

(2) The grouping bonded to the "3" carbon (circled here) came from the alkyllithium reagent.

(a)

H_2C=CHCCH$_3$ $\xrightarrow[\text{2. } H_3O^+]{\text{1. Li(CH}_3\text{CH}_2\text{CH}_2)_2\text{Cu}}$ CH$_3$CH$_2$CH$_2$CH$_2$CH$_2$CCH$_3$

2-Heptanone

(b)

$\xrightarrow[\text{2. } H_3O^+]{\text{1. Li(CH}_3)_2\text{Cu}}$

3,3-Dimethylcyclohexanone

(c)

$\xrightarrow[\text{2. } H_3O^+]{\text{1. Li(CH}_3\text{CH}_2)_2\text{Cu}}$

4-*tert*-Butyl-3-ethylcyclohexanone

(d)

$\xrightarrow[\text{2. } H_3O^+]{\text{1. Li(H}_2\text{C}=\text{CH})_2\text{Cu}}$

19.23

2-Cyclohexenone   1-Methyl-2-cyclohexen-1-ol   3-Methylcyclohexanone

2-Cyclohexenone is a cyclic α,β-unsaturated ketone whose carbonyl IR absorption occurs at 1685 cm^{-1}. If direct addition product **A** is formed, the carbonyl absorption vanishes and a hydroxyl absorption appears at 3300 cm^{-1}. If conjugate addition produces **B**, the carbonyl absorption shifts to 1715 cm^{-1}, where 6-membered-ring saturated ketones absorb.

19.24 Find the type of aldehyde or ketone and check Table 19.2 for absorptions.

(a) $H_2C=CHCH_2COCH_3$ absorbs at 1715 cm^{-1}. (4-Penten-2-one is not an α,β-unsaturated ketone.)

(b $CH_3CH=CHCOCH_3$ is an α,β-unsaturated ketone and absorbs at 1685 cm^{-1}.

(c)   2,2-Dimethylcyclopentanone, a five-membered-ring ketone, absorbs at 1750 cm^{-1}.

(d)   *m*-Chlorobenzaldehyde shows an absorption at 1705 cm^{-1} and two absorptions at 2720 cm^{-1} and 2820 cm^{-1}.

(e)   3-Cyclohexenone absorbs at 1715 cm^{-1}.

(f) $CH_3CH_2CH_2CH=CHCHO$ is an α,β-unsaturated aldehyde and absorbs at 1705 cm^{-1}.

19.25 In mass spectra, only charged particles are detected. The McLafferty rearrangement produces an uncharged alkene (not detected) and an oxygen-containing fragment, which is a cation radical and is detected. Alpha cleavage produces a neutral radical (not detected) and an oxygen-containing cation, which is detected. Since alpha cleavage occurs primarily on the more substituted side of the aldehyde or ketone, only this cleavage is shown.

(a)

4-Methyl-2-hexanone
$m/z = 114$

$m/z = 58$

$m/z = 43$

Both isomers exhibit peaks at $m/z = 43$ due to α-cleavage. The products of McLafferty rearrangement, however, occur at different values of m/z and can be used to identify each isomer.

(b)

3-Heptanone $m/z = 114$

$m/z = 72$

$m/z = 57$

The isomers can be distinguished on the basis of both α-cleavage products ($m/z = 57$ vs $m/z = 71$) and McLafferty rearrangement products ($m/z = 72$ vs $m/z = 86$).

(c)

The fragments from McLafferty rearrangement, which occur at different values of m/z, serve to distinguish the two isomers.

3-Methylpentanal
$m/z = 100$

19.26   **IR:** The only important IR absorption for the compound is seen at 1750 cm^{-1}, where 5-membered ring ketones absorb.

Mass spectrum: The products of alpha cleavage, which occurs in the ring, have the same mass as the molecular ion.

The charged fragment resulting from the McLafferty rearrangement appears at $m/z = 84$.

Visualizing Chemistry

19.27   It helps to know that all of these substances were prepared from aldehydes or ketones. Look for familiar groupings of atoms to identify the starting materials.

(a)   Notice that the substance pictured is a cyclic acetal. The starting materials were a diol (because cyclic acetals are prepared from diols) and an aldehyde (because an –H is bonded to the acetal carbon). Replace the two –OR groups with =O to identify the aldehyde starting material (acetaldehyde).

acetal

(b) We know that the product is an imine because it contains a carbon-nitrogen double bond. The carbon that is part of the C=N bond came from a ketone, and the nitrogen came from a primary amine.

imine

(c) The product is an enamine, formed from a ketone and a secondary amine. Nitrogen is bonded to the carbon that once bore the carbonyl oxygen.

bonded to the carbon that once bore the carbonyl oxygen.

enamine

(d) The secondary alcohol product might have been formed by either of two routes – by reduction of a ketone or by Grignard addition to an aldehyde.

19.28 The intermediate results from the addition of an amine to a ketone. The product is an enamine because the amine nitrogen in the carbinolamine intermediate comes from a secondary amine.

3-Methyl-2-butanone

19.29 (a) The nitrogen atom is sp^2-hybridized, and the geometry is planar.

(b) A p orbital holds the lone-pair electrons of nitrogen.

(c) The p orbital holding the lone-pair electrons of nitrogen is aligned for overlap with the π electrons of the enamine double bond. With this geometry, the nitrogen lone-pair electrons can be conjugated with the double bond, thus lowering energy.

Mechanism Problems

19.30 Note: While there may be several resonance forms for a particular intermediate, these solutions will always use the most stable form.

(a)

Mechanism:

(b)

Mechanism:

and

(c)

Mechanism:

19.31 Note: Each mechanism involves nucleophilc addition to a carbonyl followed by protonation.

(a)

Mechanism:

(b)

(c)

(d)

19.32 Note: Each reaction is an example of acetal formation and follows the same pattern: 1) protonation of the carbonyl oxygen; 2) nucleophilic addition of an alcohol oxygen; 3) removal of a proton from the oxygen now bearing the positive charge to give a hemiacetal intermediate. This is often accomplished with the solvent. In the problems where no solvent is given, a general base has been used for these solutions; 4) protonation of the hemiacetal OH converting it to water, a good leaving group; 5) loss of water to form a highly electrophilic oxonium ion; 6) nucleophilic addition of a second alcohol oxygen to the carbon adjacent to the oxonium ion; 7) deprotonation to yield the acetal product.

(a)

Mechanism:

(b)

Mechanism:

(c)

Mechanism:

(d)

Mechanism:

19.33 Note: Each mechanism is the acid catalyzed hydrolysis of an acetal and follow the same steps. Fundamentally, acetal hydrolysis is the mechanism for acetal formation in reverse. The steps are: 1) Protonation of one of the acetal oxygens; 2) Loss of an alcohol molecule with simultaneous formation of an oxonium ion; 3) Nucleophilic addition of water to the carbon that is double bonded to the oxonium ion; 4) deprotonation of the oxygen with the positive charge to give a hemiacetal intermediate; 5) protonation of the ether oxygen of the hemiacetal. 6) Loss of a second alcohol molecule with simultaneous formation of a carbon-oxygen double double bond; 7) deprotonation of the carbonyl oxygen to give the carbonyl.

(a)

CH₃O   OCH₃ $\xrightarrow[\text{H}_2\text{O}]{\text{H}^+, \text{cat}}$ +   2 CH₃OH

Mechanism:

H₃CO: :OCH₃   H₃CO: :OCH₃   :O: CH₃ HOCH₃   H₃CO: :OH   OH₂

:O: :O: OH₂   H₃CO: :OH   H₂O H   H₃CO: :OH
          CH₃OH

(b)

$\xrightarrow[\text{H}_2\text{O}]{\text{H}^+, \text{cat}}$ -OH ... OH   +

Mechanism:

CH₃ CH₃   CH₃ CH₃   OH CH₃ OH₂   OH H₃C CH₃ OH H H₂O

OH H₃C CH₃ OH   OH H₃C CH₃ OH H₂O H

(c)

Mechanism:

(d)

Mechanism:

19.34 Note: Each mechanism each mechanism involves the following steps for iminium ion formation: 1) Nucleophilic addition of the amine nitrogen to the carbonyl carbon to form a dipolar tetrahedral intermediate; 2) Proton transfer to yield a neutral carbonolamine; 3) Protonation of the OH group of the carbonolamine to convert it into a water, a good leaving group; 4) Loss of water through use of the lone pair of electrons on the nitrogen of the carbinolamine to form an iminium ion. At this point, the mechanism differs if the amine is primary the iminium ion is deprotonated to yield and imine. If the amine is secondary, an alpha hydrogen is removed in an E2-like process to yield an enamine.

(a)

Mechanism:

(b)

Mechanism:

(c)

Mechanism:

(d)

Mechanism:

19.35 Note: Each mechanism is an example of imine or enamine hydrolysis. The entire mechanism for each is the same with the exception of the first step. In problems (a) and (b) the first step is protonation of the imine nitrogen to give an iminium ion. In problems (c) and (d), the first step is protonation of the b-carbon. The remaining steps, which are identical for both mechanisms follow: 2) Nucleophilic addtion of water to the iminium carbon; 3) Deptotonation of the positively charged oxygen to form an intermediate carbinolamine; 4) Protonation of the carbinolamine nitrogen; 5) Expulsion of a neutral amine with simultaneous formation of a carbony; 6) Deprotonation of the carbonyl and protonation of the free amine.

(a)

Mechanism:

(b)

Mechanism:

(c)

Mechanism:

(d)

Mechanism:

19.36 (a)

Mechanism:

(b)

Mechanism:

19.37

Mechanism:

19.38 (a)

Mechanism:

(b)

Mechanism:

(c)

Mechanism:

(d)

Mechanism:

19.39 (a)

$$(Ph)_3P: + BrCH_2OCH_3 \longrightarrow (Ph)_3 \overset{+}{P}CH_2OCH_3 \ Br^- \xrightarrow{\ BuLi\ } (Ph)_3 \overset{+}{P} - \overset{+}{C}HOCH_3 + LiBr$$

(b)

Steps 1,4: Protonation.

Step 2: Addition of water.

Steps 3,6: Deprotonation.

Step 4: Loss of CH₃OH.

19.40

Step 1: Conjugate addition of water.

Step 2: Proton shift.

19.41 The same series of steps used to form an acetal is followed in this mechanism.

Step 1:   Protonation.

Step 2:   Addition of RSH.

Step 3:   Loss of proton.

Step 1: Protonation.

Step 2: Loss of water.

Step 3: Addition of RSH.

Step 4: Loss of proton.

19.42 Even though the product looks unusual, this reaction is made up of steps with which you are familiar.

Step 1: Addition of ylide.

Step 2: S$_N$2 displacement of dimethyl sulfide by O$^-$.

19.43

Step 1: Deprotonation by $^-$OH.

Step 2: Elimination of $^-$CN.

This sequence is the reverse of the mechanism shown in Section 19.6.

Step 3: Nucleophilic addition of cyanide

Step 4: Protonation of tetrahedral intermediate.

This step is a nucleophilic addition of cyanide.

19.44

Paraldehyde

Step 1:   Protonation makes the carbonyl carbon more electrophilic.

Steps 2,3,4:   Three successive additions of the carbonyl oxygen of acetaldehyde to the electrophilic carbonyl carbon, followed by loss of a proton (**Step 5**), give the cyclic product.

19.45

Step 1:   Aluminum, a Lewis acid, complexes with the carbonyl oxygen.

Step 2:   Complexation with aluminum makes the carbonyl group more electrophilic and facilitates hydride transfer from isopropoxide.

Step 3:   Treatment of the reaction mixture with aqueous acid cleaves the aluminum–oxygen bond and produces cyclohexanol.

Both the Meerwein–Ponndorf–Verley reaction and the Cannizzaro reaction are hydride transfers in which a carbonyl group is reduced by an alkoxide group, which is oxidized. Note that each aluminum triisopropoxide molecule is capable of reducing three ketone molecules.

19.46 (a) Nucleophilic addition of one nitrogen of hydrazine to one of the carbonyl groups, followed by elimination of water, produces a hydrazone.

(b) In a similar manner, the *other* nitrogen of hydrazine can add to the *other* carbonyl group of 2,4-pentanedione to form the pyrazole.

3,5-Dimethylpyrazole

The driving force behind this reaction is the formation of an aromatic ring. The reactions in both parts of this problem are nucleophilic addition of a primary amine (Step 2), followed by elimination of water to yield an imine or enamine (Step 5). All of the other steps are protonations (Steps 1,4) and deprotonations (Steps 3,6).

19.47 The same sequence of steps used in the previous problem leads to the formation of 3,5-dimethylisoxazole when hydroxylamine is the reagent. Loss of a proton in the last step of (b) results in a ring that is aromatic.

(a)

(b)

3,5-Dimethylisoxazole

19.48

Step 1: Addition of the phosphine nucleophile.

Step 2: Rotation of C–C bond.

Step 3: Elimination of triphenylphosphine oxide.

The final step is the same as the last step in a Wittig reaction. The same series of steps converts a trans alkene to a cis alkene.

19.49

Hydrogen peroxide and hydroxide react to form water and hydroperoxide anion.

Conjugate addition of hydroperoxide anion (Step 1) is followed by elimination of hydroxide ion, with formation of the epoxide ring (Step 2).

19.50

19.51

In this series of reactions, conjugate addition of an amine to an α,β-unsaturated ester (Step 1) is followed by nucleophilic addition of the amine to a second ester (Step 3), with loss of methanol (Step 5). The other two steps are proton transfers.

19.52 Some hydrogens have been omitted to make the drawing less cluttered.

In this series of equilibrium steps, the hemiacetal ring of α-glucose opens to yield the free aldehyde. Bond rotation is followed by formation of the cyclic hemiacetal of β-glucose. The reaction is catalyzed by both acid and base.

19.53 The first step of this reaction is the familiar nucleophilic addition to form a carbinolamine, which loses water to form an imine. This mechanism has been illustrated many times in Chapter 19. The remaining steps are deprotonations and protonations, plus loss of N₂, that lead to formation of the allylic alcohol.

Additional Problems

Naming Aldehydes and Ketones

19.54

(a)

$$CH_3CCH_2Br$$

Bromoacetone

(b)

(S)-2-Hydroxypropanal

(c)

$$CH_3CH_2CH_2CH_2CCH(CH_3)_2$$

2-Methyl-3-heptanone

(d)

(2S,3R)-2,3,4-Tri-hydroxybutanal

(e)

$$(CH_3)_3CCC(CH_3)_3$$

2,2,4,4-Tetramethyl-3-pentanone

(f)

$$CH_3C=CHCCH_3$$

4-Methyl-3-penten-2-one

(g) OHCCH₂CH₂CHO

Butanedial

(h)   CH=CHCHO

3-Phenyl-2-propenal

(i) 6,6-Dimethyl-2,4-cyclohexadienone

(j) O₂N—⟨ ⟩—C(=O)CH₃

p-Nitroacetophenone

19.55 Only 2-methylbutanal is chiral.

CH₃CH₂CH₂CH₂CHO   CH₃CH₂CHCHO   CH₃CHCH₂CHO   (CH₃)₃CCHO
Pentanal   2-Methylbutanal   3-Methylbutanal   2,2-Dimethylpropanal

CH₃CH₂CH₂CCH₃   CH₃CH₂CCH₂CH₃   CH₃CHCCH₃

2-Pentanone   3-Pentanone   3-Methyl-2-butanone

19.56

(a) 3-Methyl-3-cyclohexenone

(b) (R)-2,3-Dihydroxypropanal (D-Glyceraldehyde)

(c) 5-Isopropyl-2-methyl-2-cyclohexenone

(d) CH₃CHCCH₂CH₃ with CH₃
2-Methyl-3-pentanone

(e) CH₃CHCH₂CH with OH and O
3-Hydroxybutanal

(f) *p*-Benzenedicarbaldehyde

19.57 (a)   The α,ß-unsaturated ketone C₆H₈O contains one ring. Possible structures include:

Cyclobutenones and cyclopropenones are also possible.

(b)

$$CH_3C-CCH_3$$ and many other structures.

(c)

(d)

and many other structures.

Reactions of Aldehydes and Ketones

19.58   Reactions of phenylacetaldehyde:

(a)

(b)

(c)

(d)

(e)

(f)

(g)

(h)

Reactions of acetophenone:

(a)

(b)   no reaction

(c)

(d)

(e)

(f)

(g)

(h)

19.59 Remember:

$$RCH_2-X \quad + \quad (C_6H_5)_3P: \quad \longrightarrow \quad (C_6H_5)_3\overset{+}{P}CH_2R \ X^-$$
alkyl halide   Triphenyl   phosphonium salt
                     phosphine

$$(C_6H_5)_3\overset{+}{P}CH_2R \ X^- \quad + \quad CH_3CH_2CH_2CH_2^- \ Li^+ \quad \longrightarrow \quad (C_6H_5)_3\overset{+}{P}-\overset{-}{C}HR$$
phosphonium salt   Butyllithium   ylide

$$(C_6H_5)_3\overset{+}{P}-\overset{-}{C}HR \quad + \quad O=C \quad \longrightarrow \quad C=C$$
      ylide   aldehyde   alkene
                      or ketone

Alkyl halide   *Aldehyde/ketone*   *Product*

(a)

(b)

19.60   Remember from Chapter 17:
Primary alcohols are formed from formaldehyde + Grignard reagent. Secondary alcohols are formed from an aldehyde + Grignard reagent. Tertiary alcohols are formed from a ketone (or an ester) + Grignard reagent.

| *Aldehyde/Ketone* | *Grignard reagent* | *Product (after acidic workup)* |

(a), (b) reactions

| *Aldehyde/Ketone* | *Grignard reagent* | *Product (after acidic workup)* |

(c), (d) reactions

19.61   In general, ketones are less reactive than aldehydes for both steric (excess crowding) and electronic reasons. If the keto aldehyde in this problem were reduced with one equivalent of NaBH₄, the aldehyde functional group would be reduced in preference to the ketone.

For the same reason, reaction of the keto aldehyde with one equivalent of ethylene glycol selectively forms the acetal of the aldehyde functional group. The ketone can then be reduced with NaBH₄ and the acetal protecting group can be removed.

19.62

(a)

(b)

(c

(d)

or

19.63

(a)

The product resembles the starting material in having an aldehyde group, but a – CH_2– group lies between the aldehyde and the aromatic ring. The product aldehyde results from oxidation of an alcohol that is the product of a Grignard reaction between formaldehyde and benzylmagnesium bromide. The Grignard reagent is formed from benzyl bromide, which results from treatment of benzyl alcohol with PBr₃. Reduction of benzaldehyde yields the alcohol.

An alternate route:

In this scheme, the intermediate alcohol results from hydroboration of a double bond that is introduced by a Wittig reaction between benzaldehyde and methylenetriphenylphosphorane.

(b)

When you see a secondary amine and a double bond, you should recognize an enamine. The enamine is formed from the amine and acetophenone. Acetophenone, in turn, results from reaction of benzaldehyde with methylmagnesium bromide, followed by oxidation.

(c)

from (a)

The trisubstituted double bond suggests a Wittig reaction. Reaction of cyclopentanone with the Wittig reagent formed from benzyl bromide (formed from benzaldehyde in (a)) yields the desired product.

19.64

(a)   (b)   (c)

(d)   (e)   (f)   no reaction

(g)   +   (h)

19.65

(a)

1-Methylcyclohexene

The methyl group is introduced by a Grignard reaction with methylmagnesium bromide. Dehydration of the resulting tertiary alcohol produces 1-methylcyclohexene.

(b)

2-Phenylcyclohexanone

Reaction with phenylmagnesium bromide yields a tertiary alcohol that can be dehydrated. The resulting double bond can be treated with BH₃ to give an alcohol that can be oxidized to produce the desired ketone.

(c)

cis-1,2-Cyclohexanediol

Reduction, dehydration and hydroxylation yield the desired product.

(d)

from (c)   1-Cyclohexylcyclohexanol

A Grignard reaction forms 1-cyclohexylcyclohexanol.

Spectroscopy

19.66 Use Table 19.2 if you need help. Only carbonyl absorptions are noted.

Absorption:	*Due to:*
(a) 1750 cm^{-1}	5-membered ring ketone
1685 cm^{-1}	A,β-unsaturated ketone
(b) 1720 cm^{-1}	5-membered ring *and* aromatic ketone
(c) 1750 cm^{-1}	5-membered ring ketone
(d) 1705 cm^{-1}, 2720 cm^{-1},	aromatic aldehyde
2820 cm^{-1} 1715 cm^{-1}	aliphatic ketone

Compounds in parts (b)–(d) also show aromatic ring IR absorptions in the range 1450 cm^{-1}–1600 cm^{-1} and in the range 690–900 cm^{-1}.

19.67

3-Hydroxy-3-phenyl-
cyclohexanone

Compound **A** is a cyclic, nonconjugated enone whose carbonyl infrared absorption should occur at 1715 cm^{-1}. Compound **B** is an α,β-unsaturated, cyclic ketone; additional conjugation with the phenyl ring should lower its IR absorption below 1685 cm^{-1}. Because the actual IR absorption occurs at 1670 cm^{-1}, **B** is the correct structure.

19.68 undecylenic aldehyde

19.69 (a) IR shows that the unknown is a ketone, and ^{13}C NMR indicates that the carbonyl group is flanked by a secondary carbon and a tertiary carbon.

(b) The unknown is an aldehyde and contains an isopropyl group.

(c) The IR absorption shows that this compound is an α,β-unsaturated ketone, and the molecular formula shows 3 degrees of unsaturation. The ^{13}C NMR spectrum indicates 3 sp^2-hybridized carbons and 3 secondary carbons.

(a) $(CH_3)_2CHCCH_2CH_3$ (with C=O)   (b) $(CH_3)_2CHCH_2CHO$ (c)

19.70 Compound A has 4 degrees of unsaturation and is a five-membered ring ketone. The ^{13}C NMR spectrum has only three peaks and indicates that A is very symmetrical.

Compound A

19.71 As always, calculate the degree of unsaturation first, then use the available IR data to assign the principal functional groups.

(a)
$$a \quad c \overset{O}{\underset{\|}{}} b$$
$$CH_3CHCCH_3$$
$$|$$
$$Cl$$

a = 1.62 δ
b = 2.33 δ
c = 4.32 δ

(b)
$$a \quad c \overset{O}{\underset{\|}{}} b$$
$$(CH_3)_3CCH_2CCH_3$$

a = 1.02 δ
b = 2.12 δ
c = 2.33 δ

General Problems

19.72 4-Hydroxybutanal forms a cyclic hemiacetal when the hydroxyl oxygen adds to the aldehyde group.

Step 1:   Protonation.

Step 2:   Addition of –OH.

Step 3:   Loss of proton.

Methanol reacts with the cyclic hemiacetal to form 2-methoxytetrahydrofuran.

Step 1: Protonation.

Step 2: Loss of water.

Step 3: Addition of methanol.

Step 4: Loss of proton.

2-Methoxytetrahydrofuran is a cyclic acetal. The hydroxyl oxygen of 4-hydroxybutanal reacts with the aldehyde to form the cyclic ether linkage.

19.73

S_N2 substitution of hydroxide ion on $C_6H_5CHBr_2$ yields an unstable bromoalcohol intermediate, which loses Br^- to give benzaldehyde.

19.74

Attack can occur with equal probability on either side of the planar carbonyl group to yield a racemic product mixture that is optically inactive.

19.75

Step 1: Elimination to form the unsaturated imine.

Step 2: Conjugate addition of cysteine to the imine.

19.76 (a) Grignard addition to a conjugated ketone yields the 1,2 product, not the 1,4 product. LiAlH₄ reduces a ketone to an alcohol. The correct scheme:

(b) Oxidation of an alcohol with acidic CrO_3 converts primary alcohols to carboxylic acids, not to aldehydes. The correct scheme:

$$C_6H_5CH=CHCH_2OH \xrightarrow[\text{CH}_2\text{Cl}_2]{\text{Periodinane}} C_6H_5CH=CHCHO \xrightarrow[\text{H}^+]{\text{CH}_3\text{OH}} C_6H_5CH=CHCH(OCH_3)_2$$

(c) Treatment of a cyanohydrin with H_3O^+ produces a carboxylic acid, not an amine. The correct scheme:

19.77

$$\text{CH}_3\text{CCH}_2\text{CH}_2\text{CH}=\text{C(CH}_3)_2$$
6-Methyl-5-hepten-2-one

The reaction sequence involves protecting the ketone, converting the ester to an aldehyde, using a Wittig reaction to introduce a substituted double bond, and deprotecting the ketone.

19.78

1-Phenyl-1-propanol

(CH₃)₂NCH₂CH₂O   Tamoxifen

When you see a product that contains a double bond, and you also know that one of the starting materials is a ketone, it is tempting to use a Wittig reaction for synthesis. In this case, however, the tetrasubstituted double bond can't be formed by a Wittig reaction because of steric hindrance. The coupling step is achieved by a Grignard reaction between the illustrated ketone and a Grignard reagent, followed by dehydration. The Grignard reagent is synthesized from 1-phenyl-1-propanol. which can be prepared from benzene by either of two routes.

19.79   The molecular weight of Compound **A** shows that the molecular formula of A is C₅H₁₀O (one degree of unsaturation), and the IR absorption shows that A is an aldehyde. The uncomplicated ¹H NMR is that of 2,2-dimethylpropanal.

$$(CH_3)_3CCH \longleftarrow 9.7\ \delta$$

1.2 δ   Compound **A**

19.80   The IR of Compound **B** shows a ketone absorption. The splitting pattern of the ¹H NMR spectrum indicates an isopropyl group and indicates that the compound is a methyl ketone.

$$CH_3CHCCH_3 \longleftarrow 2.1\ \delta$$

2.4 δ

CH₃

1.2 δ   Compound **B**

19.81 Before looking at the ^{1}H NMR spectrum, we know that the compound of formula $C_9H_{10}O$ has 5 degrees of unsaturation, and we know from the IR spectrum that the unknown is an aromatic ketone. The splitting pattern in the ^{1}H NMR spectrum shows an ethyl group next to a ketone, according to chemical shift values.

19.82 The IR absorption is that of an aldehyde that isn't conjugated with the aromatic ring. The two triplets in the ^{1}H NMR spectrum are due to two adjacent methylene groups.

19.83

(a)

a = 1.44 δ
b = 4.08 δ
c,d = 6.98 δ,
         7.81 δ
e = 9.87 δ

(b)

a = 1.86 δ
b,c = 6.00 δ, 6.31 δ
d = 9.57 δ

19.84

(a)

$-CH_2CCH_2CH_3$

a = 1.01 δ
b = 2.47 δ
c = 3.66 δ
d = 7.28 δ

(b)

$(CH_3O)_2CHCH_2CCH_3$

a = 2.18 δ
b = 2.74 δ
c = 3.37 δ
d = 4.79 δ

19.85 The free aldehyde form of glucose (Problem 19.73) is reduced in the same manner described in the text for other aldehydes to produce the polyalcohol sorbitol.

Glucose   Sorbitol

19.86 (a)   The lettering here is large case and not in parenthesis but not in other places. The author should be consistent.   4-Heptanone

(b)   The lettering here is large case and not in parenthesis but not in other places. The author should be consistent.2,4-Dimethyl-3-pentanone

(c)   The lettering here is large case and not in parenthesis but not in other places. The author should be consistent.4,4-Dimethyl-2-pentanone

19.87 4-Methoxyphenylacetone

Chapter 20 – Carboxylic Acids and Nitriles

Chapter Outline

I. General information about carboxylic acids and nitriles (Sections 20.1–20.4).
   A. Naming carboxylic acids (Section 20.1).
      1. Noncyclic carboxylic acids are named by replacing the *-e* of the corresponding alkane by *-oic acid*. The $-CO_2H$ carbon is numbered C1.
      2. Compounds that have a carboxylic acid bonded to a ring are named by using the suffix *-carboxylic acid*. The $-CO_2H$ carbon is bonded to C1.
      3. Many carboxylic acids have historical, nonsystematic names.
   B. Naming nitriles.
      1. Simple nitriles are named by adding *-nitrile* to the alkane name.
         a. The nitrile carbon is C1.
      2. More complex nitriles are named as derivatives of carboxylic acids by replacing *–oic acid* by *-onitrile* or by replacing *-carboxylic acid* by *-carbonitrile*.
         a. The nitrile carbon is bonded to C1.
   C. Structure and properties of carboxylic acids (Section 20.2).
      1. The carbonyl group of carboxylic acids is sp^2-hybridized and planar.
      2. Carboxylic acids are strongly associated because of hydrogen bonding, and their boiling points are elevated.
   D. Carboxylic acid acidity (Sections 20.2–20.4).
      1. Dissociation of carboxylic acids (Section 20.2).
         a. Carboxylic acids react with bases to form salts that are water-soluble.
         b. Carboxylic acids dissociate slightly in dilute aqueous solution to give H_3O^+ and carboxylate anions.
            i. The K_a values for carboxylic acids are near 10^{-5}, making them weaker than mineral acids but stronger than alcohols.
         c. The relative strength of carboxylic acids is due to resonance stabilization of the carboxylate anion.
            i. Both carbon–oxygen bonds of carboxylate anions are the same length.
            ii. The bond length is intermediate between single and double bonds.
      2. Biological acids: the Henderson–Hasselbalch equation (Section 20.3).
         a. The pH of biological fluids (7.3) determines the ratio of dissociated (A^-) to nondissociated (HA) forms of carboxylic acids.
         b. This ratio can be calculated by using the Henderson–Hasselbalch equation.

$$\log \frac{[A^-]}{[HA]} = pH = pK_a$$

         c. At physiological pH, carboxylic acids are almost completely dissociated.
      3. Substituent effects on acidity (Section 20.4).
         a. Carboxylic acids differ in acid strength.
            i. Electron-withdrawing groups stabilize carboxylate anions and increase acidity.
            ii. Electron-donating groups decrease acidity.

   b. These inductive effects decrease with increasing distance from the carboxyl group.

   c. Substituent effects in substituted benzoic acids.

      i. Groups that are deactivating in electrophilic aromatic substitution reactions increase the acidity of substituted benzoic acids.

      ii. The acidity of benzoic acids can be used to predict electrophilic reactivity.

II. Carboxylic acids (Sections 20.5–20.6).

   A. Preparation of carboxylic acids (Section 20.5).

      1. Methods already studied.

         a. Oxidation of substituted alkylbenzenes.

         b. Oxidation of primary alcohols and aldehydes.

      2. Nitrile hydrolysis.

         a. Nitriles can be hydrolyzed by strong aqueous acids or bases to yield carboxylic acids.

         b. The sequence nitrile formation → nitrile hydrolysis can be used to prepare a carboxylic acid from a halide.

         c. This method is generally limited to compounds that can undergo S_N2 reactions.

      3. Carboxylation of Grignard reagents.

         a. A Grignard reagent can be treated with CO_2 and protonated to form a carboxylic acid.

         b. This method is limited to compounds that don't have other functional groups that interfere with Grignard reagent formation.

   B. Reactions of carboxylic acids (Section 20.6).

      1. Carboxylic acids can undergo some reactions typical of alcohols and ketones.

      2. Other types of reactions of carboxylic acids:

         a. Alpha substitution.

         b. Deprotonation.

         c. Nucleophilic acyl substitution.

         d. Reduction.

III. Chemistry of nitriles (Section 20.7).

   A. Preparation of nitriles.

      1. Nitriles can be prepared by S_N2 reaction of ^-CN with a primary alkyl halide.

      2. They can also be prepared by $SOCl_2$ dehydration of primary amides.

   B. Reactions of nitriles.

      1. Nitriles can react with nucleophiles via sp^2-hybridized imine intermediates.

      2. Aqueous base hydrolyzes nitriles to carboxylates, plus an amine/ammonia.

         a. The reaction involves formation of a hydroxy imine that isomerizes to an amide, which is further hydrolyzed to the carboxylate.

         b. Milder conditions allow isolation of the amide.

      3. Nitriles can also be hydrolyzed to carboxylic acids under acidic conditions.

      4. $LiAlH_4$ reduces nitriles to primary amines.

      5. Reaction of a nitrile with Grignard reagents yields a ketone.

IV. Spectroscopy of carboxylic acids and nitriles (Section 20.8).
   A. Infrared spectroscopy.
      1. The O–H absorption occurs at 2500–3300 cm^{-1} that is broad and easy to identify.
      2. The C=O absorption occurs at 1710–1760 cm^{-1}.
         a. The position of this absorption depends on whether the acid is free (1760 cm^{-1}) or associated (1710 cm^{-1}).
      3. Nitriles have an intense absorption at 2250 cm^{-1} that readily identifies them.
   B. NMR spectroscopy.
      1. ^{13}C NMR spectroscopy.
         a. Carboxylic acids absorb between 165–185 δ.
         b. Saturated acids absorb downfield from α,β-unsaturated acids.
         c. Nitrile carbons absorb in the range 115–130 δ.
      2. ^{1}H NMR spectroscopy.
         a. The carboxylic acid proton absorbs as a singlet at around 12 δ.

Solutions to Problems

20.1 Carboxylic acids are named by replacing *-e* of the corresponding alkane with *-oic acid*. The carboxylic acid carbon is C1.

When -CO$_2$H is a substituent of a ring, the suffix *-carboxylic acid* is used; the carboxyl carbon is not numbered in this system.

(a)
$$\underset{\text{3-Methylbutanoic acid}}{CH_3CHCH_2COH}$$

with CH$_3$ and O substituents

3-Methylbutanoic acid

(b)
$$CH_3CHCH_2CH_2COH$$
with Br and O
4-Bromopentanoic acid

(c)
$$CH_3CH_2CHCH_2CH_2CH_3$$
with CO$_2$H
2-Ethylpentanoic acid

(d)
C=C structure with H, H, H$_3$C, CH$_2$CH$_2$COH
(Z)-4-Hexenoic acid

(e)
$$CH_3CHCH_2CHCH_3$$
with CH$_3$ and CN
2,4-Dimethylpentanenitrile

(f)
cyclopentane with HO$_2$C and CO$_2$H
cis-1,3-Cyclopentane-dicarboxylic acid

20.2

(a)
$$CH_3CH_2CH_2CHCHCO_2H$$
with CH$_3$ and CH$_3$
2,3-Dimethylhexanoic acid

(b)
$$CH_3CHCH_2CH_2CO_2H$$
with CH$_3$
4-Methylpentanoic acid

(c)
cyclobutane with HO$_2$C, H, H, CO$_2$H
trans-1,2-Cyclobutanedicarboxylic acid

(d)
benzene ring with CO$_2$H and OH
o-Hydroxybenzoic acid

(e)
long chain with CO$_2$H
(9Z,12Z)-9,12-Octadecadienoic acid

(f) CH$_3$CH$_2$CH=CHCN
2-Pentenenitrile

20.3 Naphthalene is insoluble in water and benzoic acid is only slightly soluble. The *salt* of benzoic acid is very soluble in water, however, and we can take advantage of this solubility in separating naphthalene from benzoic acid.

Dissolve the mixture in an organic solvent, and extract with a dilute aqueous solution of sodium hydroxide or sodium bicarbonate, which will neutralize benzoic acid. Naphthalene remains in the organic layer, and all the benzoic acid, now converted to the benzoate salt, is in the aqueous layer. To recover benzoic acid, remove the aqueous layer, acidify it with dilute mineral acid, and extract with an organic solvent.

20.4

$$Cl_2CHCO_2H + H_2O \xrightleftharpoons{K_a} Cl_2CHCO_2^- H_3O^+$$

$$K_a = \frac{\left[Cl_2CHCO_2^-\right][H_3O^+]}{[Cl_2CHCO_2H]} = 3.32 \times 10^{-2}$$

	Initial molarity	*Molarity after dissociation*
Cl_2CHCO_2H	0.10 M	0.10 M $-$ y
Cl_2CHCO_2	0	y
H_3O^+	0	y

$$K_a = \frac{y \cdot y}{0.10 - y} = 3.32 \times 10^{-2}$$

Using the quadratic formula to solve for y, we find that y = 0.0434 M

$$\text{Percent dissociation} = \frac{0.0434 \text{ M}}{0.1000 \text{ M}} \times 100\% = 43.4\%$$

20.5 Use the Henderson–Hasselbalch equation to calculate the ratio.

(a)

$$\log\frac{[A^-]}{[HA]} = pH - pK_a = 4.50 - 3.83 = 0.67$$

$$\frac{[A^-]}{[HA]} = \text{antilog}(0.67) = 4.68 : [A^-] = 4.68\,[HA]$$

$$[HA] + [A^-] = 100\%$$
$$[HA] + 4.68\,[HA] = 5.68\,[HA] = 100\%$$
$$[HA] = 100\% \div 5.68 = 18\%$$
$$[A^-] = 100\% - 18\% = 82\%$$

82% of 0.0010 M glycolic acid is dissociated at pH = 4.50

(b)

$$\log\frac{[A^-]}{[HA]} = pH - pK_a = 5.30 - 4.87 = 0.43$$

$$\frac{[A^-]}{[HA]} = \text{antilog}(0.43) = 2.69 : [A^-] = 2.69\,[HA]$$

$$[HA] + [A^-] = 100\%$$
$$[HA] + 2.69\,[HA] = 3.69\,[HA] = 100\%$$
$$[HA] = 100\% \div 3.69 = 27\%$$
$$[A^-] = 100\% - 27\% = 73\%$$

73% of 0.0020 M propanoic acid is dissociated at pH = 5.30.

20.6   You would expect lactic acid to be a stronger acid because the electron-withdrawing inductive effect of the hydroxyl group can stabilize the lactate anion.

20.7

The pK_{a1} of oxalic acid is lower than that of a monocarboxylic acid because the carboxylate anion is stabilized both by resonance and by the electron-withdrawing inductive effect of the nearby second carboxylic acid group.

The pK_{a2} of oxalic acid is higher than pK_{a1} because an electrostatic repulsion between the two adjacent negative charges destabilizes the dianion.

20.8   A pK_a of 4.45 indicates that *p*-cyclopropylbenzoic acid is a weaker acid than benzoic acid. This, in turn, indicates that a cyclopropyl group must be electron-donating. Since electrondonating groups increase reactivity in electrophilic substitution reactions, *p*–cyclopropylbenzene should be more reactive than benzene toward electrophilic bromination.

20.9   Remember that electron-withdrawing groups increase carboxylic acid acidity, and electron donating groups decrease carboxylic acid acidity. Benzoic acid is more acidic than acetic acid.

20.10   In part (a), Grignard carboxylation must be used because the starting materials can't undergo S$_N$2 reactions. In (b), either method can be used.

(a) $(CH_3)_3 CCl \xrightarrow[\substack{\text{1. Mg, ether} \\ \text{2. } CO_2, \text{ ether} \\ \text{3. } H_3O^+}]{} (CH_3)_3 CCO_2H$

(b) $CH_3CH_2CH_2Br \xrightarrow[\substack{\text{1. Mg, ether} \\ \text{2. } CO_2, \text{ ether} \\ \text{3. } H_3O^+}]{} \quad or \quad \xrightarrow[\substack{\text{1. NaCN} \\ \text{2. } H_3O^+}]{} CH_3CH_2CH_2CO_2H$

20.11

The alcohol product can be formed by reduction of a carboxylic acid with LiAlH₄. The carboxylic acid can be synthesized either by Grignard carboxylation or by nitrile hydrolysis. The product can also be formed by a Grignard reaction between benzyl bromide and formaldehyde.

20.12

After treating the initial alcohol with PBr₃, the same steps as used in the previous problem can be followed. A Grignard reaction between the cycloalkylmagnesium bromide and formaldehyde also yields the desired product.

20.13

(a)

This symmetrical ketone can be synthesized by a Grignard reaction between propanenitrile and ethylmagnesium bromide.

(b)

p-Nitroacetophenone can be synthesized by either of two Grignard routes.

20.14

1-Phenyl-2-butanone

Once you realize that the product results from a Grignard reaction with a nitrile, this synthesis is easy.

20.15

The positions of the carbonyl absorptions are too similar to be useful. The –OH absorptions, however, are sufficiently different for distinguishing between the compounds; the broad band of the carboxylic acid hydroxyl group is especially noticeable.

20.16

^{1}H NMR:

The distinctive peak at 12 δ serves to identify the carboxylic acid. For the hydroxyketone, the absorption of the hydrogen on the oxygen-bearing carbon (3.5–4.5 δ) is significant. The position of absorption of the hydroxyl hydrogen is unpredictable, but addition of D$_2$O to the sample can be used to identify this peak.

^{13}C NMR:

The positions of the carbonyl carbon absorptions can be used to distinguish between these two compounds. The hydroxyketone also shows an absorption in the range 50–60 δ due to the hydroxyl group carbon.

Visualizing Chemistry

20.17

(a)

HO₂C

OCH₃

Br

3-Bromo-4-methoxybenzoic acid

(b)

H

H₃C   C   C   CO₂H

CH₃

3-Methyl-2-butenoic acid

(c)

CO₂H

1,3-Cyclopentadiene-carboxylic acid

(d)

OH

C=O

H₃C H

(S)-3-Cyclopentyl-2-methylpropanoic acid

20.18

(a)

CO₂H

Br

(b)

CO₂H

(CH₃)₂N

(a) *p*-Bromobenzoic acid is more acidic than benzoic acid because the electron-withdrawing bromine stabilizes the carboxylate anion.

(b) This *p*-substituted aminobenzoic acid is less acidic than benzoic acid because the electron-donating group destabilizes the carboxylate anion.

20.19

H H H₃C CH₃

HO   C   C   C   CO₂H

H H

H H H₃C CH₃

HO   C   C   C   Br

H H

Nitrile hydrolysis can't be used to synthesize the above carboxylic acid because the tertiary halide precursor (shown on the right) doesn't undergo S$_N$2 substitution with cyanide. Grignard carboxylation also can't be used because the acidic hydroxyl hydrogen interferes with formation of the Grignard reagent. If the hydroxyl group is protected, however, Grignard carboxylation can take place.

20.20 The electrostatic potential maps show that the aromatic ring of anisole is more electron-rich (red) than the aromatic ring of thioanisole, indicating that the methoxyl group is more strongly electron-donating than the methylthio group. Since electron-donating groups decrease acidity, *p*-(methylthio)benzoic acid is likely to be a stronger acid than *p*-methoxybenzoic acid.

Mechanism Problems

20.21 (a)

Mechanism:

(b)

Mechanism:

(c)

Mechanism:

20.22 (a)

Mechanism:

(b)

Mechanism:

(c)

Mechanism:

(d)

Mechanism:

20.23 (a)

Mechanism:

(b)

Mechanism:

(c)

Mechanism:

(d)

Mechanism:

20.24 **(a)**

Mechanism:

(b)

Mechanism:

(c)

Mechanism:

(d)

Mechanism:

20.25 (a)

Mechanism:

(b)

Mechanism:

(c)

Mechanism:

(d)

Mechanism:

20.26

Mechanism:

20.27 (a)

Step 1:   Protonation of acetal oxygen.

Step 2:   Loss of cyanohydrin.

Step 3:   Addition of water, followed by deprotonation.

(b)

Deprotonation of the cyanohydrin hydroxyl group is followed by loss of ⁻CN, forming 2-butanone.

20.28

Nucleophilic addition (1), alkyl shift (2), and displacement of bromide (3) lead to the observed product.

20.29

3-Phosphomevalonate 5-diphosphate   Isopentenyl diphosphate

20.30

The following steps take place in the Ritter reaction:

Step 1:   Protonation of the alkene double bond;

Step 2:   Attack of the nitrogen lone pair electrons on the carbocation;

Step 3:   Attack of water on the nitrile carbon;

Step 4:   Deprotonation;

Step 5:   Tautomerization to the ketone.

Additional Problems

Naming Carboxylic Acids and Nitriles

20.31

(a)

2,5-Dimethylhexanedioic acid

(b)

2,2-Dimethylpropanoic acid

(c) NC⌬CO$_2$H

m-Cyanobenzoic acid

(d) ⌬CO$_2$H

(*E*)-2-Cyclodecenecarboxylic acid

(e)

$$\underset{\underset{CH_3}{|}}{\overset{\overset{CH_3}{|}}{CH_3CCN}}$$

2,2-Dimethylpropanenitrile

(f)

$$CH_3CH_2CH_2\underset{\underset{CH_2CO_2H}{|}}{CH}CH_2CH_3$$

3-Ethylhexanoic acid

(g)

$$\underset{\underset{Br}{|}}{BrCH_2CHCH_2CH_2CO_2H}$$

4,5-Dibromopentanoic acid

(h) ⌬CN

2-Cyclopentenecarbonitrile

20.32

(a)

cis-1,2-Cyclohexanedicarboxylic acid

(b) HO$_2$CCH$_2$CH$_2$CH$_2$CH$_2$CH$_2$CO$_2$H
Heptanedioic acid

(c) CH$_3$C≡CCH=CHCO$_2$H

2-Hexen-4-ynoic acid

(d)

$$CH_3CH_2CH_2CH_2\underset{\underset{CH_2CH_3}{|}}{CH}CH_2\underset{\underset{CH_2CH_2CH_3}{|}}{CH}CO_2H$$

4-Ethyl-2-propyloctanoic acid

(e)

CO$_2$H
CO$_2$H
Cl 3-Chlorophthalic acid

(f) (C$_6$H$_5$)$_3$CCO$_2$H
Triphenylacetic acid

(g) ⌬—CN

2-Cyclobutenecarbonitrile

(h)

m-Benzoylbenzonitrile

20.33 (a)

$CH_3CH_2CH_2CH_2CH_2CO_2H$
Hexanoic acid

$CH_3CH_2CH_2\overset{\overset{\displaystyle CH_3}{|}}{C}HCO_2H$
2-Methylpentanoic acid

$CH_3CH_2\overset{\overset{\displaystyle CH_3}{|}}{C}HCH_2CO_2H$
3-Methylpentanoic acid

$CH_3\overset{\overset{\displaystyle CH_3}{|}}{C}HCH_2CH_2CO_2H$
4-Methylpentanoic acid

$CH_3CH_2\overset{\overset{\displaystyle CH_2CH_3}{|}}{C}HCO_2H$
2-Ethylbutanoic acid

$CH_3CH_2\overset{\overset{\displaystyle CH_3}{|}}{\underset{\underset{\displaystyle CH_3}{|}}{C}}CO_2H$
2,2-Dimethylbutanoic acid

$CH_3\overset{\overset{\displaystyle CH_3}{|}}{C}H\underset{\underset{\displaystyle CH_3}{|}}{C}HCO_2H$
2,3-Dimethylbutanoic acid

$CH_3\overset{\overset{\displaystyle CH_3}{|}}{\underset{\underset{\displaystyle CH_3}{|}}{C}}CH_2CO_2H$
3,3-Dimethylbutanoic acid

(b)

Cyclobutanecarbonitrile

$H_2C{=}CHCH_2CH_2CN$
4-Pentenenitrile

$H_2C{=}\overset{\overset{\displaystyle CH_3}{|}}{C}CH_2CN$
3-Methyl-3-butenenitrile

Other nitriles with the formula C_5H_7N can also be drawn.

20.34

Pregabalin

20.35

(2*R*,3*S*)-3-Carboxy-2-hydroxypentanedioic acid

Acidity of Carboxylic Acids

20.36

Less acidic —————————————→ *More acidic*

(a) CH_3CO_2H < HCO_2H < HO_2C-CO_2H

Acetic acid   Formic acid   Oxalic acid

(b)   *p*-Bromobenzoic acid < *p*-Nitrobenzoic acid (strongly < 2,4-Dinitrobenzoic acid
(weakly electron-   electron-withdrawing   (two strongly electron-
withdrawing   substituent)   withdrawing
substituent)   substituents)

(c) $FCH_2CH_2CO_2H < ICH_2CO_2H < FCH_2CO_2H$

In (c), the strongest acid has the most electronegative atom next to the carboxylic
acid group. The next strongest acid has a somewhat less electronegative atom next
to the carboxylic acid group. The weakest acid has an electronegative atom two
carbons away from the carboxylic acid group.

20.37   Remember that the conjugate base of a weak acid is a strong base. In other words, the
stronger the acid, the weaker the base derived from that acid.

Less basic —————————————→ *More basic*

(a) $Mg(OAc)_2 < Mg(OH)_2 < H_3C^-\ MgBr^+$

Acetic acid is a much stronger acid than water, which is a much, much stronger acid
than methane. The order of base strength is just the reverse.

(b)   Sodium p^- nitrobenzoate < Sodium benzoate < $HC\equiv C^-\ Na^+$

p-Nitrobenzoic acid is stronger than benzoic acid, which is much stronger than
acetylene.

(c) $HCO_2^-\ Li^+ < HO^-Li^+ < CH_3CH_2O^-\ Li^+$

LiOH and $LiOCH_2CH_3$ are very similar in basicity

20.38   (a) $K_a = 8.4 \times 10^{-4}$ for lactic acid

$pK_a = -\log(8.4 \times 10^{-4}) = 3.08$

(b) $K_a = 5.6 \times 10^{-6}$ for acrylic acid

$pK_a = -\log(5.6 \times 10^{-6}) = 5.25$

20.39   (a) $pK_a = 3.14$ for citric acid

$K_a = 10^{-3.14} = 7.2 \times 10^{-4}$

(b) $pK_a = 2.98$ for tartaric acid

$K_a = 10^{-2.98} = 1.0 \times 10^{-3}$

20.40

$$\log \frac{[A^-]}{[HA]} = pH - pK_a = 3.00 - 3.42 = -0.42$$

$$\frac{[A^-]}{[HA]} = \text{antilog}(-0.42) = 0.38 : [A^-] = 0.38\,[HA]$$

$$[HA] + 0.38\,[HA] = 1.38\,[HA] = 100\%$$

$$[HA] = 100\% \div 1.38 = 72\%$$

$$[A^-] = 100\% - 72\% = 28\%$$

At pH = 3.00, thioglycolic acid is 28% dissociated.

20.41

$$\log \frac{[A^-]}{[HA]} = pH - pK_a = 6.00 - 5.61 = 0.39$$

$$\frac{[A^-]}{[HA]} = \text{antilog}(0.39) = 2.45 : [A^-] = 2.45\,[HA]$$

$$[HA] + 2.45\,[HA] = 3.45\,[HA] = 100\%$$

$$[HA] = 100\% \div 3.45 = 29\%$$

$$[A^-] = 100\% - 27\% = 71\%$$

At pH = 6.00, uric acid is 71% dissociated.

Uric acid is acidic because the anion formed by dissociation of any of the three hydrogens is stabilized by resonance. An example of a resonance form:

20.42 Inductive effects of functional groups are transmitted through σ bonds. For oxalic acid, the electron-withdrawing inductive effect of one carboxylic acid group decreases the acidity of the remaining group. However, as the length of the carbon chain increases, the effect of one functional group on another decreases. In this example, the influence of the second carboxylic acid group on the ionization of the first is barely felt by succinic and adipic acids.

Reactions of Carboxylic Acids and Nitriles

20.43 (a) $CH_3CH_2CH_2CO_2H \xrightarrow[\text{2. } H_3O^+]{\text{1. } LiAlH_4} CH_3CH_2CH_2CH_2OH$

1-Butanol

(b) $CH_3CH_2CH_2CH_2OH \xrightarrow{PBr_3} CH_3CH_2CH_2CH_2Br$

1-Bromobutane

(c) $CH_3CH_2CH_2CH_2Br \xrightarrow[\text{2. } H_3O^+]{\text{1. } NaCN} CH_3CH_2CH_2CH_2CO_2H$

from (b)   Pentanoic acid

Grignard carboxylation can also be used.

(d) $CH_3CH_2CH_2CH_2Br \xrightarrow{K^{+\,-}OC(CH_3)_3} CH_3CH_2CH=CH_2$

from (b)   1-Butene

(e)

$CH_3CH_2CH_2CH_2Br \xrightarrow{2\,Li} 2\,CH_3CH_2CH_2CH_2{}^-Li^+ \xrightarrow{CuI} (CH_3CH_2CH_2CH_2)_2Cu^-Li^+$

from (b)

$(CH_3CH_2CH_2CH_2)_2Cu^-Li^+ + CH_3CH_2CH_2CH_2Br \longrightarrow CH_3(CH_2)_6CH_3$

Octane

20.44

(a) $CH_3CH_2CH_2CH_2OH \xrightarrow[H_3O^+]{CrO_3} CH_3CH_2CH_2CO_2H$

(b) $CH_3CH_2CH_2CH_2Br \xrightarrow{NaOH} CH_3CH_2CH_2CH_2OH \xrightarrow[H_3O^+]{CrO_3} CH_3CH_2CH_2CO_2H$

(c) $CH_3CH_2CH=CH_2 \xrightarrow[\text{2. } H_2O_2,\ ^-OH]{\text{1. } BH_3, THF} CH_3CH_2CH_2CH_2OH \xrightarrow[H_3O^+]{CrO_3} CH_3CH_2CH_2CO_2H$

(d)

$CH_3CH_2CH_2Br \xrightarrow[\text{2. } H_3O^+]{\text{1. } NaCN} CH_3CH_2CH_2CO_2H$

$CH_3CH_2CH_2Br \xrightarrow[\text{3. } H_3O^+]{\substack{\text{1. Mg, ether} \\ \text{2. } CO_2,\ \text{ether}}} CH_3CH_2CH_2CO_2H$

(e) $CH_3CH_2CH_2CH=CHCH_2CH_2CH_3 \xrightarrow[H_3O^+]{KMnO_4} 2\,CH_3CH_2CH_2CO_2H$

20.45 (a) $CH_3CH_2CH_2CN \xrightarrow[H_2]{NaOH} CH_3CH_2CH_2CO_2H \xrightarrow[\text{2. } H_3O^+]{\text{1. } LiAlH_4} CH_3CH_2CH_2CH_2OH$

(b) $CH_3CH_2CH_2CN \xrightarrow[\text{2. } H_2O]{\text{1. } LiAlH_4} CH_3CH_2CH_2CH_2NH_2$

(c)

$CH_3CH_2CH_2CN \xrightarrow[\text{2. } H_3O^+]{\text{1. } (CH_3)_2CHMgBr} CH_3CH_2CH_2\overset{\displaystyle O}{\overset{\|}{C}}\underset{\underset{\displaystyle CH_3}{|}}{C}HCH_3$

20.46

(a)

m-Chlorobenzoic acid

(b)

p-Bromobenzoic acid

(c)

Phenylacetic acid

Alternatively, benzyl bromide can be converted to a Grignard reagent, poured over CO$_2$, and the resulting mixture can be treated with aqueous acid in the last step.

20.47

p-Methylbenzoic acid

(a) 1. LiAlH$_4$
     2. H$_3$O$^+$

(b) NBS
     CCl$_4$

(c) 1. CH$_3$MgBr
     2. H$_3$O$^+$

(d) KMnO$_4$
     H$_2$O

In (c), the acidic proton reacts with the Grignard reagent to form methane, for no net reaction.

20.48 (a) $CH_3CH_2Br \xrightarrow{\text{Mg, ether}} CH_3CH_2MgBr \xrightarrow[\text{2. } H_3O^+]{\text{1. } ^{13}CO_2, \text{ ether}} CH_3CH_2^{13}CO_2H$

(b)

20.49 (a)

Grignard carboxylation can also be used to form the carboxylic acid.

(b)

Only Grignard carboxylation can be used because ¯CN brings about elimination of the tertiary bromide to form a double bond.

20.50 (a)   Grignard carboxylation can't be used to prepare the carboxylic acid because of the acidic hydroxyl group. Use nitrile hydrolysis.

(b)   Either method produces the carboxylic acid. Grignard carboxylation is a better reaction for preparing a carboxylic acid from a secondary bromide. Nitrile hydrolysis produces an optically active carboxylic acid from an optically active bromide.

(c)   Neither method of acid synthesis yields the desired product. Any Grignard reagent formed will react with the carbonyl functional group present in the starting material. Reaction with cyanide occurs at the carbonyl functional group, producing a cyanohydrin, as well as at halogen. However, if the ketone is first protected by forming an acetal, either method can be used.

(d)   Since the hydroxyl proton interferes with formation of the Grignard reagent, nitrile hydrolysis must be used to form the carboxylic acid.

20.51

$$H_2C=CHCH=CH_2 \xrightarrow[\text{1,4-addition}]{Br_2, CH_2Cl_2} BrCH_2CH=CHCH_2Br$$

$$\downarrow \text{2 NaCN, HCN}$$

$$H_2NCH_2CH_2CH=CHCH_2CH_2NH_2 \xleftarrow[\text{2. H}_2O]{\text{1. LiAlH}_4} NCCH_2CH=CHCH_2CN + 2 \text{ NaBr}$$

$$\downarrow H_2, \text{Pd/C}$$

$$H_2NCH_2CH_2CH_2CH_2CH_2CH_2NH_2$$
1,6-Hexanediamine

20.52

As in all of these more complex syntheses, other routes to the target compound are possible. This route was chosen because the Grignard reaction introduces a double bond without removing functionality at carbon 3. Dehydration occurs in the desired direction to produce a double bond conjugated with the carboxylic acid carbonyl group.

Spectroscopy

20.53 The peak at 1.08 δ is due to a *tert*-butyl group, and the peak at 11.2 δ is due to a carboxylic acid group. The compound is 3,3-dimethylbutanoic acid, $(CH_3)_3CCH_2CO_2H$.

20.54 Either ^{13}C NMR or ^{1}H NMR can be used to distinguish among these three isomeric carboxylic acids.

Compound	Number of ^{13}C NMR absorptions	Number of ^{1}H NMR absorptions	Splitting of ^{1}H NMR signals
CH$_3$(CH$_2$)$_3$CO$_2$H	5	5	1 triplet, peak area 3, 1.0 δ 1 triplet, peak area 2, 2.4 δ 2 multiplets, peak area 4, 1.5 δ 1 singlet, peak area 1, 12.0 δ
(CH$_3$)$_2$CHCH$_2$CO$_2$H	4	4	1 doublet, peak area 6, 1.0 δ 1 doublet, peak area 2, 2.4 δ 1 multiplet, peak area 1, 1.6 δ 1 singlet, peak area 1, 12.0 δ
(CH$_3$)$_3$CCO$_2$H	3	2	1 singlet, peak area 9, 1.3 δ 1 singlet, peak area 1, 12.1 δ

20.55 In all of these pairs, different numbers of peaks occur in the spectra of each isomer. (a), (b) Use either ^{1}H NMR or ^{13}C NMR to distinguish between the isomers.

Compound	Number of ^{13}C NMR absorptions	Number of ^{1}H NMR absorptions
(a)	3	2
(b)	5	4

(c) Use ^{1}H NMR to distinguish between these two compounds. The carboxylic acid proton of CH$_3$CH$_2$CH$_2$CO$_2$H absorbs near 12 δ, and the aldehyde proton of HOCH$_2$CH$_2$CH$_2$CHO absorbs near 10 δ and is split into a triplet.

(d) Cyclopentanecarboxylic acid shows four absorptions in both its ^{1}H NMR and ^{13}C NMR spectra. (CH$_3$)$_2$C=CHCH$_2$CO$_2$H shows six absorptions in its ^{13}C NMR and five in its ^{1}H NMR spectrum; one of the ^{1}H NMR signals occurs in the vinylic region (4.5 – 6.5 δ) of the spectrum. The ^{13}C NMR spectrum of the unsaturated acid also shows two absorptions in the C=C bond region (100–150 δ).

20.56 The compound has one degree of unsaturation, which is due to the carboxylic acid absorption seen in the IR spectrum.

$$CH_3 \overset{a}{C}H_2 \overset{b}{O}CH_2 \overset{c}{C}O_2 \overset{d}{H}$$

$$a = 1.26\ \delta$$
$$b = 3.64\ \delta$$
$$c = 4.14\ \delta$$
$$d = 11.12\ \delta$$

General Problems

20.57 2-Chloro-2-methylpentane is a tertiary alkyl halide and ⁻CN is a base. Instead of the desired S_N2 reaction of cyanide with a halide, E2 elimination occurs and yields 2-methyl-2-pentene.

20.58

20.59 (a)   Use CO2 instead of NaCN to form the carboxylic acid, or eliminate Mg from this reaction scheme and form the acid by nitrile hydrolysis.

(b)   Reduction of a carboxylic acid with LiAlH₄ yields an alcohol, not an alkyl group.

(c)   Acidic hydrolysis of the nitrile will also dehydrate the tertiary alcohol. Use basic hydrolysis to form the carboxylic acid.

20.60

Notice that the order of the reactions is very important. If toluene is oxidized first, the nitro group will be introduced in the meta position. If the nitro group is reduced first, oxidation to the carboxylic acid will reoxidize the –NH₂ group.

20.61

Other routes to this compound are possible. The illustrated route was chosen because it introduced the potential benzylic functional group and the potential carboxylic acid in one step. Notice that the aldehyde functional group and the cyclohexyl group both serve to direct the aromatic chlorination to the correct position. Also, reaction of the hydroxy acid with SOCl₂ converts –OH to –Cl and –CO₂H to –COCl. Treatment with H_3O^+ regenerates the carboxylic acid.

20.62

Substituent	pKₐ	Acidity	*E.A.S. reactivity
—PCl₂	3.59	Most acidic	Least reactive (most deactivating)
—OSO₂CH₃	3.84		
—CH=CHCN	4.03		
—HgCH₃	4.10		
—H	4.19		
—Si(CH₃)₃	4.27	Least acidic	Most reactive (least deactivating)

*Electrophilic aromatic substitution

Recall from Section 20.4 that substituents that increase acidity also decrease reactivity in electrophilic aromatic substitution reactions. Of the above substituents, only –Si(CH₃)₃ is an activator.

20.63 (a)

Again, other routes to this compound are possible. The above route was chosen because it has relatively few steps and because the Grignard reagent can be prepared without competing reactions. Notice that nitrile hydrolysis is not a possible route to this compound because the halide precursor is tertiary and doesn't undergo S_N2 substitution.

(b)

The product results from two Grignard reactions. As in (a), nitrile hydrolysis is not a route to this compound.

20.64   As we have seen throughout this book, the influence of substituents on reactions can be due to inductive effects and/or resonance effects. For *m*-hydroxybenzoic acid, the negative charge of the carboxylate anion is stabilized by the electron-withdrawing *inductive* effect of –OH, making this isomer more acidic. For *p*-hydroxybenzoic acid, the negative charge of the anion is destabilized by the electron-donating *resonance* effect of –OH that acts over the π electron system of the ring but is not important for *m*-substituents.

20.65

(a) BH_3, THF, then H_2O_2, OH⁻; (b) PBr_3; (c) Mg, then CO_2, then H_3O^+ (or ⁻CN, then H_3O^+); (d) $LiAlH_4$, then H_3O^+; (e) Dess–Martin periodinane, CH_2Cl_2; (f) H_2NNH_2,KOH

20.66 A compound with the formula C₄H₇N has two degrees of unsaturation. The IR absorption at 2250 cm⁻¹ identifies this compound as a nitrile.

$$a = 1.06\ \delta$$

$$\overset{a}{C}H_3\ \overset{b}{C}H_2\ \overset{c}{C}H_2\ C \equiv N \quad b = 1.68\ \delta$$

$$c = 2.31\ \delta$$

20.67 Both compounds contain four different kinds of protons (the H₂C= protons are nonequivalent). The carboxylic acid proton absorptions are easy to identify; the other three absorptions in each spectrum are more complex.

It is possible to assign the spectra by studying the methyl group absorptions. The methyl group peak of crotonic acid is split into a doublet by the geminal (CH₃CH=) proton, while the methyl group absorption of methacrylic acid is a singlet. The first spectrum (a) is that of crotonic acid, and the second spectrum (b) is that of methacrylic acid.

20.68 2-Methyl-2-pentenoic acid

20.69 (a) From the formula, we know that the compound has 2 degrees of unsaturation, one of which is due to the carboxylic acid group that absorbs at 183.0 δ. The ¹³C NMR spectrum also shows that no other sp² carbons are present in the sample and indicates that the other degree of unsaturation is due to a ring, which is shown to be a cyclohexane ring by symmetry and by the types of carbons in the structure.

(b) The compound has 5 degrees of unsaturation, and is a methyl-substituted benzoic acid. The symmetry shown by the aromatic absorptions identifies the compound as *p*-methylbenzoic acid.

20.70

Step 1:   Deprotonation

Step 2:   Decarboxylation

Step 3:   Protonation.

This reaction proceeds because of the loss of CO_2 and the stability of the enolate anion.

Chapter 21 – Carboxylic Acid Derivatives: Nucleophilic Acyl Substitution Reactions

Chapter Outline

I.   Introduction to carboxylic acid derivatives (Sections 21.1–21.2).
  A. Naming carboxylic acid derivatives (Section 21.1).
    1. Acid halides.
      a. The acyl group is named first, followed by the halide.
      b. For acyclic compounds, the *-ic acid* or *-oic acid* of the carboxylic acid name is replaced by *-oyl,* followed by the name of the halide.
        i.   There are eight exceptions, in which *-yl* is used
      c. For cyclic compounds, the *-carboxylic acid* ending is replaced by *-carbonyl,* followed by the name of the halide.
    2. Acid anhydrides.
      a. Symmetrical anhydrides are named by replacing *acid* by *anhydride.*
      b. Unsymmetrical anhydrides are named by citing the two acids alphabetically, followed by *anhydride.*
    3. Esters.
      a. Esters are named by first identifying the alkyl group and then the carboxylic acid group, replacing *-oic acid* by *-ate.*
    4. Amides.
      a. Amides with an unsubstituted –NH2 group are named by replacing *-oic acid* or *–ic acid* by *-amide* or by replacing *-carboxylic acid* with *-carboxamide.*
      b. If nitrogen is substituted, the nitrogen substituents are named in alphabetical order, and an *N-* is put before each.
    5. Thioesters.
      a. Thioesters are named like esters, using the prefix *thio-* before the name of the ester derivative of the carboxylic acid.
    6. Acyl phosphates.
      a. Acyl phosphates are named by citing the acyl group and adding the word *phosphate.*
  B. Nucleophilic acyl substitution reactions (Section 21.2).
    1. Mechanism of nucleophilic acyl substitution reactions.
      a. A nucleophile adds to the polar carbonyl group.
      b. The tetrahedral intermediate eliminates one of the two substituents originally bonded to it, resulting in a net substitution reaction.
      c. Reactions of carboxylic acid derivatives take this course because one of the groups bonded to the carbonyl carbon is a good leaving group.
      d. The addition step is usually rate-limiting.
    2. Relative reactivity of carboxylic acid derivatives.
      a. Both steric and electronic factors determine relative reactivity.
        i.   Steric hindrance in the acyl group decreases reactivity.
        ii. More polarized acid derivatives are more reactive than less polarized derivatives.

          iii. The effect of substituents on reactivity is similar to their effect on electrophilic aromatic substitution reactions.

      b. It is possible to convert more reactive derivatives into less reactive derivatives.

          i.   In order of decreasing reactivity: acid chlorides > acid anhydrides > thioesters > esters > amides.

          ii. Only esters, amides, and carboxylic acids are found in nature.

    3. Kinds of reactions of carboxylic acid derivatives:

      a. Hydrolysis: reaction with water to yield a carboxylic acid.

      b. Alcoholysis: reaction with an alcohol to yield an ester.

      c. Aminolysis: reaction with ammonia or an amine to yield an amide.

      d. Reduction.

          i.   Reaction with a hydride reducing agent yields an aldehyde or an alcohol.

          ii. Amides are reduced to yield amines.

      e. Reaction with an organometallic reagent to yield a ketone or alcohol.

II.   Reactions of carboxylic acids and their derivatives (Section 21.3–21.9).

  A. Nucleophilic acyl substitution reactions of carboxylic acids (Section 21.3).

    1. Carboxylic acids can be converted to acid chlorides by reaction with $SOCl_2$.

      a. The reaction proceeds through a chlorosulfite intermediate.

    2. Acid anhydrides are usually formed by heating the corresponding carboxylic acid to remove 1 equivalent of water.

    3. Conversion to esters.

      a. Conversion can be effected by the S_N2 reaction of a carboxylate and an alkyl halide.

      b. Esters can be produced by the acid-catalyzed reaction of a carboxylic acid and an alcohol.

          i.   This reaction is known as a Fischer esterification.

          ii. Mineral acid makes the acyl carbon more reactive toward the alcohol.

          iii. All steps are reversible.

          iv. The reaction can be driven to completion by removing water or by using a large excess of alcohol.

          v.   Isotopic labelling studies have confirmed the mechanism.

    4. Conversion to amides.

      a. Amides are difficult to form from carboxylic acids because amines convert carboxylic acids to carboxylate salts that no longer have electrophilic carbons.

      b. The reagent DCC (dicyclohexylcarbodiimide) can be used; it is used in the laboratory to form peptide bonds.

    5. Reduction of carboxylic acids.

      a. Reduction to alcohols can be achieved by use of $LiAlH_4$.

      b. BH_3 in THF easily reduces carboxylic acids to alcohols.

  B. Chemistry of carboxylic acid halides (Section 21.4).

    1. Carboxylic acid halides are prepared by reacting carboxylic acids with either $SOCl_2$ or PBr_3 to form the corresponding acyl halide.

2. Acyl halides are very reactive.
   a. Most reactions occur by nucleophilic acyl substitution mechanisms.
3. Hydrolysis.
   a. Acyl halides react with water to form carboxylic acids.
   b. The reaction mixture usually contains a base to scavenge the HCl produced.
3. Anhydride formation.
   a. Acid halides react with carboxylate ions to form anhydrides.
4. Alcoholysis.
   a. Acyl halides react with alcohols to form esters.
   b. Base is usually added to scavenge the HCl produced.
   c. Primary alcohols are more reactive than secondary or tertiary alcohols.
      i. It's often possible to esterify a less hindered alcohol selectively.
5. Aminolysis.
   a. Acid chlorides react with ammonia and amines to give amides.
   b. Either two equivalents of ammonia/amine must be used, or NaOH must be present, in order to scavenge HCl.
6. Reduction.
   a. LiAlH$_4$ reduces acid halides to alcohols.
      i. The reaction is a substitution of H$^-$ for Cl$^-$ that proceeds through an intermediate aldehyde, which is then reduced.
7. Reaction with organometallic reagents.
   a. Reaction with Grignard reagents yields tertiary alcohols and proceeds through an intermediate ketone.
   b. Reaction with diorganocopper (Gilman) reagents yields ketones.
      i. Reaction occurs by a radical mechanism.
      ii. This reaction doesn't occur with other carboxylic acid derivatives.
C. Chemistry of carboxylic acid anhydrides (Section 21.5).
   1. Acid anhydrides can be prepared by reaction of carboxylate anions with acid chlorides.
      a. Both symmetrical and unsymmetrical anhydrides can be prepared by this route.
   2. Acid anhydrides react more slowly than acid chlorides.
      a. Acid anhydrides undergo most of the same reactions as acid chlorides.
      b. Acetic anhydride is often used to prepare acetate esters.
      c. In reactions of acid anhydrides, half of the molecule is unused, making anhydrides inefficient to use.
D. Chemistry of esters (Section 21.6).
   1. Esters can be prepared by:
      a. S$_N$2 reaction of a carboxylate anion with an alkyl halide.
      b. Fischer esterification.
      c. Reaction of an acid chloride with an alcohol, in the presence of base.

2. Esters are less reactive than acid halides and anhydrides but undergo the same types of reactions.

3. Hydrolysis.

   a. Basic hydrolysis (saponification) occurs through a nucleophilic acyl substitution mechanism.

      i. Loss of alkoxide ion yields a carboxylic acid which is deprotonated to give a carboxylate anion.

      ii. Isotope-labelling studies confirm this mechanism.

   b. Acidic hydrolysis can occur by more than one mechanism.

      i. The usual route is by the reverse of Fischer esterification.

4. Aminolysis.

   a. Esters can be converted to amides by heating with ammonia/amines, but it's easier to start with an acid chloride.

5. Reduction.

   a. LiAlH$_4$ reduces esters to primary alcohols by a route similar to that described for acid chlorides.

   b. If DIBAH at −78 °C is used, reduction yields an aldehyde.

6. Reaction with Grignard reagents.

   a. Esters react twice with Grignard reagents to produce tertiary alcohols containing two identical substituents.

  E. Chemistry of amides (Section 21.7).

   1. Amides are prepared by the reaction of acid chlorides with ammonia/amines.

   2. Hydrolysis.

     a. Hydrolysis occurs under more severe conditions than needed for hydrolysis of other acid derivatives.

     b. Acid hydrolysis occurs by addition of water to a protonated amide, followed by loss of ammonia or an amine.

     c. Basic hydrolysis occurs by attack of HO$^-$, followed by loss of $^-$NH$_2$.

   3. Reduction.

     a. LiAlH$_4$ reduces amides to amines.

  F. Thiol esters and acyl phosphates (Section 21.8).

   1. Nature uses thiol esters and acyl phosphates in nucleophilic acyl substitution reactions because they are intermediate in reactivity between acid anhydrides and esters.

   2. Acetyl CoA is used as an acylating agent.

III. Polyamides and polyesters (Section 21.9).

  A. Formation of polyesters and polyamides.

   1. When a diamine and a diacid chloride react, a polyamide is formed.

   2. When a diacid and a diol react, a polyester is formed.

   3. These polymers are called step-growth polymers because each bond is formed independently of the others.

    B. Types of polymers.
       1. Nylons are the most common polyamides.
       2. The most common polyester, Dacron, is formed from dimethylterephthalate and ethylene glycol.
       3. Biodegradable polymers are usually polyesters of naturally-occurring hydroxycarboxylic acids.

IV. Spectroscopy of carboxylic acid derivatives and nitriles (Section 21.10).
    A. Infrared spectroscopy.
       1. All of these compounds have characteristic carbonyl absorptions that help identify them; these are listed in Table 21.3.
    B. NMR spectroscopy is of limited usefulness in distinguishing carboxylic acid derivatives.
       1. Hydrogens next to carbonyl groups absorb at around 2.1 δ in a ^{1}H NMR spectrum, but this absorption can't be used to distinguish among carboxylic acid derivatives.
       2. Carbonyl carbons absorb in the range 160–180 δ, but, again, this absorption can't be used to distinguish among carboxylic acid derivatives.

Solutions to Problems

21.1   Table 21.1 lists the suffixes for naming carboxylic acid derivatives. The suffixes used when the functional group is part of a ring are in parentheses.

(a)

$$CH_3CHCH_2CH_2CCl$$

with CH_3 and O substituents shown above

4-Methylpentanoyl
chloride

(b)

CH_2CNH_2

Cyclohexylacetamide

(c)

$$CH_3CHCOCHCH_3$$

with O above and CH_3 CH_3 below

Isopropyl
2-methylpropanoate

(d)

Benzoic anhydride

(e)

Isopropyl
cyclopentanecarboxylate

(f)

Cyclopentyl
2-methylpropanoate

(g)

$$H_2C={}CHCH_2CH_2CNHCH_3$$

with O above

N-Methyl-4-pentenamide

(h)

(*R*)-2-Hydroxypro-
panoyl phosphate

(i)

Ethyl 2,3-dimethyl-
2-butenethioate

21.2

(a)

Phenyl benzoate

(b)

$$CH_3CH_2CH_2\overset{\overset{\displaystyle O}{\|}}{C}NCH_2CH_3$$
$$\overset{|}{CH_3}$$

N-Ethyl-*N*-methylbutanamide

(c)

$$\overset{\overset{\displaystyle CH_3}{|}}{CH_3CHCH_2CH}\overset{\overset{\displaystyle O}{\|}}{C}Cl$$
$$\overset{|}{CH_3}$$

2,4-Dimethylpentanoyl
chloride

(d)

Methyl 1-methylcyclo-
hexanecarboxylate

(e)

$$CH_3CH_2\overset{\overset{\displaystyle O}{\|}}{C}CH_2\overset{\overset{\displaystyle O}{\|}}{C}OCH_2CH_3$$

Ethyl-3-oxopentanoate

(f)

Methyl *p*-bromobenzenethioate

(g)

$$H\overset{\overset{\displaystyle O}{\|}}{C}\underset{O}{}\overset{\overset{\displaystyle O}{\|}}{C}CH_2CH_3$$

Formic propanoic
anhydride

(h)

cis-2-Methylcyclopentanecarbonyl bromide

21.3

addition of methoxide
to form a tetrahedral
intermediate

elimination of Cl⁻

21.4   Use Figure 21.2 if you need help.

Most reactive ———————————→ *Least reactive*

(a)

$$\underset{\text{CH}_3\overset{\displaystyle O}{\overset{\|}{C}}\text{Cl}}{} \quad > \quad \underset{\text{CH}_3\overset{\displaystyle O}{\overset{\|}{C}}\text{OCH}_3}{} \quad > \quad \underset{\text{CH}_3\overset{\displaystyle O}{\overset{\|}{C}}\text{NH}_2}{}$$

(b)

$$\underset{\text{CH}_3\overset{\displaystyle O}{\overset{\|}{C}}\text{OCH(CF}_3)_2}{} \quad > \quad \underset{\text{CH}_3\overset{\displaystyle O}{\overset{\|}{C}}\text{OCH}_2\text{CCl}_3}{} \quad > \quad \underset{\text{CH}_3\overset{\displaystyle O}{\overset{\|}{C}}\text{OCH}_2\text{CH}_3}{}$$

The most reactive acyl derivatives contain strongly electron-withdrawing groups in the part of the structure that is to be the leaving group.

21.5   Identify the nucleophile (boxed) and the leaving group (circled), and replace the leaving group by the nucleophile in the product.

(a)

(b)

(c)

(d)

21.6   The structure represents the tetrahedral intermediate in the reaction of methyl cyclopentylacetate with hydroxide, a nucleophile. The products are cyclopentylacetate anion and methanol.

Methyl cyclopentylacetate   Cyclopentylacetate

21.7   In Fischer esterification, an alcohol undergoes a nucleophilic acyl substitution with a carboxylic acid to yield an ester. The mineral acid catalyst makes the carboxyl group of the acid more electrophilic. Predicting the products is easier if the two reagents are positioned so that the reacting functional groups point towards each other.

(a)

Acetic acid   1-Butanol   Butyl acetate

(b)

Butanoic acid   Methanol   Methyl butanoate

(c)

Cyclopentane-carboxylic acid   2-Propanol   Isopropyl cyclopentanecarboxylate

21.8 Under Fischer esterification conditions, many hydroxycarboxylic acids can form intramolecular esters (lactones).

5-Hydroxypentanoic acid   a lactone

21.9 Pyridine neutralizes the HCl byproduct by forming pyridinium chloride. This neutralization removes from the product mixture acid that might cause side reactions. As mentioned previously, positioning the reacting groups so that they face each other makes it easier to predict the products.

(a)

Propanoyl chloride   Methanol   Methyl propanoate

(b)

Acetyl chloride   Ethanol   Ethyl acetate

(c)

Benzoyl chloride   Ethanol   Ethyl benzoate

21.10 As explained in the text, only simple, low boiling alcohols are convenient to use in the Fischer esterification reaction. Thus, reaction of cyclohexanol with benzoyl chloride is the preferred method for preparing cyclohexyl benzoate.

Benzoyl chloride   Cyclohexanol   Cyclohexyl benzoate

21.11

nucleophilic
addition of
morpholine

deprotonation
by hydroxide

Trimetozine   + NaCl   elimination
of chloride

21.12 An extra equivalent of base must be added to neutralize the acid produced in these reactions. In (a) and (b), two equivalents of the amine may be used in place of NaOH.

(a)

Propanoyl chloride   Methylamine   N-Methylpropanamide

(b)

Benzoyl chloride   Diethylamine   *N,N*-Diethylbenzamide

(c)

Propanoyl chloride   Propanamide

21.13 Two combinations of acid chloride and organocopper reagent are possible.

(a)

(b)

21.14

Step 1:   Nucleophilic addition of *p*-hydroxyaniline.

Step 2:   Deprotonation by hydroxide.

Step 3:   Loss of acetate ion.

21.15

Phthalic anhydride

The second half of the anhydride becomes a carboxylic acid.

21.16   Acidic hydrolysis of an ester is a reversible reaction because the products are an alcohol and a carboxylic acid. Basic hydrolysis of an ester is irreversible because its products are an alcohol and a carboxylate anion, which has a negative charge and does not react with nucleophiles.

21.17

21.18 Lithium aluminum hydride reduces an ester to form two alcohols.

(a)

$$CH_3CH_2CH_2CH(CH_3)COCH_3 \xrightarrow[\text{2. H}_3\text{O}^+]{\text{1. LiAlH}_4} CH_3CH_2CH_2CH(CH_3)CH_2OH + CH_3OH$$

2-Methyl-1-pentanol   Methanol

(b)

$$\xrightarrow[\text{2. H}_3\text{O}^+]{\text{1. LiAlH}_4}$$

Benzyl alcohol   Phenol

21.19 Remember that Grignard reagents can only be used with esters to form a tertiary alcohol that has two identical substituents. Identify these two substituents, which come from the Grignard reagent, and work backward to select the ester (the alkyl group of the ester is unimportant).

Tertiary Alcohol ⟵ *Grignard Reagent* + *Ester*

(a)

2 CH₃MgBr +

(b)

2 [phenyl]MgBr + H₃C–C(=O)–OR

(c)

2 CH₃CH₂MgBr + CH₃CH₂CH₂CH₂–C(=O)–OR

21.20

N-Ethylbenzamide

21.21

The product is a *N,N*-disubstituted amine, which can be formed by reduction of an amide. The amide results from treatment of an acid chloride with the appropriate amine. The acid chloride is the product of the reaction of SOCl₂ with a carboxylic acid that is formed by carboxylation of the Grignard reagent synthesized from the starting material.

21.22 Even though the entire molecule of coenzyme A is biologically important, we are concerned in this problem only with the –SH group. The remainder of the structure is represented here as "R".

Acetyl CoA

Step 1:   Nucleophilic addition of the –SR group of CoA (after deprotonation) to acetyl adenylate to form a tetrahedral intermediate.

Step 2:   Loss of adenosine monophosphate.

21.23   In each example, if n molecules of one component react with n molecules of the other component, a polymer with n repeating units is formed, and $2n$ small molecules are formed as byproducts; these are shown in each reaction.

(a)

$$BrCH_2CH_2CH_2Br + HOCH_2CH_2CH_2OH \xrightarrow{\text{Base}} \left(CH_2CH_2CH_2OCH_2CH_2CH_2O\right)_n$$

$$+ \ 2n \ HBr$$

(b)

$$HOCH_2CH_2OH + HO_2C(CH_2)_6CO_2H \xrightarrow[\text{catalyst}]{H_2SO_4} \left(OCH_2CH_2O-\overset{O}{\underset{\|}{C}}(CH_2)_6\overset{O}{\underset{\|}{C}}\right)_n$$

$$+ \ 2n \ H_2O$$

(c)

$$H_2N(CH_2)_6NH_2 + \overset{O}{\underset{\|}{Cl C}}(CH_2)_4\overset{O}{\underset{\|}{CCl}} \longrightarrow \left(HN\overset{O}{\underset{\|}{C}}(CH_2)_6NH-\overset{O}{\underset{\|}{C}}(CH_2)_4\overset{O}{\underset{\|}{C}}\right)_n$$

$$+ \ 2n \ HCl$$

21.24

1,4-Benzenedicarboxylic acid   1,4-Benzenediamine

Kevlar $+ \ 2n \ H_2O$

21.25 Use Table 21.3 if you need help.

Absorption	Functional group present
(a) 1735 cm^{-1}	Saturated ester *or* 6-membered ring lactone
(b) 1810 cm^{-1}	Saturated acid chloride
(c) $2500 - 3300$ cm^{-1} and 1710 cm^{-1}	Carboxylic acid
(d) 1715 cm^{-1}	Saturated ketone *or* 6-membered ring ketone

21.26 (a)   IR 1735 cm^{-1} corresponds to a saturated ester.

The remaining five carbons and twelve hydrogens can be arranged in a number of ways to produce a structure for this compound. For example:

$$\underset{\text{O}}{\overset{\text{O}}{\|}}$$

CH₃CH₂COCH₂CH₂CH₃   *or*   HCOCH₂CH₂CH₂CH₂CH₃

The structural formula indicates that this compound can't be a lactone.

(b)

CH₃CN(CH₃)₂

(c)

CH₃CH=CHCCl   *or*   H₂C=C(CH₃)CCl

Visualizing Chemistry

21.27

(a)

N,N-Dimethyl-3-methylbutanamide

(b)

3-Methylbutyl benzoate

21.28 (a)

o-Bromobenzoic acid   2-Propanol   Isopropyl
                                                  o-bromobenzoate

This compound can also be synthesized by Fischer esterification of *o*-bromobenzoic acid with 2-propanol and an acid catalyst.

(b)

Cyclopentylacetic acid   Cyclopentylacetamide

21.29

3-Methyl-4-pentenoyl chloride   3-Methyl-4-pentenamide

The starting material is 3-methyl-4-pentenoyl chloride, as indicated by the –Cl in the tetrahedral intermediate. Ammonia adds to give the observed tetrahedral intermediate, which eliminates Cl⁻ to yield the above amide.

21.30 According to the electrostatic potential maps, the carbonyl carbon of acetyl azide is more electron-poor (less red) and therefore more reactive in nucleophilic acyl substitution reactions. Resonance donation of nitrogen lone-pair electrons to the carbonyl group is greater in an amide than in an acyl azide.

Acetamide

Acetyl azide

Mechanism Problems

21.31 (a)

Mechanism:

(b)

Mechanism:

(c)

Mechanism:

(d)

Mechanism:

21.32 (a)

Mechanism:

(b)

Mechanism:

(c)

Mechanism:

(d)

Mechanism:

21.33 (a)

Mechanism:

(b)

Mechanism:

(c)

Mechanism:

(d)

Mechanism:

21.34 **(a)**

Mechanism:

(b)

Mechanism:

(c)

Mechanism:

not simultaneous

(d)

Mechanism:

not simultaneous

21.35   Note: Two factors are responsible for the observed regioselectivity. First, the sterics of the tert-butyl group blocks the nucleophile from reacting at the adjacent carbonyl. Second, the nitrogen of the amino acid has an inductive effect increasing the electrophilic character of the amino acid carbonyl.

pavalyl - amino acid
mixed anhydride

Mechanism:

21.36

Mechanism:

21.37

Glycerol
3-phosphate

1-Acylglycerol
3-phosphate

Addition of –OH to the fatty acyl CoA (Step 1), followed by loss of –SCoA from the tetrahedral intermediate (Step 2), produces 1-acylglycerol 3-phosphate.

21.38 Formation of the dipeptide:

Step 1:   The carboxylate group from one amino acid adds to DCC to form a reactive intermediate.

Step 2:   The amino group of the second amino acid adds to the carbonyl group to yield a tetrahedral intermediate.

Step 3:   The intermediate loses dicyclohexylurea to produce the amide.

Proton transfers occur in steps 1 and 3.

Formation of the 2,5-diketopiperazine:

Step 1: Addition of carboxylate to DCC.

Step 2: Intramolecular nucleophilic attack of the amino terminal end of the amide on the acylating agent.

Step 3: Loss of dicyclohexylurea.

Proton transfers occur in steps 1 and 2.

21.39

A summary of steps:

Step 1: Protonation

Step 2: Nucleophilic addition of NH_3

Steps 3,5,7,9: Proton transfers

Step 4: Ring opening

Step 6: Nucleophilic addition of $-NH_2$

Step 8 Loss of H_2O

This reaction requires high temperatures because the intermediate amide is a poor nucleophile and the carboxylic acid carbonyl group is unreactive.

21.40 Both steps involve nucleophilic acyl substitutions.

Formation of acyl phosphate:

Step 1:   Reaction of the phosphate oxygen with the carbonyl carbon of succinyl CoA.

Step 2:   Loss of ⁻SCoA from the tetrahedral intermediate, yielding acyl phosphate.

Conversion of acyl phosphate to succinate:

Step 1:   Reaction of the diphosphate oxygen of GDP with the phosphorus of the acyl phosphate to produce an intermediate similar to the intermediates formed in nucleophilic acyl substitutions of carboxylic acid derivatives.

Step 2:   Loss of phosphate to form GTP and succinate.

21.41 In all of these reactions, a nucleophile adds to either carbon or phosphorus to form an intermediate that expels a leaving group to give the desired product.

Formation of 1,3-bisphosphoglycerate:

Route to enzyme-bound thioester:

Reduction to glyceraldehyde 3-phosphate:

21.42

In (c), the imine rearranges to an α,β-unsaturated ester, to which the nucleophile adds to give the trapped β-lactamate.

21.43

21.44 This is a nucleophilic acyl substitution reaction whose mechanism is similar to others we have studied.

$I_3C:^-$ can act as a leaving group because the electron-withdrawing iodine atoms stabilize the carbanion.

Additional Problems

Naming Carboxylic Acid Derivatives

21.45

(a)

p-Methylbenzamide

(b)

4-Ethyl-2-hexenoyl chloride

(c)

Dimethyl butanedioate
or
Dimethyl succinate

(d)

Isopropyl
3-phenylpropanoate

(e)

N-Methyl-3-bromo-
butanamide

(f)

Methyl 1-cyclopentene-
carboxylate

(g)

Phenyl benzoate

(h)

Isopropyl thiobenzoate

21.46

(a)

p-Bromophenylacetamide

(b)

O
‖
C

O
‖
CNH₂

m-Benzoylbenzamide

(c)

CH₃ O
| ‖
CH₃CH₂CH₂CH₂C—CNH₂
|
CH₃

2,2-Dimethylhexanamide

(d)

O
‖
C
O

Cyclohexyl cyclo-
hexanecarboxylate

(e)

O
‖
C
OCH₂CH₃

Ethyl 2-cyclobutene-
carboxylate

(f)

O   O   O

Succinic anhydride

21.47 Many structures can be drawn for each part of this problem.

(a)

O
‖
C
Cl

Cyclopentanecarbonyl
chloride

CH₃CH₂   CH₃
     \   /
      C=C
     /   \
    H   C=O
           |
           Cl

(*E*)-2-Methyl-2-
pentenoyl chloride

O
‖
CH₂C
Cl
H₂C=C
CH₂CH₃

3-Ethyl-3-butenoyl
chloride

(b)

O
‖
C
NH₂

1-Cyclohexenecarboxamide

O
‖
CH₃CH₂CH₂C≡CCH₂C
NH₂

3-Heptynamide

O
‖
H₂C=CHCH=CHC
N(CH₃)₂

N,N-Dimethyl-
2,4-pentadienamide

Nucleophilic Acyl Substitution Reactions

21.48

(a)

$$CH_3CH_2CCl \xrightarrow[\text{ether}]{(Ph)_2CuLi} CH_3CH_2C\text{—}C_6H_5$$

(b)

$$CH_3CH_2CCl \xrightarrow[\text{2. } H_3O^+]{\text{1. LiAlH}_4} CH_3CH_2CH_2OH$$

(c)

$$CH_3CH_2CCl \xrightarrow[\text{2. } H_3O^+]{\text{1. 2 CH}_3\text{MgBr}} CH_3CH_2C(CH_3)_2\text{—}OH$$

(d)

$$CH_3CH_2CCl \xrightarrow{H_3O^+} CH_3CH_2COH + HCl$$

(e)

$$CH_3CH_2CCl + HO\text{—}C_6H_{11} \xrightarrow{\text{Pyridine}} CH_3CH_2C\text{—}O\text{—}C_6H_{11}$$

(f)

$$CH_3CH_2CCl + H_2N\text{—}C_6H_5 \xrightarrow{\text{NaOH}} CH_3CH_2C\text{—}N(H)\text{—}C_6H_5$$

(g)

$$CH_3CH_2CCl \xrightarrow{CH_3CO_2^-\ Na^+} CH_3CH_2C\text{—}OCCH_3$$

21.49 The reagents in parts (a), (e), and (g) don't react with methyl propanoate.

(b)

$$CH_3CH_2COCH_3 \xrightarrow[\text{2. } H_3O^+]{\text{1. LiAlH}_4} CH_3CH_2CH_2OH + CH_3OH$$

(c)

$$CH_3CH_2COCH_3 \xrightarrow[\text{2. } H_3O^+]{\text{1. 2 CH}_3\text{MgBr}} CH_3CH_2C(CH_3)_2\text{—}OH + CH_3OH$$

(d)

$$CH_3CH_2COCH_3 \xrightarrow{H_3O^+} CH_3CH_2COH + CH_3OH$$

(f)

$$CH_3CH_2COCH_3 + H_2N\text{—}C_6H_5 \longrightarrow CH_3CH_2C\text{—}N(H)\text{—}C_6H_5$$

21.50 The reagents in parts (a), (e), (f), and (g) don't react with propanamide.

(b)

$$CH_3CH_2\overset{\displaystyle O}{\overset{\|}{C}}NH_2 \quad \xrightarrow[\text{2. H}_2\text{O}]{\text{1. LiAlH}_4} \quad CH_3CH_2CH_2NH_2$$

(c)

$$CH_3CH_2\overset{\displaystyle O}{\overset{\|}{C}}NH_2 \quad \xrightarrow[\text{2. H}_3\text{O}^+]{\text{1. CH}_3\text{MgBr}} \quad CH_3CH_2\overset{\displaystyle O}{\overset{\|}{C}}NH_2 \ + \ CH_4$$

(d)

$$CH_3CH_2\overset{\displaystyle O}{\overset{\|}{C}}NH_2 \quad \xrightarrow{\text{H}_3\text{O}^+} \quad CH_3CH_2\overset{\displaystyle O}{\overset{\|}{C}}OH \ + \ NH_4^+$$

21.51 Dimethyl carbonate is a diester. Use your knowledge of the Grignard reaction to work your way through this problem.

Triphenylmethanol

The overall reaction consists of three additions of phenylmagnesium bromide, two eliminations of methoxide and one protonation.

21.52

(a) $CH_3CH_2CH_2CO_2H \xrightarrow[\text{2. } H_3O^+]{\text{1. } BH_3} CH_3CH_2CH_2CH_2OH$

(b) $CH_3CH_2CH_2CH_2OH \xrightarrow[CH_2Cl_2]{\text{Periodinane}} CH_3CH_2CH_2CHO$

   from (a)

(c) $CH_3CH_2CH_2CH_2OH \xrightarrow{PBr_3} CH_3CH_2CH_2CH_2Br$

   from (a)

(d) $CH_3CH_2CH_2CH_2Br \xrightarrow{NaCN} CH_3CH_2CH_2CH_2CN$

   from (c)

(e) $CH_3CH_2CH_2CH_2Br \xrightarrow{K^+ \ ^-OC(CH_3)_3} CH_3CH_2CH=CH_2$

   from (c)

(f) $CH_3CH_2CH_2CH_2CN \xrightarrow{H_3O^+} CH_3CH_2CH_2CH_2CO_2H$

   from (d)

$CH_3CH_2CH_2CH_2CO_2H \xrightarrow{SOCl_2}$

$CH_3CH_2CH_2CH_2CONHCH_3 \xleftarrow{2\ CH_3NH_2} CH_3CH_2CH_2CH_2COCl$

(g) $CH_3CH_2CH_2CH_2CN \xrightarrow[\text{2. } H_3O^+]{\text{1. } CH_3MgBr} CH_3CH_2CH_2CH_2\overset{\displaystyle O}{\overset{\displaystyle \|}{C}}CH_3$

   from (d)

(h) $CH_3CH_2CH_2CO_2H \xrightarrow{SOCl_2} CH_3CH_2CH_2COCl$

$\underset{\text{benzene ring } CH_2CH_2CH_2CH_3}{} \xleftarrow{H_2,\ Pd/C} \underset{\text{benzene ring } \overset{O}{\overset{\|}{C}}CH_2CH_2CH_3}{} \xleftarrow[AlCl_3]{C_6H_6}$

(i) $CH_3CH_2CH_2CO_2H \xrightarrow{SOCl_2} CH_3CH_2CH_2COCl \xrightarrow{2\ NH_3} CH_3CH_2CH_2CONH_2$

$\xrightarrow{SOCl_2}$

$SO_2 + 2\ HCl + CH_3CH_2CH_2CN$

21.53

(a)

(b)

(c)

(d)

(e)

(f)

(g)

(h)

21.54 The reactivity of esters in saponification reactions is influenced by steric factors. Branching in both the acyl and alkyl portions of an ester hinders attack of the hydroxide nucleophile. This effect is less dramatic in the alkyl portion of the ester than in the acyl portion because alkyl branching is one atom farther away from the site of attack, but it is still significant.

Most reactive ⟶ *Least reactive*

$$CH_3COCH_3 > CH_3COCH_2CH_3 > CH_3COCHCH_3 > CH_3COCCH_3$$

21.55

2,4,6-Trimethylbenzoic acid

2,4,6-Trimethylbenzoic acid has two methyl groups ortho to the carboxylic acid functional group. These bulky methyl groups block the approach of the alcohol and prevent esterification from occurring under Fischer esterification conditions. A possible route to the methyl ester:

This route succeeds because reaction occurs farther away from the site of steric hindrance. It is also possible to form the acid chloride of 2,4,6-trimethylbenzoic acid and react it with methanol and pyridine.

21.56

(a)

(b)

(c)

Reaction of an ester with Grignard reagent produces a tertiary alcohol, not a ketone.

(d)

(e)

21.57

Step 1: The carboxylic acid protonates DCC.

Step 2: The carboxylate oxygen adds to DCC to form a reactive intermediate.

Step 3: The amine nitrogen adds to the carbonyl group to yield a tetrahedral intermediate.

Step 4: The intermediate loses dicyclohexylurea to produce the lactam.

21.58

Step 1: Protonation.

Step 2: Addition of methanol.

Step 3: Proton transfer.

Step 4: Loss of ethanol.

Step 5: Deprotonation.

In acidic methanol, the ethyl ester reacts by a nucleophilic acyl substitution mechanism to yield a methyl ester. The equilibrium favors the methyl ester because of the large excess of methanol present.

21.59

This reaction is a typical nucleophilic acyl substitution reaction, with azide as the nucleophile and chloride as the leaving group.

Step-Growth Polymers

21.60 Step 1:   Water opens the caprolactam ring to form the amino acid intermediate.

Step 2:   Reaction of the intermediate with a second molecule of caprolactam forms a dimer.

Steps 3 and beyond: Reaction of the dimer with caprolactam. This process repeats itself many, many times until the polymer stops growing. Remember that each new bond is formed in a discrete step. Heat forces the equilibrium in the direction of polymer formation.

nylon 6

21.61 Look for the monomer units, which are difunctional compounds, in the polymer.

21.62 Hydroxide opens the lactone ring, and the resulting anion can add to a second lactone molecule to produce a polyester.

21.63 The polyimide pictured is a step-growth polymer of a benzene tetracarboxylic acid and an aromatic diamine.

Spectroscopy

21.64 In some of these pairs, IR spectroscopy alone can differentiate between the isomers. For others, either ^{1}H NMR or a combination of ^{1}H NMR and IR data is necessary.

(a)

$$CH_3CH_2\overset{\overset{\displaystyle O}{\|}}{C}NHCH_3$$
N-Methylpropanamide

$$CH_3\overset{\overset{\displaystyle O}{\|}}{C}N(CH_3)_2$$
N,N-Dimethylacetamide

IR:	1680 cm^{-1}	1650 cm^{-1}
	(*N*-substituted amide)	(*N,N*-disubstituted amide)
^{1}H NMR:	one methyl group one ethyl group	three methyl groups

(b)

$$HOCH_2CH_2CH_2CH_2C\equiv N$$
5-Hydroxypentanenitrile

$$\overset{\overset{\displaystyle O}{\|}}{\text{—}}CNH_2$$
Cyclobutanecarboxamide

| *IR:* | 3300–3400 cm^{-1}
(hydroxyl)
2250 cm^{-1}
(nitrile) | 1690 cm^{-1}
(amide) |

(c)

$$ClCH_2CH_2CH_2\overset{\overset{\displaystyle O}{\|}}{C}OH$$
4-Chloropentanoic acid

$$CH_3OCH_2CH_2\overset{\overset{\displaystyle O}{\|}}{C}Cl$$
3-Methoxypropanoyl chloride

| *IR:* | 2500–3300 cm^{-1}
(hydroxyl)
1710 cm^{-1}
(carboxylic acid) | 1810 cm^{-1}
(carboxylic acid chloride) |

(d)

$$CH_3CH_2\overset{\overset{\displaystyle O}{\|}}{C}OCH_2CH_3$$
Ethyl propanoate

$$CH_3\overset{\overset{\displaystyle O}{\|}}{C}OCH_2CH_2CH_3$$
Propyl acetate

| *^{1}H NMR:* | two triplets
two quartets | one singlet
one triplet
one quartet
one multiplet |

21.65 The IR spectrum indicates that this compound has a carbonyl group.

$$\underset{a\quad c\quad\ \ b}{CH_3\overset{\overset{\displaystyle Cl}{|}}{C}H\overset{\overset{\displaystyle O}{\|}}{C}OCH_3}$$

$a = 1.69\ \delta$
$b = 3.79\ \delta$
$c = 4.41\ \delta$

21.66

(a)

$$N \equiv CCH_2COCH_2CH_3$$

with O double bonded at the central carbon, labeled b, c, a

a = 1.32 δ

b = 3.51 δ

c = 4.27 δ

(b)

$$CH_3CH_2CH_2CCl$$

with O double bonded at the carbonyl carbon, labeled a, b, c

a = 1.00 δ

b = 1.75 δ

c = 2.86 δ

General Problems

21.67 A negatively charged tetrahedral intermediate is formed when the nucleophile ^-OH attacks the carbonyl carbon of an ester. An electron-withdrawing substituent can stabilize this negatively charged tetrahedral intermediate and increase the rate of reaction. (Contrast this effect with substituent effects in electrophilic aromatic substitution, in which positive charge developed in the intermediate is stabilized by electron-*donating* substituents.) Substituents that are deactivating in electrophilic aromatic substitution are activating in ester hydrolysis, as the observed reactivity order shows. The substituents –CN and –CHO are electron-withdrawing; –NH₂ is strongly electron-donating.

Most reactive —————————————————→ *Least reactive*

$$Y = -NO_2 > -C \equiv N > -CHO > -Br > -H > -CH_3 > -OCH_3 > -NH_2$$

21.68

The tetrahedral intermediate **T** can eliminate any one of the three –OH groups to reform either the original carboxylic acid or labeled carboxylic acid. Further reaction of water with mono-labeled carboxylic acid leads to the doubly labeled product.

21.69

Ethyl propanoate

Remember that the ^{18}O label appears in both oxygens of the acetic acid starting material.

21.70

(a)

addition of carboxylic acid

loss of proton

elimination of trifluoroacetate

(b) The electron-withdrawing fluorine atoms polarize the carbonyl group, making it more reactive toward nucleophiles.

(c) Because trifluoroacetate is a better leaving group than other carboxylate anions, the reaction proceeds as indicated.

21.71 This synthesis requires a nucleophilic aromatic substitution reaction, explained in Section 16.7.

Nucleophilic aromatic substitution

Reduction of nitro group

Aminolysis

Butacetin

The amide can be formed by the reaction of acetyl chloride with the appropriate amine, which is produced by reduction of the nitro group of the starting material. A nucleophilic aromatic substitution of –F by –OC(CH₃)₃ can take place because the ring has an electron-withdrawing nitro group para to the site of substitution. Acetic anhydride can also be used to acetylate the amine.

21.72

Phenyl 4-aminosalicylate

21.73

N,N-Diethyl-*m*-toluamide   formation of amide

Grignard carboxylation yields *m*-methylbenzoic acid, which can be converted to an acid chloride and treated with diethylamine to produce the amide.

21.74

Tranexamic acid

Using a rhodium catalyst, the aromatic ring is hydrogenated to form the cis-substituted cyclohexane, which is converted to the trans isomer by heating to 300°. The nitrile starting material is hydrolyzed to form a carboxylic acid, and the methyl group is brominated and treated with ammonia to form the amine.

21.75 (a)

Resonance forms show that the carbon of diazomethane is basic, and reaction with an acid can occur to form a methyldiazonium ion.

(b)

An S_N2 reaction takes place in which the carboxylate ion displaces N_2 as the leaving group to form the methyl ester.

21.76

Dimethyl terephthalate   +   Glycerol

The product of the reaction of dimethyl terephthalate with glycerol has a high degree of cross-linking and is more rigid than Dacron.

21.77

(a)

$$CH_3COCH(CH_3)_2$$
  b   c   a

with O double bonded above the first carbon

a = 1.22 δ
b = 2.01 δ
c = 4.99 δ

(b)

$$CH=CHCOCH_2CH_3$$
 e   c   b   a

with O double bonded above the C, attached to a benzene ring (d)

a = 1.32 δ
b = 4.24 δ
c = 6.41 δ
d = 7.36, 7.49 δ
e = 7.68 δ

21.78

(a)

$$CH_3CH_2OCCH_2COCH_2CH_3$$
 a   c   b   c   a

with two O double bonded

a = 1.27 δ
b = 3.34 δ
c = 4.20 δ

(b)

$$ClCH_2CH_2COCH_2CH_3$$
  c   b   d   a

with O double bonded

a = 1.26 δ
b = 2.77 δ
c = 3.76 δ
d = 4.19 δ

21.79   Ethyl 4-cyanobenzoate

21.80   Ethyl levulinate (ethyl 4-oxopentanoate)

21.81   Note: The final step in the formation of an amide in each case is deprotonation. The amine that serves as the nucleophile in the reaction will also react as a base and form an unreactive ammonium salt. If a base strong enough to deprotonate the ammonium salt is present, then the amine will be deprotoned and free to react as a nucleophile. The pK_a an ammonium ion is approximately 10. For an acid chloride, chloride ion is not a strong enough base to deprotonate an ammonium ion (the pK_a of HCl is -5) so a second mole of base is needed to complete the reaction. In the case of an ester, an alkoxide leaving group is generated. An alkoxide is a strong enough base to deprotonate an ammonium ion (the pK_a of an alcohol is approximately 16). Thus, the reaction does not require any additional base.

The first reaction scheme shows an amide-type species reacting with an amine:

$$R_1\text{-C(=O)-}\overset{+}{N}(R_2)(H)(H) \;+\; R_2-\ddot{N}H_2 \longrightarrow R_1\text{-C(=O)-}\ddot{N}(R_2)(H) \;+\; R_2-\overset{+}{N}H_3$$

non-nucleophilic
ammonium salt

21.82 Addition of the triamine causes formation of cross-links between prepolymer chains.

$$\text{CH}_2\text{CH}_2\text{NCH}_2\overset{\text{OH}}{\underset{|}{\text{CH}}}\text{CH}_2\left(\text{O}-\text{C}_6\text{H}_4-\underset{\text{CH}_3}{\overset{\text{CH}_3}{\text{C}}}-\text{C}_6\text{H}_4-\text{OCH}_2\overset{\text{OH}}{\underset{|}{\text{CH}}}\text{CH}_2\right)_n$$

$$-\text{N}-\text{CH}_2-\text{CH}_2-\text{N}-\text{CH}_2\overset{\text{OH}}{\underset{|}{\text{CH}}}\text{CH}_2\left(\text{O}-\text{C}_6\text{H}_4-\underset{\text{CH}_3}{\overset{\text{CH}_3}{\text{C}}}-\text{C}_6\text{H}_4-\text{OCH}_2\overset{\text{OH}}{\underset{|}{\text{CH}}}\text{CH}_2\right)_n$$

Review Unit 8: Carbonyl Compounds 1 - Reaction at the Carbonyl Group

Major Topics Covered (with vocabulary);

Aldehydes and ketones:
-carbaldehyde   acyl group   acetyl group   formyl group   benzoyl group   hydrate

Reactions of aldehydes and ketones:
nucleophilic addition reaction   gem diol   cyanohydrin   imine   enamine   carbinolamine
2,4-dinitrophenylhydrazone   Wolff-Kishner reaction   acetal   hemiacetal   Wittig reaction
ylide betaine   Cannizzaro reaction   conjugate addition   α,β-unsaturated carbonyl compound

Carboxylic acids and their derivatives:
carboxylation   carboxylic acid derivative   acid halide   acid anhydride   amide   ester
nitrile–carbonitrile

Reactions of carboxylic acids and their derivatives:
nucleophilic   acyl substitution   hydrolysis   alcoholysis   aminolysis   Fischer
esterification reaction lactone   saponification   DIBAH   lactam   thiol   ester   acyl
phosphate   polyamide   polyester   step-growth polymer   chain-growth polymer   nylon

Types of Problems:

After studying these chapters you should be able to:

– Name and draw aldehydes, ketones, carboxylic acids and their derivatives.
– Prepare all of these compounds.
– Explain the reactivity difference between aldehydes and ketones and between carboxylic acids and all their derivatives.
– Calculate dissociation constants of carboxylic acids, and predict the relative acidities of substituted carboxylic acids.
– Formulate mechanisms for reactions related to the reactions we have studied.
– Predict the products of the reactions for all functional groups we have studied.
– Use spectroscopic techniques to identify these compounds.
– Draw representative segments of step-growth polymers.

Points to Remember:

*   In all of these reactions, a nucleophile adds to a positively polarized carbonyl carbon to form a tetrahedral intermediate. There are three possible fates for the tetrahedral intermediate: (1) The intermediate can be protonated, as occurs in Grignard reactions, reductions, and cyanohydrin formation. (2) The intermediate can lose water (or ⁻OH), as happens in imine and enamine formation. (3) The intermediate can lose a leaving group, as occurs in most reactions of carboxylic acid derivatives.

*   Many of the reactions in these three chapters require acid or base catalysis. An acid catalyst, protonates the carbonyl oxygen, making the carbonyl carbon more reactive toward nucleophiles, and/or protonates the tetrahedral intermediate, making loss of a leaving group easier. A base catalyst deprotonates the nucleophile, making it more nucleophilic. The pH optimum for these reactions is a compromise between the two needs.

*   Here are a few reminders for drawing the mechanisms of nucleophilic addition and substitution reactions. (1) When a reaction is acid-catalyzed, none of the intermediates are negatively charged, although, occasionally, a few may be neutral. Check your mechanisms for charge balance. (2) Make sure you have drawn arrows correctly. The point of the arrow shows the new location of the electron pair at the base of the arrow. (3) In a polar reaction, two arrows never point at each other. If you find two arrows pointing at each other, redraw the mechanism.

*   Reactions of acyl halides are almost always carried out with an equivalent of base present. The base is used to scavenge the protons produced when a nucleophile adds to an acyl halide. If base were not present, hydrogen ions would protonate the nucleophile and make it unreactive.

*   The products of acidic cleavage of an amide are a carboxylic acid and a protonated amine. The products of basic cleavage of an amide are a carboxylate anion and an amine.

*   In some of the mechanisms shown in the answers, a series of protonations and deprotonations occur. These steps convert the initial tetrahedral intermediate into an intermediate that more easily loses a leaving group. These deprotonations may be brought about by the solvent, by the conjugate base of the catalyst, by other molecules of the carbonyl compound or may occur intramolecularly. When a "proton transfer" is shown as part of a mechanism, the base that removes the proton has often not been shown. However, it is implied that the proton transfer is assisted by a base: the proton doesn't fly off the intermediate unassisted.

*   The most useful spectroscopic information for identifying carbonyl compounds comes from IR spectroscopy and ^{13}C NMR spectroscopy. Carbonyl groups have distinctive identifying absorptions in their infrared spectra. ^{13}C NMR is also useful for identifying aldehydes, ketones, and nitriles, although other groups are harder to distinguish. The ^{1}H NMR absorptions of aldehydes and carboxylic acids are also significant. Look at mass spectra for McLafferty rearrangements and alpha-cleavage reactions of aldehydes and ketones.

Self-Test:

A
Jasmone
(used in perfumery)

B
Erythrocentaurin
(a bitter tonic)

Predict the products of the reaction of **A** with: (a) LiAlH₄, then H₃O⁺; (b) C₆H₅MgBr, then H₃O⁺; (c) (CH₃)₂NH, H₃O⁺; (d) CH₃OH, H⁺ catalyst (e) (C₆H₅)₃P⁺—CH₂–; (f) 1 equiv. CH₃CH₂NH₂, H₃O⁺. How would you reduce **A** to yield a saturated hydrocarbon? Where would you expect the carbonyl absorption of **A** to occur in its IR spectrum?

Predict the products of **B** with the reagents (a) – (d) above. What product(s) would be formed if **B** was treated with Br₂, FeBr₃? Where do the carbonyl absorptions occur in the IR spectrum of **B**? Describe the ¹³C NMR spectrum of **B**.

C
Kethoxal
(antiviral)

D
Julocrotine

E
Xanthoxylin

Kethoxal (**C**) exists in solution as an equilibrium mixture. With what compound is it in equilibrium. Why does the equilibrium lie on the side of kethoxal?

Identify the carboxylic acid derivatives present in **D**. Show the products of treatment of **D** with (a) ⁻OH, H₂O (b) LiAlH₄, then H₂O.

Name **E**. Describe the IR spectrum and ¹H NMR spectrum of **E**.

Multiple Choice:

1. In which of the following nucleophilic addition reactions does the equilibrium lie on the side of the products?
   (a) Propanal + HCN
   (b) Acetone + H_2O
   (c) Acetaldehyde + HBr
   (d) 2,2,4,4-Tetramethyl-3-pentanone + HCN

2. Which alcohol can be formed by three different combinations of carbonyl compound + Grignard reagent?
   (a) 2-Butanol
   (b) 3-Methyl-3-hexanol
   (c) Triphenylmethanol
   (d) 1-Phenylethanol

3. A nitrile can be converted to all of the following except:
   (a) an aldehyde
   (b) an amide
   (c) an amine
   (d) A nitrile can be converted to all of the above compounds.

4. Which of the following *p*- substituted benzoic acids is the least acidic?
   (a) $CH_3COC_6H_5CO_2H$
   (b) $CH_3OC_6H_5CO_2H$
   (c) $BrC_6H_5CO_2H$
   (d) $NCC_6H_5CO_2H$

5. A carboxylic acid can be reduced by all of the following except:
   (a) $LiAlH_4$, then H_3O^+
   (b) BH_3, THF, then H_3O^+
   (c) $NaBH_4$, then H_3O^+
   (d) All of these reagents can reduce a carboxylic acid.

6. Which of the following carboxylic acids can be formed by both Grignard carboxylation and by nitrile hydrolysis?
   (a) Phenylacetic acid
   (b) Benzoic acid
   (c) Trimethylacetic acid
   (d) 3-Butynoic acid

7. Acid anhydrides are used mainly for:
   (a) synthesizing carboxylic acids
   (b) forming alcohols
   (c) introducing acetyl groups
   (d) forming aldehydes

8.   A ketone is formed from an acid halide by reaction with:
      (a)   DIBAH
      (b)   LiAlH$_4$
      (c)   RMgBr
      (d)   (CH$_3$CH$_2$)CuLi

9.   From which carboxylic acid derivative can you form a ketone as the product of a Grignard reaction?
      (a)   acid chloride
      (b)   ester
      (c)   nitrile
      (d)   amide

10.   An infrared absorption at 1650 cm^{-1} indicates the presence of:
      (a)   aromatic acid chloride
      (b)   *N,N*-disubstituted amide
      (c)   α,β-unsaturated ketone
      (d)   aromatic ester

Chapter 22 – Carbonyl Alpha-Substitution Reactions

Chapter Outline

I. Keto–enol tautomerism (Section 22.1).
   A. Nature of tautomerism.
      1. Carbonyl compounds with hydrogens bonded to their α carbons equilibrate with their corresponding enols.
      2. This rapid equilibration is called tautomerism, and the individual isomers are tautomers.
      3. Unlike resonance forms, tautomers are isomers.
      4. Despite the fact that very little of the enol isomer is present at room temperature, enols are very important because they are reactive.
   B. Mechanism of tautomerism.
      1. In acid-catalyzed enolization, the carbonyl carbon is protonated to form an intermediate that can lose a hydrogen from its α carbon to yield a neutral enol.
      2. In base-catalyzed enol formation, an acid-base reaction occurs between a base and an α hydrogen.
         a. The resultant enolate ion is protonated to yield an enol.
         b. Protonation can occur either on carbon or on oxygen.
         c. Only hydrogens on the α positions of carbonyl compounds are acidic.
II. Enols (Sections 22.2–22.4).
   A. Reactivity of enols (Section 22.2).
      1. The electron-rich double bonds of enols cause them to behave as nucleophiles.
         a. The electron-donating enol –OH groups make enols more reactive than alkenes.
      2. When an enol reacts with an electrophile, the initial adduct loses –H from oxygen to give an α-substituted carbonyl compound.
   B. Reactions of enols (Sections 22.3–22.4).
      1. Alpha halogenation of aldehydes and ketones (Section 22.3).
         a. Aldehydes and ketones can be halogenated at their α positions by reaction of X_2 in acidic solution.
         b. The reaction proceeds by acid-catalyzed formation of an enol intermediate.
         c. Halogen isn't involved in the rate-limiting step: the rate doesn't depend on the identity of the halogen, but only on [carbonyl] and [H^+].
         d. α-Bromo ketones are useful in syntheses because they can be dehydrobrominated by base treatment to form α,β-unsaturated ketones.
      2. Alpha-bromination of carboxylic acids (Section 22.4).
         a. In the Hell–Volhard–Zelinskii (HVZ) reaction, a mixture of Br_2 and PBr_3 can be used to brominate carboxylic acids in the α position.
         b. The initially formed acid bromide reacts with Br_2 to form an α-bromo acid bromide, which is hydrolyzed by water to give the α-bromo carboxylic acid.
         c. The reaction proceeds through an acid bromide enol.
III. Enolates (Sections 22.5–22.7).
   A. Enolate ion formation (Section 22.5).

1. Hydrogens alpha to a carbonyl group are weakly acidic.
   a. This acidity is due to overlap of a filled *p* orbital with the carbonyl group *p* orbitals, allowing the carbonyl group to stabilize the negative charge by resonance.
   b. The two resonance forms aren't equivalent: the form with the negative charge on oxygen is of lower energy.
2. Strong bases are needed for enolate ion formation.
   a. Alkoxide ions are often too weak to use in enolate formation.
   b. Lithium diisopropylamide can be used to form the enolates of many different carbonyl compounds.
3. When a hydrogen is flanked by two carbonyl groups, it is much more acidic.
   a. Both carbonyl groups can stabilize the negative charge.
B. Reactivity of enolate ions (Section 22.6).
   1. Enolates are more useful than enols for two reasons:
      a. Unlike enols, stable solutions of enolates are easily prepared.
      b. Enolates are more reactive than enols because they are more nucleophilic.
   2. Enolates can react either at carbon or at oxygen.
      a. Reaction at carbon yields an *α*-substituted carbonyl compound.
      b. Reaction at oxygen yields an enol derivative.
C. Reactions of enolate ions (Sections 22.6–22.7).
   1. Base-promoted *α*-halogenation.
      a. Base-promoted halogenation of aldehydes and ketones proceeds readily because each halogen added makes the carbonyl compound more reactive.
      b. Consequently, polyhalogenated compounds are usually produced.
      c. This reaction is only useful with methyl ketones, which form HCX_3 when reacted with halogens.
      d. This reaction is known as the haloform reaction.
         i.   The HCX_3 is a solid that can be identified.
         ii. The last step of the reaction involves a carbanion leaving group.
   2. Alkylation reactions of enolates (Section 22.7).
      a. General features.
         i.   Alkylations are useful because they form a new C–C bond.
         ii. Alkylations have the same limitations as S_N2 reactions; the alkyl groups must be methyl, primary, allylic or benzylic.
      b. The malonic ester synthesis.
         i.   The malonic ester synthesis is used for preparing a carboxylic acid from a halide while lengthening the chain by two carbon atoms.
         ii. Diethyl malonate is useful because its enolate is easily prepared by reaction with sodium ethoxide.
         iii. Since diethyl malonate has two acidic hydrogens, two alkylations can take place.

    iv. Heating in aqueous HCl causes hydrolysis and decarboxylation of the
         alkylated malonate to yield a substituted monocarboxylic acid.
      (a). Decarboxylations are common only to β-keto acids and malonic acids.
    v. Cycloalkanecarboxylic acids can also be prepared.
  c. The acetoacetic ester synthesis.
    i.   The acetoacetic ester synthesis is used for converting an alkyl halide to a
         methyl ketone, while lengthening the carbon chain by 3 atoms.
    ii. As with malonic ester, acetoacetic ester has two acidic hydrogens which are
         flanked by a ketone and an ester, and two alkylations can take place.
    iii. Heating in aqueous HCl hydrolyzes the ester and decarboxylates the acid to
         yield the ketone.
    iv. Most β-keto esters can undergo this type of reaction.
  d. Direct alkylation of ketones, esters, and nitriles.
    i.   LDA in a nonprotic solvent can be used to convert the above compounds to
         their enolates.
    ii. Alkylation of an unsymmetrical ketone leads to a mixture of products, but the
         major product is alkylated at the less hindered position.

Solutions to Problems

22.1–22.2   Acidic hydrogens in the keto form of each of these compounds are bold. One of these hydrogens is removed by base when an enolate is formed.

	Keto Form	*Enol Form*	*Number of Acidic Hydrogens*
(a)			4
(b)			3
(c)			3
(d)			2
(e)			4
(f)			5

In (d) and (f), cis and trans enolates are possible.

22.3

equivalent;
more stable

equivalent;
less stable

The first two monoenols are more stable because the enol double bond is conjugated with the carbonyl group.

22.4

deuteration
of carbonyl
oxygen

loss of proton
at alpha position

enol

enol   deuteration
of enol
double bond

loss of deuterium
on carbonyl oxygen

22.5   Alpha-bromination, followed by dehydration using pyridine, yields the enone below.

1-Penten-3-one

22.6

Methyl 2-bromo-3-methylpentanoate

The mechanism of the ester-forming step is a nucleophilic acyl substitution, which was described in Chapter 21.

22.7   Hydrogens α to one carbonyl group (or nitrile) are weakly acidic. Hydrogens α to two carbonyl groups are much more acidic, but they are not as acidic as carboxylic acid protons.

(a)

weakly acidic

(b)

weakly acidic

(c)

weakly acidic   most acidic

(d)

weakly acidic

(e)

weakly acidic

(f)

weakly acidic

22.8   Nitriles are weakly acidic because the nitrile anion can be stabilized by resonance involving the π bonds of the nitrile group.

22.9 Halogenation in acid medium is acid-*catalyzed* because hydrogen ions are regenerated:

Halogenation in basic medium is base-*promoted* because a stoichiometric amount of base is consumed:

22.10 The malonic ester synthesis converts a primary or secondary alkyl halide into a carboxylic acid with two more carbons (a substituted acetic acid). Identify the component that originates from malonic ester (the acid component). The rest of the molecule comes from the alkyl halide, which should be primary or methyl.

(a)

from halide (PhCH$_2$ + CH$_2$CO$_2$H) from malonic ester

$$CH_2(CO_2Et)_2 \xrightarrow[\text{2. PhCH}_2\text{Br}]{\text{1. Na}^+ \text{ }^-\text{OEt}} PhCH_2-CH(CO_2Et)_2 + NaBr$$

$\downarrow$ H$_3$O$^+$, heat

⌬—CH$_2$—CH$_2$CO$_2$H + CO$_2$ + 2 EtOH

3-Phenylpropanoic acid

(b)

(CH$_3$CH$_2$CH$_2$ + CHCO$_2$H) from malonic ester

from halide (CH$_3$)

$$CH_2(CO_2Et)_2 \xrightarrow[\text{2. CH}_3\text{CH}_2\text{CH}_2\text{Br}]{\text{1. Na}^+ \text{ }^-\text{OEt}} CH_3CH_2CH_2-CH(CO_2Et)_2 + NaBr$$

$\downarrow$ 1. Na$^+$ $^-$OEt
2. CH$_3$Br

$$CH_3CH_2CH_2-\underset{\underset{CH_3}{|}}{C}HCO_2H \xleftarrow[\text{heat}]{H_3O^+} CH_3CH_2CH_2-\underset{\underset{CH_3}{|}}{C}(CO_2Et)_2 + NaBr$$

2-Methylpentanoic acid

+ CO$_2$ + 2 EtOH

(c)

from halide ⟨ (CH₃)₂CHCH₂ ⊹ CH₂CO₂H ⟩   from malonic ester

$$CH_2(CO_2Et)_2 \xrightarrow[\text{2. } (CH_3)_2CHCH_2Br]{\text{1. Na}^+ \text{ }^-OEt} (CH_3)_2CHCH_2-CH(CO_2Et)_2 + NaBr$$

$$\downarrow H_3O^+, \text{heat}$$

$$(CH_3)_2CHCH_2-CH_2CO_2H + CO_2 + 2 \text{ EtOH}$$
4-Methylpentanoic acid

22.11 Since malonic ester has only two acidic hydrogen atoms, it can be alkylated only two times. Formation of trialkylated acetic acids is thus not possible.

22.12

from malonic ester

from halides

$$CH_2(CO_2Et)_2 \xrightarrow[\substack{\text{2. } CH_3CHCH_2Br \\ | \\ CH_3}]{\text{1. Na}^+ \text{ }^-OEt} \begin{array}{c} CH_3CHCH_2-CH(CO_2Et)_2 + NaBr \\ | \\ CH_3 \end{array}$$

$$\begin{array}{c} CH_3 \\ | \end{array} \begin{array}{l} \text{1. Na}^+ \text{ }^-OEt \\ \text{2. } CH_3Br \end{array}$$

$$CH_3CHCH_2-CHCO_2H \xleftarrow[\text{heat}]{H_3O^+} CH_3CHCH_2-C(CO_2Et)_2 + NaBr$$
$$\quad | \qquad\qquad | \qquad\qquad\qquad\qquad | \qquad\quad |$$
$$\quad CH_3 \qquad CH_3 \qquad\qquad\qquad\qquad CH_3 \quad CH_3$$

2,4-Dimethylpentanoic acid
  + CO₂ + 2 EtOH

22.13 As in the malonic ester synthesis, you should identify the structural fragments of the target compound. The acetoacetic ester synthesis converts an alkyl halide to a methyl ketone ("substituted acetone"). The methyl ketone component comes from acetoacetic ester; the other component comes from a halide.

(a)

from halide ⟨(CH₃)₂CHCH₂⟩ ⟨CH₂CCH₃⟩ from acetoacetic ester

$$
\underset{\substack{|\\CO_2Et}}{CH_2CCH_3}
\quad \xrightarrow[\text{2. (CH}_3)_2\text{CHCH}_2\text{Br}]{\text{1. Na}^+ \ ^-\text{OEt}} \quad
(CH_3)_2CHCH_2 - \underset{\substack{|\\CO_2Et}}{CHCCH_3} \ + \ NaBr
$$

↓ H₃O⁺, heat

$$
(CH_3)_2CHCH_2 - CH_2CCH_3 \ + \ CO_2 \ + \ EtOH
$$

5-Methyl-2-hexanone

(b)

from halide ⟨C₆H₅CH₂CH₂⟩ ⟨CH₂CCH₃⟩ from acetoacetic ester

$$
\underset{\substack{|\\CO_2Et}}{CH_2CCH_3}
\quad \xrightarrow[\text{2. C}_6\text{H}_5\text{CH}_2\text{CH}_2\text{Br}]{\text{1. Na}^+ \ ^-\text{OEt}} \quad
C_6H_5CH_2CH_2 - \underset{\substack{|\\CO_2Et}}{CHCCH_3} \ + \ NaBr
$$

↓ H₃O⁺, heat

$$
C_6H_5CH_2CH_2 - CH_2CCH_3 \ + \ CO_2 \ + \ EtOH
$$

5-Phenyl-2-pentanone

22.14 The acetoacetic ester synthesis can only be used for certain products:

(1) Three carbons must originate from acetoacetic ester. In other words, compounds of the type RCOCH₃ can't be synthesized by the reaction of RX with acetoacetic ester.

(2) Alkyl halides must be primary or methyl.

(3) The acetoacetic ester synthesis can't be used to prepare compounds that are trisubstituted at the α position.

(a)

Phenylacetone

(b)

Acetophenone

(c)

3,3-Dimethylbutan-2-one

(a) Phenylacetone can't be produced by an acetoacetic ester synthesis because bromobenzene, the necessary halide, does not enter into S_N2 reactions. [See (2) above.]

(b)   Acetophenone can't be produced by an acetoacetic ester synthesis. [See (1) above.]

(c)   3,3-Dimethyl-2-butanone can't be prepared because it is trisubstituted at the α position. [See (3) above.]

22.15

22.16 Direct alkylation is used to introduce substituents α to an ester, ketone or nitrile. Look at the target molecule to identify these substituents. Alkylation is achieved by treating the starting material with LDA, followed by a primary halide.

(a)

3-Phenyl-2-butanone

Alkylation occurs at the carbon next to the phenyl group because the phenyl group can help stabilize the enolate anion intermediate.

(b) $CH_3CH_2CH_2CH_2C{\equiv}N$ $\xrightarrow[\text{2. } CH_3CH_2I]{\text{1. LDA}}$ $CH_3CH_2CH_2CHC{\equiv}N$
                                                              $\underset{CH_2CH_3}{|}$

2-Ethylpentanenitrile

(c)

2-Allylcyclohexanone

(d)

2,2,6,6-Tetramethylcyclohexanone

(e)

Isopropyl phenyl ketone

(f)

$$CH_3CHCH_2COCH_3 \xrightarrow[\text{2. } CH_3CH_2I]{\text{1. LDA}} CH_3CHCHCOCH_3$$

Methyl 2-ethyl-3-methylbutanoate

Visualizing Chemistry

22.17 (a) Check to see if the target molecule is a methyl ketone or a substituted carboxylic acid. (The target molecule is a methyl ketone, and the reaction is an acetoacetic ester synthesis.) Next, identify the halide or halides that react with acetoacetic ester. (The halide is 1-bromo-3-methyl-2-butene.) Formulate the reaction, remembering to include a decarboxylation step.

from halide $(CH_3)_2C{=}CHCH_2 \;\; CH_2CCH_3$ from acetoacetic ester

$$CH_2CCH_3 \xrightarrow[\text{2.}(CH_3)_2C=CHCH_2Br]{\text{1. Na}^+ \text{ }^-OEt} (CH_3)_2C=CHCH_2-CHCCH_3 + NaBr$$
$$CO_2Et \qquad\qquad\qquad\qquad\qquad CO_2Et$$

$\downarrow$ H_3O^+, heat

$$CO_2 + EtOH + (CH_3)_2C=CHCH_2-CH_2CCH_3$$

6-Methyl-5-hepten-2-one

(b) This product is formed from the reaction of malonic ester with both benzyl bromide and bromomethane.

from malonic ester

from halides

$CH_2(CO_2Et)_2$ $\xrightarrow{\text{1. } Na^+ \, {}^-OEt}{\text{2. } C_6H_5CH_2Br}$ $C_6H_5CH_2-CH(CO_2Et)_2$ + NaBr

1. $Na^+ \, {}^-OEt$
2. CH_3Br

$C_6H_5CH_2-\underset{\underset{CH_3}{|}}{C}HCO_2H$ $\xleftarrow[\text{heat}]{H_3O^+}$ $C_6H_5CH_2-\underset{\underset{CH_3}{|}}{C}(CO_2Et)_2$ + NaBr

2-Methyl-3-phenylpropanoic acid + CO_2 + 2 EtOH

22.18

Ordinarily, β-diketones are acidic because they can form enolates that can be stabilized by delocalization over both carbonyl groups. In this case, loss of the proton at the bridgehead carbon doesn't occur because the strained ring system doesn't allow formation of the bridgehead double bond. Instead, enolization takes place in the opposite direction, and the diketone resembles acetone, rather than a β-diketone, in it pK_a and degree of dissociation.

22.19

Enolization can occur on only one side of the carbonyl group because of the two methyl groups on the other side. The circled axial hydrogen is more acidic because the p orbital that remains after its removal is aligned for optimum overlap with the π electrons of the carbonyl oxygen.

Mechanism Problems

22.20 (a)

Mechanism:

(b)

Mechanism:

(c)

Mechanism:

(d)

OH⁻ catalyst

H_2O

Mechanism:

22.21 (a)

Br_2

CH_3CO_2H

Mechanism:

(b)

Mechanism:

(c)

Mechanism:

(d)

Mechanism:

22.22 (a)

1. LDA
2. CH₃CH₂Br

Mechanism:

(b)

1. LDA
2. CH₃Br

Mechanism:

(c)

1. LDA
2. PhCH₂Br

Mechanism:

(d)

1. LDA
2. CH₃CH₂I

Mechanism:

22.23 (a)

Mechanism:

Note: The product mixture *will not be optically active*, but a racemic mixture. During the reaction the chiral center becomes part of an achiral enol intermediate. When the enol tautomerizes both enantiomeric ketones are formed in equal amounts.

(b)

Mechanism:

Note: A single, *optically active* product is formed in this reaction. The chiral center in the starting material is never involved in the mechanism, so it remains unchanged.

22.24

Mechanism:

22.25 Note: Under acidic conditions, the ketone must form an enol that reacts with the halogen. The first step of enol formation is protonation of the carbonyl carbon. When there is a halogen a to the carbonyl, the inductive effect created by the halogen removes electron density from the carbonyl oxygen making it more difficult to protonante. This makes the enol more difficult to form and thus inhibits formation of an enol. When the reaction is performed under basic conditions, the inductive effect of the halogen removes electron removes electron density from the carbon to which it is attached, which in turn makes the proton attached to the same carbon more acidic. This facilitates formation of an enolate making the addition of a second halogen faster than the first.

22.26

protonation of
carbonyl oxygen

abstraction
of α proton   *enol*

protonation
at γ position

loss of proton
on oxygen

22.27

abstraction
of γ proton

abstraction
of α proton

protonation
at α position

protonation
at γ position

All protons in the five-membered ring can be exchanged by base treatment.

22.28

22.29 Decarboxylation, which takes place because of the stability of the resulting anion, is followed by protonation.

22.30

Step 1: Base-catalyzed enolization.

Step 2: Equilibration of two enolates by proton transfer.

Step 3: Protonation.

22.31

Step 1:   Loss of proton at the α position.

Step 2:   Displacement of bromide.

Step 3:   Nucleophilic addition of ⁻OH.

Step 4:   Ring opening.

Step 5:   Proton transfer.

Step 6:   Protonation.

22.32

nucleophilic addition
of diazomethane

bond migration
and loss of N_2

22.33 Laurene differs in stereochemical configuration from the observed product at the carbon
α to the methylene group. Since this position is α to the carbonyl group in the precursor to
laurene, enolization and isomerization must have occurred during the reaction.
Isomerization of the ketone precursor is brought about by a reversible reaction with the
basic Wittig reagent, which yields an equilibrium mixture of two diastereomeric ketones.
One of the ketone isomers then reacts preferentially with the Wittig reagent to give only
the observed product.

22.34

+ 2 EtOH + CO₂ + CH₃CO₂H

Acid hydrolyzes both ester bonds, as well as the amide bond, by mechanisms that were shown in Figure 21.8 and Section 21.6. Decarboxylation of the β-keto acid produces alanine.

22.35

A malonic ester synthesis is used to form 4-methylpentanoic acid. Hell–Volhard–Zelinskii bromination of the acid, followed by reaction with ammonia, yields leucine. The last reaction is an S$_N$2 displacement of bromide by ammonia.

22.36

Step 1: Protonation.

Step 2: Hydride shift.

Step 3: Deprotonation.

Step 4: Enolization to form the aromatic ring.

Additional Problems

Acidity of Carbonyl Compounds

22.37 Acidic hydrogens are bold. The most acidic hydrogens are the two between the carbonyl groups in (b) and the hydroxyl hydrogen in (c). The hydrogens in (c) that are bonded to the methyl group are acidic (draw resonance forms to prove it).

(a)

$$CH_3CH_2CHCCH_3$$
$$\underset{CH_3}{\overset{O}{||}}$$

(b)

(c)

$$HOCH_2CH_2CC\equiv CCH_3$$

(d)

CO$_2$CH$_3$

CH$_2$CN

(e)

(f)

$$CH_3CH_2CC=CH_2$$
$$\underset{CH_3}{\overset{O}{||}}$$

22.38 Check your answer by using the pK_as in Table 22.1.

Least Acidic ⟶ *Most Acidic*

$$(CH_3CH_2)_2NH < CH_3\overset{O}{\overset{||}{C}}CH_3 < CH_3CH_2OH < CH_3\overset{O}{\overset{||}{C}}CH_2\overset{O}{\overset{||}{C}}CH_3 < CH_3CH_2\overset{O}{\overset{||}{C}}OH < CCl_3\overset{O}{\overset{||}{C}}OH$$

22.39

(a)

(b)

(c)

(d)

(e)

22.40 Enolization at the γ position produces a conjugated enolate anion that is stabilized by delocalization of the negative charge over the π system of five atoms.

22.41 The illustrated compound, 1-phenyl-2-propenone, doesn't yield an anion when treated with base because the hydrogen on the α carbon is vinylic and isn't acidic (check Table 22.1 for acidity constants).

1-Phenyl-2-propenone

α-Substitution Reactions

22.42

(a)

(b)

(c)

(d)

22.43 (a)

(b)

$$CH_2(CO_2Et)_2 \xrightarrow[\text{2. } (CH_3)_2CHBr]{\text{1. Na}^+ \ ^-OEt} (CH_3)_2CH-CH(CO_2Et)_2 \ + \ NaBr$$

$$\downarrow H_3O^+, \text{heat}$$

$$(CH_3)_2CHCH_2CO_2Et \xleftarrow[\text{H}^+ \text{ catalyst}]{\text{EtOH}} (CH_3)_2CHCH_2CO_2H$$
Ethyl 3-methylbutanoate $+ \ CO_2 \ + \ 2 \ EtOH$

Some elimination product will also be formed.

(c)

$$CH_2(CO_2Et)_2 \xrightarrow[\text{2. } CH_3CH_2Br]{\text{1. Na}^+ \ ^-OEt} \begin{array}{c} CH_3CH_2CH(CO_2Et)_2 \\ + \ NaBr \end{array} \xrightarrow[\text{2. } CH_3Br]{\text{1. Na}^+ \ ^-OEt} \begin{array}{c} \overset{CH_3}{\underset{|}{\ }} \\ CH_3CH_2C(CO_2Et)_2 \\ + \ NaBr \end{array}$$

$$\downarrow H_3O^+, \text{heat}$$

$$\underset{\text{Ethyl 2-methylbutanoate}}{\overset{CH_3}{\underset{|}{CH_3CH_2CHCO_2Et}}} \xleftarrow[\text{H}^+ \text{ catalyst}]{\text{EtOH}} \overset{CH_3}{\underset{|}{CH_3CH_2CHCO_2H}} \ + \ CO_2 \ + \ 2 \ EtOH$$

(d)   The malonic acid synthesis can't be used to synthesize carboxylic acids that are trisubstituted at the alpha position.

22.44 Look back to Problem 22.14, which describes compounds that can be prepared by an acetoacetic ester synthesis. Neither (a) or (c) are products of an acetoacetic ester synthesis because the halide component that would be needed for each synthesis doesn't undergo S_N2 reactions. Compound (b) can be prepared by the reaction of acetoacetic ester with 1,5-dibromopentane.

22.45 Two alkylations are needed if the target molecule has two alkyl groups α to the carbonyl group.

(a)

$$\underset{CO_2Et}{\overset{O}{\underset{|}{CH_2\overset{\|}{C}CH_3}}} \xrightarrow[\text{2. 2 } CH_3CH_2Br]{\text{1. 2 Na}^+ \ ^-OEt} \begin{array}{c} \overset{O}{(CH_3CH_2)_2\overset{\|}{C}CH_3} \\ \underset{CO_2Et}{|} \\ + \ 2 \ NaBr \end{array} \xrightarrow[\text{heat}]{H_3O^+} \begin{array}{c} \overset{O}{(CH_3CH_2)_2CH\overset{\|}{C}CH_3} \\ \text{3-Ethyl-2-pentanone} \\ + \ CO_2 \ + \ EtOH \end{array}$$

(b)

$$\underset{\underset{CO_2Et}{|}}{CH_2\overset{\overset{O}{\|}}{C}CH_3} \xrightarrow[\text{2. } CH_3CH_2CH_2Br]{\text{1. } Na^+ \ ^-OEt} \underset{\underset{CO_2Et}{|}}{CH_3CH_2CH_2\overset{\overset{O}{\|}}{C}HCCH_3} + NaBr$$

1. Na$^+$ $^-$OEt
2. CH$_3$Br

$$\underset{\text{3-Methyl-2-hexanone}}{\underset{}{CH_3CH_2CH_2\overset{\overset{H_3C}{|}}{C}H-\overset{\overset{O}{\|}}{C}CH_3}} \xleftarrow[\text{heat}]{H_3O^+} \underset{\underset{CO_2Et}{|}}{CH_3CH_2CH_2\overset{\overset{H_3C}{|}}{\underset{}{C}}-\overset{\overset{O}{\|}}{C}CH_3} + NaBr$$

+ CO$_2$ + EtOH

22.46 Use a malonic ester synthesis if the product you want is an α-substituted carboxylic acid or derivative. Use an acetoacetic acid synthesis if the product you want is an α-substituted methyl ketone.

(a)

$$CH_2(CO_2Et)_2 \xrightarrow[\text{2. 2 } CH_3Br]{\text{1. 2 } Na^+ \ ^-OEt} \underset{\underset{CH_3}{|}}{CH_3\overset{}{C}(CO_2Et)_2} + 2 \ NaBr$$

(b)

$$\underset{\underset{CO_2Et}{|}}{CH_2\overset{\overset{O}{\|}}{C}CH_3} \xrightarrow[\text{2. } BrCH_2(CH_2)_4CH_2Br]{\text{1. 2 } Na^+ \ ^-OEt}$$

+ NaBr

H$_3$O$^+$, heat

+ CO$_2$ + EtOH

(c)

$$CH_2(CO_2Et)_2 \xrightarrow[\text{2. } BrCH_2CH_2CH_2Br]{\text{1. 2 } Na^+ \ ^-OEt}$$

+ 2 NaBr

H$_3$O$^+$, heat

+ CO$_2$ + 2 EtOH

(d)

$$CH_3CCH_3 \atop CO_2Et \quad \xrightarrow[\text{2. } H_2C=CHCH_2Br]{\text{1. } Na^+ \; ^-OEt} \quad H_2C=CHCH_2CHCCH_3 \quad + \quad NaBr$$

(with structure CH_2CCH_3 / CO_2Et on left; product $H_2C=CHCH_2CHCCH_3$ with CO_2Et)

$\downarrow H_3O^+$, heat

$$H_2C=CHCH_2CH_2CCH_3 \quad + \quad CO_2 \quad + \quad EtOH$$

22.47 The haloform reaction (Problem 22.25d) is an alpha-substitution reaction in which a methyl ketone is trihalogenated at the alpha position, and the trihalomethyl group is displaced by –OH. It is a test for methyl ketones.

Positive haloform reaction: Negative haloform reaction:

(a) CH_3CCH_3 (b) $C_6H_5CCH_3$ (c) CH_3CH_2CHO (d) CH_3CO_2H (e) CH_3CN

22.48 First, treat geraniol with PBr₃ to form $(CH_3)_2C=CHCH_2CH_2C(CH_3)=CHCH_2Br$ (geranyl bromide).

(a)

$$CH_3CO_2Et \quad \xrightarrow[\text{2. Geranyl bromide}]{\text{1. LDA}} \quad (CH_3)_2C=CHCH_2CH_2C(CH_3)=CHCH_2CH_2CO_2Et$$

Ethyl geranylacetate

Alternatively:

$$CH_2(CO_2Et)_2 \quad \xrightarrow[\text{2. Geranyl bromide}]{\text{1. } Na^+ \; ^-OEt} \quad (CH_3)_2C=CHCH_2CH_2C(CH_3)=CHCH_2CH_2(CO_2Et)_2$$

$\downarrow H_3O^+$, heat

$$CO_2 \quad + \quad 2 \; EtOH \quad + \quad (CH_3)_2C=CHCH_2CH_2C(CH_3)=CHCH_2CH_2CO_2H$$

$\downarrow$ 1. $SOCl_2$
2. EtOH, pyridine

$$(CH_3)_2C=CHCH_2CH_2C(CH_3)=CHCH_2CH_2CO_2Et$$

Ethyl geranylacetate

(b)

$$CH_2CCH_3 \atop CO_2Et \quad \xrightarrow[\text{2. Geranyl bromide}]{\text{1. } Na^+ \; ^-OEt} \quad (CH_3)_2C=CHCH_2CH_2C(CH_3)=CHCH_2CHCCH_3$$

(with CO_2Et on product)

$\downarrow H_3O^+$, heat

$$CO_2 \quad + \quad EtOH \quad + \quad (CH_3)_2C=CHCH_2CH_2C(CH_3)=CHCH_2CH_2CCH_3$$

Geranylacetone

22.49   Dialkylation of diethylmalonate:

Nucleophilic acyl substitution:

This series of steps is repeated to form the 6-membered ring.

General Problems

22.50   When a compound containing acidic hydrogen atoms is treated with NaOD in D₂O, all acidic hydrogens are gradually replaced by deuterium atoms. For each proton (atomic weight 1) lost, a deuteron (atomic weight 2) is added. Since the molecular weight of cyclohexanone increases by four after NaOD/D₂O treatment (from 98 to 102), cyclohexanone contains four acidic hydrogen atoms.

22.51   Reaction of (R)-2-methylcyclohexanone with aqueous base is shown below. Reaction with aqueous acid proceeds by a related mechanism through an enol, rather than an enolate ion, intermediate.

(R)-2-Methylcyclohexanone

Carbon 2 loses its chirality when the enolate ion double bond is formed. Protonation occurs with equal probability from either side of the planar sp^2-hybridized carbon 2, resulting in a racemic product.

22.52

(S)-3-Methylcyclohexanone

(S)-3-Methylcyclohexanone isn't racemized by base because its chirality center is not involved in the enolization reaction.

22.53 The Hell–Volhard–Zelinskii reaction involves formation of an intermediate acid bromide enol, with loss of stereochemical configuration at the chirality center. Bromination of (R)-2-phenylpropanoic acid can occur from either face of the enol double bond, producing racemic 2-bromo-2-phenylpropanoic acid. If the molecule had a chirality center that didn't take part in enolization (Problem 22.35), the product would be optically active.

22.54

(a) Na^+ $^-OCH_3$, then CH_3I; (b) H_3O^+, heat; (c) LDA, then CH_3I (some 2,2-dimethylcyclohexanone may also be formed).

22.55

abstraction
of α proton

protonation
at γ position

The enolate of 3-cyclohexenone can be protonated at three different positions. Protonation at the γ position yields the α,β-unsaturated ketone.

22.56 Protons α to a carbonyl group or γ to an enone carbonyl group are acidic (Problem 22.39). Thus for 2-methyl-2-cyclopentenone, protons at the starred positions are acidic.

Isomerization of a 2-substituted 2-cyclohexenone to a 6-substituted 2-cyclohexenone requires removal of a proton from the 5-position of the 2-substituted isomer. Since protons in this position are not acidic, double bond isomerization does not occur.

22.57 A nitroso compound is analogous to a carbonyl compound. If there are hydrogens α to the nitroso group, enolization similar to that observed for carbonyl compounds can occur, leading to formation of the more stable oxime. If no hydrogens are adjacent to the nitroso group, enolization to the oxime can't occur, and the nitroso compound is stable.

22.58

(a)

(b)

(c)

(d)

(e)

Warm aqueous acid both hydrolyzes the nitrile and dehydrates the alcohol.

(f)

22.59 Treatment of either the cis or trans isomer with base causes enolization α to the carbonyl group and results in loss of configuration at the α position. Reprotonation at carbon 2 produces either of the diastereomeric 4-*tert*-butyl-2-methylcyclohexanones. In both diastereomers the *tert*-butyl group of carbon 4 occupies the equatorial position for steric reasons. The methyl group of the cis isomer is also equatorial, but the methyl group of the trans isomer is axial. The trans isomer is less stable because of 1,3-diaxial interactions of the methyl group with the ring hydrogens.

22.60 (a) Reaction with Br$_2$ at the α position occurs only with aldehydes and ketones, not with esters.

(b) Aryl halides can't be used in malonic ester syntheses because they don't undergo S$_N$2 reactions.

(c) The product of this reaction sequence, H$_2$C=CHCH$_2$CH$_2$COCH$_3$, is a methyl ketone, not a carboxylic acid.

22.61 The reaction of cyclohexanone and *tert*-butylmagnesium bromide gives the expected carbonyl addition product. The yield of the *tert*-butylmagnesium bromide addition product is very low, however, because of the difficulty of approach of the bulky *tert*-butyl Grignard reagent to the carbonyl carbon. More favorable is the acid-base reaction between the Grignard reagent and the acidic carbonyl α proton.

When D_3O^+ is added to the reaction mixture, the deuterated ketone is produced.

22.62

Step 1: Acid-catalyzed enolization (Figure 22.1).

Step 2: Attack of enol π electrons on phenylselenyl chloride, with loss of Cl^-.

Step 3: Loss of proton.

22.63 Start at the end of the sequence of reactions and work backwards.

(a) Because the keto acid $C_9H_{13}NO_3$ loses CO_2 on heating, it must be a β-keto acid. Neglecting stereoisomerism, we can draw the structure of the β-keto acid as:

keto acid   Ecgonine   Cocaine

(b) When ecgonine ($C_9H_{15}NO_3$) is treated with CrO_3, the keto acid $C_9H_{13}NO_3$ is produced. Since CrO_3 is used for oxidizing alcohols to carbonyl compounds, ecgonine has the structure shown above. Again, the stereochemistry is unspecified.

(c) Ecgonine contains carboxylic acid and alcohol functional groups. The other products of hydroxide treatment of cocaine are a carboxylic acid (benzoic acid) and an alcohol (methanol). Cocaine thus contains two ester functional groups, which are saponified on reaction with hydroxide.

The complete reaction sequence:

Cocaine

$^-$OH, H$_2$O

Ecgonine

+ CH$_3$OH

+ C$_6$H$_5$CO$_2$H

CrO$_3$, H$_3$O$^+$

keto acid

heat

Tropinone + CO$_2$

22.64   The key step is an intramolecular alkylation reaction of the ketone α-carbon, with the tosylate in the second ring serving as the leaving group.

Base

abstraction of α proton

alkylation, with –OTos as a leaving group

+ $^-$OTos

H$_2$C—PPh$_3$

Wittig reaction

Sativene

22.65 This sequence resembles the one shown in Problem 22.49.

Sodium Pentothal

The series of steps is repeated to form the 6-membered ring.

Chapter 23 – Carbonyl Condensation Reactions

Chapter Outline

I. The aldol reaction (Sections 23.1–23.6).
  A. Characteristics of the aldol reaction (Sections 23.1–23.2).
    1. The aldol condensation is a base-catalyzed dimerization of two aldehydes or ketones.
    2. The reaction can occur between two components that have α hydrogens.
    3. One component (the nucleophilic donor) is converted to its enolate and undergoes an α-substitution reaction.
    4. The other component (the electrophilic acceptor) undergoes nucleophilic addition.
    5. For simple aldehydes, the equilibrium favors the products, but for other aldehydes and ketones, the equilibrium favors the reactants.
    6. Carbonyl condensation reactions require only a catalytic amount of base (Section 23.2).
      a. Alpha-substitution reactions, on the other hand, use one equivalent of base.
  B. Dehydration of aldol products (Section 23.3).
    1. Aldol products are easily dehydrated to yield α,β-unsaturated aldehydes and ketones.
      a. Dehydration is catalyzed by both acid and base.
      b. Reaction conditions for dehydration are only slightly more severe than for condensation.
      c. Often, dehydration products are isolated directly from condensation reactions.
    2. Conjugated enones are more stable than nonconjugated enones.
    3. Removal of the water byproduct drives the aldol equilibrium towards product formation.
  C. Aldol products (Sections 23.4–23.5).
    1. Using aldol reactions in synthesis (Section 23.4).
      a. Obvious aldol products are:
        i. α,β-Unsaturated aldehydes/ketones.
        ii. β-Hydroxy aldehydes/ketones.
      b. Often, it's possible to work backwards from a compound that doesn't seem to resemble an aldol product and recognize aldol components.
    2. Mixed aldol reactions (23.5).
      a. If two similar aldehydes/ketones react under aldol conditions, 4 products may be formed – two self-condensation products and two mixed products.
      b. A single product can be formed from two different components :
        i.   If one carbonyl component has no α-hydrogens.
        ii. If one carbonyl compound is much more acidic than the other.
  D. Intramolecular aldol condensations (Section 23.6).
    1. Treatment of certain dicarbonyl compounds with base can lead to cyclic products.
    2. A mixture of cyclic products may result, but the more strain-free ring usually predominates.
II. The Claisen condensation (Sections 23.7–23.9).
  A. Features of the Claisen condensation (Section 23.7).

1. Treatment of an ester with 1 equivalent of base yields a β-keto ester.
2. The reaction is reversible and has a mechanism similar to that of the aldol reaction.
3. A major difference from the aldol condensation is the expulsion of an alkoxide ion from the tetrahedral intermediate of the initial Claisen adduct.
4. Because the product is often acidic, one equivalent of base is needed; addition of this amount of base drives the reaction to completion.
5. Addition of acid yields the final product.

B   Mixed Claisen condensations (Section 23.8).

1. Mixed Claisen condensations of two different esters can succeed if one component has no α hydrogens.
2. Mixed Claisen condensations between a ketone and an ester with no α hydrogens are also successful.

C.   Intramolecular Claisen condensations: the Dieckmann cyclization (Section 23.9).

1. The Dieckmann cyclization is used to form cyclic β-keto esters.
   a.   1,6-Diesters form 5-membered rings.
   b.   1,7-Diesters form 6-membered rings.
2. The mechanism is similar to the Claisen condensation mechanism.
3. The product β-keto esters can be further alkylated.
   a.   This is a good route to 2-substituted cyclopentanones and cyclohexanones.

III. Other carbonyl condensation reactions (Sections 23.10–23.13).

A. The Michael reaction (Section 23.10).

1. The Michael reaction is the conjugate addition of an enolate to an α,β-unsaturated carbonyl compound.
   a.   The highest-yielding reactions occur between stable enolates and unhindered α,β-unsaturated carbonyl compounds.
2. Stable enolates are Michael donors, and α,β-unsaturated compounds are Michael acceptors.

B. The Stork reaction (Section 23.11).

1. A ketone that has been converted to an enamine can act as a Michael donor in a reaction known as the Stork reaction.
2. The sequence of reactions in the Stork reaction:
   a.   Enamine formation from a ketone.
   b.   Michael-type addition to an α,β-unsaturated carbonyl compound.
   c.   Enamine hydrolysis back to a ketone.
3. This sequence is equivalent to the Michael addition of a ketone to an α,β-unsaturated carbonyl compound and yields a 1,5 diketone product.

C. The Robinson annulation reaction (Section 23.12).

1. The Robinson annulation reaction combines a Michael reaction with an intramolecular aldol condensation to synthesize substituted ring systems.
2. The components are a nucleophilic donor, such as a β-keto ester, and an α,β-unsaturated ketone acceptor.

3. The intermediate 1,5-diketone undergoes an intramolecular aldol condensation to yield a cyclohexenone.

D. Biological carbonyl condensation reactions (Section 23.13).

1. Many biomolecules are synthesized by carbonyl condensation reactions.

2. The enzyme aldolase catalyzes the addition of a ketone enolate to an aldehyde.

    a. This mixed aldol reaction is successful because of the selectivity of enzyme catalysis.

3. Acetyl CoA is the major building block for the synthesis of biomolecules.

    a. Acetyl CoA can act as an electrophilic acceptor by being attacked at its carbonyl group.

    b. Acetyl CoA can act as a nucleophilic donor by loss of its acidic α hydrogen.

Solutions to Problems

23.1 (1) Form the enolate of one molecule of the carbonyl compound.

(2) Have the enolate attack the electrophilic carbonyl of the second molecule.

(3) Protonate the alkoxide oxygen.

Practice writing out these steps for the other aldol condensations.

(a) See above.

(b)

(c)

23.2

4-Hydroxy-4-methyl-2-pentanone

The steps for the reverse aldol are the reverse of those described in Problem 23.1.

(1)   Deprotonate the alcohol oxygen.

(2)   Eliminate the enolate anion.

(3)   Reprotonate the enolate anion.

23.3   As in Problem 23.1, align the two carbonyl compounds so that the location of the new bond is apparent. After drawing the addition product, form the conjugated enone product by dehydration. In parts (b) and (c), a mixture of E,Z isomers may be formed.

(a)

(b)

(c)

23.4 Including double bond isomers, 4 products can be formed. The major product is formed by reaction of the enolate formed by abstraction of a proton at position "a" because position "b" has more steric hindrance.

major (less hindered)   minor (more hindered)

23.5 (a)

CH₃CH₂CH₂C—CH   2-Hydroxy-2-methylpentanal

This is not an aldol product. The hydroxyl group in an aldol product must be β, not α, to the carbonyl group.

(b)

5-Ethyl-4-methyl-4-hepten-3-one

This product results from the aldol self-condensation of 3-pentanone, followed by dehydration.

23.6

1-Butanol $CH_3CH_2CH_2CH_2OH$

23.7

Cyclopropylacetaldehyde

23.8   (a)

4-Phenyl-3-buten-2-one

This mixed aldol will succeed because one of the components, benzaldehyde, is a good acceptor of nucleophiles, yet has no α-hydrogen atoms. Although it is possible for acetone to undergo self-condensation, the mixed aldol reaction is much more favorable.

(b)

Four products result from the aldol condensation of acetone and acetophenone. The two upper compounds are mixed aldol products, and the bottom two are self-condensation products.

(c)

As in (b), a mixture of products is formed because both carbonyl partners contain α-hydrogen atoms. The upper two products result from mixed aldol condensations; the lower two are self-condensation products.

23.9

2,4–Pentanedione is in equilibrium with two enolate ions after treatment with base. Enolate **A** is stable and unreactive, while enolate **B** can undergo internal aldol condensation to form a cyclobutenone product. But, because the aldol reaction is reversible and the cyclobutenone product is highly strained, there is little of this product present when equilibrium is reached. At equilibrium, only the stable, diketone enolate ion **A** is present.

23.10 This intramolecular aldol condensation gives a product with a seven-membered ring fused to a five-membered ring.

23.11 As in the aldol condensation, writing the two Claisen components in the correct orientation makes it easier to predict the product.

(a)

(b)

(c)

23.12

Hydroxide ion can react at two different sites of the β-keto ester. Abstraction of the acidic α-proton is more favorable but is reversible and does not lead to product. Addition of hydroxide ion to the carbon of the carbonyl group, followed by irreversible elimination of ethyl acetate anion, accounts for the observed product.

23.13 As shown in Worked Example 23.4, dimethyl oxalate is a very effective reagent in mixed Claisen reactions.

23.14

Diethyl 4-methylheptanedioate

23.15

C1–C6 bond formation

C2–C7 bond formation

Unlike diethyl 4-methylheptanedioate, shown in the previous problem, diethyl 3-methylheptanedioate is unsymmetrical. Two different enolates can form, and each can cyclize to a different product.

23.16 A Michael reaction takes place when a stable enolate ion (Michael donor) adds to the double bond of an α,β-unsaturated carbonyl compound (Michael acceptor). The enolate adds to the double bond of the conjugated system. Predicting Michael products is easier when the donor and acceptor are positioned so that the product can be visualized.

	Michael Donor	Michael Acceptor	Product

(a)

(b)

$H_2C=CHC\equiv N$

(c)

23.17

	Michael Donor	Michael Acceptor	Product

(a)

$H_2C=CHCCH_3$

(b)

$H_2C=CHCCH_3$

23.18 Find the carbonyl group of the Michael acceptor, and count three carbons away from the carbonyl group. The Michael donor forms the new bond to this carbon. Break this bond to identify the Michael donor and Michael acceptor.

new bond formed   Michael acceptor   Michael donor

23.19 An enamine is formed from a ketone when it is necessary to synthesize a 1,5-diketone or a 1,5-dicarbonyl compound containing an aldehyde or ketone. The ketone starting material is converted to an enamine in order to increase the reactivity of the ketone and to direct the regiochemistry of addition. The process, as described in Section 23.11, is: (1) conversion of a ketone to its enamine; (2) Michael addition to an α,β-unsaturated carbonyl compound; (3) hydrolysis of the substituted enamine to the diketone product.

23.20 Analyze the product for the Michael acceptor and the ketone. In (a), the Michael acceptor is propenenitrile. The ketone is cyclopentanone, which is treated with pyrrolidine to form the enamine.

(a)

(b)

23.21 The Robinson annulation is a combination of two reactions covered in this chapter. First, a Michael reaction takes place between a nucleophilic donor (the diketone in this problem) and an α,β-unsaturated carbonyl compound (the enone shown). The resulting product can cyclize in an aldol reaction. The base catalyzes both reactions.

23.22

This is one of the more complicated-looking syntheses that we have seen. First, analyze the product for the two Michael components. The carbon–carbon double bond arises from dehydration of the aldol addition product, and is located where one of the two C=O groups of the original diketone used to be. The Michael addition takes place at the carbon between these ketone groups. The Michael acceptor is an enone that can also enter into the aldol condensation and furnishes the methyl group attached to the double bond.

Visualizing Chemistry

23.23

(a)

2 $CH_3CH_2CCH_2CH_3$

3-Pentanone

(b)

2 $CCH_2CH(CH_3)_2$

3-Methylbutanal

23.24 The enolate of methyl phenylacetate adds to a second molecule of methyl phenylacetate to form the Claisen intermediate that is pictured. Elimination of methoxide (circled) and acidification give the product shown.

Methyl phenylacetate

23.25

4-Oxoheptanal

23.26 Remember that the new carbon–carbon double bond in the product connects one of the carbonyl carbons of the Michael donor with an α carbon of the Michael acceptor.

Michael acceptor   Michael donor

Mechanism Problems

23.27 (a)

Mechanism:

(b)

Mechanism:

(c)

Mechanism:

(d)

Mechanism:

23.28 (a)

Mechanism:

(b)

Mechanism:

(c)

Mechanism:

(d)

Mechanism:

23.29 (a)

Mechanism:

(b)

Mechanism:

(c)

Mechanism:

(d)

Mechanism:

23.30 (a)

Mechanism:

(b)

Mechanism:

(c)

Mechanism:

(d)

Mechanism:

23.31 (a)

Mechanism:

(b)

Mechanism:

(c)

Mechanism:

(d)

Mechanism:

23.32

Mechanism:

23.33

Azlactone

intermediate

dehydro α-amino acid

Mechanism:

23.34 This sequence is a reverse aldol reaction.

: Base

Acetyl CoA

Acetoacetate

Step 1:   Deprotonation by base.

Step 2:   Elimination of acetyl CoA enolate.

Step 3:   Protonation of enolate.

23.35 In contrast to the previous problem, this sequence is a reverse Claisen reaction. The first step (not illustrated) is the reaction of HSCoA with a base to form ⁻SCoA.

2-Methyl-3-keto-
butyryl CoA

Propionyl CoA
(propanoyl CoA)

Acetyl CoA

Step 1:   Addition of ⁻SCoA to form a tetrahedral intermediate.

Step 2:   Elimination of propionyl CoA anion.

Step 3:   Protonation.

23.36 *Formation of (S)-citryl CoA:*

(S)-Citryl CoA

Step 1:   Formation of acetyl CoA enolate.

Step 2:   Aldol-like nucleophilic addition of acetyl CoA to the carbonyl group of oxaloacetate and protonation.

Loss of CoA to form citrate:

Step 1: Nucleophilic addition of hydroxyl to the carbonyl group of (*S*)-citryl CoA.

Step 2: Loss of ⁻SCoA and protonation of the leaving group.

23.37

Leucine

Step 1: Addition of ⁻SCoA.

Step 2: Hydrolysis

Step 3: Elimination

Step 4: Conjugate addition.

Step 5:   Oxidation.

Step 6:   Decarboxylation of a β-keto ester.

Step 7:   Nucleophilic acyl substitution, loss of water, reduction.

23.38 Formation of the enolate of diethyl malonate is the first step:

$$CH_2(CO_2Et)_2 \; + \; {}^-OEt \; \rightleftharpoons \; {}^-{:}CH(CO_2Et)_2 \; + \; HOEt$$

+ 2 EtOH + CO₂

Cinnamic acid

Step 1:   Nucleophilic addition of the enolate to the carbonyl group.

Step 2:   Protonation, dehydration

Step 3:   Ester cleavage and decarboxylation of a β-keto ester.

23.39

$$ClCH_2CO_2Et \; + \; Na^+ \; {}^-OEt \; \rightleftharpoons \; Cl\overset{..}{\overset{..}{C}}HCO_2Et \; + \; EtOH$$

nucleophilic
addition

S_N2 displacement

+ Cl⁻

23.40 This mechanism divides into two sequences of steps. In the first part of the mechanism, an acetal is hydrolyzed to acetone and a dihydroxycarboxylic acid.

In the second set of steps, the dihydroxycarboxylic acid forms a cyclic ester (a lactone).

23.41 This problem becomes easier if you draw the starting material so that it resembles the product.

23.42

Step 1:   Conjugate addition of H₃C⁻.

Step 2:   Claisen condensation.

Step 3:   Loss of methoxide.

23.43

Step 1: Enolate formation.

Step 2: Intramolecular aldol condensation.

Step 3: Retro-aldol condensation.

Step 4: Protonation.

23.44

Step 1: Deprotonation and retro aldol reaction.

Step 2: Equilibration between two enolates.

Step 3: Internal aldol condensation.

Step 4: Protonation.

23.45 (a)

(b)

Step 1:   Protonation.

Step 2:   Addition of amine

Step 3:   Proton transfer.

Step 4:   Elimination of water.

Step 5:   Aldol-like addition of enol to iminium ion.

Step 6:   Loss of proton.

23.46   The Mannich reaction occurs between the diester, butanedial, and methylamine.

23.47

Step 1:   S$_N$2 displacement of Br$^-$ by enamine.

Step 2:   Tautomerization of ketoester.

Step 3:   Attack of enol on iminium ion.

Additional Problems

Aldol Reactions

23.48   (a)   (CH$_3$)$_3$CCHO has no α hydrogens and does not undergo aldol self-condensation.

(b)

(c)

Benzophenone doesn't undergo aldol self condensation because it has no α hydrogens

(d)

(e)

(f)   C₆H₅CH=CHCHO does not undergo aldol reactions because its α proton isn't acidic.

23.49 As always, analyze the product for the carbon–carbon double bond that is formed by dehydration of the initial aldol adduct. Break the bond, and add a carbonyl oxygen to the appropriate carbon to identify the carbonyl reactant(s).

(a)

(b)

(c)

(d)

23.50

23.51

23.52   Product **A**, which has two singlet methyl groups and no vinylic protons in its ^{1}H NMR, is the major product of the intramolecular cyclization of 2,5-heptanedione.

23.53

Because all steps in the aldol reaction are reversible, the more stable product is formed at equilibrium.

23.54

An aldol condensation involves a series of reversible equilibrium steps. In general, formation of product is favored by the dehydration of the β-hydroxy ketone to form a conjugated enone. Here, dehydration to form conjugated product can't occur. In addition, the **B ⇌ C** equilibrium favors **B** because of steric hindrance.

23.55 The reactive nucleophile in the acid-catalyzed aldol condensation is the *enol* of one of the reactants. The electrophile is a reactant with a protonated carbonyl group.

Step 1: Enol formation.

Step 2: Addition of the enol nucleophile to the protonated carbonyl compound.

electrophile   nucleophile

Step 3: Loss of proton from the carbonyl oxygen.

23.56

1. NaOH, EtOH
2. heat

Cinnamaldehyde

Although self-condensation of acetaldehyde can take place, the mixed aldol product predominates.

23.57 The first step of an aldol condensation is enolate formation. The ketone shown here does not enolize because double bonds at the bridgehead of small bicyclic ring systems are too strained to form. Since the bicyclic ketone does not enolize, it doesn't undergo aldol condensation.

Claisen Condensations

23.58 (a)

Approximately equal amounts of each product will form if the two esters are of similar reactivity.

(b)

The mixed condensation product predominates.

(c)

This is the only *Claisen* monocondensation product (aldol self-condensation of cyclohexanone also occurs to a small extent).

(d)

The mixed Claisen product is the major product.

23.59 If cyclopentanone and base are mixed first, aldol self-condensation of cyclopentanone can occur before ethyl formate is added. If both carbonyl components are mixed together before adding base, the more favorable mixed Claisen condensation occurs with less competition from the aldol self-condensation reaction.

23.60

addition of ethoxide

elimination of ethyl dimethyl-acetate anion

$$\overset{..}{\overset{..}{C}}(CH_3)_2COEt + EtOH \rightleftharpoons (CH_3)_2CHCOEt + \,^-OEt$$

This is a reverse Claisen reaction.

23.61 Two different reactions are possible when ethyl acetoacetate reacts with ethoxide anion. In one reaction, attack of ethoxide ion on the carbonyl carbon is followed by elimination of the anion of ethyl acetate—a reverse Claisen reaction similar to the one illustrated in 23.39. More likely, however, is the acid–base reaction of ethoxide ion and a doubly activated α-hydrogen of ethyl acetoacetate.

The resonance-stabilized acetoacetate anion is no longer reactive toward nucleophiles, and no further reaction occurs at room temperature. Elevated temperatures are required to make the cleavage reaction proceed. This complication doesn't occur with ethyl dimethylacetoacetate because it has no acidic hydrogens between its two carbonyl groups.

Michael and Enamine Reactions

23.62 Michael reactions occur between stabilized enolate anions and α,β-unsaturated carbonyl compounds. Learn to locate these components in possible Michael products. Usually, it is easier to recognize the enolate nucleophile; in (a), the nucleophile is the ethyl acetoacetate anion.The rest of the compound is the Michael acceptor. Draw a double bond in conjugation with the electron-withdrawing group in this part of the molecule.

Michael donor

Michael acceptor

CH₃CCH₂ (CO₂Et) + H₂C=CHCC₆H₅

$$\begin{array}{c}\text{1. NaOEt,}\\\text{EtOH}\\\xrightarrow{}\\\text{2. H}_3\text{O}^+\end{array}$$

Michael donor   Michael acceptor

(b)   When the Michael product has been decarboxylated after the addition reaction, it is more difficult to recognize the original enolate anion.

Michael donor   Michael acceptor

(c)

Michael donor   Michael acceptor

(d)

Michael donor   Michael acceptor

(e)

Michael donor   Michael acceptor

(f)

CH₃NO₂   +

Michael
donor

Michael acceptor

1. NaOEt,
   EtOH
2. H₃O⁺

23.63

1. Na⁺ ⁻OEt,
   EtOH
2. CH₃I

1. Na⁺ ⁻OEt, EtOH
2. H₂C=CHCOCH₃
3. H₃O⁺

1. Na⁺ ⁻OEt,
   EtOH
2. heat

This sequence of reactions consists of an alkylation of a 1,3-diketone, followed by a Robinson annulation. The carbon–carbon double bond appears where the second carbonyl group of the diketone used to be and is the site of the ring-forming aldol reaction. A Michael reaction between the diketone and the Michael acceptor 3-buten-2-one adds the carbon atoms used to form the second ring, and an alkylation with CH₃I adds the methyl group.

23.64

Step 1:   Michael addition of enamine.

Step 2:   Tautomerization.

Step 3:   Enamine hydrolysis.

Step 4:   Abstraction of proton by hydroxide.

Step 5:   Internal aldol condensation.

Step 6:   Dehydration.

23.65 (a)

Step 1:   Enamine formation.

Step 2:   Michael addition of enamine.

Step 3:   Enamine hydrolysis.

Step 4:   Internal aldol condensation.

Step 5:   Dehydration.

(b)

Notice that the desired enamine is formed (see Problem 23.73).

Step 1:   Enamine formation.

Step 2:   Michael addition of enamine.

Step 3:   Enamine hydrolysis.

Step 4:   Internal aldol condensation.

Step 5:   Dehydration.

(c)

The enamine double bond is conjugated with the aromatic ring.

Step 1:   Enamine formation.

Step 2:   Michael addition of enamine.

Step 3:   Enamine hydrolysis.

Step 4:   Internal aldol condensation.

Step 5:   Dehydration.

23.66

General Problems

23.67

(a)

$$CH_3CH_2CH_2COEt \ + \ \underset{\underset{CH_2CH_3}{|}}{CH_2COEt} \xrightarrow[\text{2. } H_3O^+]{\text{1. NaOEt, EtOH}} \ CH_3CH_2CH_2\underset{}{C}-\underset{\underset{CH_2CH_3}{|}}{CHCOEt} \ + \ HOEt$$

(b)

(c)

(d)

$$C_6H_5CH_2CH_2CH \ + \ \underset{\underset{CH_2C_6H_5}{|}}{CH_2CH} \xrightarrow[\text{2. heat}]{\text{1. NaOH, EtOH}} \ C_6H_5CH_2CH_2CH = \underset{\underset{CH_2C_6H_5}{|}}{CCH} \ + \ H_2O$$

23.68 (a) Several other products are formed in addition to the one pictured. Self-condensation of acetaldehyde and acetone (less likely) can occur, and an additional mixed product is formed.

(b) There are two problems with this reaction. (1) Michael reactions occur in low yield with mono-ketones. Formation of the enamine, followed by the Michael reaction, gives a higher yield of product. (2) Addition can occur on either side of the ketone to give a mixture of products.

(c) Internal aldol condensation of 2,6-heptanedione can produce a four-membered ring or a six-membered ring. The six-membered ring is more likely to form because it is less strained.

23.69

(a) LiAlH₄, then H₃O⁺;

(b) POCl₃, pyridine

(c) KMnO₄, H₃O⁺;

(d) CH₃CH₂OH, H⁺;

(e) Na⁺ ⁻OEt;

(f) H₃O⁺;

(g) Na⁺ ⁻OEt, then CH₃Br;

(h) H₃O⁺, heat

23.70

(a)

(b)

(c)

(d)

23.71 (a)

Step 1: Formation of enolate.

Step 2: Aldol condensation.

Step 3: Loss of hydroxide.

(b)

The protonated form of the above structure is the product.

23.72 (c)

(d)

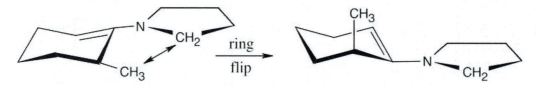

Step 1: Internal aldol condensation.

Step 2: Dehydration

Step 3: Ester cleavage and decarboxylation of a β-keto ester.

23.73

Crowding between the methyl group and the pyrrolidine ring disfavors this enamine.

The crowding in this enamine can be relieved by a ring-flip, which puts the methyl group in an axial position. This enamine is the only one formed.

Chapter 24 – Amines and Heterocycles

Chapter Outline

I. Facts about amines (Section 24.1–24.5).
  A. Naming amines (Section 24.1).
    1. Amines are classified as primary (RNH_2), secondary (R_2NH), tertiary (R_3N) or quaternary ammonium salts (R_4N^+).
    2. Primary amines are named in several ways:
      a. For simple amines, the suffix *-amine* is added to the name of the alkyl substituent.
      b. The suffix *-amine* can replace the final *-e* of the parent compound.
      c. For more complicated amines, the $-NH_2$ group is an amino substituent on the parent molecule.
    3. Secondary and tertiary amines:
      a. Symmetrical amines are named by using the prefixes *di-* and *tri-* before the name of the alkyl group.
      b. Unsymmetrical amines are named as *N*-substituted primary amines.
        i. The largest group is the parent.
    4. The simplest arylamine is aniline.
    5. Heterocyclic amines (nitrogen is part of a ring) have specific parent names.
      a. The nitrogens receive the lowest possible numbers.
  B. Structure and properties of amines (Section 24.2).
    1. The three amine bonds and the lone pair occupy the corners of a tetrahedron.
    2. An amine with three different substituents is chiral.
      a. The two amine enantiomers interconvert by pyramidal inversion.
      b. This process is rapid at room temperature.
    3. Amines with fewer than 5 carbons are water-soluble and form hydrogen bonds.
    4. Amines have higher boiling points than alkanes of similar molecular weight.
    5. Amines smell nasty!
  C. Basicity of amines (Sections 24.3–24.5).
    1. The lone pair of electrons makes amines both nucleophilic and basic (Section 24.3).
    2. The basicity constant K_b is the measure of the equilibrium of an amine with water.
      a. The larger the value of K_b (smaller pK_b), the stronger the base.
    3. More often, K_a is used to describe amine basicity.
      a. K_a is the dissociation constant of the conjugate acid of an amine.
      b. $pK_a + pK_b = 14$ (for aqueous media).
      c. The smaller the value of K_a (larger pK_a), the stronger the base.
    4. Base strength.
      a. Primary, secondary, and tertiary alkylamines have similar basicities.
      b. Arylamines and heterocyclic amines are less basic than alkylamines.
        i. The sp^2 electrons of the pyridine lone pair are less available for bonding.
        ii. The pyrrole lone pair electrons are part of the aromatic ring π system.
      c. Amides are nonbasic.

  d. Amine basicity can be used as a means of separating amines from a mixture.

    i. An amine can be converted to its salt, extracted from an organic solution with water, neutralized, and re-extracted with an organic solvent.

  e. Some amines are very weak acids.

    i. LDA is formed from diisopropylamine and acts as a strong base.

 5. Basicity of substituted arylamines (Section 24.4).

  a. Arylamines are less basic than alkylamines for two reasons:

    i. Arylamine lone-pair electrons are delocalized over the aromatic ring and are less available for bonding.

    ii. Arylamines lose resonance stabilization when they are protonated.

  b. Electron-donating substituents increase arylamine basicity.

 6. Biological amines and the Henderson–Hasselbalch equation (Section 24.5).

  a. The Henderson–Hasselbalch equation (Section 20.3) can be used to calculate the percent of protonated vs. unprotonated amines.

  b. At physiological pH (7.3), most amines exist in the protonated form.

II. Synthesis of amines (Section 24.6).

 A. Reduction of amides, nitriles and nitro groups.

  1. S_N2 displacement with ^-CN, followed by reduction, turns a primary alkyl halide into an amine with one more carbon atom.

  2. Amide reduction converts an amide or nitrile into an amine with the same number of carbons.

  3. Arylamines can be prepared by reducing aromatic nitro compounds.

   a. Catalytic hydrogenation can be used if no other interfering groups are present.

   b. $SnCl_2$ can also be used.

 B. S_N2 reactions of alkyl halides.

  1. It is possible to alkylate ammonia or an amine with RX.

   a. Unfortunately, it is difficult to avoid overalkylation.

  2. An alternative is displacement of ^-X by azide, followed by reduction.

  3. Also, reaction of an alkyl halide with phthalimide anion, followed by hydrolysis, gives a primary amine (Gabriel amine synthesis).

 C. Reductive amination of aldehydes and ketones.

  1. Treatment of an aldehyde or ketone with ammonia or an amine in the presence of a reducing agent yields an amine.

   a. The reaction proceeds through an imine, which is reduced.

   b. $NaBH_4$ or $NaBH(OAc)_3$ are the reducing agents most commonly used.

   c. Tertiary amines do not undergo reductive amination.

 D. Rearrangements.

  1. Hofmann rearrangement.

   a. When a primary amide is treated with Br_2 and base, CO_2 is eliminated, and an amine with one less carbon is produced.

   b. The mechanism is lengthy and proceeds through an isocyanate intermediate.

  c. In the rearrangement step, the –R group migrates at the same time as the Br⁻ ion leaves.

 2. The Curtius rearrangement starts with an acyl azide and occurs by a mechanism very similar to that of the Hofmann rearrangement.

III. Reactions of amines (Sections 24.7–24.9).

 A. Alkylation and acylation (Section 24.7).

  1. Alkylation of primary and secondary amines occurs but is hard to control.

  2. Primary and secondary amines can also be acylated.

 B. Hofmann elimination.

  1. Alkylamines can be converted to alkenes by the Hofmann elimination reaction.

   a. The amine is treated with an excess of methyl iodide to form a quaternary ammonium salt.

   b. Treatment of the quaternary salt with Ag_2O, followed by heat, gives the alkene.

  2. The elimination is an E2 reaction.

  3. The less substituted double bond is formed because of the bulk of the leaving group.

  4. The reaction was formerly used for structure determination and is rarely used today.

 C. Reactions of arylamines (Section 24.8).

  1. Electrophilic aromatic substitution.

   a. Electrophilic aromatic substitutions are usually carried out on *N*-acetylated amines, rather than on unprotected amines.

    i. Amino groups are *o,p*-activators, and polysubstitution sometimes occurs.

    ii. Friedel–Crafts reactions don't take place with unprotected amines.

   b. Aromatic amines are acetylated by treatment with acetic anhydride.

   c. The *N*-acetylated amines are *o,p*-directing activators, but are less reactive than unprotected amines.

   d. Synthesis of sulfa drugs was achieved by electrophilic aromatic substitution reactions on *N*-protected aromatic amines.

  2. The Sandmeyer reaction.

   a. When a primary arylamine is treated with HNO_2 (nitrous acid), an arenediazonium salt is formed.

   b. The diazonio group of arenediazonium salts can be replaced by many types of nucleophiles in radical substitution reactions.

    i. Aryl halides are formed by treatment with CuCl, CuBr or NaI.

    ii. Aryl nitriles are formed by treatment with CuCN.

    iii. Phenols are formed by treatment with Cu_2O and $Cu(NO_3)_2$.

    iv. H_3PO_2 (hypophosphorous acid) converts a diazonium salt to an arene, and is used when a substituent must be introduced and then removed.

    v. These reactions occur through radical, rather than polar, pathways.

3. Diazonium coupling reactions.

    a. Diazonium salts can react with activated aromatic rings to form colored azo compounds.

    b. The reaction is an electrophilic aromatic substitution that usually occurs at the *p*-position of the activated ring.

    c. The extended π system of the azo ring system makes these compounds brightly colored due to absorption in the visible region of the spectrum.

IV. Heterocyclic amines (Section 24.9).

  A. Pyrrole, imidazole and other 5-membered ring unsaturated heterocycles.

    1. Structures of pyrrole, furan and thiophene.

       a. All are aromatic because they have six π electrons in a cyclic conjugated system.

       b. Pyrrole is nonbasic because all 5 nitrogen electrons are used in bonding.

       c. The carbon atoms in pyrrole are electron-rich and are reactive toward electrophiles.

    2. Electrophilic substitution reactions.

       a. All three compounds undergo electrophilic aromatic substitution reactions readily.

       b. Halogenation, nitration, sulfonation and Friedel–Crafts alkylation can take place if reaction conditions are modified.

       c. Reaction occurs at the 2-position because the reaction intermediate from attack at that position is more stable.

    3. Imidazole and thiazole.

       a. A nitrogen in each of these compounds is basic.

  B. Pyridine and pyrimidine.

    1. Structure of pyridine.

       a. Pyridine is the nitrogen-containing analog of benzene.

       b. The nitrogen lone pair isn't part of the π electron system .

       c. Pyridine is a stronger base than pyrrole but a weaker base than alkylamines.

    2. Electrophilic substitution of pyridine.

       a. Electrophilic substitutions take place with great difficulty.

          i. The pyridine ring is electron-poor due to the electron-withdrawing inductive effect of nitrogen.

          ii. Acid-base complexation between nitrogen and an electrophile puts a positive charge on the ring.

    3. Pyrimidine has two nitrogens in the 1 and 3 positions of a six-membered ring.

       a. Pyrimidine is less basic than pyridine.

  C. Polycyclic heterocycles.

    1. The reactivity of polycyclic heterocyclic compounds is related to the type of heteroatom and to the size of the ring.

    2. Indole has a pyrrole-like nitrogen and undergoes electrophilic aromatic substitutions in the heterocyclic ring.

    3. Purines have 4 nitrogens (3 pyridine-like, and one pyrrole-like) in a fused-ring structure.

IV. Spectroscopy of amines (Section 24.10).
    A. IR spectroscopy.
        1. Primary and secondary amines absorb in the region 3300–3500 cm^{-1}.
            a. Primary amines show a pair of bands at 3350 cm^{-1} and 3450 cm^{-1}.
            b. Secondary amines show a single band at 3350 cm^{-1}.
            c. These absorptions are sharper and less intense than alcohol absorptions, which also occur in this range.
    B. NMR spectroscopy.
        1. ^{1}H NMR.
            a. Amine protons are hard to identify because they appear as broad signals.
            b. Exchange with D$_2$O causes the amine signal to disappear and allows identification.
            c. Hydrogens on the carbon next to nitrogen are somewhat deshielded.
        2. ^{13}C NMR.
            a. Carbons next to nitrogen are slightly deshielded.
    C. Mass spectrometry.
        1. The nitrogen rule: A compound with an odd number of nitrogens has an oddnumbered molecular weight (and molecular ion).
        2. Alkylamines undergo α-cleavage and show peaks that correspond to both possible modes of cleavage.

Solutions to Problems

24.1   Facts to remember about naming amines:

(1)   Primary amines are named by adding the suffix -*amine* to the name of the alkyl substituent.

(2)   The prefix *di-* or *tri-* is added to the names of symmetrical secondary and tertiary amines.

(3)   Unsymmetrical secondary and tertiary amines are named as *N*-substituted primary amines. The parent amine has the largest alkyl group.

(4)   Heterocyclic amines have unique parent names; the heteroatoms have the lowest possible numbers.

(a)   CH₃NHCH₂CH₃

N-Methylethylamine

(b)   Tricyclohexylamine

(c)   *N*-Ethyl-*N*-methyl-cyclohexylamine

(d)   *N*-Methylpyrrolidine

(e)   [(CH₃)₂CH]₂NH

Diisopropylamine

(f)   H₂NCH₂CH₂CHNH₂

1,3-Butanediamine

24.2

(a)   [(CH₃)₂CH]₃N

Triisopropylamine

(b)   (H₂C=CHCH₂)₃N

Triallylamine

(c)   *N*-Methylaniline

(d)   *N*-Ethyl-*N*-methyl-cyclopentylamine

(e)   *N*-Isopropylcyclohexylamine

(f)   *N*-Ethylpyrrole

24.3 The numbering of heterocyclic rings is described in Section 24.1.

(a)

5-Methoxyindole

(b)

1,3-Dimethylpyrrole

(c)

4-(*N,N*-Dimethylamino)-pyridine

(d)

5-Aminopyrimidine

24.4 Amines are less basic than hydroxide but more basic than amides. The pK_a values of the conjugate acids of the amines in (c) are shown. The larger the pK_a, the stronger the base.

	More Basic	*Less Basic*
(a)	$CH_3CH_2NH_2$	$CH_3CH_2CONH_2$
(b)	NaOH	CH_3NH_2
(c)	CH_3NHCH_3 $pK_a = 10.73$	pyridine $pK_a = 5.25$

24.5

$CH_2NH_3^+$

$pK_a = 9.33$
stronger acid (smaller pK_a)

$CH_3CH_2CH_2NH_3^+$

$pK_a = 10.71$
weaker acid (larger pK_a)

CH_2NH_2

$pK_b = 14 - 9.33 = 4.67$
weaker base

$CH_3CH_2CH_2NH_2$

$pK_b = 14 - 10.71 = 3.29$
stronger base

The stronger base (propylamine) holds a proton more tightly than the weaker base (benzylamine). Thus, the propylammonium ion is less acidic (larger pK_a) than the benzylammonium ion (smaller pK_a).

To calculate pK_b: $K_a \cdot K_b = 10{-}14$, $pK_a + pK_b = 14$ and $pK_b = 14 - pK_a$.

24.6 The basicity order of substituted arylamines is the same as their reactivity order in electrophilic aromatic substitution reactions because, in both cases, electron-withdrawing substituents make the site of reaction more electron-poor and destabilize a positive charge.

Least Basic ⟶ *Most Basic*

(a)

O₂N—⟨ ⟩—NH₂ < (O=)C(H)—⟨ ⟩—NH₂ < Br—⟨ ⟩—NH₂

(b)

CH₃C(=O)—⟨ ⟩—NH₂ < Cl—⟨ ⟩—NH₂ < H₃C—⟨ ⟩—NH₂

(c)

F₃C—⟨ ⟩—NH₂ < H₂FC—⟨ ⟩—NH₂ < H₃C—⟨ ⟩—NH₂

24.7   Use the expressions shown in Section 24.5.

$$\log \frac{[RNH_2]}{[RNH_3^+]} = pH - pK_a = 7.3 - 1.3 = 6.0$$

$$\frac{[RNH_2]}{[RNH_3^+]} = \text{antilog}\,(6.0) = 10^6 : [RNH_2] = 10^6 [RNH_3^+]$$

At pH = 7.3, virtually 100% of the pyrimidine molecules are in the neutral form.

24.8   Amide reduction can be used to synthesize most amines, but nitrile reduction can be used to synthesize only primary amines. Thus, the compounds in (b) and (d) can be synthesized only by amide reduction.

Amine   *Nitrile Precursor*   *Amide Precursor*

(a)

CH₃CH₂CH₂NH₂   CH₃CH₂C≡N   CH₃CH₂C(=O)NH₂

(b)

(CH₃CH₂CH₂)₂NH   CH₃CH₂C(=O)NHCH₂CH₂CH₃

(c)

⟨ ⟩—CH₂NH₂   ⟨ ⟩—C≡N   ⟨ ⟩—C(=O)NH₂

(d)

⟨ ⟩—NHCH₂CH₃   ⟨ ⟩—NHC(=O)CH₃

The compounds in parts (b) and (d) can't be prepared by reduction of a nitrile.

24.9

Step 1: Addition of hydroxide.

Step 2: Ring opening.

Step 3: Proton transfer.

Step 4: Addition of hydroxide.

Step 5: Elimination of amine.

Step 6: Proton transfer.

24.10 The upper reaction is the azide synthesis, and the lower reaction is the Gabriel synthesis.

24.11 Look at the target molecule to find the groups bonded to nitrogen. One group comes from the aldehyde/ketone precursor, and the other group comes from the amine precursor. In most cases, two combinations of amine and aldehyde/ketone are possible.

	Amine	*Amine Precursor*	*Carbonyl Precursor*

(a)

$$CH_3CH_2NHCHCH_3 \quad (\overset{CH_3}{|})$$

$CH_3CH_2NH_2$

$$H_2NCHCH_3 \quad (\overset{CH_3}{|})$$

or

$$CH_3\overset{O}{\overset{||}{C}}CH_3$$

CH_3CHO

(b)

⬡—NHCH₂CH₃

⬡—NH₂

CH_3CHO

(c)

⬠—NHCH₃

⬠—NH₂

or

H_2NCH_3

CH_2O

⬠=O

24.12

$$\underset{H_3C}{\overset{\overset{CH_3}{|}}{N}}\underset{H}{} \;+\; O{=}C\!\!-\!\!(aryl)CH_3 \xrightarrow{NaBH_4} \underset{H_3C}{\overset{\overset{CH_3}{|}}{N}}CH_2{-}(aryl)CH_3$$

24.13 In both of these reactions, the product amine is formed from a carboxylic acid derivative precursor that has one more carbon than the amine. In the Hofmann rearrangement, the precursor is an amide, which is treated with Br_2, NaOH and H_2O. In the Curtius rearrangement, the precursor is an acid chloride, which is treated with NaN_3, then with H_2O and heat.

(a)

$$(CH_3)_3CCH_2CH_2CO_2H \xrightarrow{SOCl_2} (CH_3)_3CCH_2CH_2COCl \xrightarrow{NH_3} (CH_3)_3CCH_2CH_2CONH_2$$

$$\downarrow NaN_3 \qquad\qquad\qquad\qquad \downarrow \begin{array}{c} Br_2,\ NaOH \\ H_2O \end{array}$$

$$(CH_3)_3CCH_2CH_2CON_3 \xrightarrow[\text{heat}]{H_2O} (CH_3)_3CCH_2CH_2NH_2$$

(b)

24.14 The Hofmann elimination yields alkenes and amines from larger amines. The major alkene product has the less substituted double bond, but all possible products may be formed. The hydrogens that can be eliminated are starred. When possible, cis and trans double bond isomers are both formed.

(a)

Amine	*Alkene Products*	*Amine products*

$CH_3CH_2CH_2CHCH_2CH_2CH_2CH_3$ (NH$_2$) $CH_3CH_2CH=CHCH_2CH_2CH_2CH_3$ $(CH_3)_3N$

or

$CH_3CH_2CH_2CH=CHCH_2CH_2CH_3$ $(CH_3)_3N$

Both hydrogens that might be eliminated are secondary, and both possible products should form in approximately equal amounts.

(b)

$(CH_3)_3N$

(c)

$CH_3CH_2CH_2CHCH_2CH_2CH_3$ (NH$_2$) $CH_3CH_2CH=CHCH_2CH_2CH_3$ $(CH_3)_3N$

In each of the above reactions, only one product can form.

(d)

$H_2C=CH_2$ *(major)*   +   N(CH$_3$)$_2$

or

(minor)   + $(CH_3)_2NCH_2CH_3$

The first pair of products in (d) results from elimination of a primary hydrogen and are the major products. The second pair of products results from elimination of a secondary hydrogen.

24.15

The product, which contains both the double bond and the tertiary amine in an ring-opened structure, can undergo a second Hoffmann elimination.

24.16 This reaction sequence is similar to the sequence used to synthesize sulfanilamide. Key steps are: (1) treatment of aniline with acetic anhydride to modulate reactivity, (2) reaction of acetanilide with chlorosulfonic acid, (3) treatment of the chlorosulfonate with the heterocyclic base, and (4) removal of the acetyl group.

24.17 In all of these reactions, benzene is nitrated and the nitro group is ultimately reduced, but the timing of the reduction step is important in arriving at the correct product. In (a), nitrobenzene is immediately reduced and alkylated. In (c), chlorination occurs before reduction so that chlorine can be introduced in the *m*-position. In (b) and (d), nitrobenzene is reduced and then acetylated in order to overcome amine basicity and to control reactivity. In both cases, the acetyl group is removed in the last step.

Either method of nitro group reduction ($SnCl_2$, H_2) can be used in all parts of this problem; both methods are shown.

(a)

Mono- and trialkylated anilines are also formed.

(b)

(problem 24.16)

(c)

(d)

(problem 24.16)

24.18 (a)

p-Bromobenzoic acid

The route shown above is one of several ways to synthesize *p*- bromobenzoic acid and is definitely <u>not</u> the simplest way. (The simplest route is Friedel–Crafts alkylation → bromination → oxidation). The illustrated synthesis shows the use of the diazonium replacement reaction that substitutes bromine for a nitro group. Oxidation of the methyl group yields the substituted benzoic acid.

(b)

m-Bromobenzoic acid

Again, this isn't the easiest route to this compound. In this case, nitration is followed by bromination, then by diazotization, treatment with CuCN, and hydrolysis of the nitrile.

(c)

from (b)   *m*-Bromochlorobenzene

(d)

from (a)   *p*-Methylbenzoic acid

(e)

(Problem 24.16)   1,2,4-Tribromobenzene

24.19

Problem 24.17(a)

p-(*N*,*N*-Dimethylamino)azobenzene

Coupling takes place between *N*,*N*-dimethylaniline and a benzenediazonium salt to yield the desired product.

24.20

Thiazole

Thiazole contains six π electrons. Each carbon contributes one electron, nitrogen contributes one electron, and sulfur contributes two electrons to the ring π system. Both sulfur and nitrogen have lone electron pairs in sp^2 orbitals that lie in the plane of the ring.

24.21

$$\log \frac{[RNH_2]}{[RNH_3{}^+]} = pH - pK_a = 7.37 - 6.00 = 1.37$$

$$\frac{[RNH_2]}{[RNH_3{}^+]} = \text{antilog } (1.37) = 23.4 : [RNH_2] = 23.4[RNH_3{}^+]$$

$$[RNH_3{}^+] + 23.4[RNH_3{}^+] = 24.4[RNH_3{}^+] = 100\%$$

$$[RNH_3{}^+] = 100\% \div 24.4 = 4.1\%$$

$$[RNH_2] = 100\% - 4.1 = 95.9\%$$

4.1% of histidine molecules have the imidazole nitrogen in the protonated form at physiological pH.

24.22

Attack at C2:

Pyridine

unfavorable

Attack at C3:

Attack at C4:

unfavorable

Reaction at C3 is favored over reaction at C2 or C4. The positive charge of the cationic intermediate of reaction at C3 is delocalized over three carbon atoms, rather than over two carbons and the electronegative pyridine nitrogen as occurs in reaction at C2 or C4.

24.23

N,N-Dimethyltryptamine

The side chain nitrogen atom of *N,N*-dimethyltryptamine is more basic than the ring nitrogen atom because its lone electron pair is more available for donation to a Lewis acid. The aromatic nitrogen electron lone pair is part of the ring π electron system.

24.24

Attack at C2:

Attack at C3:

Positive charge can be stabilized by the nitrogen lone-pair electrons in reaction at both C2 and C3. In reaction at C2, however, stabilization by nitrogen destroys the aromaticity of the fused benzene ring. Reaction at C3 is therefore favored, even though the cationic intermediate has fewer resonance forms, because the aromaticity of the six-membered-ring is preserved.

24.25

The IR spectrum (pair of bands at around 3300 cm^{-1}) shows that **B** is a primary amine, and the ^{1}H NMR spectrum shows a 9-proton singlet, a one-proton quartet, and a 3-proton doublet. An absorption due to the amine protons is not visible.

Visualizing Chemistry

24.26

(a)

N-Methylisopropylamine

secondary amine

(b)

trans-(2-Methyl-cyclopentyl)amine

primary amine

(c)

N-Isopropylaniline

secondary amine

24.27

pyridine
nitrogen → (less basic)

alkylamine (more basic)

amide (not basic)

24.28

(1*S*,2*S*)-(1,2-Diphenylpropyl)amine

1. excess CH$_3$I
2. Ag$_2$O, H$_2$O, heat

(*Z*)-1,2-Diphenyl-1-propene

Hofmann elimination is an E2 elimination, in which the two groups to be eliminated must be 180° apart. The product that results from this elimination geometry is the *Z* isomer.

24.29

⟵ most basic

The indicated nitrogen is most basic because it is more electron-rich. The electrons of the other nitrogens are part of the fused-ring π system and are less available for donation to a Lewis acid.

Mechanism Problems

24.30 (a)

Mechanism:

not simultaneos, either order

not simultaneos, either order

(b)

Mechanism:

not simultaneos,
either order

not simultaneos,
either order

(c)

Mechanism:

not simultaneos,
either order

not simultaneos,
either order

(d)

Mechanism:

not simultaneos,
either order

not simultaneos,
either order

24.31 **(a)**

Mechanism:

(b)

Mechanism:

(c)

Mechanism:

(d)

Mechanism:

addition from top

and

addition from bottom

24.32 (a)

Mechanism:

(b)

(c)

(d)

Mechanism:

24.33 (a)

Mechanism:

842   Chapter 24

(b)

Mechanism:

(c)

Mechanism:

(d)

Mechanism:

24.34

24.35

Steps 1,6: Protonation.

Steps 2,7: Nucleophilic addition.

Steps 3,8: Proton transfer.

Steps 4,9: Loss of water.

Steps 5,10: Loss of proton.

The mechanism consists of the nucleophilic addition of ammonia, first to one of the ketones, and then to the other, with loss of two equivalents of water.

24.36

3,5-Dimethylisoxazole

Steps 1,6: Protonation.

Steps 2,7: Nucleophilic addition of hydroxylamine.

Steps 3,8: Proton transfer.

Steps 4,9: Loss of water.

Steps 5,10: Loss of proton.

This mechanism is virtually identical to the mechanism illustrated in the previous problem and involves two nucleophilic additions to carbonyl groups, with loss of water. These addition-eliminations are nucleophilic acyl substitution reactions.

24.37 Benzaldehyde first reacts with methylamine and NaBH₄ in the usual way to give the reductive amination product *N*-methylbenzylamine. This product then reacts further with benzaldehyde in a second reductive amination to give *N*-methyldibenzylamine.

N-Methyldibenzylamine

24.38

Step 1:   Nucleophilic addition.

Step 2:   Cyclization.

Step 3:   Elimination of lysine.

Step 4:   Elimination of the other lysine.

Step 5:   Tautomerization.

24.39 The reaction of trimethylamine with ethylene oxide is an S$_N$2 reaction that opens the epoxide ring.

24.40

Mitomycin C

enamine

Steps 1 – 3:   E1 elimination (protonation, loss of HOCH$_3$, deprotonation).

Steps 4 – 6:   S$_N$2 substitution (protonation, substitution, deprotonation).

Step 7:   E1 elimination of carbamate.

Steps 8 – 9:   conjugate addition of DNA (addition, deprotonation).

Notice that five of the nine steps are either protonations or deprotonations.

24.41

Step 1: Addition of NH₃.

Step 2: Elimination of ⁻OH.

Step 3: Addition of ⁻CN.

Step 4: Hydrolysis of nitrile.

The mechanism of acid-catalyzed nitrile hydrolysis is shown in Problem 20.25.

24.42

Step 1: Conjugate addition of hydrazine.

Step 2: Proton transfer.

Step 3: Nucleophilic acyl substitution, forming the cyclic amide.

Step 4: Proton transfer.

24.43

Hofmann rearrangement (the mechanism is shown in Section 24.6) of an α-hydroxy amide produces a carbinolamine intermediate that expels ammonia to give an aldehyde.

24.44

Step 1: Conjugate addition of amine.

Step 2: Proton transfer.

Step 3: Nucleophilic addition of amine.

Step 4: Proton transfer.

Step 5: Elimination of methanol.

24.45

Step 1: S$_N$2 displacement of Br$^-$ by amine.

Step 2: Deprotonation.

Step 3: Conjugate addition of alcohol.

Step 4: Proton transfer.

24.46

(S)-Norcoclaurine

Step 1: Nucleophilic addition of the amine to the protonated aldehyde.

Step 2: Proton transfer.

Step 3: Loss of water.

Step 4: Electrophilic aromatic substitution.

Step 5: Loss of proton.

Additional Problems

Naming Amines

24.47

(a)

2,4-Dibromoaniline

(b)

(2-Cyclopentylethyl)amine

(c)

—NHCH₂CH₃

N-Ethylcyclopentylamine

(d)

N,N-Dimethylcyclopentylamine

(e)

N—CH₂CH₂CH₃

N-Propylpyrrolidine

(f) H₂NCH₂CH₂CH₂CN

4-Aminobutanenitrile

24.48

(a)

—N(CH₃)₃

N,N-Dimethylaniline

(b)

—CH₂NH₂

(Cyclohexylmethyl)amine

(c)

—NHCH₃

N-Methylcyclohexylamine

(d)

(2-Methylcyclohexyl)amine

(e) (H₃C)₂NCH₂CH₂CO₂H

3-(*N,N*-dimethylaminopropano

24.49

(a)

secondary amine

(b)

secondary amine

(c)

tertiary amide

tertiary amine

secondary amine

Lysergic acid diethylamide

Amine Basicity

24.50 The pyrrole anion, $C_4H_4N{:}^-$, is a 6 π electron species that has the same electronic structure as the cyclopentadienyl anion. Both of these anions possess the aromatic stability of 6 π electron systems.

24.51

Histamine

The "a" nitrogen is most basic because its electron pair is most available to Lewis acids. The "c" nitrogen is the least basic because the lone-pair electrons of the pyrrole nitrogen are part of the ring π electron system.

24.52

The inductive effect of the electron-withdrawing nitro group makes the amine nitrogens of both *m*-nitroaniline and *p*-nitroaniline less electron-rich and less basic than aniline.

When the nitro group is para to the amino group, conjugation of the amino group with the nitro group can also occur. *p*-Nitroaniline is thus even less basic than *m*-nitroaniline.

Synthesis of Amines

24.53

(a)

$$CH_2CH_2CH_2CH_2OH \xrightarrow{PBr_3} CH_2CH_2CH_2CH_2Br \xrightarrow{NaN_3} CH_2CH_2CH_2CH_2N_3$$

$$\downarrow \begin{array}{l} 1.\ LiAlH_4 \\ 2.\ H_2O \end{array}$$

$$CH_2CH_2CH_2CH_2NH_2$$
Butylamine

(b)

$$CH_2CH_2CH_2CH_2OH \xrightarrow[H_3O^+]{CrO_3} CH_2CH_2CH_2CO_2H \xrightarrow{SOCl_2} CH_2CH_2CH_2COCl$$

$$\downarrow \begin{array}{l} CH_2CH_2CH_2CH_2NH_2 \\ from\ (a) \\ NaOH \end{array}$$

$$\overset{O}{\overset{\|}{CH_3CH_2CH_2CNHCH_2CH_2CH_2CH_3}}$$

$$CH_3CH_2CH_2CH_2NHCH_2CH_2CH_2CH_3 \xleftarrow[2.\ H_2O]{1.\ LiAlH_4}$$
Dibutylamine

(c)

$$CH_2CH_2CH_2CH_2OH \xrightarrow{CrO_3,\ H_3O^+} CH_2CH_2CH_2CO_2H$$

$$\xrightarrow[H_2O]{Br_2,\ NaOH} CH_2CH_2CH_2CONH_2 \xleftarrow{2\ NH_3}$$

$$CH_2CH_2CH_2NH_2$$
Propylamine

$$CH_2CH_2CH_2CON_3 \xleftarrow{NaN_3}$$
$$H_2O,\ heat$$

$$\downarrow SOCl_2$$
$$CH_2CH_2CH_2COCl$$

(d)

$$CH_2CH_2CH_2CH_2OH \xrightarrow{PBr_3} CH_2CH_2CH_2CH_2Br \xrightarrow{NaCN} CH_2CH_2CH_2CH_2CN$$

$$\downarrow \begin{array}{l} 1.\ LiAlH_4 \\ 2.\ H_2O \end{array}$$

$$CH_2CH_2CH_2CH_2CH_2NH_2$$
Pentylamine

(e) $CH_2CH_2CH_2CH_2OH$ $\xrightarrow[CH_2Cl_2]{Periodinane}$ $CH_2CH_2CH_2CHO$ $\xrightarrow[(CH_3)_2NH]{NaBH_4}$ $CH_2CH_2CH_2CH_2N(CH_3)_2$

N,N-Dimethylbutylamine

(f) $CH_2CH_2CH_2NH_2$ $\xrightarrow[CH_3I]{excess}$ $CH_2CH_2CH_2\overset{+}{N}(CH_3)I^-$ $\xrightarrow[2.\ heat]{1.\ Ag_2O,\ H_2O}$ $CH_3CH=CH_2$

from (c)

Propene   + $(CH_3)_3N$

24.54

(a) $CH_3CH_2CH_2CH_2CO_2H$ $\xrightarrow{SOCl_2}$ $CH_3CH_2CH_2CH_2COCl$ $\xrightarrow{2\ NH_3}$ $CH_3CH_2CH_2CH_2CONH_2$

Pentanamide

(b) $CH_3CH_2CH_2CH_2CONH_2$ $\xrightarrow[H_2O]{Br_2,\ NaOH}$ $CH_3CH_2CH_2CH_2NH_2$

from (a)   Butylamine

(c) $CH_3CH_2CH_2CH_2CONH_2$ $\xrightarrow[2.\ H_2O]{1.\ LiAlH_4}$ $CH_3CH_2CH_2CH_2CH_2NH_2$

from (a)   Pentylamine

(d) $CH_3CH_2CH_2CH_2CO_2H$ $\xrightarrow[2.\ H_2O]{1.\ Br_2,\ PBr_3}$ $CH_3CH_2CH_2\overset{\overset{\displaystyle Br}{|}}{C}HCO_2H$

2-Bromopentanoic acid

(e) $CH_3CH_2CH_2CH_2CO_2H$ $\xrightarrow[2.\ H_3O^+]{1.\ BH_3}$ $CH_3CH_2CH_2CH_2CH_2OH$

$\downarrow PBr_3$

$CH_3CH_2CH_2CH_2CH_2CN$ $\xleftarrow{NaCN}$ $CH_3CH_2CH_2CH_2CH_2Br$

Hexanenitrile

(f) $CH_3CH_2CH_2CH_2CH_2CN$ $\xrightarrow[2.\ H_2O]{1.\ LiAlH_4}$ $CH_3CH_2CH_2CH_2CH_2CH_2NH_2$

from (e)   Hexylamine

24.55

(a) $\xrightarrow[H_2SO_4]{HNO_3}$ (benzene with NO$_2$) $\xrightarrow[2.\ NaOH,\ H_2O]{1.\ SnCl_2,\ H_3O^+}$ (benzene with NH$_2$)

(b) (benzene with CONH$_2$) $\xrightarrow[H_2O]{Br_2,\ NaOH}$ (benzene with NH$_2$)

(c) (toluene CH$_3$) $\xrightarrow[H_2O]{KMnO_4}$ (benzene CO$_2$H) $\xrightarrow{SOCl_2}$ (benzene COCl) $\xrightarrow{2\ NH_3}$ (benzene CONH$_2$) $\xrightarrow[H_2O]{Br_2,\ NaOH}$ (benzene NH$_2$)

24.56 First, synthesize aniline from benzene, as shown in Problem 24.38 (a).

24.57

(a) $CH_3CH_2CH_2CH_2CONH_2$ $\xrightarrow[\text{2. H}_2\text{O}]{\text{1. LiAlH}_4}$ $CH_3CH_2CH_2CH_2CH_2NH_2$

(b) $CH_3CH_2CH_2CH_2CN$ $\xrightarrow[\text{2. H}_2\text{O}]{\text{1. LiAlH}_4}$ $CH_3CH_2CH_2CH_2CH_2NH_2$

(c) $CH_3CH_2CH{=}CH_2$ $\xrightarrow[\text{2. H}_2\text{O}_2, \ ^-\text{OH}]{\text{1. BH}_3, \text{THF}}$ $CH_3CH_2CH_2CH_2OH$ $\xrightarrow{\text{PBr}_3}$ $CH_3CH_2CH_2CH_2Br$

$\downarrow$ NaCN

$CH_3CH_2CH_2CH_2CH_2NH_2$ $\xleftarrow[\text{2. H}_2\text{O}]{\text{1. LiAlH}_4}$ $CH_3CH_2CH_2CH_2CN$

(d) $CH_3CH_2CH_2CH_2CH_2CONH_2$ $\xrightarrow[\text{H}_2\text{O}]{\text{Br}_2, \text{NaOH}}$ $CH_3CH_2CH_2CH_2CH_2NH_2$

(e) $CH_3CH_2CH_2CH_2CH{=}CHCH_2CH_2CH_2CH_3$ $\xrightarrow[\text{2. Zn, H}_3\text{O}^+]{\text{1. O}_3}$ $2 \ CH_3CH_2CH_2CH_2CHO$

$CH_3CH_2CH_2CH_2CHO$ $\xrightarrow[\text{NaBH}_4]{\text{NH}_3}$ $CH_3CH_2CH_2CH_2CH_2NH_2$

(f) $CH_3CH_2CH_2CH_2OH$ $\xrightarrow{\text{PBr}_3}$ $CH_3CH_2CH_2CH_2Br$ $\xrightarrow{\text{NaCN}}$ $CH_3CH_2CH_2CH_2CN$

$\downarrow$ 1. LiAlH$_4$
$\downarrow$ 2. H$_2$O

$CH_3CH_2CH_2CH_2CH_2NH_2$

(g) $CH_3CH_2CH_2CH_2CO_2H$ $\xrightarrow{\text{SOCl}_2}$ $CH_3CH_2CH_2CH_2COCl$

$\downarrow$ 2 NH$_3$

$CH_3CH_2CH_2CH_2CH_2NH_2$ $\xleftarrow[\text{2. H}_2\text{O}]{\text{1. LiAlH}_4}$ $CH_3CH_2CH_2CH_2CONH_2$

24.58

Reactions of Amines

24.59

(a)

(b)

(c)

24.60

(a)

m Toluidine

(b)

(c)

(d)

24.61

(a)

(b)

(c)

(d)

Br—⟨aryl⟩—NH_2 $\xrightarrow{CH_3COCl}$ Br—⟨aryl⟩—$NHCOCH_3$

(e)

Br—⟨aryl⟩—NH_2 $\xrightarrow{CH_3MgBr}$ Br—⟨aryl⟩—$NH\ MgBr$ $+$ CH_4

(f)

Br—⟨aryl⟩—NH_2 $\xrightarrow[AlCl_3]{CH_3CH_2Cl}$ Br—⟨aryl⟩—$\overset{H}{\underset{H}{N}}:AlCl_3$

(g)

Br—⟨aryl⟩—$\overset{+}{N_2}\ HSO_4^{-}$ $\xrightarrow{CuCl}$ Br—⟨aryl⟩—Cl

(h)

Br—⟨aryl⟩—$NHCOCH_3$ $\xrightarrow[AlCl_3]{CH_3CH_2Cl}$ Br—⟨aryl⟩—$NHCOCH_3$ with CH_2CH_3

In (e), the amine reacts with the Grignard reagent and inactivates it.

24.62 Hydrogens that can be eliminated are starred. In cases where more than one alkene can form, the alkene with the less substituted double bond is the major product.

Amine $\xrightarrow[\substack{\text{1. excess } CH_3I \\ \text{2. } Ag_2O,\ H_2O \\ \text{3. heat}}]{}$ Alkene $+$ Amine

(a)

cyclopentyl—$NHCH_3$ → cyclopentene $+$ $N(CH_3)_3$

(b)

$\overset{*}{C}H_3$
$HN\overset{*}{C}HCH_2CH_2CH_2CH_3$ (with phenyl)

$H_2C{=}CHCH_2CH_2CH_2CH_3$ major
$CH_3CH{=}CHCH_2CH_2CH_3$ minor

$+$ phenyl—$N(CH_3)_2$

(c)

$$CH_3$$
$$|$$
$$CH_3CHCHCH_2CH_2CH_3$$
$$\quad * | \quad *$$
$$NH_2$$

$$CH_3$$
$$|$$
$$CH_3CHCH=CHCH_2CH_3 \quad + \quad N(CH_3)_3$$

major

$$CH_3$$
$$|$$
$$CH_3C=CHCH_2CH_2CH_3$$

minor

24.63

(a)

(b)

from (a)

(c)

24.64

(a)

(b)

(c)

(d)

Spectroscopy

24.65   The ^{1}H NMR of the amine shows 5 peaks. Two are due to an ethyl group bonded to an electronegative element (oxygen), two are due to 4 aromatic ring hydrogens, and the peak at 3.40 δ is due to 2 amine hydrogens.

$C_{10}H_{13}NO_2$ $C_8H_{11}NO$ C_6H_7NO
Phenacetin   *p*-Ethoxyaniline   *p*-Aminophenol

24.66

(a)   HOCH$_2$CH$_2$CH$_2$NH$_2$ a = 1.68 δ   (b)   (CH$_3$O)$_2$CHCH$_2$NH$_2$ a = 1.29 δ
     e   d   a   c   b   b = 2.69 δ   c   d b a   b = 2.78 δ

                                  c = 2.89 δ   c = 3.39 δ

                                  d = 3.72 δ   d = 4.31 δ

(c)   2-Phenylethylamine

24.67   Dibutylamine

General Problems

24.68

(a)   NH₃, NaBH₄;

(b)   excess CH₃I;

(c)   Ag₂O, H₂O, heat;

(d)   RCO₃H

(e)   (CH₃)₂NH.

Step (e) is an S$_N$2 ring opening of the epoxide by nucleophilic substitution of the amine at the primary carbon.

24.69

Oxazole

Oxazole is an aromatic 6 π electron heterocycle. Two oxygen electrons and one nitrogen electron are in *p* orbitals that are part of the π electron system of the ring, along with one electron from each carbon. An oxygen lone pair and a nitrogen lone pair are in *sp*² orbitals that lie in the plane of the ring. Since the nitrogen lone pair is available for donation to acids, oxazole is more basic than pyrrole.

24.70

N-Protonation
(no resonance stabilization)

O-Protonation
(resonance stabilization)

Protonation occurs on oxygen because an *O*-protonated amide is stabilized by resonance.

24.71 *N*-Ethylaniline

24.72

(a)   CH₂=CHCH₂Cl, AlCl₃;

(b)   Hg(OAc)₂, H₂O; NaBH₄;

(c)   CrO₃, H₃O⁺

(d)   CH₃NH₂, NaBH₄.

24.73

The last step of the synthesis is a reductive amination of a ketone that is formed by oxidation of the corresponding alcohol. The alcohol results from the Grignard reaction between cyclopentylmagnesium bromide and propylene oxide.

24.74

(a)

Tetracaine

(b)

1. SOCl$_2$
2. HOCH$_2$CH$_2$N(CH$_3$)$_2$, Pyridine

(c) Benzene

CH$_3$Cl
AlCl$_3$

The synthesis in (a) is achieved by a reductive amination reaction. Reactions in (b) include formation of an acid chloride, esterification, and reduction of the nitro group.

24.75

Atropine   Tropine   Tropidene

+

HO$_2$CCHC$_6$H$_5$
|
CH$_2$OH   Tropic acid

We know the location of the –OH group of tropine because it is stated that tropine is an optically inactive alcohol. This hydroxyl group results from basic hydrolysis of the ester that is composed of tropine and tropic acid.

24.76

Tropidene

1. CH$_3$I
2. Ag$_2$O, H$_2$O
3. heat

1. CH$_3$I
2. Ag$_2$O, H$_2$O
3. heat

+ N(CH$_3$)$_3$

Tropilidene

Tropilidene results from two cycles of Hofmann elimination on tropidene.

24.77 The formula $C_9H_{17}N$ indicates two degrees of unsaturation in the product. Both are probably due to rings since the product results from catalytic reduction.

Step 1:   Reduction of nitrile

Step 2:   Nucleophilic addition.

Step 3:   Dehydration.

Step 4:   Reduction of double bond.

24.78 The molecular formula indicates that coniine has one double bond or ring, and the Hofmann elimination product shows that the nitrogen atom is part of a ring.

Coniine   5-(N,N-Dimethylamino)-1-octene

24.79

Step 1: Michael addition.

Step 2: Ester hydrolysis; decarboxylation.

Step 3: Reduction.

Step 4: Cyclization.

Step 5: Reduction.

24.80

Tyramine

When you see –CH₂NH₂, think of the reduction of a nitrile. The nitrile comes from substitution of a benzylic bromide by ⁻CN.

24.81

(a)

(b)

The reactive intermediate is benzyne, which undergoes a Diels–Alder reaction with cyclopentadiene to yield the observed product.

24.82

Two successive cycles of Hofmann elimination lead to formation of cyclooctatriene.

Allylic bromination followed by elimination yield cyclooctatetraene.

24.83

(a)

a = 2.25 δ

b = 2.89 δ

c = 6.66 δ, 7.03 δ

(b)

a = 1.14 δ

b = 3.40 δ

c = 4.47 δ

d = 6.65 δ

e = 7.24 δ

24.84 Note: Although there are several resonance forms, with the exception of the completely neutral form, the most stable one is the charge separated form with a negative charge on the pyridine nitrogen. The increased electron density on the pyridine nitrogen increases its ability to react as a nucleophile.

Review Unit 9: Carbonyl Compounds II − Reaction at the α Carbon; Amines

Major Topics Covered (with vocabulary):

Carbonyl α-substitution reactions:

α-substitution reaction   tautomerism   tautomer   enolate ion   Hell-Volhard-Zelinskii reaction   β-diketone   β-keto eater   malonic ester synthesis   acetoacetic eater synthesis   LDA

Carbonyl condensation reactions:

carbonyl condensation reactions   aldol reaction   enone   mixed aldol reaction   Claisen condensation reaction   Dieckmann cyclization   Michael reaction   Michael acceptor   Michael donor   Stork enamine reaction   Robinson annulation reaction

Amines:

primary, secondary, tertiary amine   quaternary ammonium salt   arylamine   heterocyclic amine   pyramidal inversion K_b azide synthesis   Gabriel amine synthesis   reductive amination   Hofmann rearrangement   Curtius rearrangement   Hofmann elimination reaction   arenediazonium salt   diazotization   Sandmeyer reaction   azo compound   diazonium coupling reaction   pyrrole   thiophene   furan   pyridine   fused-ring heterocycle   pyrimidine   purine   nitrogen rule

Types of Problems:

After studying these chapters, you should be able to:

- Draw keto-enol tautomers of carbonyl compounds, identify acidic hydrogens, and draw the resonance forms of enolates.
- Formulate the mechanisms of acid- and base-catalyzed enolization and of other α-substitution reactions.
- Predict the products of α-substitution reactions.
- Use α-substitution reactions in synthesis.
- Predict the products of carbonyl condensation reactions.
- Formulate the mechanisms of carbonyl condensation reactions.
- Use carbonyl condensation reactions in synthesis.
- Name and draw amines, and classify amines as primary, secondary, tertiary , quaternary, arylamines, or heterocyclic amines.
- Predict the basicity of alkylamines, arylamines and heterocyclic amines.
- Synthesize alkylamines and arylamines by several routes.
- Predict the products of reactions involving alkylamines and arylamines.
- Use diazonium salts in reactions involving arylamines, including diazo coupling reactions.
- Draw orbital pictures of heterocycles and explain their acid-base properties.
- Explain orientation and reactivity in heterocyclic reactions, and predict the products of reactions involving heterocycles.
- Propose mechanisms for reactions involving alkylamines, arylamines, and heterocycles.
- Identify amines by spectroscopic techniques.

Points to Remember:

*   It is unusual to think of a carbonyl compound as an acid, but the protons α to a carbonyl group can be removed by a strong base. Protons α to two carbonyl groups are even more acidic: in some cases, acidity approaches that of phenols. This acidity is the basis for α-substitution reactions of compounds having carbonyl groups. Abstraction by base of an α proton produces a resonance-stabilized enolate anion that can be used in alkylations involving alkyl halides and tosylates.

*   Alkylation of an unsymmetrical LDA-generated enolate generally occurs at the less hindered α carbon.

*   When you need to synthesize a β-hydroxy ketone or aldehyde or an α,β-unsaturated ketone or aldehyde, use an aldol reaction. When you need to synthesize a β-diketone or β-keto ester, use a Claisen reaction. When you need to synthesize a 1,5-dicarbonyl compound, use a Michael reaction. The Robinson annulation is used to synthesize polycyclic molecules by a combination of a Michael reaction with an aldol condensation.

*   In many of the mechanisms in this group of chapters, the steps involving proton transfer are not explicitly shown. The proton transfers occur between the proton and the conjugate base with the most favorable pK of those present in the solution. These steps have been omitted at times to simplify the mechanisms.

*   In the Claisen condensation, the enolate of the β-dicarbonyl compound is treated with H_3O^+ to yield the neutral product.

*   For an amine, the larger the value of pK_a of its ammonium ion, the stronger the base. The smaller the value of pK_b of the amine, the stronger the base.

*   The Sandmeyer reaction allows the synthesis of substituted benzenes that can't be formed by electrophilic aromatic substitution reactions. These reactions succeed because N_2 is a very good leaving group.

Self-Test:

A

Pentymal

(a sedative)

B

Dypnone

(sunscreen)

The six-membered ring in **A** is formed by the cyclization of two difunctional compounds. What are they? What type of reaction occurs to form the ring? The two alkyl groups are introduced into one of the difunctional compounds prior to cyclization. What type of reaction is occurring, and how is it carried out? What type of reaction occurs in the formation of Dypnone (**B**)? Why might **B** be effective as a sunscreen?

C

Benzphetamine

(an appetite suppressant)

D

Butralin

(an herbicide)

What type of amine is **C**? Do you expect it to be more or less basic than ammonia? Than aniline? What product do you expect from Hofmann elimination of **C**? What significant absorptions might be seen in the IR spectrum of **C**? What information can be obtained from the mass spectrum? Plan a synthesis of **D** from benzene.

Multiple Choice:

1. Which of the following compounds has four acidic hydrogens?
    (a)   2-Pentanone
    (b)   3-Pentanone
    (c)   Acetophenone
    (d)   Phenylacetone

2. In which of the following reactions is an enol, rather than an enolate, the reacting species?
    (a)   acetoacetic acid synthesis
    (b)   malonic ester synthesis
    (c)   LDA alkylation
    (d)   Hell–Volhard–Zelinskii reaction

3. Cyclobutanecarboxylic acid is probably the product of a:
    (a)   malonic ester synthesis
    (b)   acetoacetic ester synthesis
    (c)   LDA alkylation
    (d)   Hell–Volhard–Zelinskii reaction

4. An LDA alkylation can be used to alkylate all of the following, except:
    (a)   aldehydes
    (b)   ketones
    (c)   esters
    (d)   nitriles

5. If you want to carry out a carbonyl condensation, and you don't want to form α-substitution product, you should:
    (a)   lower the temperature
    (b)   use one equivalent of base
    (c)   use a catalytic amount of base
    (d)   use a polar aprotic solvent

6. Which reaction forms a cyclohexenone?
    (a)   Dieckmann cyclization
    (b)   Michael reaction
    (c)   Claisen condensation
    (d)   intramolecular aldol condensation

7. All of the following molecules are good Michael donors except:
    (a)   Ethyl acetoacetate
    (b)   Nitroethylene
    (c)   Malonic ester
    (d)   Ethyl 2-oxocyclohexanecarboxylate

8.   The ammonium ion of which of the following amines has the smallest value of pK_a?
     (a)   Methylamine
     (b)   Trimethylamine
     (c)   Aniline
     (d)   p-Bromoaniline

9.   All of the following methods of amine synthesis are limited to primary amines, except:
     (a)   Curtius rearrangement
     (b)   reductive amination
     (c)   Hofmann rearrangement
     (d)   azide synthesis

10. To form an azo compound, an aryldiazonium salt should react with:
     (a)   CuCN
     (b)   benzene
     (c)   nitrobenzene
     (d)   phenol

Chapter 25 – Biomolecules: Carbohydrates

Chapter Outline

I. Classification of carbohydrates (Section 25.1).
   A. Simple *vs.* complex:
      1. Simple carbohydrates (monosaccharides) can't be hydrolyzed to smaller units.
      2. Complex carbohydrates are made up of two or more simple sugars linked together.
         a. A disaccharide is composed of two monosaccharides.
         b. A polysaccharide is composed of three or more monosaccharides.
   B. Aldoses *vs.* ketoses:
      1. A monosaccharide with an aldehyde carbonyl group is an aldose.
      2. A monosaccharide with a ketone carbonyl group is a ketose.
   C. *Tri-, tetr-, pent-*, etc. indicate the number of carbons in the monosaccharide.
II. Monosaccharides (Sections 25.2–25.7).
   A. Configurations of monosaccharides (Section 25.2–25.4).
      1. Fischer projections (Section 25.2).
         a. Each chirality center of a monosaccharide is represented by a pair of crossed lines.
            i. The horizontal line represents bonds coming out of the page.
            ii. The vertical line represents bonds going into the page.
         b. Allowed manipulations of Fischer projections:
            i. A Fischer projection can be rotated on the page by 180°, but not by 90° or 270°.
            ii. Holding one group steady, the other three groups can be rotated clockwise or counterclockwise.
         c. Rules for assigning *R,S* configurations.
            i. Assign priorities to the substituents in the usual way (Section 5.5).
            ii. Perform one of the two allowed motions to place the lowest priority group at the top of the Fischer projection.
            iii. Determine the direction of rotation of the arrow that travels from group 1 to group 2 to group 3, and assign *R* or *S* configuration.
         d. Carbohydrates with more than one chirality center are shown by stacking the centers on top of each other.
            i. The carbonyl carbon is placed at or near the top of the Fischer projection.
      2. D,L sugars (Section 25.3).
         a. (*R*)-Glyceraldehyde is also known as D-glyceraldehyde.
         b. In D sugars, the –OH group farthest from the carbonyl group points to the right in a Fischer projection.
            i. Most naturally-occurring sugars are D sugars.
         c. In L sugars, the –OH group farthest from the carbonyl group points to the left in a Fischer projection.
         d. D,L designations refer only to the configuration farthest from the carbonyl carbon and are unrelated to the direction of rotation of plane-polarized light.

3. Configurations of aldoses (Section 25.4).
    a. There are 4 aldotetroses – D and L erythrose and threose.
    b. There are 4 D,L pairs of aldopentoses: ribose, arabinose, xylose and lyxose.
    c. There are 8 D,L pairs of aldohexoses : allose, altrose, glucose, mannose, gulose, idose, galactose, and talose.
    d. A scheme for drawing and memorizing the D-aldohexoses:
        i.   Draw all –OH groups at C5 pointing to the right.
        ii. Draw the first four –OH groups at C4 pointing to the right and the second four pointing to the left.
        iii. Alternate –OH groups at C3: two right, two left, two right, two left.
        iv. Alternate –OH groups at C2: right, left, etc.
        v.   Use the mnemonic "<u>All</u> <u>al</u>truists <u>g</u>ladly <u>m</u>ake <u>g</u>um *in* <u>gal</u>lon <u>t</u>anks" to assign names.

B. Cyclic structures of monosaccharides (Section 25.5).
    1. Hemiacetal formation.
        a. Monosaccharides are in equilibrium with their internal hemiacetals.
            i.   Glucose exists primarily as a six-membered pyranose ring, formed between the –OH group at C5 and the aldehyde group at C1.
            ii. Fructose exists primarily as a five-membered furanose ring.
        b. Structure of pyranose rings.
            i.   Pyranose rings have a chair-like geometry.
            ii. The hemiacetal oxygen is at the right rear for D-sugars.
            iii. An –OH group on the right in a Fischer projection is on the bottom face in a pyranose ring, and an –OH group on the left is on the top face.
            iv. For D sugars, the –CH₂OH group is on the top.
    2. Mutarotation.
        a. When a monosaccharide cyclizes, a new chirality center is generated.
            i.   The two diastereomers are anomers.
            ii. The form with the anomeric –OH group trans to the –CH₂OH group is the α anomer (minor anomer).
            iii. The form with the anomeric –OH group cis to the –CH₂OH group is the β anomer (major anomer).
        b. When a solution of either pure anomer is dissolved in water, the optical rotation of the solution reaches a constant value.
            i.   This process is called mutarotation.
            ii. Mutarotation is due to the reversible opening and recyclizing of the hemiacetal ring and is catalyzed by both acid and base.

C. Reactions of monosaccharides (Section 25.6).
    1. Ester and ether formation.
        a. Esterification occurs by treatment with an acid anhydride or acid chloride.
        b. Ethers are formed by treatment with methyl iodide and Ag₂O.
        c. Ester and ether derivatives are crystalline and easy to purify.

2. Glycoside formation.
    a. Treatment of a hemiacetal with an alcohol and an acid catalyst yields an acetal.
        i. Acetals aren't in equilibrium with an open-chain form and do exhibit mutarotation.
        ii. Aqueous acid reconverts the acetal to a monosaccharide.
    b. These acetals, called glycosides, occur in nature.
    c. Glycosides are named by first citing the alkyl group and then replacing the *–ose* suffix of the sugar with *-oside*.
    d. The laboratory synthesis of glycosides is achieved by the Koenigs–Knorr reaction.
        i. Treatment of the acetylpyranose with HBr, followed by treatment with the appropriate alcohol and Ag_2O, gives the acetylglycoside.
        ii. Both anomers give the same product.
        iii. The reaction involves neighboring-group participation by acetate.
3. Phosphorylation.
    a. Monosaccharides can be phosphorylated by ATP to form a glycosyl monophosphate.
    b. The resulting glycosyl monophosphate can react with a second nucleoside triphosphate to produce a glycosyl diphosphate.
    c. This product can react with a lipid or a protein to form a glycoconjugate.
4. Reduction of monosaccharides.
    a. Reaction of a monosaccharide with $NaBH_4$ yields an alditol (a polyalcohol).
5. Oxidation of monosaccharides.
    a. Several mild reagents can oxidize the carbonyl group to a carboxylic acid (aldonic acid).
        i. In the laboratory, aqueous Br_2 is used to oxidize aldoses (not ketoses).
        ii. Historically, Tollens reagent, Fehling's reagent and Benedict's reagent have served as tests for reducing sugars.
        iii. All aldoses and some ketoses are reducing sugars, but glycosides are nonreducing.
    b. The more powerful oxidizing agent, dilute HNO_3, oxidizes aldoses to dicarboxylic acids (aldaric acids).
    c. Enzymes can oxidize the $-CH_2OH$ of a monosaccharide (with oxidizing the aldehyde) to form a uronic acid.
6. Chain-lengthening: the Kiliani–Fischer synthesis.
    a. In the Kiliani–Fischer synthesis, an aldehyde group becomes C2 of a chainlengthened monosaccharide and the added carbon is the new C1.
    b. The reaction involves cyanohydrin formation, reduction and hydrolysis.
    c. The products are two diastereomeric aldoses that differ in configuration at C2.
7. Chain-shortening: the Wohl degradation.
    a. The Wohl degradation shortens an aldose by one carbon.

b. The reaction involves treatment of the aldose with hydroxylamine, dehydration and loss of HCN from the resulting cyanohydrin.

D. Eight essential monosaccharides (Section 25.7).

1. Glucose, galactose, mannose and xylose are aldoses.

2. Fucose is a deoxy sugar.

3. *N*-Acetylglucosamine and *N*-acetylgalactosamine are amino sugars.

4. *N*-Acetylneuraminic acid is the parent compound of the sialic acids.

5. All of the essential monosaccharides arise from glucose.

III. Other carbohydrates (Sections 25.8–25.11).

A. Disaccharides (Section 25.8).

1. Cellobiose and maltose.

a. Cellobiose and maltose contain a $1 \rightarrow 4$-glycosidic acetal bond between two glucose monosaccharide units.

b. Maltose consists of two glucopyranose units joined by a $1 \rightarrow 4$-α-glycosidic bond.

c. Cellobiose consists of two glucopyranose units joined by a $1 \rightarrow 4$-β-glycosidic bond.

d. Both maltose and cellobiose are reducing sugars and exhibit mutarotation.

e. Humans can't digest cellobiose but can digest maltose.

2. Lactose.

a. Lactose consists of a unit of galactose joined by a β-glycosidic bond between C1 and C4 of a glucose unit.

b. Lactose is a reducing sugar found in milk.

3. Sucrose.

a. Sucrose is a disaccharide that yields glucose and fructose on hydrolysis.

i. Sucrose is called "invert sugar" because the sign of rotation changes when sucrose is hydrolyzed.

ii. Sucrose is one of the most abundant pure organic chemicals in the world.

b. The two monosaccharides are joined by a glycosidic link between C1 of glucose and C2 of fructose.

c. Sucrose isn't a reducing sugar and doesn't exhibit mutarotation.

B. Polysaccharides and their synthesis (Section 25.9).

1. Polysaccharides have a reducing end and undergo mutarotation, but aren't considered to be reducing sugars because of their size.

2. Important polysaccharides.

a. Cellulose.

i. Cellulose consists of thousands of D-glucose units linked by $1 \rightarrow 4$-β-glycosidic bonds.

ii. In nature, cellulose is used as structural material.

b. Starch.

i. Starch consists of thousands of D-glucose units linked by $1 \rightarrow 4$-α-glycosidic bonds.

ii. Starch can be separated into amylose (water-soluble) and amylopectin (water-insoluble) fractions.

(a) Amylopectin contains $1\rightarrow6$-α-glycosidic branches.

iii. Starch is digested in the mouth by glycosidase enzymes, which only cleave α-glycosidic bonds.

c. Glycogen.

i. Glycogen is an energy-storage polysaccharide.

ii. Glycogen contains both $1\rightarrow4$- and $1\rightarrow6$-links.

3. An outline of the glycan assembly method of polysaccharide synthesis.

a. A glycal (a monosaccharide with a C1–C2 double bond) is protected at C6 by formation of a silyl ether and at C3–C4 by formation of a cyclic carbonate ester.

b. The protected glycal is epoxidized.

c. Treatment of the glycal epoxide (in the presence of $ZnCl_2$) with a second glycal having a free C6 hydroxyl group forms a disaccharide.

d. The process can be repeated.

C. Other important carbohydrates (Section 25.10).

1. Deoxy sugars have an –OH group missing and are components of nucleic acids.

2. In amino sugars, an –OH is replaced by a $-NH_2$.

a. Amino sugars are found in chitin and in antibiotics.

D. Cell surface carbohydrates and influenza viruses (Section 25.11).

1. Polysaccharides are involved in cell surface recognition.

a. Polysaccharide markers on the surface of influenza viruses are variants of two types of glycoproteins –hemagglutinin (H – Type 5 or Type 1), and neuraminidase (N – Type 1).

b. Infection occurs when a virus binds to a receptor on a target cell and is engulfed by the cell.

c. New viral particles are produced, pass out of the cell, and are held to surface receptors.

d. A neuraminidase enzyme cleaves the receptor–virus bond, allowing the virus to invade a new cell.

2. Antiviral vaccines block the neuraminidase enzyme, limiting the spread of the virus.

Solutions to Problems

25.1

(a) Threose
an aldotetrose

(b) Ribulose
a ketopentose

(c) Tagatose
a ketohexose

(d) 2-Deoxyribose
an aldopentose

25.2 Horizontal bonds of Fischer projections point out of the page, and vertical bonds point into the page.

(a)

(b)

(c)

25.3 To decide if two Fischer projections are identical, use the two allowable rotations to superimpose two groups of each projection. If the remaining groups are also superimposed after rotation, the projections represent the same enantiomer.

(a) Since –H is in the same position in both A and B, keep it steady, and rotate the other three groups. If, after rotation, all groups are superimposed, the two projections are identical. If only two groups are superimposed, the projections are enantiomers. Thus, A is identical to B.

A

B

D

Projections A, B and C are identical, and D is their enantiomer.

25.4   Rotate the structure 180° around the horizontal axis to arrive at a drawing having the hydrogen at the rear. Assign the *R,S* configuration as usual, and draw the Fischer projection

25.5   Draw the skeleton of the Fischer projection and add the –CHO and –CH₂OH groups to the top and bottom, respectively. Look at each carbon from the direction in which the –H and –OH point out of the page, and draw what you see on the Fischer projection.

View C3 from this side:
–OH is on the right.

View C2 from this side:
–OH is on the right.

25.6   The hydroxyl group bonded to the chiral carbon farthest from the carbonyl group points to the right in a D sugar, and points to the left in an L sugar.

(a)
```
      CHO
      |
HO----S----H
      |
HO----S----H
      |
    CH2OH
```
L-Erythrose

(b)
```
      CHO
      |
 H----R----OH
      |
HO----S----H
      |
 H----R----OH
      |
    CH2OH
```
D-Xylose

(c)
```
    CH2OH
      |
     C=O
      |
HO----S----H
      |
 H----R----OH
      |
    CH2OH
```
D-Xylulose

25.7

```
      CHO
      |
 H----R----OH
      |
HO----S----H
      |
HO----S----H
      |
    CH2OH
```
L-(+)-Arabinose

25.8

(a)
```
      CHO
      |
HO---------H
      |
 H---------OH
      |
HO---------H
      |
    CH2OH
```
L-Xylose

(b)
```
      CHO
      |
HO---------H
      |
 H---------OH
      |
 H---------OH
      |
HO---------H
      |
    CH2OH
```
L-Galactose

(c)
```
      CHO
      |
HO---------H
      |
HO---------H
      |
HO---------H
      |
HO---------H
      |
    CH2OH
```
L-Allose

25.9   An aldoheptose has 5 chirality centers. Thus, there are $2^5 = 32$ aldoheptoses – 16 D aldoheptoses and 16 L aldoheptoses.

25.10   See Problem 25.5 for the method of solution.

View C3 from this side;
–OH is on the right.

```
           HO H
            \ |
OHC   2  C   4  CH2OH
     C   3  C
     / \     / \
    H OH   H OH
```

View C2,C4 from this side;
–OH is on the right.

=

```
      CHO
      |
 H◄━━━━OH
      |
 H◄━━━━OH
      |
 H◄━━━━OH
      |
    CH2OH
```

=

```
      CHO
      |
 H----R----OH
      |
 H----R----OH
      |
 H----R----OH
      |
    CH2OH
```
D-Ribose

25.11 The steps for drawing a furanose are similar to the steps for drawing a pyranose. Ring formation occurs between the –OH group at C4 and the carbonyl carbon. A groups on the right in a Fischer projection is on the bottom face of the ring.

D-Ribose   (Furanose form)

25.12 The furanose of fructose results from ring formation between the –OH group at C5 and the ketone at C2. In the α anomer, the anomeric –OH group is trans to the C6 –CH$_2$OH group, and in the β anomer the two groups are cis. In the pyranose form, cyclization occurs between the –OH group at C6 and the ketone. The more stable chair conformations are shown.

α-D-Fructopyranose β-D-Fructopyranose

α-D-Fructofuranose β-D-Fructofuranose

25.13 There are two ways to draw these anomers: (1) Following the steps in Worked Example 25.3, draw the Fischer projection, lay it on its side, form the pyranose ring, and convert it to a chair, remembering that the anomeric –OH group is cis to the C6 group; (2) Draw β-D-glucopyranose, and exchange the hydroxyl groups that differ between glucose and the other two hexoses.

β-D-Galactopyranose   β-D-Mannopyranose

β-D-Galactopyranose and β-D-mannopyranose each have one hydroxyl group in the axial position and are therefore of similar stability.

25.14 In the previous problem we drew β-D-galactopyranose. In this problem, invert the configuration at each chirality center of the D enantiomer and perform a ring-flip to arrive at the structure of the L enantiomer.

β-D-Galactopyranose   β-L-Galactopyranose

All substituents, except for the –OH at C4, are equatorial in the more stable conformation of β-L-galactopyranose.

25.15 From the model, we can see that the monosaccharide is the pyranose form of a D-hexose. It is an α-anomer because the anomeric hydroxyl group is trans to the group at C6. Comparing the model with α-D-glucopyranose, we see that all groups have the same axial/equatorial relationship, except for the hydroxyl group at C3, which is axial in the model and equatorial in α-D-glucopyranose. The monosaccharide is α-D-allopyranose. Use Figure 25.3 as a reference.

α-D-Allopyranose   D-Allose

25.16

β-D-Ribofuranose

(a) CH₃I, Ag₂O

(b) (CH₃CO)₂O, pyridine

25.17

D-Glucose   1. NaBH₄   D-Glucitol   D-Galactose   1. NaBH₄   Galactitol
             2. H₂O   2. H₂O

Reaction of D-galactose with NaBH₄ yields an alditol that has a plane of symmetry and is a meso compound.

25.18

D-Glucose   1. NaBH₄   D-Glucitol   ≡   1. NaBH₄   L-Gulose
             2. H₂O   2. H₂O

Reaction of an aldose with NaBH₄ produces a polyol (alditol). Because an alditol has the same functional group at both ends, two different aldoses can yield the same alditol. Here, L-gulose and D-glucose form the same alditol (rotate the Fischer projection of L-gulitol 180° to see the identity).

25.19

CHO
H——OH
HO——H
H——OH
H——OH
CH₂OH
D-Glucose

dil. HNO₃, heat →

CO₂H
H——OH
HO——H
H——OH
H——OH
CO₂H
Glucaric acid

CHO
H——OH
H——OH
H——OH
H——OH
CH₂OH
D-Allose

dil. HNO₃, heat →

CO₂H
H——OH
H——OH
- - - - -
H——OH
H——OH
CO₂H
Allaric acid

Allaric acid has a plane of symmetry and is an optically inactive meso compound. Glucaric acid has no symmetry plane.

25.20 D-Allose and D-galactose yield meso aldaric acids. All other D-hexoses produce optically active aldaric acids on oxidation because they lack a plane of symmetry.

25.21 The products of Kiliani–Fischer reaction of D-ribose have the same configuration at C3, C4 and C5 as D-ribose. Use Figure 25.3 as a reference.

CHO
H——OH
H——OH
H——OH
CH₂OH
D-Ribose

1. HCN
2. H₂, Pd catalyst
3. H₃O⁺ →

CHO
H——OH
H——OH
H——OH
H——OH
CH₂OH
D-Allose

+

CHO
HO——H
H——OH
H——OH
H——OH
CH₂OH
D-Altrose

25.22 The aldopentose, L-xylose has the same configuration as the configuration at C3, C4 and C5 of L-idose and L-gulose. Use Figure 25.3 as a reference.

CHO
HO——H
H——OH
HO——H
CH₂OH
L-Xylose

1. HCN
2. H₂, Pd catalyst
3. H₃O⁺ →

CHO
H——OH
HO——H
H——OH
HO——H
CH₂OH
L-Idose

+

CHO
HO——H
HO——H
H——OH
HO——H
CH₂OH
L-Gulose

25.23   The aldopentoses have the same configurations at C3 and C4 as D-threose.

D-Xylose   D-Lyxose

1. H$_2$NOH
2. (CH$_3$CO)$_2$O, CH$_3$CO$_2$$^-Na^+$
3. Na$^+$ $^-$OCH$_3$

D-Threose

25.24

N-Acetylmannosamine

N-Acetyl-D-neur-aminic acid

25.25

(a) $\xrightarrow[\text{2. H}_2\text{O}]{\text{1. NaBH}_4}$

Cellobiose   (b) $\xrightarrow[\text{H}_2\text{O}]{\text{Br}_2}$

(c) $\xrightarrow[\text{pyridine}]{\text{CH}_3\text{COCl}}$

$$\text{Ac} = \text{CH}_3\overset{\overset{\displaystyle O}{\displaystyle \|}}{\text{C}}—$$

Visualizing Chemistry

25.26 (a) Convert the model to a Fischer projection, remembering that the aldehyde group is on top, pointing into the page, and that the groups bonded to the carbons below point out of the page. The model represents a D-aldose because the –OH group at the chiral carbonfarthest from the aldehyde points to the right.

D-Threose

(b) Break the hemiacetal bond and uncoil the aldohexose. Notice that all hydroxyl groups point to the right in the Fischer projection. The model represents the β anomer of D-allopyranose.

β-D-Allopyranose

25.27 The hints in the previous problem also apply here. Molecular models are helpful.

(a)

L-Glyceraldehyde

(b)

D-Erythrose

25.28 The structure represents an α anomer because the anomeric –OH group and the –CH$_2$OH group are trans. The compound is α-L-mannopyranose because the –OH group at C2 is the only non-anomeric axial hydroxyl group.

α-L-Mannopyranose

25.29 (a)

| L-Mannose | D-Mannose (enantiomer) | D-Glucose (diastereomer) |

(b) The model represents an L-aldohexose because the hydroxyl group on the chiral carbon farthest from the aldehyde group points to the left.

β-L-Mannofuranose

(c) This is tricky! The furanose ring of an aldohexose is formed by connecting the –OH group at C4 to the aldehyde carbon. The best way to draw the anomer is to lie L-mannose on its side and form the ring. All substituents point down in the furanose, and the anomeric –OH and the –CH(OH)CH$_2$OH group are cis.

Mechanism Problems

25.30

UDP-Glucose   UDP-Galactose

Oxidation of the C4 hydroxyl group by NAD^+ forms a ketone plus NADH, and reduction of the ketone by NADH yields UDP-galactose. The result is an epimerization at carbon 4 of the pyranose ring.

25.31

Fructose 6-phosphate

Mannose 6-phosphate

All of these reactions are acid/base catalyzed enolizations or hemiacetal openings/formations.

25.32

Fructose 6-phosphate

Glucosamine 6-phosphate

Nucleophilic acyl substitution and tautomerization lead to the formation of glucosamine 6-phosphate from fructose 6-phosphate. The mechanism of opening of the fructofuranose ring was shown in the previous problem.

25.33 Glucopyranose is in equilibrium with glucofuranose

Reaction with two equivalents of acetone occurs by the mechanism we learned for acetal formation (Sec 19.10)

A five-membered acetal ring forms much more readily when the hydroxyl groups are cis to one another. In glucofuranose, the C3 hydroxyl group is trans to the C2 hydroxyl group, and acetal formation occurs between acetone and the C1 and C2 hydroxyls of glucofuranose. Since the C1 hydroxyl group is part of the acetone acetal, the furanose is no longer in equilibrium with the free aldehyde, and the diacetone derivative is not a reducing sugar.

25.34 Cleavage of fructose 1,6-bisphosphate occurs by a retro aldol reaction.

Fructose 1,6-bisphosphate

Glyceraldehyde
3-phosphate

Dihydroxyacetone
3-phosphate

25.35 (a) Oxidation by NADP⁺, elimination, and conjugate reduction by NADPH give the observed product. Notice that there is no net consumption of NADP⁺. The mechanism of NADP⁺ oxidations and reductions has been shown many times in this book and also appears in part (c).

GDP-D-Mannose

(b) Two epimerizations, both α to the carbonyl group, cause a change in stereochemistry.

(c)   Reduction at C4 by NADPH forms GDP-L-fucose.

GDP-L-Fucose

Additional Problems

Carbohydrate Structure

25.36

(a)
```
CH2OH
|
C=O
|
CH2OH
```
a ketotriose

(b)
```
    CH2OH
 H——OH
    C=O
 H——OH
    CH2OH
```
a ketopentose

(c)
```
    CHO
 H———OH
HO———H
 H———OH
HO———H
 H———OH
    CH2OH
```
an aldoheptose

25.37

(a)
```
    CH2OH
    C=O
 H——OH
    CH2OH
```
a ketotetrose

(b)
```
    CH2OH
    C=O
 H——OH
HO——H
    CH2OH
```
a ketopentose

(c)
```
    CHO
 H——H
 H——OH
HO——H
 H——OH
    CH2OH
```
a deoxyaldohexose

(d)
```
    CHO
 H——NH2
 H——OH
 H——OH
    CH2OH
```
a five-carbon amino sugar

25.38 D-Ribose and L-xylose are diastereomers and differ in all physical properties (or if they have identical physical properties in any category, it is a coincidence).

25.39–25.40

Ascorbic acid has an L configuration because the hydroxyl group at the lowest chirality center points to the left.

L-Ascorbic acid

25.41

(a)

(b)

(c)

25.42

(a)

(S)-2-Bromobutane

CH₃CH₂CH(Br)CH₃

(b)

(R)-Alanine

CH₃CH(NH₂)CO₂H

(c)

(R)-2-Hydroxypropanoic acid

CH₃CH(OH)CO₂H

(d)

(S)-3-Methylhexane

CH₃CH₂CH₂CH(CH₃)CH₂CH₃

25.43

25.44

β-D-Allopyranose

This structure is a pyranose (6-membered ring) and is a β anomer (the C1 hydroxyl group and the –CH2OH groups are cis). It is a D sugar because the -O- at C5 is on the right in the uncoiled form.

25.45

cis

5 OH

HOCH2   O   OH

                OH

OH

β-L-Gulopyranose

This sugar is a β-pyranose. It is an L sugar because the -O- at C5 points to the left in the uncoiled form. It's also possible to recognize this as an L sugar by the fact that the configuration at C5 is S.

25.46

(a)

HOCH2 OH

HO   O

                OH

OH   H

β-D-Altropyranose

CHO

HO—————H

H—————OH

H—————OH

H—————OH

CH2OH

(b)

α-D-Fructofuranose

(c)

α-L-Mannopyranose

25.47

β-D-Ribulofuranose

25.48

β-D-Talopyranose

Carbohydrate Reactions

25.49

In (f), –Ac represents an acetate.

25.50-25.52 Four D-2-ketohexoses are possible.

	CH$_2$OH			CH$_2$OH			CH$_2$OH			CH$_2$OH	
	C=O			C=O			C=O			C=O	
H—		—OH	HO—		—H	H—		—OH	HO—		—H
H—		—OH	H—		—OH	HO—		—H	HO—		—H
H—		—OH	H—		—OH	H—		—OH	H—		—OH
	CH$_2$OH			CH$_2$OH			CH$_2$OH			CH$_2$OH	
	D-Psicose			D-Fructose			D-Sorbose			D-Tagatose	

1. NaBH$_4$
2. H$_2$O

1. NaBH$_4$
2. H$_2$O

	CH$_2$OH			CH$_2$OH			CH$_2$OH			CH$_2$OH	
H—		—OH	HO—		—H	H—		—OH	HO—		—H
H—		—OH	H—		—OH	H—		—OH	H—		—OH
H—		—OH	H—		—OH	HO—		—H	HO—		—H
H—		—OH	H—		—OH	H—		—OH	H—		—OH
	CH$_2$OH			CH$_2$OH			CH$_2$OH			CH$_2$OH	
	Allitol		+	Altritol			Gulitol		+	Iditol	

25.53 The two lactones are formed between a carboxylic acid and a hydroxyl group 4 carbons away. When the lactones are reduced with sodium amalgam, the resulting hexoses have an aldehyde at one end and a hydroxyl group at the other end.

D-Glucose   L-Gulose

25.54 The hard part of this problem is determining where the glycosidic bond occurs on the second glucopyranose ring. Treatment with iodomethane, followed by hydrolysis, yields a tetra-*O*-methyl glucopyranose and a tri-*O*-methyl glucopyranose. The oxygen in the tri-*O*-methylated ring that is not part of the hemiacetal group and is not methylated is the oxygen that forms the acetal bond. In this problem, the C6 oxygen forms the glycosidic link.

Gentiobiose

6-*O*-(β-D-Glucopyranosyl)-β-glucopyranose

General Problems

25.55

CHO
H——OH
HO——H
HO——H
H——OH
CH₂OH
D-Galactose

α-D-Galactopyranose
$[α]_D = +150.7°$

β-D-Galactopyranose
$[α]_D = +52.8°$

Let x be the percent of D-galactose present as the $α$ anomer and y be the percent of D-galactose present as the $β$ anomer.

$150.7°x + 52.8°y = 80.2°$ $x + y = 1;$ $y = 1 - x$

$150.7°x + 52.8°(1-x) = 80.2°$

$97.9°x = 27.4°$

$x = 0.280$

$y = 0.720$

28.0% of D-galactose is present as the $α$ anomer, and 72.0% is present as the $β$ anomer.

25.56

CHO
HO——H
HO——H 1. NaBH₄
HO——H 2. H₂O
H——OH
CH₂OH
D-Talose

CH₂OH
HO——H
HO——H   rotate
HO——H   180°
H——OH   ≡
CH₂OH

CH₂OH
HO——H
H——OH
H——OH
H——OH
CH₂OH

1. NaBH₄
2. H₂O

CHO
HO——H
H——OH
H——OH
H——OH
CH₂OH
D-Altrose

25.57

CHO
H——OH
H——OH   dil.
H——OH   HNO₃
H——OH
CH₂OH
D-Allose

CO₂H
H——OH   rotate
H——OH   180°
H——OH   ≡
H——OH
CO₂H

CO₂H
HO——H
HO——H
HO——H
HO——H
CO₂H

dil.
HNO₃

CHO
HO——H
HO——H
HO——H
HO——H
CH₂OH
L-Allose

D-Galactose   **L-Galactose**

25.58

D-Lyxose   **D-Arabinose**

25.59 (a)   D-Galactose gives the same aldaric acid as L-galactose.

L-Galactose   **D-Galactose**

(b)   The other aldohexose is a D-sugar.

(c)

β-D-Galactopyranose

25.60 Amygdalin has the same carbohydrate skeleton as gentiobiose (Problem 25.54). Draw the cyanohydrin of benzaldehyde, and form a bond between the hemiacetal oxygen and the carbonyl carbon of benzaldehyde, with elimination of water.

Amygdalin

25.61 Since trehalose is a nonreducing sugar, the two glucose units must be connected through an oxygen atom at the anomeric carbon of each glucose. There are three possible structures for trehalose: The two glucopyranose rings can be connected (α,α), (β, β), or (α,β).

25.62 Since trehalose is not cleaved by β-glycosidases, it must have an α,α glycosidic linkage.

Trehalose
1-*O*-(α-D-Glucopyranosyl)-α-glucopyranose

25.63

Neotrehalose
1-*O*-(β-D-Glucopyranosyl)-β-glucopyranose

Isotrehalose
1-*O*-(α-D-Glucopyranosyl)-β-glucopyranose

25.64

2,3:4,6-Diacetone mannopyranoside

Acetone forms an acetal with the hydroxyl groups at C2 and C3 of D-mannopyranoside because the hydroxyl groups at these positions are cis to one another. The pyranoside ring is still a hemiacetal that is in equilibrium with free aldehyde, which is reducing toward Tollens' reagent.

25.65 Dilute base abstracts a proton α to the carbonyl carbon, forming an enolate. The enolate double bond can be protonated from either side, giving either mannose or glucose as the product.

25.66

CO$_2$H

H——OH

HO——H

H——OH

H——OH

CH$_2$OH

D-Gluconic acid

lactone formation

O=C

H——OH

HO——H O

H——OH

H——

CH$_2$OH

+ H$_2$O

≡

HOCH$_2$ H

HO——

HO——

OH O

enolization ⇅ pyridine

HOCH$_2$

HO——

HO——

OH

OH

enediol

enolization ⇅ pyridine

CO$_2$H

HO——H

HO——H

H——OH

H——OH

CH$_2$OH

D-Mannonic acid

lactone formation

O=C

HO——H

HO——H O

H——OH

H——

CH$_2$OH

+ H$_2$O

≡

HOCH$_2$ OH

HO——

HO——

H O

Isomerization at C2 occurs because the enediol can be reprotonated on either side of the double bond.

25.67 There are eight diastereomeric cyclitols.

25.68

Because **A** is oxidized to an optically inactive aldaric acid, the possible structures are Dribose and D-xylose. Chain extension of D-xylose, however, produces two hexoses that, when oxidized, yield optically active aldaric acids.

25.69 (a)

(b)

phenylhydrazone

(c)

enol   keto imine

(d)

In these last steps, two nucleophilic addition reactions take place to yield imine products. The mechanism has been worked out in greater detail in Section 19.8, but the essential steps are additions of phenylhydrazine, first to the imine, then to the ketone. Proton transfers are followed by eliminations, first of ammonia, then of H_2O.

25.70   (a)

less stable
β-D-Idopyranose

more stable

less stable
α-D-Idopyranose

more stable

(b)   *α*-D-Idopyranose is more stable than *β*-D-idopyranose because only one group is axial in its more stable chair conformation, whereas *β*-D-idopyranose has two axial groups in its more stable conformation.

(c)

1,6-Anhydro-D-idopyranose is formed from the *β* anomer because the axial hydroxyl groups on carbons 1 and 6 are close enough for the five-membered ring to form.

(d)   The hydroxyl groups at carbons 1 and 6 of D-glucopyranose are equatorial in the most stable conformation and are too far apart for a ring to form.

25.71

D-Ribofuranose is the sugar present in acetyl CoA.

Chapter 26 – Biomolecules: Amino Acids, Peptides and Proteins

Chapter Outline

I.   Amino acids (Sections 26.1–26.3).
   A. Structure of amino acids (Section 26.1).
      1. Amino acids exist in solution as zwitterions.
         a. Zwitterions are internal salts and have many of the properties associated with salts.
            i.   They have large dipole moments.
            ii. They are soluble in water.
            iii. They are crystalline and high-melting.
         b. Zwitterions can act either as acids or as bases.
            i.   The $-CO_2^-$ group acts as a base.
            ii. The ammonium group acts an acid.
      2. All natural amino acids are α-amino acids: the amino group and the carboxylic acid group are bonded to the same carbon.
      3. All but one (proline) of the 20 common amino acids are primary amines.
      4. All of the amino acids are represented by both a three-letter code and a one-letter code. See Table 26.1.
      5. All amino acids except glycine are chiral.
         a. Only one enantiomer (L) of each pair is naturally-occurring.
         b. In Fischer projections, the carboxylic acid is at the top, and the amino group points to the left.
         c. α-Amino acids are referred to as L-amino acids.
      6. Side chains can be neutral, acidic, or basic.
         a. Fifteen of the 20 amino acids are neutral.
         b. Two (aspartic acid and glutamic acid) are acidic.
             i.   At pH = 7.3, their side chains exist as carboxylate ions.
         c. Three (lysine, arginine and histidine) are basic.
             i.   At pH = 7.3, the side chains of lysine and arginine exist as ammonium ions.
            ii. Histidine is not quite basic enough to be protonated at pH = 7.3.
            iii. The double-bonded nitrogen in the histidine ring is basic.
         d. Cysteine and tyrosine are weakly acidic but are classified as neutral.
      7. Humans are able to synthesize only 11 of the 20 amino acids.
         a. These are nonessential amino acids.
         b. The 9 essential amino acids must be supplied in the diet.
   B. The Henderson–Hasselbalch equation and isoelectric points (Section 26.2).
      1. The Henderson–Hasselbalch equation.
         a. If we know the values of pH and pK_a, we can calculate the percentages of protonated, neutral and deprotonated forms of an amino acid.
         b. If we do these calculations at several pH values, we can construct a titration curve for each amino acid.

2. The isoelectric point (p*I*) is the pH at which an amino acid exists as a neutral, dipolar zwitterion.

    a. p*I* is related to side chain structure.

       i. The 15 amino acids that are neutral have p*I* near neutrality.

       ii. The two acidic amino acids have p*I* at a lower pH.

       iii. The 3 basic amino acids have p*I* at a higher pH.

    b. For neutral amino acids, p*I* is the average of the two pK_a values.

    c. For acidic amino acids, p*I* is the average of the two lowest pK_a values.

    d. For basic amino acids, p*I* is the average of the two highest pK_a values.

    e. Proteins have an overall p*I*.

3. Electrophoresis allows the separation of amino acids by differences in their p*I*.

    a. A buffered solution of amino acids is placed on a paper or gel.

    b. Electrodes are connected to the solution, and current is applied.

    c. Negatively charged amino acids migrate to the positive electrode, and positively charged amino acids migrate to the negative electrode.

    d. Amino acids can be separated because the extent of migration depends on p*I*.

C. Synthesis of *α*-amino acids (Section 26.3).

1. The Hell–Volhard–Zelinskii method and the phthalimide method.

    a. An *α*-bromo acid is produced from a carboxylic acid by *α*-bromination.

    b. Displacement of –Br by ammonia gives the *α*-amino acid.

2. The amidomalonate synthesis.

    a. An alkyl halide reacts with the anion of diethyl amidomalonate.

    b. Hydrolysis of the adduct yields the *α*-amino acid.

3. Reductive amination.

    a. Reductive amination of an *α*-keto carboxylic acid gives an *α*-amino acid.

    b. This method is related to the biosynthetic pathway for synthesis of amino acids.

4. All of the methods listed above produce a racemic mixture of amino acids.

D. Enantioselective synthesis of amino acids.

1. Resolution of racemic mixtures:

    a. The mixture can react with a chiral reagent, followed by separation of the diastereomers and reconversion to amino acids.

    b. Enzymes selectively catalyze reactions that form one of the enantiomers, but not the other.

2. Enantioselective synthesis.

    a. Enantioselective hydrogenation of *Z*-enamido acids produces chiral *α*-amino acids.

    b. The most effective catalysts are complexes of rhodium (I), cyclooctadiene and a chiral diphosphine.

II. Peptides (Sections 26.4–26.8).

    A. Peptide structure (Section 26.4).

        1. Peptide bonds.

           a. A peptide is an amino acid polymer in which the amine group of one amino acid forms an amide bond with the carboxylic acid group of a second amino acid.

           b. The sequence of –N–CH–CO– is known as the backbone of the peptide or protein.

           c. Rotation about the amide bond is restricted.

        2. The N-terminal amino acid of the polypeptide is always drawn on the left.

        3. The C-terminal amino acid of the polypeptide is always drawn on the right.

        4. Peptide structure is described by using three-letter codes, or one-letter codes, for the individual amino acids, starting with the N-terminal amino acid on the left.

        5. Disulfide bonds.

           a. Two cysteines can form a disulfide bond (–S–S–).

           b. Disulfide bonds can link two polypeptides or introduce a loop within a polypeptide chain.

    B. Structure determination of peptides (Sections 26.5–26.6).

        1. Amino acid analysis (Section 26.5).

           a. Amino acid analysis provides the identity and amount of each amino acid present in a protein or peptide.

           b. First, all disulfide bonds are reduced and all peptide bonds are hydrolyzed.

           c. The mixture is placed on a chromatography column, and the residues are eluted.

               i. As each amino acid elutes, it undergoes reaction with ninhydrin, which produces a purple color that is detected and measured spectrophotometrically.

           d. Alternatively, the mixture can be analyzed by HPLC.

           e. Amino acid analysis is reproducible on properly maintained equipment; residues always elute at the same time, and only small sample sizes are needed.

        2. The Edman degradation (peptide sequencing) (Section 26.6).

           a. The Edman degradation removes one amino acid at a time from the –NH$_2$ end of a peptide.

               i. The peptide is treated with phenylisothiocyanate (PITC), which reacts with the amino-terminal residue.

               ii. The PITC derivative is split from the peptide.

               iii. The residue undergoes acid-catalyzed rearrangement to a PTH, which is identified chromatographically.

               iv. The shortened chain undergoes another round of Edman degradation.

           b. Since the Edman degradation can only be used on peptides containing fewer than 50 amino acids, a protein must be cleaved into smaller fragments.

               i. Partial acid hydrolysis is unselective and therefore is of limited usefulness.

               ii. The enzyme trypsin cleaves proteins at the carboxyl side of Arg and Lys residues.

               iii. The enzyme chymotrypsin cleaves proteins at the carboxyl side of Phe, Tyr and Trp residues.

   c. The complete amino acid sequence of a protein results from determining the individual sequences of peptides and overlapping them.
   C. Synthesis of peptides (Sections 26.7–26.8).
      1. Laboratory synthesis of peptides (Section 26.7).
         a. Groups that are not involved in peptide bond formation are protected.
            i.   Carboxyl groups are often protected as methyl or benzyl esters.
            ii. Amino groups are protected as Boc or Fmoc derivatives.
         b. The peptide bond is formed by coupling with DCC (dicyclohexylcarbodiimide).
         c. The protecting groups are removed.
            i.   Boc groups are removed by brief treatment with trifluoroacetic acid.
            ii. Fmoc groups are removed by treatment with base.
            iii. Esters are removed by mild hydrolysis or by hydrogenolysis (benzyl).
      2. Automated peptide synthesis – Merrifield solid-phase method (Section 26.8).
         a. The carboxyl group of a Boc-protected amino acid is attached to a polystyrene resin by formation of an ester bond.
         b. The resin is washed with trifluoroacetic acid, and the Boc group is removed.
         c. A second Boc-protected amino acid is coupled to the first, and the resin is washed.
         d. The cycle (deprotecting, coupling, washing) is repeated as many times as needed.
         e. Finally, treatment with anhydrous HF removes the final Boc group and frees the polypeptide.
         f. Robotic peptide synthesizers have improved yield and preparation time.
III. Proteins (Section 26.9).
   A. Classification of proteins.
      1. Fibrous proteins consist of long, filamentous polypeptide chains.
      2. Globular proteins are compact and roughly spherical.
   B. Protein structure.
      1. Levels of protein structure.
         a. Primary structure refers to the amino acid sequence of a protein.
         b. Secondary structure refers to the organization of segments of the peptide backbone into a regular pattern, such as a helix or sheet.
         c. Tertiary structure describes the overall three-dimensional shape of a protein.
         d. Quaternary structure describes how protein subunits aggregate into a larger structure.
      2. Examples of structural features.
         a. α-Helix.
            i.   An α-helix is a right-handed coil; each turn of the coil contains 3.6 amino acids.
            ii. The structure is stabilized by hydrogen bonds between amide N–H groups and C=O groups four residues away.

  b. β-Pleated sheet.
    i. In a β-pleated sheet, hydrogen bonds occur between residues in adjacent chains.
    ii. In a β-pleated sheet, the peptide chain is extended, rather than coiled.
  c. Tertiary structure.
    i. The nonpolar amino acid side chains congregate in the center of a protein to avoid water.
    ii. The polar side chain residues are on the surface, where they can take part in hydrogen bonding and salt bridge formation.
    iii. Other important features of tertiary structure are disulfide bridges and hydrogen bonds between amino acid side chains.
 3. Denaturation of proteins.
  a. Modest changes in temperature and pH can disrupt a protein's tertiary structure.
    i. This process is known as denaturation.
    ii. Denaturation doesn't affect protein primary structure.
  b. Denaturation affects both physical and catalytic properties of proteins.
  c. Occasionally, spontaneous renaturation can occur.
C. Enzymes (Sections 26.10–26.11).
 1. Description of enzymes and cofactors (Section 26.10).
  a. An enzyme is a substance (usually protein) that catalyzes a biochemical reaction.
  b. An enzyme is specific and usually catalyzes the reaction of only one substrate.
    i. Some enzymes, such as papain, can operate on a range of substrates.
  c. How enzymes function.
    i. Enzymes form an enzyme–substrate complex, within which the conversion to product takes place.
    ii. Enzymes accelerate the rate of reaction by lowering the energy of the transition state.
    iii. The rate constant for the conversion of $E \cdot S$ to $E + P$ is the turnover number.
  d. Enzymes are grouped into 6 classes according to the reactions they catalyze.
    i. Oxidoreductases catalyze oxidations and reductions.
    ii. Transferases catalyze the transfer of a group from one substrate to another.
    iii. Hydrolases catalyze hydrolysis reactions.
    iv. Lyases catalyze the addition or loss of a small molecule to or from a substrate.
    v. Isomerases catalyze isomerizations.
    vi. Ligases catalyze bond formation between two molecules, often coupled with hydrolysis of ATP
  e. The name of an enzyme has two parts, ending with -*ase*.
    i. The first part identifies the substrate.
    ii. The second part identifies the enzyme's class.
  f. Most enzymes are globular proteins, and many consist of a protein portion (apoenzyme) and a cofactor.
    i. Cofactors may be small organic molecules (coenzymes) or inorganic ions.

       ii. Many coenzymes are derived from vitamins.

2. How enzymes work – citrate synthase (Section 26.11).

    a. Citrate synthase catalyzes the aldol-like addition of acetyl CoA to oxaloacetate to produce citrate.

    b. Functional groups in a cleft of the enzyme bind oxaloacetate.

    c. Functional groups in a second cleft bind acetyl CoA.

       i. The two reactants are now in close proximity.

    d. Two enzyme amino acid residues generate the enol of acetyl CoA.

    e. The enol undergoes nucleophilic addition to the ketone carbonyl group of oxaloacetate.

    f. Two enzyme amino acid residues deprotonate the enol and protonate the carbonyl oxygen.

    g. Water hydrolyzes the thiol ester, releasing citrate and CoA.

Solutions to Problems

26.1 Amino Acids with aromatic rings: Phe, Tyr, Trp, His.

Amino acids containing sulfur: Cys, Met.

Amino acids that are alcohols: Ser, Thr (Tyr is a phenol.)

Amino acids having hydrocarbon side chains: Ala, Ile, Leu, Val, Phe.

26.2

A Fischer projection of the α-carbon of an L-amino acid is pictured above.

For most L-amino acids:

Group   Priority
–NH$_2$ 1
–CO$_2$H   2
–R   3
–H   4

For cysteine:

Group   Priority
–NH$_2$ 1
–CH$_2$SH   2
–CO$_2$H   3
–H   4

26.3

L-Threonine

Diastereomers of Threonine

26.4 On the low pH (acidic) side of p*I*, a protein has a net positive charge, and on the high pH (basic) side of p*I*, a protein has a net negative charge. Thus, hemoglobin (p*I* = 6.8) has a net positive charge at pH =5.3 and a net negative charge at pH = 7.3.

26.5 This method of amino acid synthesis is simple and uses methods we have already studied. The phthalimide synthesis can also be used to introduce the amino group. Remember that only racemic amino acids are produced by this method.

(a)

$$C_6H_5CH_2CH_2CO_2H \xrightarrow[\text{2. H}_2\text{O}]{\text{1. Br}_2, \text{ PBr}_3} C_6H_5CH_2\overset{\overset{\displaystyle Br}{|}}{C}HCO_2H \xrightarrow[\text{excess}]{\text{NH}_3} C_6H_5CH_2\overset{\overset{\displaystyle NH_2}{|}}{C}HCO_2^-$$

3-Phenylpropanoic acid

Phenylalanine

(b)

$$(CH_3)_2CHCH_2CO_2H \xrightarrow[\text{2. H}_2\text{O}]{\text{1. Br}_2, \text{ PBr}_3} (CH_3)_2CH\overset{\overset{\displaystyle Br}{|}}{C}HCO_2H \xrightarrow[\text{excess}]{\text{NH}_3} (CH_3)_2CH\overset{\overset{\displaystyle NH_2}{|}}{C}HCO_2^-$$

3-Methylbutanoic acid

Valine

26.6

In the amidomalonate synthesis, shown above, an alkyl halide RX is converted to RCH(NH₃⁺)CO₂H. Choose an alkyl halide that completes the structure of the target amino acid.

(a)   *Amino Acid*   *Halide*

$$(CH_3)_2CHCH_2\overset{\overset{\displaystyle NH_3^+}{|}}{C}HCO_2^-$$

(CH₃)₂CHCH₂Br

Leucine

(b)   *Amino Acid*   *Halide*

Histidine

(c)
$$\overset{\overset{\displaystyle NH_3^+}{|}}{CH_2CHCO_2^-}$$

CH₂Br

Tryptophan

(d)

$$CH_3SCH_2CH_2 \overset{\overset{\displaystyle NH_3^+}{\displaystyle |}}{C}HCO_2^-$$

Methionine

$$CH_3SCH_2CH_2Br$$

26.7 The precursor to an amino acid prepared by enantioselective hydrogenation has a *Z* double bond conjugated with a carboxylic acid carbonyl group.

26.8 Val–Tyr–Gly (VYG)   Tyr–Gly–Val (YGV)   Gly–Val–Tyr (GVY)

Val–Gly–Tyr (VGY)   Tyr–Val–Gly (YVG)   Gly–Tyr–Val (GYV)

26.9 The N-terminal group is on the right, and the C-terminal group is on the left.

26.10

The cysteine sulfur is a good nucleophile, and iodide is a good leaving group.

26.11 One product of the reaction of an amino acid with ninhydrin is the extensively conjugated purple ninhydrin product. The other major product is the aldehyde derived from the side chain of the amino acid. When valine reacts, the resulting aldehyde is 2-methylpropanal. The other products are carbon dioxide and water. The identity of the aldehyde is determined by the amino acid side chain.

26.12 Trypsin cleaves peptide bonds at the carboxyl (right) side of lysine and arginine. Chymotrypsin cleaves peptide bonds at the carboxyl side of phenylalanine, tyrosine and tryptophan.

26.13 The part of the PTH derivative that lies to the right of the indicated dotted lines comes from the N-terminal residue. Complete the structure to identify the amino acid, which in this problem is methionine.

26.14 The N-terminal residue of angiotensin II is aspartic acid. Replace the –R group of the PTH derivative in Figure 26 .4 with –CH₂CO₂H to arrive at the correct structure.

26.15   Line up the fragments so that the amino acids overlap.

(a)   Arg–Pro
            Pro–Leu–Gly
                       Gly–Ile–Val V–L

The complete sequence:

Arg–Pro–Leu–Gly–Ile–Val

(b)   V–M–W
                W–N-V

The complete sequence:

V–M–W–N–V–L

26.16

26.17

$$R = CH_2CH(CH_3)_2$$

1. Protect the amino group of leucine.

2. Protect the carboxylic acid group of alanine.

3. Couple the protected amino acids with DCC.

4. Remove the leucine protecting group.

$$(CH_3)_3COCNHCHC—NHCHCOCH_3 \xrightarrow{CF_3CO_2H} H_3\overset{+}{N}CHC—NHCHCOCH_3$$

$$+ \quad (CH_3)_2C{=}CH_2 \quad + \quad CO_2$$

(with R and CH₃ substituents shown)

5. Remove the alanine protecting group.

$$H_3\overset{+}{N}CHC—NHCHCOCH_3 \xrightarrow[\text{2. H}_3O^+]{\text{1. NaOH, H}_2O} H_3\overset{+}{N}CHC—NHCHCOH + CH_3OH$$

(R, CH₃, (CH₃)₂CHCH₂ substituents shown)

Leu——Ala

26.18 (a) Pyruvate decarboxylase is a lyase.

(b) Chymotrypsin is a hydrolase.

(c) Alcohol dehydrogenase is an oxidoreductase.

Visualizing Chemistry

26.19

(a)
Isoleucine

(b)
Histidine

(c)
Glutamine

26.20

Cys——Lys——Ala——Asp
C——K——A——D

26.21

26.22 It's possible to identify this representation of valine as the D enantiomer by noting the configuration at the chirality center. The configuration is R, and thus the structure is D-valine.

^-O_2C $\overset{R}{\underset{C}{\diagdown}}$ $CH(CH_3)_2$

$H_3\overset{+}{N}$ H

D-Valine
(R)-Valine

26.23 After identifying the amino acid residues, notice that the tetrapeptide has been drawn with the amino terminal residue on the right. To name the sequence correctly, the amino terminal residue must be cited first. Thus, the tetrapeptide should be named Ser–Leu–Phe–Ala.

Ala	Phe	Leu	Ser
A	F	L	S

Mechanism Problems

26.24 (a)

Step 1: Dehydration.

Step 2: Nucleophilic addition of the amino group of the amino acid.

Step 3: Proton transfer.

Step 4:   Loss of water.

(b)

Decarboxylation produces a different imine.

(c)

Step 1:   Addition of water.

Step 2:   Proton transfer.

Step 3:   Bond cleavage to yield an aldehyde and an amine.

(d)

Step 1: Addition of the amine to a carbonyl carbon of a second ninhydrin molecule.

Step 2: Proton shift.

Step 3: Loss of water to form the purple anion.

Notice that the amino nitrogen is all that remains of the original amino acid.

26.25

Formation of cation:

Electrophilic aromatic substitution:

Protonation of the ether oxygen, followed by displacement of methanol by Cl⁻.

26.26

Step 1:   NaOH brings about elimination of the carboxylated peptide.

Step 2:   Loss of CO_2.

The Fmoc group is acidic because the Fmoc anion is similar to the cyclopentadienyl anion, which is resonance-stabilized and is aromatic.

26.27 (a)

The first step is a substitution similar to the nucleophilic acyl substitution reactions that we studied in Chapter 21.

(b)

Internal S$_N$2 displacement of sulfide results in formation of a 5-membered ring containing an iminium group.

(c)

In this sequence of steps, water adds to the imine double bond, and the peptide bond is cleaved.

(d)

Water opens the lactone ring to give the product shown.

26.28

Additional Problems

Amino Acid Structures and Chirality

26.29 Both (*R*)-serine and (*R*)-alanine are D-amino acids. In a D-amino acid, the –NH$_2$ group is on the right.

(*R*)-Serine   (*R*)-Alanine

26.30

L-Bromoalanine
(*R*)-Bromoalanine

This L "amino acid" also has an *R* configuration because the –CH$_2$Br "side chain" is higher in priority than the –CO$_2$H group.

26.31

(*S*)-Proline

26.32

(a)

Tryptophan (Trp)

Isoleucine (Ile)

Cysteine (Cys)

(d)

Histidine (His)

26.33

H_2A^+

protonated

HA

neutral

A^-

deprotonated

At pH = 2.50:

$$\log \frac{[HA]}{[H_2A^+]} = pH - pk_{a1} = 2.50 - 1.99 = 0.51; \frac{[HA]}{[H_2A^+]} = 3.24$$

At pH = 2.50, approximately three times as many proline molecules exist in the neutral form as exist in the protonated form.

At pH = 9.70:

$$\log \frac{[A^-]}{[HA]} = pH - pk_{a2} = 9.70 - 10.60 - 0.90; \frac{[A^-]}{[HA]} = 0.126$$

At pH = 9.70, the ratio of deprotonated proline to neutral proline is approximately 1:8.

26.34

(a)	Val–Leu–Ser	V-L-S		Ser–Val–Leu	S-V-L
	Val–Ser–Leu	V-S-L		Leu–Val–Ser	L-V-S
	Ser–Leu–Val	S-L-V		Leu–Ser–Val	L-S-V

(b)

Ser–Leu–Leu–Pro	S-L-L-P	Leu–Leu–Ser–Pro	L-L-S-P
Ser–Leu–Pro–Leu	S-L-P-L	Leu–Leu–Pro–Ser	L-L-P-S
Ser–Pro–Leu–Leu	S-P-L-L	Leu–Ser–Leu–Pro	L-S-L-P
Pro–Leu–Leu–Ser	P-L-L-S	Leu–Ser–Pro–Leu	L-S-P-L
Pro–Leu–Ser–Leu	P-L-S-L	Leu–Pro–Leu–Ser	L-P-L-S
Pro–Ser–Leu–Leu	P-S-L-L	Leu–Pro–Ser–Leu	L-P-S-L

26.35 Aldehydes and ketones can undergo nucleophilic addition reactions. In particular, aldehydes and ketones can react with amines to form imines and enamines, reactions that might compete with formation of amide bonds between amino acids. Because of this reactivity, aldehydes and ketones are unlikely to be found in amino acid side chains.

Amino Acid Synthesis and Reactions

26.36 The diethylamidomalonate anion is formed by treating diethylamidomalonate with sodium ethoxide. Choose the appropriate halide based on the amino acid side chain.

(a)

$$CH_3CO_2H + CO_2 + 2\ EtOH + (CH_3)_2CHCH_2CHCO_2H \quad\quad Leucine$$
$$\overset{+}{N}H_3$$

(b)

$$CH_3CO_2H + CO_2 + 2\ EtOH +$$

Tryptophan

26.37

(a)

$$CH_3SCH_2CH_2\overset{O}{\overset{||}{C}}CO_2H \xrightarrow[NaBH_4]{NH_3} CH_3SCH_2CH_2\overset{NH_2}{\overset{|}{C}HCO_2H}$$

Methionine

(b)

$$CH_3CH_2\overset{H_3C}{\overset{|}{C}H}\overset{O}{\overset{||}{C}}CO_2H \xrightarrow[NaBH_4]{NH_3} CH_3CH_2\overset{H_3C}{\overset{|}{C}H}\overset{NH_2}{\overset{|}{C}HCO_2H}$$

Isoleucine

26.38

(a)

1. H$_2$, [Rh(DiPAMP)(COD)]$^+$ BF$_4^-$
2. NaOH, H$_2$O

Valine

(b)

1. H$_2$, [Rh(DiPAMP)(COD)]$^+$ BF$_4^-$
2. NaOH, H$_2$O

Proline

26.39

$\xrightarrow{H_3O^+}$

$$HOCH_2\overset{+}{\underset{NH_3}{C}}HCO_2H$$

Serine

+ CH$_3$CO$_2$H

+ CO$_2$ + 2 EtOH

26.40

(a)

$$(CH_3)_2CH\overset{}{\underset{\overset{|}{\overset{+}{N}H_3}}{C}}HCO_2^- \xrightarrow[H^+\text{ catalyst}]{CH_3CH_2OH} (CH_3)_2CH\overset{}{\underset{\overset{|}{\overset{+}{N}H_3}}{C}}HCO_2CH_2CH_3$$

Valine

(b)

$$(CH_3)_2CH\overset{}{\underset{\overset{|}{\overset{+}{N}H_3}}{C}}HCO_2^- \xrightarrow[Et_3N]{\overset{O\ \ O}{\overset{||\ \ ||}{(CH_3)_3COCOCOC(CH_3)_3}}} (CH_3)_2CH\overset{}{\underset{\overset{|}{\underset{\overset{||}{O}}{NHCOC(CH_3)_3}}}{C}}HCO_2^- + HOC(CH_3)_3 + CO_2$$

(c)

$(CH_3)_2CHCHCO_2^-$
|
$^+NH_3$
$\xrightarrow{\text{KOH, H}_2\text{O}}$
$(CH_3)_2CHCHCO_2^-$ K^+ $+$ H_2O
|
NH_2

(d)

$(CH_3)_2CHCHCO_2^-$
|
$^+NH_3$
$\xrightarrow[\text{2. H}_2\text{O}]{\text{1. CH}_3\text{COCl, pyridine}}$
$(CH_3)_2CHCHCO_2^-$
|
$NHCCH_3$
$\underset{\text{O}}{\|}$

26.41

It is also possible to draw many other resonance forms that involve the π electrons of the aromatic 6-membered rings.

Peptides and Enzymes

26.42 (a)

Cys——His———Glu———Met
C———H———E———M

(b)

P————E—P————T————I———D———E
Pro———Glu—Pro———Thr———Ile———Asp———Glu

26.43 The tripeptide is cyclic.

26.44

Step 1: Valine is protected as its Boc derivative.

$$(CH_3)_3COCOCOC(CH_3)_3 \ + \ Val \ \xrightarrow{Et_3N} \ (CH_3)_3COC\text{—}Val\text{—}OH$$
$$(Boc\text{—}Val\text{—}OH)$$

Step 2: Boc–Val bonds to the polymer in an S_N2 reaction.

$$Boc\text{—}Val\text{—}OH \ + \ ClCH_2\text{—}\boxed{Polymer} \ \xrightarrow{Base} \ Boc\text{—}Val\text{—}OCH_2\text{—}\boxed{Polymer}$$

Step 3: The polymer is first washed, then is treated with CF₃CO₂H to cleave the Boc group.

Boc — Val — OCH₂ — (Polymer) →[1. wash / 2. CF₃CO₂H] Val — OCH₂ — (Polymer)

Step 4: A Boc-protected Ala is coupled to the polymer-bound valine by reaction with. DCC. The polymer is washed.

Boc — Ala + Val — OCH₂ — (Polymer) →[1. DCC / 2. wash] Boc — Ala — Val — OCH₂ — (Polymer)

Step 5: The polymer is treated with CF₃CO₂H to remove Boc.

Boc — Ala — Val — OCH₂ — (Polymer) →[CF₃CO₂H] Ala — Val — OCH₂ — (Polymer)

Step 6: A Boc-protected Phe is coupled to the polymer by reaction with DCC. The polymer is washed.

Boc — Phe + Ala — Val — OCH₂ — (Polymer) →[1. DCC / 2. wash]

Boc — Phe — Ala — Val — OCH₂ — (Polymer)

Step 7: Treatment with anhydrous HF removes the Boc group and cleaves the ester bond between the peptide and the polymer.

Boc — Phe — Ala — Val — OCH₂ — (Polymer) →[HF]

Phe — Ala — Val + (CH₃)₂C=CH₂ + CO₂ + HOCH₂ — (Polymer)

26.45

Peptide →[PITC] *Phenylthiohydantoin* + *Shortened Peptide*

(a)

Ile–Leu–Pro–Phe
I — L — P — F

Leu–Pro–Phe
L — P — F

(b)

Asp–Thr–Ser–Gly–Ala
D — T — S — G — A

Thr–Ser–Gly–Ala
T — S — G — A

26.46

$$\text{Phe} \overset{?}{\underset{?}{\sim}} \text{Leu} - \text{Met} - \text{Lys} \overset{|}{\underset{|}{+}} \text{Tyr} \overset{?}{\underset{?}{\sim}} \text{Asp} - \text{Gly} - \text{Gly} - \text{Arg} \overset{|}{\underset{|}{+}} \text{Val} - \text{Ile} - \text{Pro} - \text{Tyr}$$

Cleaved by trypsin = - - - -
Cleaved by chymotrypsin = ⌁⌁⌁

26.47 (a) Hydrolases catalyze the cleavage of bonds by addition of water (hydrolysis).

(b) Lyases catalyze the elimination of a small molecule (H_2O, CO_2) from a molecule.

(c) Transferases catalyze the transfer of a functional group between substrates.

26.48 Amino acids with polar side chains are likely to be found on the outside of a globular protein, where they can form hydrogen bonds with water and with each other. Amino acids with nonpolar side chains are found on the inside of a globular protein, where they can avoid water. Thus, aspartic acid (b) and lysine (d) are found on the outside of a globular protein, and valine (a) and phenylalanine (c) are likely to be found on the inside. Refer to Table 26.1.

26.49

Leuprolide

Glu	His	Trp	Ser	Tyr	Leu	Leu	Arg	Pro—NHEt
E	H	W	S	Y	L	L	R	P —NHEt

(a) The N-terminal glutamic acid is a cyclic lactam. The C-terminal proline is an *N*-ethyl amide.

(b) One of the leucines (indicated above) has D stereochemistry.

(c) See above.

(d) The charge on a peptide is due to the side chains. According to Table 26.1, the only side chain that is charged at neutral pH is arginine. Thus, leuprolide has a charge of +1 at neutral pH.

General Problems

26.50   A proline residue in a polypeptide chain interrupts α-helix formation because the amide nitrogen of proline has no hydrogen that can contribute to the hydrogen-bonded structure of an α-helix.

26.51

The protonated guanidino group can be stabilized by resonance.

26.52   100 g of cytochrome *c* contains 0.43 g iron, or 0.0077 mol Fe:

$$0.43 \text{ g Fe} \times \frac{1 \text{ mol Fe}}{55.8 \text{ g Fe}} = 0.0077 \text{ mol Fe}$$

Assuming that each mole of protein contains 1 mol Fe, then mol Fe = mol protein.

$$\frac{100 \text{ g Cytochrome } c}{0.0077 \text{ mol Fe}} = \frac{13,000 \text{ g Cytochrome } c}{1 \text{ mol Fe}}$$

Cytochrome *c* has a minimum molecular weight of 13,000 g/ mol.

26.53 ^{1}H NMR shows that the two methyl groups of *N,N*-dimethylformamide are nonequivalent at room temperature. If rotation around the CO–N bond were unrestricted, the methyl groups would be interconvertible, and their ^{1}H NMR absorptions would coalesce into a single signal.

The presence of two methyl absorptions shows that there is a barrier to rotation around the CO–N bond. This barrier is due to the partial double-bond character of the CO–N bond, as indicated by the two resonance forms below. Rotation to interconvert the two methyl groups is slow at room temperature, but heating to 180° supplies enough energy to allow rapid rotation and to cause the two NMR absorptions to merge.

26.54 Gly

Gly–Asp–Phe–Pro

Phe–Pro–Val

Val–Pro–Leu

The complete sequence:

Gly–Gly–Asp–Phe–Pro–Val–Pro–Leu

26.55

Cleaved by Trypsin = - - - -

Cleaved by Chymotrypsin = ⌇⌇⌇

26.56 Ser–Ile–Arg–Val–Val–Pro–Tyr–Leu–Arg

26.57 Reduced oxytocin: Cys–Tyr–Ile–Gln–Asn–Cys–Pro–Leu–Gly–NH$_2$

Oxidized oxytocin: Cys–Tyr–Ile–Gln–Asn–Cys–Pro–Leu–Gly–NH$_2$

with S—S bridge connecting the two Cys residues

The C-terminal end of oxytocin is actually an amide, but this can't be determined from the information given.

26.58

(a)

Aspartame (nonzwitterionic form)

(b)

Aspartame at pH = 5.9, 7.3

At pH = 7.3, aspartame exists in the zwitterionic form, as it does at pH = 5.9.

26.59

Step 1:   Protonation

Step 3:   Proton transfer.

Step 5:   Bond rotation, addition of amine.

Step 7:   Loss of water

Step 2:   Addition of water

Step 4:   Ring opening.

Step 6:   Proton transfer.

Step 8:   Deprotonation.

26.60

$$H_3C \quad NH_2$$
$$CH_3CH_2CHCHCO_2H \quad + \quad \alpha\text{-keto acid} \quad \xrightarrow{\text{transamination}} \quad CH_3CH_2CHCCO_2H \quad + \quad \text{amino acid}$$
$$\qquad\qquad\qquad\qquad\qquad\qquad\qquad\qquad\qquad\qquad\qquad H_3C \quad O$$

26.61

4-Methylideneimidazol-5-one (MIO)

Step 1: Nucleophilic addition of the amino group of the amino acid.

Step 2: Proton transfer.

Step 3: Elimination.

Step 4: Elimination.

26.62

Step 1:   Nucleophilic addition of the amine to α-ketoglutarate.

Step 2:   Loss of water.

Step 3:   Reduction by NADPH/H$^+$

Review Unit 10: Biomolecules I – Carbohydrates, Amino Acids, Peptides

Major Topics Covered (with vocabulary):

Monosaccharides:
carbohydrate   monosaccharide   aldose   ketose   Fischer projection   D,L sugars   pyranose furanose   anomer   anomeric center α anomer β anomer   mutarotation   glycoside   Koenigs- Knorr reaction   aldonic acid   alditol   reducing sugar   aldaric acid   Kiliani-Fischer synthesis   Wohl degradation   fucose   glucosamine   galactosamine   neuraminic acid

Other sugars:
disaccharide   1,4' link   cellobiose   maltose   lactose   sucrose   polysaccharide   cellulose amylose   amylopectin   glycogen   glycal assembly method   deoxy sugar   amino sugar   cell-surface carbohydrate   hemagglutinin   neuraminidase

Amino acids:
amino acid   zwitterion   amphoteric α-amino acid   side chain   isoelectric point (pI)   electrophoresis   Henderson-Hasselbalch equation   amidomalonate synthesis   reductive   amination   resolution   enantioselective synthesis

Peptides:
residue   backbone   *N*-terminal amino acid   C-terminal amino acid   disulfide link   amino   acid analysis   Edman degradation   phenylthiohydantoin   trypsin   chymotrypsin   peptide   synthesis protection   Boc derivative   Fmoc derivative   DCC   Merrifield solid-phase   technique

Proteins:
simple protein   conjugated protein   primary structure   secondary structure   tertiary   structure   quaternary structure α -helix β-pleated sheet   salt bridge   prosthetic group   enzyme   cofactor   apoenzyme   holoenzyme   coenzyme   vitamin   isomerase   hydrolase   ligase   lyase   oxidoreductase   transferase   denaturation

Types of Problems:

After studying these chapters, you should be able to:

- Classify carbohydrates as aldoses, ketoses, D or L sugars, monosaccharides, or polysaccharides.
- Draw monosaccharides as Fischer projections or chair conformations.
- Predict the products of reactions of monosaccharides and disaccharides.
- Deduce the structures of monosaccharides and disaccharides.
- Formulate mechanisms for reactions involving carbohydrates.
- Identify the common amino acids and draw them with correct stereochemistry in dipolar form.
- Explain the acid-base behavior of amino acids.
- Synthesize amino acids.
- Draw the structure of simple peptides.
- Deduce the structure of peptides and proteins.
- Outline the synthesis of peptides.

– Explain the classification of proteins and the levels of structure of proteins.
– Draw structures of reaction products of amino acids and peptides.

Points to Remember:

* Aldohexoses, ketohexoses and aldopentoses can all exist in both pyranose forms and furanose forms.

* A reaction that produces the same functional group at both ends of a monosaccharide halves the number of possible stereoisomers of the monosaccharide.

* The reaction conditions that form a glycoside are different from those that form a polyether, even though both reactions, technically, form –OR bonds.

* At physiological pH, the side chains of the amino acids aspartic acid and glutamic acid exist as anions, and the side chains of the amino acids lysine and arginine exist as cations. The imidazole ring of histidine exists as a mixture of protonated and neutral forms.

* Since the amide backbone of a protein is neutral and uncharged, the isoelectric point of a protein or peptide is determined by the relative numbers of acidic and basic amino acid residues present in the peptide.

* In the Merrifield technique of protein synthesis, a protecting group isn't needed for the carboxyl group because it is attached to the polymer support.

Self-Test:

A

Digitalose
(hydrolysis product of
digitoxigenin, a heart
medication)

$$H_3\overset{+}{N}CH_2CH_2CH_2\underset{\underset{NH_3^+}{|}}{C}HCO_2^-$$

C

Ornithine

Digitalose (**A**) is related to which D-aldohexose? Provide a name for **A**, including the configuration at the anomeric carbon. Predict the products of the reaction of **A** with: (a) CH_3OH, H^+ catalyst; (b) CH_3I, Ag_2O.

Vicianose (**B**) is a disaccharide associated with a natural product found in seeds. Treatment of **B** with CH3I and Ag_2O, followed by hydrolysis, gives 2,3,4-tri-O-methyl-D-glucose and 2,3,4-tri-O-methyl-L-arabinose. What is the structure of **B**? Is **B** a reducing sugar?

Ornithine (**C**) is a nonstandard amino acid that occurs in metabolic processes. Which amino acid does it most closely resemble? Estimate pK_a values and pI for ornithine, and draw the major form present at pH = 2, pH = 6, and pH = 11. If ornithine were a component of proteins, how would it affect the tertiary structure of a protein?

Tyr–Gly–Gly–Phe–Leu–Arg–Arg–Ile–Arg–Pro–Lys–Leu–Lys–Trp–Asp–Asn–Gln
Porcine Dynorphin (**D**)

Dynorphin (**D**) is a neuropeptide. Indicate the *N*-terminal end and the C-terminal end. Show the products of cleavage with: (a) trypsin; (b) chymotrypsin. Show the *N*-phenylthiohydantoin that results from treatment of **D** with phenyl isothiocyanate. Do you expect **D** to be an acidic, a neutral or a basic peptide? Kallidin (**E**) is a decapeptide that serves as a vasodilator. The composition of **E** is Arg$_2$ Gly Lys Phe$_2$ Pro$_3$ Ser. The C-terminal residue is Arg. Partial acid hydrolysis yields the following fragments: Pro–Gly–Phe, Lys–Arg–Pro, Pro–Phe–Arg, Pro–Pro–Gly, Phe–Ser–Pro What is the structure of **E**.

Multiple choice:

1.   The enantiomer of α-D-glucopyranose is:
     (a) β-D-Glucopyranose
     (b) α-L-Glucopyranose
     (c) β-L-Glucopyranose
     (d)   none of these

2.   All of the following reagents convert an aldose to an aldonic acid except:
     (a)   dilute HNO_3
     (b)   Fehling's reagent
     (c)   Benedict's reagent
     (d)   aqueous Br_2

3.   Which two aldoses yield D-lyxose after Wohl degradation?
     (a)   D-Glucose and D-Mannose
     (b)   D-Erythrose and D-Threose
     (c)   D-Galactose and D-Altrose
     (d)   D-Galactose and D-Talose

4.   All of the following disaccharides are reducing sugars except:
     (a)   Cellobiose
     (b)   Sucrose
     (c)   Maltose
     (d)   Lactose

5.   Which of the following polysaccharides contains β-glycosidic bonds?
     (a)   Amylose
     (b)   Amylopectin
     (c)   Cellulose
     (d)   Glycogen

6.   To find the pI of an acidic amino acid:|
     (a)   find the average of the two lowest pK_a values
     (b)   find the average of the two highest pK_a values
     (c)   find the average of all pK_a values
     (d)   use the value of the pK_a of the side chain.

7.   Which of the following techniques can synthesize a single enantiomer of an amino acid?
     (a)   Hell–Volhard–Zelinskii reaction
     (b)   reductive amination
     (c)   amidomalonate synthesis
     (d)   hydrogenation of a Z enamido acid

8. The purple product that results from the reaction of ninhydrin with an amino acid contains which group of the amino acid?
    (a) the amino group
    (b) the amino nitrogen
    (c) the carboxylic acid group
    (d) the side chain

9. Which of the following reagents is not used in peptide synthesis?
    (a) Phenylthiohydantoin
    (b) Di-*tert*-butyl dicarbonate
    (c) Benzyl alcohol
    (d) Dicyclohexylcarbodiimide

10. Which structural element is not present in myoglobin?
    (a) a prosthetic group
    (b) regions of α-helix
    (c) hydrophobic regions
    (d) quaternary structure

Chapter 27 – Biomolecules: Lipids

Chapter Outline

I. Esters (Sections 27.1–27.3).
   A. Waxes, fats and oils (Section 27.1).
      1. Waxes are esters of long-chain carboxylic acids with long-chain alcohols.
      2. Fats and oils are triacylglycerols.
         a. Hydrolysis of a fat yields glycerol and three fatty acids.
         b. The fatty acids need not be the same.
      3. Fatty acids.
         a. Fatty acids are even-numbered, unbranched long-chain (C_{12}–C_{20}) carboxylic acids.
         b. The most abundant saturated fatty acids are palmitic (C_{16}) and stearic (C_{18}) acids.
         c. The most abundant unsaturated fatty acids are oleic and linoleic acids (both C_{18}).
            i. Linoleic and arachidonic acids are polyunsaturated fatty acids.
         d. Unsaturated fatty acids are lower-melting than saturated fatty acids because the double bonds keep molecules from packing closely.
         e. The C=C bonds can be catalytically hydrogenated to produce higher-melting fats.
            i. Occasionally, cis–trans bond isomerization takes place.
   B. Soap (Section 27.2).
      1. Soap is a mixture of the sodium and potassium salts of fatty acids produced by hydrolysis (saponification) of animal fat.
      2. Soap acts as a cleanser because the two ends of a soap molecule are different.
         a. The hydrophilic carboxylate end dissolves in water.
         b. The hydrophobic hydrocarbon tails solubilize greasy dirt.
         c. In water, the hydrocarbon tails aggregate into spherical clusters (micelles), in which greasy dirt can accumulate in the interior.
      3. Soaps can form scum when a fatty acid anion encounters Mg^{2+} or Ca^{2+} cations.
         a. This problem is circumvented by detergents, which don't form insoluble metal salts.
   C. Phospholipids (Section 27.3).
      1. Glycerophospholipids.
         a. Glycerophospholipids consist of glycerol, two fatty acids (at C1 and C2 of glycerol), and a phosphate group bonded to an amino alcohol at C3 of glycerol.
      2. Sphingomyelins.
         a. Sphingomyelins have sphingosine or a related dihydroxyamine as their backbone.
         b. They are abundant in brain and nerve tissue.
      3. Phospholipids comprise the major lipids in cell membranes.
         a. The phospholipid molecules are organized into a lipid bilayer, which has polar groups on the inside and outside, and nonpolar tails in the middle.

II. Prostaglandins and other eicosanoids (Section 27.4).

    A. Prostaglandins.

        1. Prostaglandins are C_{20} lipids that contain a C_5 ring and two side chains.

        2. Prostaglandins are present in small amounts in all body tissues and fluids.

        3. Prostaglandins have many effects: they lower blood pressure, affect blood platelet aggregation, affect kidney function and stimulate uterine contractions.

    B. Eicosanoids.

        1. Prostaglandins, thromboxanes, and leukotrienes make up the eicosanoid class of compounds.

        2. Eicosanoids are named by their ring system, substitution pattern and number of double bonds.

        3. Eicosanoids are biosynthesized from arachidonic acid, which is synthesized from linoleic acid.

            a. The transformation from arachidonic acid is catalyzed by the cyclooxygenase (COX) enzyme.

            b. One form of the COX enzyme catalyzes the usual functions, and a second form produces additional prostaglandin as a result of inflammation.

III. Terpenoids (Section 27.5).

    A. Facts about terpenoids.

        1. Terpenoids occur as essential oils in lipid extractions of plants.

        2. Terpenoids are small organic molecules with diverse structures.

        3. All terpenoids are structurally related.

            a. Terpenoids arise from head-to-tail bonding of isopentenyl diphosphate units.

            b. Carbon 1 is the head, and carbon 4 is the tail.

        4. Terpenoids are classified by the number of five-carbon multiples they contain.

            a. Monoterpenoids are synthesized from two five-carbon units.

            b. Sesquiterpenoids are synthesized from three five-carbon units.

            c. Larger terpenoids occur in both animals and plants.

    B. Biosynthesis of terpenoids.

        1. Nature uses the isoprene equivalent isopentenyl diphosphate (IPP) to synthesize terpenoids.

            a. IPP is biosynthesized by two routes that depend on the organism and the structure of the terpenoid.

                i. The mevalonate pathway produces sesquiterpenoids and triterpenoids in most animals and plants.

                ii. The 1-deoxyxylulose 5-phosphate pathway gives monoterpenoids, diterpenoids, and tetraterpenoids.

        2. The mevalonate pathway.

            a. Acetyl CoA undergoes Claisen condensation to form acetoacetyl CoA.

            b. Another acetyl CoA undergoes an aldol-like addition to acetoacetyl CoA to give (3S)-3-hydroxy-3-methylglutaryl CoA (HMG-CoA).

    c. HMG CoA is reduced by NADPH, yielding (*R*)-mevalonate.

    d. Phosphorylation and decarboxylation convert (*R*)-mevalonate to IPP.

  3. Conversion of IPP to terpenoids.

    a. IPP is isomerized to dimethylallyl diphosphate (DMAPP) by a carbocation pathway.

    b. The C=C bond of IPP displaces the PPO⁻ group of dimethallyl diphosphate, to form geranyl diphosphate (GPP), the precursor to all monoterpenoids.

    c. Geranyl diphosphate reacts with IPP to yield farnesyl diphosphate (FPP), the precursor to sesquiterpenoids.

    d. GPP is isomerized and cyclizes on the way to yielding many monoterpenoids.

IV. Steroids (Sections 27.6–27.7).

  A. Steroids are derived from the triterpenoid lanosterol.

    1. Steroids have a tetracyclic fused ring system, whose rings are designated A, B, C, and D.

    2. The three six-membered rings adopt chair geometry and do not undergo ring-flips.

  B. Stereochemistry of steroids (Section 27.6).

    1. Two cyclohexane rings can be joined either cis or trans.

      a. In a trans-fused ring, the groups at the ring junction are trans.

      b. In cis-fused rings, the groups at the ring junction are cis.

      c. Cis ring fusions usually occur between rings A and B.

    2. In both kinds of ring fusions, the angular methyl groups usually protrude above the rings.

    3. Steroids with A–B trans fusions are more common.

    4. Substituents can be either axial or equatorial.

      a. Equatorial substituents are more favorable for steric reasons.

  C. Types of steroid hormones.

    1. Sex hormones.

      a. Androgens (testosterone, androsterone) are male sex hormones.

      b. Estrogens (estrone, estradiol) and progestins are female sex hormones.

    2. Adrenocortical hormones.

      a. Mineralocorticoids (aldosterone) regulate cellular Na^+ and K^+ balance.

      b. Glucocorticoids (hydrocortisone) regulate glucose metabolism and control inflammation.

    3. Synthetic steroids.

      a. Oral contraceptives and anabolic steroids are examples of synthetic steroids.

  D. Biosynthesis of steroids (Section 27.7).

    1. All steroids are biosynthesized from lanosterol.

    2. Lanosterol is formed from squalene, which is the product of dimerization of farnesyl diphosphate (FPP).

    3. Squalene is first epoxidized to form 2,3-oxidosqualene.

4. Nine additional steps are needed to form lanosterol.
   a. The first several steps are cyclization reactions.
   b. The last steps are hydride and methyl shifts involving carbocations.
5. Other enzymes convert lanosterol to cholesterol.

Solutions to Problems

27.1

$$CH_3(CH_2)_{18}\overset{\displaystyle O}{\overset{\displaystyle \|}{C}}\diagdown$$

from C_{20} acid $O(CH_2)_{31}CH_3$

from C_{32} alcohol

Carnauba wax

27.2

$$CH_2O\overset{O}{\overset{\|}{C}}(CH_2)_{14}CH_3$$
$$|\quad\overset{O}{\overset{\|}{}}$$
$$CHO\overset{O}{\overset{\|}{C}}(CH_2)_{14}CH_3$$
$$|\quad\overset{O}{\overset{\|}{}}$$
$$CH_2O\overset{O}{\overset{\|}{C}}(CH_2)_{14}CH_3$$

Glyceryl tripalmitate

$$CH_2O\overset{O}{\overset{\|}{C}}(CH_2)_7CH=CH(CH_2)_7CH_3 \quad (cis)$$
$$|\quad\overset{O}{\overset{\|}{}}$$
$$CHO\overset{O}{\overset{\|}{C}}(CH_2)_7CH=CH(CH_2)_7CH_3 \quad (cis)$$
$$|\quad\overset{O}{\overset{\|}{}}$$
$$CH_2O\overset{O}{\overset{\|}{C}}(CH_2)_7CH=CH(CH_2)_7CH_3 \quad (cis)$$

Glyceryl trioleate

Glyceryl tripalmitate is higher melting because it is saturated.

27.3

$$CH_3(CH_2)_7CH=CH(CH_2)_7\overset{O}{\overset{\|}{C}}O^- \; Mg^{2+} \; {}^-O\overset{O}{\overset{\|}{C}}(CH_2)_7CH=CH(CH_2)_7CH_3$$

Magnesium oleate

The double bonds are cis.

27.4

$$CH_2O\overset{O}{\overset{\|}{C}}(CH_2)_{14}CH_3$$
$$|\quad\overset{O}{\overset{\|}{}}$$
$$CHO\overset{O}{\overset{\|}{C}}(CH_2)_7CH=CH(CH_2)_7CH_3 \xrightarrow[\text{H}_2\text{O}]{\text{NaOH}}$$
$$|\quad\overset{O}{\overset{\|}{}}$$
$$CH_2O\overset{O}{\overset{\|}{C}}(CH_2)_7CH=CH(CH_2)_7CH_3$$

Glyceryl dioleate mono-
palmitate (cis double bonds)

$$CH_2OH$$
$$|$$
$$CHOH$$
$$|$$
$$CH_2OH$$

Glycerol

$+$

$$Na^+ \; {}^-O\overset{O}{\overset{\|}{C}}(CH_2)_{14}CH_3$$
Sodium palmitate

$$2 \; Na^+ \; {}^-O\overset{O}{\overset{\|}{C}}(CH_2)_7CH=CH(CH_2)_7CH_3$$
Sodium oleate   cis

27.5

Prostaglandin E$_2$

27.6 The *pro-S* hydrogen (blue) ends up cis to the methyl group, and the *pro-R* hydrogen (red) ends up trans.

27.7 As described in Worked Example 27.1, draw the diphosphate precursor so that it resembles the product. Often, the precursor is linalyl diphosphate, which results from isomerization of geranyl diphosphate (the mechanism is shown in Figure 27.10). In (a), it's not easy to see the relationship, but once you've arrived at the product, rotate the structure.

(a)

Linalyl
diphosphate

+ ⁻OPP

α-Pinene

(b)

γ-Bisabolene

27.8   Both ring systems are trans-fused, and both hydrogens at the ring junctions are axial. Refer back to Chapter 4 if you have trouble remembering the relationships of substituents on a cyclohexane ring.

(a)

equatorial

(b)

axial

27.9 Draw the three-dimensional structure and note the relationship of the hydroxyl group to groups whose orientation is known.

Lithocholic acid

OH ← equatorial

27.10

Lanosterol

Lanosterol

Cholesterol

Cholesterol

	Lanosterol		*Cholesterol*
1.	Two methyl groups at C4.	1.	Two hydrogens at C4.
2.	One methyl group at C14.	2.	One hydrogen at C14.
3.	C5–C6 single bond.	3.	C5–C6 double bond.
4.	C8–C9 double bond.	4.	C8–C9 single bond.
5.	Double bond in side chain	5.	Saturated side chain.

Visualizing Chemistry

27.11

Cholic acid

Cholic acid is an A–B cis steroid because the groups at the fusion of ring A and ring B have a cis relationship.

27.12

Helminthogermacrene

Draw farnesyl diphosphate in the configuration that resembles the product, then draw its allylic isomer (the mechanism for the formation of the isomer is shown in Problem 27.7). In this reaction, a cyclization, followed by loss of a proton to form the double bond, gives helminthogermacrene.

27.13

Linoleic acid

A polyunsaturated fat such as linoleic acid is more likely to be found in peanut oil.

Mechanism Problems

27.14

Caryophyllene

Draw farnesyl diphosphate in the correct orientation in order to make this problem much easier. Internal displacement of –OPP by the electrons of one double bond is followed by attack of the electrons of the second double bond on the resulting carbocation. Loss of a proton from the carbon next to the resulting carbocation produces the double bond.

27.15

ψ–Ionone

β–Ionone

Acid protonates a double bond, and the electrons of a second double bond attack the carbocation. Deprotonation yields β-ionone.

27.16

Isoborneol

Camphene

Step 1:   Protonation.   **Step 2:**   Loss of water.

Step 3:   Carbocation rearrangement.   **Step 4:**   Loss of proton.

The key step is the carbocation rearrangement, which occurs by the migration of one of the ring bonds.

Additional Problems

Fats, Oils, and Related Lipids

27.17

Eicosa-5,8,11,14,17-pentaenoic acid
(all cis)

27.18

$$CH_2O\overset{\displaystyle O}{\overset{\|}{C}}(CH_2)_{16}CH_3$$

*$CH O\overset{\displaystyle O}{\overset{\|}{C}}(CH_2)_{16}CH_3$

$CH_2O\overset{\displaystyle O}{\overset{\|}{C}}(CH_2)_7CH\text{=}CH(CH_2)_7CH_3$
(cis)

or

$CH_2O\overset{\displaystyle O}{\overset{\|}{C}}(CH_2)_{16}CH_3$

$CHO\overset{\displaystyle O}{\overset{\|}{C}}(CH_2)_7CH\text{=}CH(CH_2)_7CH_3$
(cis)

$CH_2O\overset{\displaystyle O}{\overset{\|}{C}}(CH_2)_{16}CH_3$

optically active

optically inactive

1. $^-$OH, H_2O
2. H_3O^+

CH_2OH
$CHOH$ + $HO\overset{\displaystyle O}{\overset{\|}{C}}(CH_2)_7CH\text{=}CH(CH_2)_7CH_3$ (cis)   +   2 $HO\overset{\displaystyle O}{\overset{\|}{C}}(CH_2)_{16}CH_3$
CH_2OH
　　　　　　　　　Oleic acid　　　　　　　　　　　　　Stearic acid

Four different groups are bonded to the central glycerol carbon atom in the optically active fat.

27.19

$CH_3(CH_2)_{14}\overset{\displaystyle O}{\overset{/\!/}{C}}\underset{OCH_2(CH_2)_{14}CH_3}{\diagdown}$　　　　Cetyl palmitate

27.20

$CH_2O\overset{\displaystyle O}{\overset{\|}{C}}(CH_2)_7CH\text{=}CH(CH_2)_7CH_3$ (cis)

$CHO\overset{\displaystyle O}{\overset{\|}{C}}(CH_2)_7CH\text{=}CH(CH_2)_7CH_3$ (cis)

$CH_2O\overset{\displaystyle O}{\overset{\|}{C}}(CH_2)_7CH\text{=}CH(CH_2)_7CH_3$ (cis)

Glyceryl trioleate

(a)

Glyceryl trioleate $\xrightarrow[\text{CH}_2\text{Cl}_2]{\text{Br}_2}$

$CH_2O\overset{\displaystyle O}{\overset{\|}{C}}(CH_2)_7CH(Br)CH(Br)(CH_2)_7CH_3$

$CHO\overset{\displaystyle O}{\overset{\|}{C}}(CH_2)_7CH(Br)CH(Br)(CH_2)_7CH_3$

$CH_2O\overset{\displaystyle O}{\overset{\|}{C}}(CH_2)_7CH(Br)CH(Br)(CH_2)_7CH_3$

(b)

Glyceryl trioleate $\xrightarrow{\text{H}_2/\text{Pd}}$

$$\begin{array}{c} \text{O} \\ \parallel \\ \text{CH}_2\text{OC(CH}_2)_{16}\text{CH}_3 \\ \quad \text{O} \\ \quad \parallel \\ \text{CHOC(CH}_2)_{16}\text{CH}_3 \\ \quad \text{O} \\ \quad \parallel \\ \text{CH}_2\text{OC(CH}_2)_{16}\text{CH}_3 \end{array}$$

(c)

Glyceryl trioleate $\xrightarrow[\text{H}_2\text{O}]{\text{NaOH}}$

$$\begin{array}{c} \text{CH}_2\text{OH} \\ | \\ \text{CHOH} \\ | \\ \text{CH}_2\text{OH} \end{array} + 3 \ \text{Na}^+ \ {}^-\text{O}_2\text{C(CH}_2)_7\text{CH}=\text{CH(CH}_2)_7\text{CH}_3$$

(d)

Glyceryl trioleate $\xrightarrow[\text{2. Zn, CH}_3\text{CO}_2\text{H}]{\text{1. O}_3}$

$$\begin{array}{c} \text{O} \quad\quad \text{O} \\ \parallel \quad\quad \parallel \\ \text{CH}_2\text{OC(CH}_2)_7\text{CH} \\ \quad \text{O} \quad\quad \text{O} \\ \quad \parallel \quad\quad \parallel \\ \text{CHOC(CH}_2)_7\text{CH} \\ \quad \text{O} \quad\quad \text{O} \\ \quad \parallel \quad\quad \parallel \\ \text{CH}_2\text{OC(CH}_2)_7\text{CH} \end{array} + 3 \ \begin{array}{c} \text{O} \\ \parallel \\ \text{HC(CH}_2)_7\text{CH}_3 \end{array}$$

(e)

Glyceryl trioleate $\xrightarrow[\text{2. H}_3\text{O}^+]{\text{1. LiAlH}_4}$

$$\begin{array}{c} \text{CH}_2\text{OH} \\ | \\ \text{CHOH} \\ | \\ \text{CH}_2\text{OH} \end{array} + 3 \ \text{HOCH}_2(\text{CH}_2)_7\text{CH}=\text{CH(CH}_2)_7\text{CH}_3$$

(f)

Glyceryl trioleate $\xrightarrow[\text{2. H}_3\text{O}^+]{\text{1. CH}_3\text{MgBr}}$

$$\begin{array}{c} \text{CH}_2\text{OH} \\ | \\ \text{CHOH} \\ | \\ \text{CH}_2\text{OH} \end{array} + 3 \ \begin{array}{c} \text{CH}_3 \\ | \\ \text{HOC(CH}_2)_7\text{CH}=\text{CH(CH}_2)_7\text{CH}_3 \\ | \\ \text{CH}_3 \end{array}$$

27.21

$$\text{CH}_3(\text{CH}_2)_7\text{CH}=\text{CH}(\text{CH}_2)_7\text{CO}_2\text{H} \quad \text{(cis)}$$

Oleic Acid

(a)   Oleic acid $\xrightarrow[\text{HCl}]{\text{CH}_3\text{OH}}$ $\text{CH}_3(\text{CH}_2)_7\text{CH}=\text{CH(CH}_2)_7\text{CO}_2\text{CH}_3$
Methyl oleate

(b)   Methyl oleate from (a) $\xrightarrow{\text{H}_2/\text{Pd}}$ $\text{CH}_3(\text{CH}_2)_{16}\text{CO}_2\text{CH}_3$
Methyl stearate

(c)   Oleic acid $\xrightarrow[\text{2. Zn, CH}_3\text{CO}_2\text{H}]{\text{1. O}_3}$ $\underset{\text{Nonanal}}{\text{CH}_3(\text{CH}_2)_7\text{CHO}}$ + $\underset{\text{9-Oxononanoic acid}}{\text{OHC(CH}_2)_7\text{CO}_2\text{H}}$

(d)

9-Oxononanoic acid
from (c)
$\xrightarrow[\text{H}_3\text{O}^+]{\text{CrO}_3}$
$\text{HO}_2\text{C(CH}_2)_7\text{CO}_2\text{H}$
Nonanedioic acid

(e)

Oleic acid $\xrightarrow[\text{CH}_2\text{Cl}_2]{\text{Br}_2}$ $\text{CH}_3(\text{CH}_2)_7\text{CH(Br)CH(Br)(CH}_2)_7\text{CO}_2\text{H}$

$\downarrow$ 1. 3 NaNH$_2$, NH$_3$
2. H$_3$O$^+$

$\text{CH}_3(\text{CH}_2)_7\text{C}\equiv\text{C(CH}_2)_7\text{CO}_2\text{H}$
Stearolic acid

Three equivalents of the base are needed because one of them is neutralized by the carboxylic acid.

(f)

Oleic acid $\xrightarrow{\text{H}_2/\text{Pd}}$ $\text{CH}_3(\text{CH}_2)_{15}\text{CH}_2\text{CO}_2\text{H}$ $\xrightarrow[\text{2. H}_2\text{O}]{\text{1. Br}_2, \text{PBr}_3}$ $\overset{\overset{\text{Br}}{|}}{\text{CH}_3(\text{CH}_2)_{15}\text{CHCO}_2\text{H}}$

Stearic acid

2-Bromostearic acid

(g)

2 $\text{CH}_3(\text{CH}_2)_{16}\text{CO}_2\text{CH}_3$
from (b)
$\xrightarrow[\text{2. H}_3\text{O}^+]{\text{1. Na}^+ \text{ }^-\text{OCH}_3}$
$\text{CH}_3(\text{CH}_2)_{16}\overset{\overset{\text{O}}{||}}{\text{C}}\underset{\underset{\text{CO}_2\text{CH}_3}{|}}{\text{CH}}(\text{CH}_2)_{15}\text{CH}_3$ + HOCH$_3$

$\downarrow$ H$_3$O$^+$, heat

$\text{CH}_3(\text{CH}_2)_{16}\overset{\overset{\text{O}}{||}}{\text{C}}\text{CH}_2(\text{CH}_2)_{15}\text{CH}_3$ + CO$_2$ + HOCH$_3$
18-Pentatriacontanone

This synthesis uses a Claisen condensation, followed by a β-keto ester decarboxylation.

27.22 Fats and plasmalogens are both esters of a glycerol molecule that has carboxylic acid ester groups at C2 and C3. The third group bonded to glycerol, however, differs with the type of lipid: a fat has a carboxylic acid ester at C1, and a plasmalogen has a vinyl ether in that location.

27.23

Basic hydrolysis cleaves the carboxylic acid ester bonds but doesn't affect the ether bond. Acidic hydrolysis cleaves all three groups bonded to glycerol and produces an aldehyde from the vinyl ether group.

27.24

a cardiolipin

Saponification of a cardiolipin yields 4 different carboxylates, 3 equivalents of glycerol and two equivalents of phosphate.

27.25

$$CH_3(CH_2)_7C \equiv C(CH_2)_7CO_2H \xrightarrow[\text{2. Zn, } CH_3CO_2H]{\text{1. } O_3} CH_3(CH_2)_7CO_2H + HO_2C(CH_2)_7CO_2H$$

Stearolic acid   Nonanoic acid   Nonanedioic acid

Stearolic acid contains a triple bond because the products of ozonolysis are carboxylic acids.

27.26

$$CH_3(CH_2)_7C{\equiv}CH \xrightarrow[NH_3]{NaNH_2} \left[CH_3(CH_2)_7C{\equiv}C{:}^-Na^+\right] + NH_3$$

$$I{-}CH_2(CH_2)_5CH_2{-}Cl$$

$$CH_3(CH_2)_7C{\equiv}C(CH_2)_6CH_2CN \xleftarrow{NaCN} CH_3(CH_2)_7C{\equiv}C(CH_2)_6CH_2{-}Cl$$

$$\downarrow H_3O^+$$

$$CH_3(CH_2)_7C{\equiv}C(CH_2)_6CH_2CO_2H$$

Stearolic acid

I$^-$, rather than Cl$^-$, is displaced
by acetylide because iodide is a
better leaving group than chloride.

Terpenoids and Steroids

27.27–27.29

Remember that a compound with n chirality centers can have a maximum of 2^n stereoisomers. Not all the possible stereoisomers of these compounds are found in nature or can be synthesized. Some stereoisomers have highly strained ring fusions; others contain 1,3-diaxial interactions.

(a)

Guaiol

(8 possible stereoisomers)

(b)

Sabinene

(4 possible stereoisomers)

(c)

(c)

Cedrene

(16 possible stereoisomers)

If carbon 1 of each diphosphate were isotopically labeled, the labels would appear at the circled positions of the terpenoids.

27.30

(R)-Mevalonate   + $^-$SCoA

27.31   First, mevalonate 5-diphosphate is converted to isopentenyl diphosphate (IPP) and dimethallyl diphosphate (DMAPP).

IPP is isomerized to DMAPP.

DMAPP and IPP couple to give geranyl diphosphate (GPP).

A second molecule of IPP adds to GPP to give farnesyl diphosphate, the precursor to α-cadinene.

Notice that the ^{14}C labels are located at two different positions: (1) at the carbon to which –OPP was bonded; (2) at the carbon bonded to the methyl group.

Now, arrange farnesyl diphosphate to resemble the skeleton of α-cadinene. The first step in the reaction sequence is formation of the allylic isomer of FPP; the mechanism was shown in Problem 27.7.

α-Cadinol

27.32   Farnesyl diphosphate (from the previous problem) dimerizes to form squalene.

Farnesyl diphosphate

Squalene

27.33   Squalene is converted to lanosterol by the series of steps pictured in Figure 27.14.

Squalene

Lanosterol

General Problems

27.34

Farnesyl diphosphate   Isopentenyl diphosphate

Geranylgeranyl diphosphate

The precursor to flexibilene is formed from the reaction of farnesyl diphosphate and isopentenyl diphosphate.

Flexibilene

The precursor cyclizes by the now-familiar mechanism to produce flexibilene.

27.35

Dihydrocarvone

The two hydrocarbon substituents are equatorial in the most stable chair conformation.

27.36

Menthol

All ring substituents are equatorial in the most stable conformation of menthol.

27.37

(a)

(b)

As always, use the stereochemistry of the groups at the ring junction to label the other substituents as equatorial or axial and then esterify the appropriate –OH group.

27.38

Linalyl
diphosphate

Isoborneol

The initial addition is followed by a carbocation rearrangement to produce a secondary carbocation, which reacts with water to yield the secondary alcohol.

27.39

Digitoxigenin

The hydroxyl group in ring A is axial, and the hydroxyl group at the ring C–D fusion is equatorial to ring C and axial to ring D. Notice that digitoxigenin has both an A–B cis ring fusion and a C–D cis ring fusion.

27.40

Lithium aluminum hydride reduces the lactone ring to a diol. Periodinane oxidizes only one hydroxyl group because the second group is tertiary.

27.41

CH₃(CH₂)₄CH₂   CH₂(CH₂)₈CO₂H

C=C

H   H

Vaccenic acid

$$\xrightarrow[\text{2. Zn, CH}_3\text{CO}_2\text{H}]{\text{1. O}_3}$$

CH₃(CH₂)₅CHO

Heptanal

+

OHC(CH₂)₉CO₂H

11-Oxoundecanoic acid

$$\xrightarrow[\text{Zn/Cu}]{\text{CH}_2\text{I}_2}$$

CH₃(CH₂)₄CH₂   CH₂(CH₂)₈CO₂H   (and enantiomer)

H   H

C—C

H   H

C

H   H

Lactobacillic acid

27.42

CH₃CH₂CH₂CH₂

H   H   H

C=C   C=C   C=C

H   H   H

CH₂(CH₂)₆CO₂H

(9Z,11E,13E)-9,11,13-Octadecatrienoic acid

(Eleostearic acid)

$$\downarrow \begin{array}{l}\text{1. O}_3\\\text{2. Zn, CH}_3\text{CO}_2\text{H}\end{array}$$

CH₃CH₂CH₂CH₂CHO + OHC—CHO + OHC—CHO + OHC(CH₂)₇CO₂H

The stereochemistry of the double bonds can't be determined from the information given.

27.43 This mechanism also appears in Problem 27.32

Farnesyl diphosphate

Isopentenyl diphosphate

GGDP

27.44

Estradiol   Diethylstilbestrol

Estradiol and diethylstilbestrol resemble each other in having similar carbon skeletons, in having a phenolic ring, and in being diols.

27.45

Diethylstilbestrol

The key reaction is a Grignard reaction between two molecules that are both synthesized from phenol. Phenol is first converted to anisole, in order to avoid problems with acidic hydrogens interfering with the Grignard reaction. Next, anisole undergoes Friedel–Crafts acylation with propanoyl chloride.The resulting ketone is one of the Grignard components. The other component is prepared by reduction, bromination and treatment with magnesium of a quantity of the ketone. After the Grignard reaction, HI serves to both dehydrate the alcohol and cleave the methyl ether groups.

27.46

(a)

Estradiol $\xrightarrow[\text{2. CH}_3\text{I}]{\text{1. NaH}}$

(b)

Estradiol $\xrightarrow[\text{pyridine}]{\text{CH}_3\text{COCl}}$

(c)

Estradiol $\xrightarrow[\text{FeBr}_3]{\text{Br}_2}$

(d)

Estradiol $\xrightarrow[\text{CH}_2\text{Cl}_2]{\text{Periodinane}}$

27.47

One equivalent of H₂ hydrogenates the least substituted double bond. Dihydrocembrene has no ultraviolet absorption because it is not conjugated.

27.48

The mechanism follows the usual path: cyclization of linalyl diphosphate, followed by attack of the π electrons of the second double bond, produces an intermediate carbocation. A carbocation rearrangement occurs, and the resulting carbocation reacts with water to form an alcohol that is oxidized to give α-fenchone.

27.49

3-Ketobutyryl–protein

Step 1: Attack of malonyl–protein anion on acetyl–protein (Claisen condensation).
Step 2: Loss of S–protein.
Step 3: Decarboxylation.
Step 4: Protonation and tautomerization.

27.50

In this series of steps, dissociation of diphosphate ion allows bond isomerization to take place, making it possible for ring formation to occur. This mechanism is very similar to the mechanism shown in Figure 27.10.

Two cyclizations produce the trichodiene ring skeleton and a secondary carbocation.

A hydride shift, two methyl shifts, and loss of $-H^+$ yield trichodiene.

Chapter 28 – Biomolecules: Nucleic Acids

Chapter Outline

I. Nucleic acids (Sections 28.1–28.2).
    A. Nucleotides (Section 28.1).
        1. Nucleotides are composed of a heterocyclic purine or pyrimidine base, an aldopentose, and a phosphate group.
            a. In RNA, the purines are adenine and guanine, the pyrimidines are uracil and cytosine, and the sugar is ribose.
            b. In DNA, thymine replaces uracil, and the sugar is 2'-deoxyribose.
        2. Positions on the base receive non-prime superscripts, and positions on the sugar receive prime superscripts.
        3. The heterocyclic base is bonded to C1' of the sugar.
        4. DNA is vastly larger than RNA and is found in the cell nucleus.
    B. Nucleic acids.
        1. Nucleic acids are composed of nucleotides connected by a phosphodiester bond between the 5' ester of one nucleotide and the 3' hydroxyl group of another.
            a. One end of the nucleic acid polymer has a free hydroxyl group and is called the 3' end.
            b. The other end has a free phosphate group and is called the 5' end.
        2. The structure of a nucleic acid depends on the order of bases.
        3. The sequence of bases is described by starting at the 5' end and listing the bases by their one-letter abbreviations in order of occurrence.
    C. Base-pairing in DNA (Section 28.2).
        1. DNA consists of two polynucleotide strands coiled in a double helix.
            a. Adenine and thymine hydrogen-bond with each other, and cytosine and guanine hydrogen-bond with each other.
        2. Because the two DNA strands are complementary, the amount of A equals the amount of T, and the amount of C equals the amount of G.
        3. The double helix is 20 Å wide, there are 10 bases in each turn, and each turn is 34 Å in height.
        4. The double helix has a major groove and a minor groove into which polycyclic aromatic molecules can intercalate.
    D. The "central dogma" of molecular genetics.
        1. The function of DNA is to store genetic information and to pass it on to RNA, which, in turn, uses it to make proteins.
        2. Replication, transcription and translation are the three processes that are responsible for carrying out the central dogma.
II. The transfer of genetic information (Sections 28.3–28.5).
    A. Replication of DNA (Section 28.3).
        1. Replication is the enzyme-catalyzed process whereby DNA makes a copy of itself.
        2. Replication is semiconservative: each new strand of DNA consists of one old strand and one newly synthesized strand.

3. How replication occurs:

    a. The DNA helix partially unwinds.

        i. This process is catalyzed by the enzyme helicase.

    b. New nucleotides form base pairs with their complementary partners.

    c. Formation of new bonds is catalyzed by DNA polymerase and takes place in the $5' \rightarrow 3'$ direction.

        i. Bond formation occurs by attack of the 3' hydroxyl group on the 5' triphosphate, with loss of a diphosphate leaving group.

    d. Both new chains are synthesized in the $5' \rightarrow 3'$ direction.

        i. One chain is synthesized continuously (the leading strand).

        ii. The other strand is synthesized in small pieces, which are later joined by DNA ligase enzymes (the lagging strand).

B. Transcription – synthesis of RNA (Section 28.4).

  1. There are 3 main types of RNA:

    a. Messenger RNA (mRNA) carries genetic information to ribosomes when protein synthesis takes place.

    b. Ribosomal RNA (rRNA), complexed with protein, comprises the physical makeup of the ribosomes.

    c. Transfer RNA (tRNA) brings amino acids to the ribosomes, where they are joined to make proteins.

    d. There are also small RNAs, which carry out a variety of cellular functions.

  2. DNA contains "promoter sites", which indicate where mRNA synthesis is to begin, and base sequences that indicate where mRNA synthesis stops.

    a. RNA polymerase binds to the promoter sequence.

  3. mRNA is synthesized in the nucleus by transcription of DNA.

    a. The DNA partially unwinds, forming a "bubble".

    b. Ribonucleotides form base pairs with their complementary DNA bases.

    c. Bond formation occurs in the $5' \rightarrow 3'$ direction.

    d. Only one strand of DNA (the antisense, or noncoding, strand) is transcribed.

    e. Thus, the synthesized mRNA is a copy of the sense (coding) strand with U replacing T.

  4. Synthesis of mRNA is not necessarily continuous.

    a. Often, synthesis begins in a region of DNA called an exon and is interrupted by a seemingly noncoding region of DNA called an intron.

    b. In the final mRNA, the noncoding sections have been removed and the remaining pieces have been spliced together by specific enzymes.

C. Translation (Section 28.5).

  1. Translation is the process in which proteins are synthesized at the ribosomes by using mRNA as a template.

2. The message delivered by mRNA is contained in "codons" – 3-base groupings that are specific for an amino acid.
    a. Amino acids are coded by 61 of the possible 64 codons.
    b. The other 3 codons are "stop" codons.
3. Each tRNA is responsible for bringing an amino acid to the growing protein chain.
    a. A tRNA has a cloverleaf-shaped secondary structure and consists of 70–100 ribonucleotides.
    b. Each tRNA contains an anticodon complementary to the mRNA codon.
4. The protein chain is synthesized by enzyme-catalyzed peptide bond formation.
5. A 3-base "stop" codon on mRNA signals when synthesis is complete.

III. DNA technology (Sections 28.6–28.8).
  A. DNA sequencing (Section 28.6).
    1. Before sequencing, the DNA chain is cleaved at specific sites by restriction endonucleases.
        a. The restriction endonuclease recognizes both a sequence on the sense strand and its complement on the antisense strand.
        b. The DNA strand is cleaved by several different restriction endonucleases, to produce fragments that overlap those from a different cleavage.
    2. Maxam–Gilbert DNA sequencing.
        a. This method uses chemical techniques.
    3. Sanger dideoxy DNA sequencing.
        a. The following mixture is assembled:
            i. The restriction fragment to be sequenced.
            ii. A primer (a small piece of DNA whose sequence is complementary to that on the 3' end of the fragment).
            iii. The 4 DNA nucleoside triphosphates.
            iv. Small amounts of the four dideoxynucleotide triphosphates, each of which is labeled with a different fluorescent dye.
        b. DNA polymerase is added to the mixture, and a strand begins to grow from the end of the primer.
        c. Whenever a dideoxynucleotide is incorporated, chain growth stops.
        d. When reaction is complete, the fragments are separated by gel electrophoresis.
        e. Because fragments of all possible lengths are represented, the sequence can be read by noting the color of fluorescence of each fragment.
  B. DNA synthesis (Section 28.7).
    1. DNA synthesis is based on principles similar to those for peptide synthesis.
    2. The following steps are needed:
        a. The nucleosides are protected and bound to a silica support.
            i. Adenine and cytosine bases are protected by benzoyl groups.
            ii. Guanine is protected by an isobutyryl group.
            iii. Thymine isn't protected.
            iv. The 5' –OH group is protected as a DMT ether.

      b. The DMT group is removed.

      c. The polymer-bound nucleoside is coupled with a protected nucleoside containing a phosphoramidite group.

         i.   One of the phosphoramidite oxygens is protected as a β-cyano ether.

         ii. Tetrazole catalyzes the coupling.

      d. The phosphite is oxidized to a phosphate with I_2.

      e. Steps b – d are repeated until the desired chain is synthesized.

      f. All protecting groups are removed and the bond to the support is cleaved by treatment with aqueous ammonia.

C. The polymerase chain reaction (Section 28.8).

    1. The polymerase chain reaction (PCR) can produce vast quantities of a DNA fragment.

    2. The key to PCR is *Taq* DNA polymerase, a heat-stable enzyme.

      a. Newer heat-stable DNA polymerase enzymes have become available.

    3. Steps in PCR:

      a. The following mixture is heated to 95 °C (a temperature at which DNA becomes single-stranded);

         i.   *Taq* polymerase.

         ii. Mg^{2+} ion.

         iii. The 4 deoxynucleotide triphosphates.

         iv. A large excess of two oligonucleotide primers, each of which is complementary to the ends of the fragment to be synthesized.

      b. The temperature is lowered to 37 °C – 50 °C, causing the primers to hydrogenbond to the single-stranded DNA.

      c. After raising the temperature to 72 °C, *Taq* catalyzes the addition of further nucleotides, yielding two copies of the original DNA.

      d. The process is repeated until the desired quantity of DNA is produced.

Solutions to Problems

28.1

5' end

2'-Deoxyadenosine 5'-phosphate (A)

2'-Deoxyguanosine 5'-phosphate (G)

3' end OH

28.2

5' end

Uridine 5'-phosphate (U)

Adenosine 5'-phosphate (A)

3' end OH   OH

28.3   DNA (5' end) GGCTAATCCGT (3' end) is complementary to

DNA (3' end) CCGATTAGGCA (5' end)

Remember that the complementary strand has the 3' end on the left and the 5' end on the right.

The complementary sequence can also be written as:

DNA (5' end) ACGGATTAGCC (3' end). Be sure that you know which format is being used (3' to 5', or 5' to 3').

28.4

Uracil ———— Adenine

28.5 DNA (5' end) GATTACCGTA (3' end) is complementary to

RNA (3' end) CUAAUGGCAU (5' end)

28.6 RNA (5' end) UUCGCAGAGU (3' end)

DNA (3' end) AAGCGTCTCA (5' end) antisense (noncoding) strand

28.7–28.8 Several different codons can code for the same amino acid (Table 28.1). The corresponding anticodon follows the slash mark after each codon. The mRNA codons are written with the 5' end on the left and the 3' end on the right, and the tRNA anticodons have the 3' end on the left and the 5' end on the right.

Amino acid:	Ala	Phe	Leu	Tyr
Codon sequence/	GCU/CGA	UUU/AAA	UUA/AAU	UAU/AUA
tRNA anticodon:	GCC/CGG	UUC/AAG	UUG/AAC	UAC/AUG
	GCA/CGU		CUU/GAA	
	GCG/CGC		CUC/GAG	
			CUA/GAU	
			CUG/GAC	

28.9–28.10

The mRNA base sequence:	(5' end)	CUU–AUG–GCU–UGG–CCC–UAA (3' end)
The amino acid sequence:		Leu—Met—Ala—Trp—Pro–(stop)
The DNA sequence: (antisense strand)	(3' end)	GAA–TAC–CGA–ACC–GGG–ATT(5' end)

28.11

Cleavage of DMT ethers proceeds by an S_N1 mechanism and is rapid because the DMT cation is unusually stable.

28.12

This is an E2 elimination reaction, which proceeds easily because the hydrogen α to the nitrile group is acidic.

Visualizing Chemistry

28.13

(a)

Guanine (G)
DNA
RNA

(b)

Uracil (U)
RNA

(c)

Cytosine (C)
DNA
RNA

All three bases are found in RNA, but only guanine and cytosine are found in DNA.

28.14

2',3'-Dideoxythymidine 5'-phosphate

The triphosphate made from 2',3'-dideoxythymidine 5' phosphate is labeled with a fluorescent dye and used in the Sanger method of DNA sequencing. Along with the restriction fragment to be sequenced, a DNA primer, and a mixture of the four dNTPs, small quantities of the four labeled dideoxyribonucleotide triphosphates are mixed together. DNA polymerase is added, and a strand of DNA complementary to the restriction fragment is synthesized. Whenever a dideoxyribonucleotide is incorporated into the DNA chain, chain growth stops. The fragments are separated by electrophoresis, and each terminal dideoxynucleotide can be identified by the color of its fluorescence. By identifying these terminal dideoxynucleotides, the sequence of the restriction fragment can be read.

28.15   According to the electrostatic potential map, the nitrogen at the 7 position of 9-methylguanine is more electron-rich (red) and should be more nucleophilic. Thus 9-methylguanine should be the better nucleophile.

9-Methylguanine   9-Methyladenine

Mechanism Problems

28.16   This reaction involves addition of a thiol residue of the enzyme to malonic semialdehyde, yielding a hemithioacetal (Step 1). Oxidation by NAD^+ (step 2), followed by nucleophilic acyl substitution by CoA (Step 3), gives malonyl CoA.

Malonic
semialdehyde

Malonyl CoA

28.17 The steps: (1) phosphorylation by ATP; (2) cyclization; (3) loss of phosphate; (4) tautomerization.

Formylglycinamidine
ribonucleotide

Aminoimidazole
ribonucleotide

28.18 The steps: (1) Addition of water; (2) Proton shift; (3) Elimination of NH₃.

Guanine

Xanthine

28.19   Both steps are nucleophilic acyl substitutions.

(a)

Carbamoyl aspartate

(b)

Dihydroorotate

Additional Problems

28.20   The DNA that codes for natriuretic peptide (32 amino acids) consists of 99 bases; 3 bases code for each of the 32 amino acids in the chain (96 bases), and a 3-base "stop" codon is also needed.

28.21   **Position 9:**

Horse amino acid = Gly   Human amino acid = Ser

mRNA codons (5' —> 3'):

GGU GGC GGA GGG   UCU UCC UCA UCG AGU AGC

DNA bases (antisense strand 3' —> 5'):

<u>CCA</u> <u>CCG</u> CCT CCC   AGA AGG AGT AGC <u>TCA</u> <u>TCG</u>

The underlined horse DNA base triplets differ from their human counterparts (also underlined) by only one base.

Position 30:

Horse amino acid = Ala   Human amino acid = Thr

mRNA codons (5' —> 3'):

GCU GCC GCA GCG   ACU ACC ACA ACG

DNA bases (antisense strand 3' —> 5'):

CGA CGG CGT CGC   TGA TGG TGT TGC

Each of the above groups of DNA bases from horse insulin has a counterpart in human insulin that differs from it by only one base. It is possible that horse insulin DNA differs from human insulin DNA by only two bases out of 159!

28.22 The percent of A always equals the percent of T, since A and T are complementary. The percent G equals the percent C for the same reason. Thus, sea urchin DNA contains about 32% each of A and T, and about 18% each of G and C.

28.23 Even though the stretch of DNA shown contains UAA in sequence, protein synthesis doesn't stop. The codons are read as 3-base individual units from start to end, and, in this mRNA sequence, the unit UAA is read as part of two codons, not as a single codon.

28.24 Restriction endonucleases cleave DNA base sequences that are palindromes, meaning that the sequence reads the same as the complement when both are read in the (5') to (3') direction. Thus, the sequence in (c), CTCGAG is recognized. The sequence in (a), GAATTC, is also a palindrome and is recognized by a restriction endonuclease. The sequence in (b) is not a palindrome and is not recognized.

28.25–28.27

		(a)	(b)	(c)	(d)
mRNA codon :	(5'–>3')	AAU	GAG	UCC	CAU
Amino acid:		Asn	Glu	Ser	His
DNA sequence:	(3'–>5')	TTA	CTC	AGG	GTA
tRNA anticodon:	(3'–>5')	UUA	CUC	AGG	GUA

The DNA sequence of the antisense (noncoding) strand is shown.

28.28–28.29 UAC is a codon for tyrosine. It was transcribed from ATG of the antisense strand of a DNA chain.

mRNA codon
5' end

DNA
3' end

U

A

C

3' end

A

T

G

5' end

28.30

Tyr		Gly		Gly		Phe		Met		(stop)		is coded by
UAC		GGU		GGU		UUU		AUG		UAA		
UAU		GGC		GGC		UUC				UAG		
		GGA		GGA						UGA		
		GGG		GGG								

A total of 2 x 4 x 4 x 2 x 1 x 3 = 194 different mRNA sequences can code for metenkephalin!

28.31   Angiotensin II: Asp——Arg——Val——Tyr——-Ile——-His——Pro——Phe (stop)

mRNA sequence: (5'–>3')	GAU	CGU	GUU	UAU	AUU	CAU	CCU	UUU	UAA
	GAC	CGC	GUC	UAC	AUC	CAC	CCC	UUC	UAG
	CGA		GUA		AUA		CCA		UGA
	CGG		GUG				CCG		
	AGA								
	AGG								

As in the previous problem, many mRNA sequences (13,824) can code for angiotensin II.

28.32   DNA coding strand (5'–>3'):   CTT— CGA—CCA— GAC—AGC—TTT
mRNA (5'–>3'):   CUU—CGA—CCA—GAC—AGC—UUU
Amino acid sequence: Leu——Arg—Pro—Asp—Ser——Phe

The mRNA sequence is the complement of the DNA noncoding (antisense) strand, which is the complement of the DNA coding (sense) strand. Thus, the mRNA sequence is a copy of the DNA coding (sense) strand, with T replaced by U.

28.33   mRNA sequence (5'–>3'):   CUA—GAC—CGU—UCC—AAG—UGA

Amino Acid: Leu——Asp—Arg—Ser—Lys   (stop)

28.34

	Original Sequence	*Miscopied Sequence*
DNA coding strand (5'–>3'):	-CAA-CCG-GAT-	-CGA-CCG-GAT-
mRNA sequence (5'–>3'):	-CAA-CCG-GAU-	-CGA-CCG-GAU-
Amino acid sequence:	-Gln—Pro—Asp-	-Arg—Pro—Asp-

If this gene sequence were miscopied in the indicated way, a glutamine in the original protein would be replaced by an arginine in the mutated protein.

28.35 1.   First, protect the nucleosides.

(a)   Bases are protected by amide formation.

Adenine

C_6H_5COCl
pyridine

Cytosine

C_6H_5COCl
pyridine

Guanine

$(CH_3)_2CHCOCl$
pyridine

Thymine does not need to be protected.

(b)   The 5' hydroxyl group is protected as its *p*-dimethoxytrityl (DMT) ether.

1. Base
2. DMTBr

2. Attach a protected 2-deoxycytidine nucleoside to the polymer support.

3. Cleave the DMT ether.

4. Couple protected 2'-deoxythymidine to the polymer-2'-deoxycytidine. (The nucleosides have a phosphoramidite group at the 3' position.)

5.   Oxidize the phosphite product to a phosphate triester, using iodine.

6.   Repeat steps 3–5 with protected 2'-deoxyadenosine and protected 2'-deoxyguanosine.

7.   Cleave all protecting groups with aqueous ammonia to yield the desired sequence.

28.36 Both of these cleavages occur by the now-familiar nucleophilic acyl substitution route. A nucleophile adds to the carbonyl group, a proton shifts location, and a second group is eliminated. Only the reacting parts of the structures are sown.
Deprotection at 1:
Deprotection at 1:

Deprotection at 2:

28.37

Cyclic AMP

28.38 (a), (b)

The –CH$_2$– group at C2 of deoxyribose is missing from ganciclovir.

(d)   The missing atoms are part of the relatively inflexible deoxyribose ring. Without the ring, the DNA chain is floppy and base pairing to form a double helix can't occur.

(e)   As mentioned in (d), the inability to form base pairs stops the replication of DNA.

Chapter 29 – The Organic Chemistry of Metabolic Pathways

Chapter Outline

I. Overview of metabolism and biochemical energy (Section 29.1).
  A. Metabolism.
    1. The reactions that take place in the cells of organisms are collectively called metabolism.
      a. The reactions that produce smaller molecules from larger molecules are called catabolism and produce energy.
      b. The reactions that build larger molecules from smaller molecules are called anabolism and consume energy.
    2. Catabolism can be divided into four stages:
      a. In digestion, bonds in food are hydrolyzed to yield monosaccharides, fats, and amino acids.
      b. These small molecules are degraded to acetyl CoA.
      c. In the citric acid cycle, acetyl CoA is catabolized to CO_2, and energy is produced.
      d. Energy from the citric acid cycle enters the electron transport chain, where ATP is synthesized.
  B. Biochemical energy.
    1. ATP, a phosphoric acid anhydride, is the storehouse for biochemical energy.
    2. The breaking of a P–O bond of ATP can be coupled with an energetically unfavorable reaction, so that the overall energy change is favorable.
    3. The resulting phosphates are much more reactive than the original compounds.
II. Lipid metabolism (Sections 29.2–29.4).
  A. Catabolism of fats (Section 29.2–29.3).
    1. Triacylglycerols are first hydrolyzed in the stomach and small intestine to yield glycerol plus fatty acids (Section 29.2).
      a. The reaction is catalyzed by a lipase.
        i. Aspartic acid, serine and histidine residues in the enzyme bring about reaction.
      b. Glycerol is phosphorylated and oxidized and enters glycolysis.
        i. The mechanism of oxidation involves a hydride transfer to NAD^+.
        ii. The addition to NAD^+ is stereospecific.
    2. β-Oxidation (in the mitochondria) (Section 29.3).
      a. Fatty acids are degraded by β-oxidation, a 4-step spiral that results in the cleavage of an n-carbon fatty acid into $n/2$ molecules of acetyl CoA.
      b. Before entering β-oxidation, a fatty acid is first converted to its fatty-acyl CoA.
    3. Steps of β oxidation.
      a. Introduction of a double bond conjugated with the carbonyl group.
        i. The reaction is catalyzed by acyl CoA dehydrogenase.
        ii. The enzyme cofactor FAD is also involved and is reduced.
        iii. The mechanism involves abstraction of the *pro-R* α and β hydrogens, resulting in formation of a trans double bond.

    b. Conjugate addition of water to form an alcohol.

        i.   The reaction is catalyzed by enoyl CoA hydratase.

    c. Alcohol oxidation.

        i.   The reaction is catalyzed by L-3-hydroxyacyl CoA dehydrogenase.

        ii. The cofactor NAD+ is reduced to NADH/H$^+$ at the same time.

        iii. Histidine deprotonates the hydroxyl group.

    d. Cleavage of acetyl CoA from the chain.

        i.   The reaction, which is catalyzed by β-keto thiolase, is a retro-Claisen reaction.

        ii. Nucleophilic addition of coenzyme A to the keto group is followed by loss of acetyl CoA enolate, leaving behind a chain-shortened fatty-acyl CoA.

  3. An *n*-carbon fatty acid yields *n*/2 molecules of acetyl CoA after (*n*/2–1) passages of β-oxidation.

    a. Since most fatty acids have an even number of carbons, no carbons are left over after β-oxidation.

    b. Those with an odd number of carbons require further steps for degradation.

B. Biosynthesis of fatty acids (Section 29.4).

  1. General principles.

    a. In most cases, the pathway of synthesis isn't the exact reverse of degradation.

        i.   If $\Delta G°$ is negative for one route, it must be positive for the exact reverse, which is thus energetically unfavorable.

        ii. The metabolic strategy is for one pathway to be related to its reverse but not to be identical.

    b. All common fatty acids have an even number of carbons because they are synthesized from acetyl CoA.

    c. In vertebrates, a large multienzyme synthase complex catalyzes all steps in the pathway.

  2. Synthetic pathway.

    a. Steps 1–2: Acyl transfers convert acetyl CoA to more reactive species.

        i.   Acetyl CoA is converted to acetyl ACP.

        ii. The acetyl group of acetyl ACP is transferred to the synthase enzyme.

    b. Steps 3–4: Carboxylation and acyl transfer.

        i.   Acetyl CoA reacts with bicarbonate to yield malonyl CoA and ADP.

           (a) The coenzyme biotin, a CO_2 carrier, transfers CO_2 in a nucleophilic acyl substitution reaction.

        ii. Malonyl CoA is converted to malonyl ACP.

        iii. At this point, both acetyl groups and malonyl groups are bound to the synthase enzyme.

    c. Step 5: Condensation.

        i.   A Claisen condensation forms acetoacetyl CoA from acetyl synthase and malonyl ACP.

        ii. The reaction proceeds by an initial decarboxylation of malonyl ACP to give an enolate that adds to acetyl synthase to form acetoacetyl CoA.

    d. Steps 6–8: Reduction and dehydrogenation.
        i. The ketone group of acetoacetyl CoA is reduced by NADPH.
        ii. The β-hydroxy thiol ester is dehydrated.
        iii. The resulting double bond is hydrogenated by NADPH to yield butyryl ACP.
    e. The steps are repeated with butyryl synthase and malonyl ACP to give a six carbon unit.
    f. Fatty acids up to palmitic acid (16 carbon atoms) are synthesized by this route.
        i. Elongation of palmitic acid and larger acids occurs with acetyl CoA units as the two-carbon donor, rather than ACP.

III. Carbohydrate metabolism (Sections 29.5–29.8).
    A. Catabolism of carbohydrates (Sections 29.5–29.7).
        1. Glycolysis (Section 29.5).
            a. Glycolysis is a 10-step series of reactions that converts glucose to pyruvate.
            b. Steps 1–2: Phosphorylation and isomerization.
                i. Glucose is phosphorylated at the 6-position by reaction with ATP.
                    (a) The enzyme hexokinase is involved.
                ii. Glucose 6-P is isomerized to fructose 6-P by glucose-6-P isomerase.
            c. Step 3: Fructose 6-P is phosphorylated to yield fructose 1,6-bisphosphate.
                i. ATP and phosphofructokinase are involved.
            d. Step 4: Cleavage.
                i. Fructose 1,6-bisphosphate is cleaved to glyceraldehyde 3-phosphate and dihydroxyacetone phosphate.
                    (a) The reaction is a reverse aldol reaction catalyzed by aldolase.
            e. Step 5: Isomerization.
                i. Dihydroxyacetone phosphate is isomerized to glyceraldehyde 3-phosphate.
                ii. The net result is production of two glyceraldehyde 3-phosphates, both of which pass through the rest of the pathway.
            f. Steps 6–7: Oxidation, phosphorylation, and dephosphorylation.
                i. Glyceraldehyde 3-phosphate is both oxidized and phosphorylated to give 1,3-bisphosphoglycerate.
                    (a) Oxidation by NAD^+ occurs via a hemithioacetal to yield a product that forms the mixed anhydride.
                ii. The mixed anhydride reacts with ADP to form ATP and 3-phosphoglycerate
                    (a) The enzyme phosphoglycerate kinase is involved.
            g. Step 8: Isomerization.
        1. 3-Phosphoglycerate is isomerized to 2-phosphoglycerate by phosphoglycerate mutase.
            h. Steps 9–10: Dehydration and dephosphorylation.
                i. 2-Phosphoglycerate is dehydrated by enolase to give phosphoenolpyruvate.
                ii. Pyruvate kinase catalyzes the transfer of a phosphate group to ADP, with formation of pyruvate.

2. The conversion of pyruvate to acetyl CoA (Section 29.6).
   a. The conversion pyruvate → acetyl CoA is catalyzed by an enzyme complex called pyruvate dehydrogenase complex.
   b. Step 1: Addition of thiamin.
      i. A nucleophilic ylide group on thiamin diphosphate adds to the carbonyl group of pyruvate to yield a tetrahedral intermediate.
   c. Step 2: Decarboxylation.
   d. Step 3: Reaction with lipoamide.
      i. The enamine product of decarboxylation reacts with lipoamide, displacing sulfur and opening the lipoamide ring.
   e. Step 4: Elimination of thiamin diphosphate ylide.
   f. Step 5: Acyl transfer.
      i. Acetyl dihydrolipoamide reacts with coenzyme A to give acetyl CoA.
      ii. The resulting dihydrolipoamide is reoxidized to lipoamide by FAD.
      iii. $FADH_2$ is reoxidized to FAD by NAD^+.
   g. Other fates of pyruvate.
      i. In the absence of oxygen, pyruvate is reduced to lactate.
      ii. In bacteria, pyruvate is fermented to ethanol.
3. The citric acid cycle (conversion of acetyl CoA to CO_2) (Section 29.7).
   a. Characteristics of the citric acid cycle.
      i. The citric acid cycle is a closed loop of eight reactions.
      ii. The intermediates are constantly regenerated.
      iii. The cycle operates as long as NAD^+ and $FADH_2$ are available, which means that oxygen must also be available.
   b. Steps 1–2: Addition to oxaloacetate.
      i. Acetyl CoA adds to oxaloacetate to form citryl CoA, which is hydrolyzed to citrate.
         (a) The reaction is catalyzed by citrate synthase.
      ii. Citrate is isomerized to isocitrate by aconitase.
         (a) The reaction is an E1cb dehydration, followed by conjugate addition of water.
   c. Steps 3–4: Oxidative decarboxylations.
      i. Isocitrate is oxidized by isocitrate dehydrogenase to give a ketone that loses CO_2 to give α-ketoglutarate.
      ii. α-Ketoglutarate is transformed to succinyl CoA in a reaction catalyzed by a multienzyme dehydrogenase complex.
   d. Steps 5–6: Hydrolysis and dehydrogenation of succinyl CoA.
      i. Succinyl CoA is converted to an acyl phosphate, which transfers a phosphate group to GDP in a reaction catalyzed by succinyl CoA synthase.
      ii. Succinate is dehydrogenated by FAD and succinate dehydrogenase to give fumarate; the reaction is stereospecific.
   e. Steps 7–8: Regeneration of oxaloacetate.

   i.   Fumarase catalyzes the addition of water to fumarate to produce (*S*)-malate.

   ii. (*S*)-malate is oxidized by NAD^+ and malate dehydrogenase to complete the cycle.

C. Carbohydrate biosynthesis: gluconeogenesis (Section 29.8).

1. Step 1: Carboxylation.

   a.   Pyruvate is carboxylated to yield oxaloacetate in a reaction that uses biotin and ATP.

2. Step 2: Decarboxylation and phosphorylation.

   a.   Concurrent decarboxylation and phosphorylation of oxaloacetate produce phosphoenolpyruvate.

3. Steps 3–4: Hydration and isomerization.

   a.   Conjugate addition of water gives 2-phosphoglycerate.

   b.   Isomerization produces 3-phosphoglycerate.

4. Steps 5–7: Phosphorylation, reduction and tautomerization.

   a.   Reaction of 3-phosphoglycerate with ATP yields an acyl phosphate.

   b.   The acyl phosphate is reduced by $NADPH/H^+$ to an aldehyde.

   c.   The aldehyde tautomerizes to dihydroxyacetone phosphate.

5. Step 8: Aldol reaction.

   a.   Dihydroxyacetone phosphate and glyceraldehyde 3-phosphate join to form fructose 1,6-bisphosphate.

   b.   This reaction involves the imine of dihydroxyacetone phosphate, which forms an enamine that takes part in the condensation.

6. Steps 9-11: Hydrolysis and isomerization.

   a.   Fructose 1,6-bisphosphate is hydrolyzed to fructose 6-phosphate.

   b.   Fructose 6-phosphate isomerizes to glucose 6-phosphate.

   c.   Glucose 6-phosphate is hydrolyzed to glucose.

7. Several of these steps are the reverse of steps of glycolysis.

IV. Protein metabolism (Section 29.9).

A. Catabolism of proteins: Deamination.

1. The pathway to amino acid catabolism:

   a.   The amino group is removed as ammonia by transamination.

   b.   The ammonia is converted to urea.

   c.   What remains is converted to a compound that enters the citric acid cycle.

      i.   Each carbon skeleton is degraded in a unique pathway.

2. Transamination.

   a.   The –NH₂ group of an amino acid adds to the aldehyde group of pyridoxal phosphate to form an imine (Schiff base).

   b.   The imine tautomerizes to a different imine.

   c.   The second imine is hydrolyzed to give an α-keto acid and an amino derivative of pyridoxal phosphate.

   d.   The pyridoxal derivative transfers its amino group to α-ketoglutarate, to regenerate pyridoxal phosphate and form glutamate.

    3. Deamination.

       a. The glutamate from transamination undergoes oxidative deamination to yield ammonium ion and α-ketoglutarate.

V. Some conclusions about biological chemistry (Section 29.10).

    1. The mechanisms of biochemical reactions are almost identical to the mechanisms of laboratory reactions.

    2. Most metabolic pathways are linear.

       a. Linear pathways make sense when a multifunctional molecule is undergoing transformation.

       b. Cyclic pathways may be more energetically feasible when a molecule is small.

Solutions to Problems

29.1 This reaction is a substitution at phosphorus, with ADP as the leaving group.

29.2

29.3   A fatty acid with n carbons yields $n/2$ acetyl CoA molecules after $(n/2 - 1)$ passages of the β-oxidation pathway

(a)

$$CH_3CH_2 - CH_2CH_2 - CH_2CH_2 - CH_2CH_2 - CH_2CH_2 - CH_2CH_2 - CH_2CH_2 - CH_2CO_2H$$

β oxidation

$$8\ CH_3CSCoA$$

Seven passages of the β-oxidation pathway are needed.

Seven passages of the β-oxidation pathway are needed.

(b)

$$CH_3CH_2 - (CH_2CH_2)_8 - CH_2CO_2H \xrightarrow{\ \beta\ oxidation\ } 10\ CH_3CSCoA$$

Nine passages of the β-oxidation pathway are needed.

Nine passages of the β-oxidation pathway are needed.

29.4 β-Hydroxybutyryl ACP resembles the β-hydroxy ketones that were described in Chapter 23 and that dehydrate readily by an E1cB mechanism.

29.5   A fatty acid synthesized from $^{13}CH_3CO_2H$ has an alternating labeled and unlabeled carbon chain. The carboxylic acid carbon is unlabeled.

$$\overset{*}{C}H_3CH_2\ \overset{*}{C}H_2CH_2\ \overset{*}{C}H_2CH_2\ \overset{*}{C}H_2CH_2\ \overset{*}{C}H_2CH_2\ \overset{*}{C}H_2CH_2\ \overset{*}{C}H_2CH_2\ \overset{*}{C}H_2CO_2H$$

29.6   The face in front of the plane of the page is the *Re* face. Since addition occurs from behind the plane of the page, it occurs at the *Si* face.

29.7   ATP is produced in step 7 (1,3-bisphosphoglycerate —> 3-phosphoglycerate) and in step 10 (phosphoenolpyruvate —> pyruvate). Refer to Figure 29.7.

29.8   **Step 1** is a nucleophilic acyl substitution at phosphorus (*phosphate transfer*) by the –OH group at C6 of glucose, with ADP as the leaving group.

   Step 2 is an *isomerization*, in which the pyranose ring of glucose 6-phosphate opens, tautomerism causes isomerization to fructose 6-phosphate, and a furanose ring is formed.

   Step 3 is a substitution, similar to the one in step 1, involving the –OH group at C1 of fructose 6-phosphate (*phosphate transfer*).

   Step 4 is a *retro-aldol reaction* that cleaves fructose 1,6-bisphosphate to glyceraldehydes 3-phosphate and dihydroxyacetone phosphate.

   Step 5 is an *isomerization* of dihydroxyacetone phosphate to glyceraldehyde 3-phosphat that occurs by keto–enol tautomerization.

   Step 6 begins with a *nucleophilic addition* reaction to the aldehyde group of glyceraldehyde 3-phosphate by a thiol group of an enzyme to form a hemithioacetal, which is *oxidized* by NAD^+ to an acyl thioester. *Nucleophilic acyl substitution* by phosphate yields the product 1,3-bisphosphoglycerate.

   Step 7 is a nucleophilic acyl substitution reaction at phosphorus, in which ADP react with 1,3-diphosphoglycerate, yielding ATP and 3-phosphoglycerate (*phosphate transfer*).

   Step 8 is an *isomerization* of 3-phosphoglycerate to 2-phosphoglycerate.

   Step 9 is an *E1cB elimination* of H_2O to form phosphoenolpyruvate.

   Step 10 is a substitution reaction at phosphorus that forms ATP and enolpyruvate, which tautomerizes to pyruvate (*phosphate transfer*).

29.9

Carbons 1 and 6 of glucose end up as –CH₃ groups of acetyl CoA. and carbons 3 and 4 of glucose end up as CO₂.

29.10 Citrate and isocitrate are tricarboxylic acids. Refer to Figure 29.12.

29.11

29.12 The *pro-R* hydrogen is removed during dehydration, and the reaction occurs with *anti* geometry.

29.13 First, 1,3-bisphosphoglycerate reacts with a cysteine residue of the enzyme in a nucleophilic acyl substitution reaction, with loss of phosphate. Then, reduction by NADH in a second nucleophilic acyl substitution reaction yields glyceraldehyde 3-phosphate.

Glyceraldehyde 3-phosphate

Glyceraldehyde 3-phosphate

29.14

PMP   α-Ketoglutarate   α-Keto acid imine tautomer

Nucleophilic acyl substitution, followed by loss of water, forms the imine tautomer.

α-Keto acid imine tautomer   α-Keto acid imine

A lysine residue deprotonates the carbon next to the ring, leading to tautomerization.

α-Keto acid imine   PLP–glutamate imine

Enzymatic protonation of the keto acid imine yields PLP glutamate imine.

PLP–glutamate imine

PLP imine   Glutamate

Addition of the enzyme, followed by loss of glutamate, regenerates PLP imine.

29.15 Position leucine and α-ketoglutarate so that the groups to be exchanged are aligned. This arrangement makes it easy to predict the products of transamination reactions.

Leucine α-Ketoglutarate   keto acid   Glutamate

29.16 As in the previous problem, redraw the α-keto acid and align it with glutamate. By exchanging the keto group and the amino group, you can identify the amino acid as asparagine.

keto acid   Glutamate α-Ketoglutarate   Asparagine

Visualizing Chemistry

29.17 The amino acid precursors are valine (a) and methionine (b).

(a)

Valine α-Ketoglutarate   Glutamate

(b)

Methionine α-Ketoglutarate   Glutamate

29.18 The intermediate is (*S*)-malate. Refer to Figure 29.12.

(*S*)-Malate

29.19

Decarboxylation of the intermediate yields lysine.

29.20 The intermediate is derived from D-erythrose.

D-Erythrose 4-phosphate

Mechanism Problems

29.21

Ribulose
5-phosphate

Ribose
5-phosphate

The isomerization of ribulose 5-phosphate to ribose 5-phosphate occurs by way of an intermediate enolate.

29.22   This is a reverse aldol reaction, similar to step 4 of glycolysis.

29.23   (a)   The first sequence of steps in this mechanism involves formation of the imine (Schiff base) of sedoheptulose 7-phosphate, followed by retro-aldol cleavage to form erythrose 4-phosphate and the enamine of dihydroxyacetone.

Erythrose
4-phosphate

(b)   The enamine of dihydroxyacetone adds to glyceraldehyde 3-phosphate to yield fructose 6-phosphate. This reaction is almost identical to the reaction pictured for Step 8 of gluconeogenesis in Section 29.8.

Fructose 6-phosphate

29.24

Xylulose
5-phosphate

(a)

(b)

Glyceraldehyde
3-phosphate

Ribose
5-phosphate

Sedoheptulose
7-phosphate

29.25

Maleoylacetoacetate

addition of nucleophile

bond rotation

expulsion of nucleophile

Fumaroylacetoacetate

+ ⁻Nu:

29.26

nucleophilic addition of water

retro-Claisen condensation

29.27 The first step in the conversion acetoacetate ——> acetyl CoA is the formation of acetoacetyl CoA. This reaction also occurs as the first step in fatty acid catabolism. Although we haven't studied the mechanism, it involves formation of a mixed anhydride.

HSCoA, ATP   AMP, PP$_i$

The final step is a retro-Claisen reaction, whose mechanism is pictured in Section 29.3 as Step 4 of β-oxidation of fatty acids.

29.28 Now is a good time to use retrosynthetic analysis, which was first encountered in Chapter 9. In this degradative pathway, what might be the precursor to acetyl CoA (the final product)? Pyruvate is a good guess, because we learned how to convert pyruvate to acetyl CoA in Section 29.6. How do we get from serine to pyruvate? A transamination reaction is a possibility. However, the immediate transamination precursor to pyruvate is the amino acid alanine, which differs from serine by one hydroxyl group. Thus, we probably have to design a pathway from serine to pyruvate that takes this difference into account.

Many routes are possible, but here's the simplest:

The coenzymes thiamin diphosphate and lipoamide are involved in the last step.

29.29

This reaction is a transamination that requires the coenzyme pyridoxal phosphate as a cofactor. The mechanism, which is described in Figure 29.14, involves two steps. The first step is the nucleophilic addition of glutamate nitrogen to the aldehyde group of pyridoxal phosphate to yield an imine intermediate, which is hydrolyzed to give α-ketoglutarate plus a nitrogen-containing pyridoxal phosphate byproduct.

Pyridoxal phosphate

Glutamate

This byproduct reacts with 3-phosphohydroxypyruvate to give 3-phosphoserine plus regenerated pyridoxal phosphate.

3-Phosphohy-
droxypyruvate

3-Phosphoserine

29.30

α-Ketoisovalerate

α-Ketoisocaproate

29.31

(a)

Cystathionine

(b)

Cysteine

(c)

α-Ketobutyrate

The product of double-bond reduction is α-ketobutyrate. $FADH_2$ is the necessary coenzyme.

Additional Problems

Enzymes and Coenzymes

29.32 Digestion is the breakdown of bulk food in the stomach and small intestine. Hydrolysis of amide, ester and acetal bonds yields amino acids, fatty acids, and simple sugars, and these processes release energy.

29.33 Metabolism refers to all reactions that take place inside cells. Digestion is a catabolic part of metabolism in which food is broken down into small organic molecules.

29.34 Metabolic processes that break down large molecules are known as catabolism. Metabolic processes that assemble larger biomolecules from smaller ones are known as anabolism.

29.35

29.36

29.37 ATP transfers a phosphate group to another molecule in anabolic reactions.

29.38 NAD^+ is a biochemical oxidizing agent that converts alcohols to aldehydes or ketones, yielding NADH and H^+ as byproducts.

29.39 FAD is an oxidizing agent that introduces a conjugated double bond into a biomolecule, yielding $FADH_2$ as the reduced byproduct.

29.40 (a)   Pyridoxal phosphate is the cofactor associated with transamination.

(b)   Biotin is the cofactor associated with carboxylation of a ketone.

(c)   Thiamin diphosphate is the cofactor associated with decarboxylation of an α-keto acid.

29.41

Lactate   Pyruvate

NAD$^+$ is needed to convert lactate to pyruvate because the reaction involves the oxidation of an alcohol.

Metabolism

29.42

Acetoacetyl CoA   Acetyl CoA

29.43

(a)

(b)

(c)

29.44   The exact reverse of an energetically favorable reaction is energetically unfavorable. Since glycolysis is energetically favorable (negative $\Delta G^{o\prime}$), its exact reverse has a positive ΔG°; and is energetically unfavorable. Instead, glucose is synthesized by gluconeogenesis, an alternate pathway that also has a negative $\Delta G^{o\prime}$.

29.45   (a)   One mole of glucose is catabolized to two moles of pyruvate, each of which yields one mole of acetyl CoA. Thus,

1.0 mol of glucose —> 2.0 mol of acetyl CoA

(b)   A fatty acid with n carbons yields $n/2$ moles of acetyl CoA per mole of fatty acid. For palmitic acid ($C_{15}H_{31}CO_2H$),

$$1.0 \text{ mol of palmitic acid} \times \frac{8 \text{ mol of acetyl CoA}}{1 \text{ mol of palmitic acid}} -> 8.0 \text{ mol of acetyl CoA}$$

(c)   Maltose is a disaccharide that yields two moles of glucose on hydrolysis. Since each mole of glucose yields two moles of acetyl CoA,

$$1.0 \text{ mol of maltose} —> 2.0 \text{ mol of glucose} —> 4.0 \text{ mol of acetyl CoA}$$

29.46

	(a) Glucose	(b) Palmitic acid	(c) Maltose
Molecular weight	180.2 amu	256.4 amu	342.3 amu
Moles in 100.0 g	0.5549 mol	0.3900 mol	0.2921 mol
Moles of acetyl CoA produced	2 x 0.5549 mol = 1.110 mol	8 x 0.3900 mol = 3.120 mol	4 x 0.2921 mol = 1.168 mol
Grams acetyl CoA produced	898.6 g	2526 g	945.6 g

Palmitic acid is the most efficient precursor of acetyl CoA on a weight basis.

29.47

| *Amino acid* | *α-Keto acid* |

(a)

$$CH_3\overset{\overset{NH_3^+}{|}}{C}H\overset{}{C}HCO_2^-$$
$$\underset{OH}{|}$$

$$CH_3\overset{\overset{O}{\|}}{C}HCCO_2^-$$
$$\underset{OH}{|}$$

(b)

$$\text{—}CH_2\overset{\overset{NH_3^+}{|}}{C}HCO_2^-$$

$$\text{—}CH_2\overset{\overset{O}{\|}}{C}CO_2^-$$

(c)

$$H_2N\overset{\overset{O}{\|}}{C}CH_2\overset{\overset{NH_3^+}{|}}{C}HCO_2^-$$

$$H_2N\overset{\overset{O}{\|}}{C}CH_2\overset{\overset{O}{\|}}{C}CO_2^-$$

29.48   As we saw in Section 29.1, formation of glucose 6-phosphate from glucose and ATP is energetically favorable (negative $\Delta G^{\circ\prime}$) The reverse reaction, transfer of a phosphate group to ADP from glucose 6-phosphate, is energetically unfavorable and doesn't occur spontaneously. have negative $\Delta G^{\circ\prime}$ values and are energetically favorable reactions. In chemical terms, the leaving groups in the reactions of 3-phosphoglyceroyl phosphate (carboxylate) and phosphoenolpyruvate (enolate) are more stable anions than the leaving group in the reaction of glucose (alkoxide), so these reactions are more favorable.

Glucose   3-Phosphoglyceroyl   Phosphoenolpyruvate
                         phosphate

29.49   The steps in the conversion of α-ketoglutarate to succinyl CoA are similar to steps in the conversion of pyruvate to acetyl CoA shown in Figure 29.11, and the same coenzymes are involved: lipoamide, thiamin diphosphate, acetyl CoA and NAD$^+$.

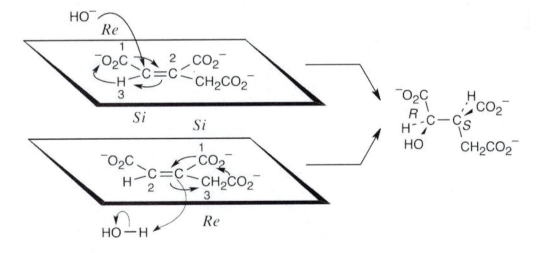

Step 1:   Nucleophilic addition of thiamin diphosphate ylid.

Step 2:   Decarboxylation.

Step 3:   Addition of double bond to lipoamide, with ring opening.

Step 4:   Elimination of thiamin diphosphate.

Step 5:   Nucleophilic addition of acetyl CoA to succinyl lipoamide and elimination of dihydrolipoamide to give succinyl CoA.

Step 6:   Reoxidation of dihydrolipoamide to lipoamide.

29.50   **Addition of –OH:** For carbon 2, the top face is the *Re* face, and –OH adds from this face to give an *R* configuration at carbon 2.
Addition of H⁺: For carbon 3, the top face is *Si*, and H⁺ adds from the bottom, or *Re*, face to give an *S* configuration at carbon 3. The reaction occurs with anti geometry.

General Problems

29.51

The face above the plane of the ring is the *Si face*. Since the problem states that addition of –H takes place from the *Re face*, the circled hydrogen comes from *sn*-glycerol. The added hydrogen has *pro-R* stereochemistry.

29.52

29.53

29.54 Addition occurs from the *Si face* to form the *R* enantiomer.

29.55 If dehydration removes the *pro-R* hydrogen and the resulting double bond is trans, as indicated, *anti* elimination must have taken place.

29.56

The reduction is a syn addition.

Review Unit 11: Biomolecules II – Lipids, Nucleic Acids, Metabolic Pathways

Major Topics Covered (with vocabulary):

Lipids:
wax   fat   oil   triacylglycerol   fatty acid   polyunsaturated fatty acid   soap
saponification   micelle   phosphoglyceride   sphingolipid   lipid   bilayer   sphingosine
sphingomyelin   prostaglandin   terpenoid   essential oil   monoterpenoid   sesquiterpenoid
isopentenyl diphosphate   steroid   hormone   sex hormone   adrenocortical hormone
androgen   estrogen   mineralocorticoid   glucocorticoid   squalene   lanosterol

Nucleic acids and nucleotides:
nucleoside   nucleotide   deoxyribonucleic acid (DNA)   ribonucleic acid (RNA)   adenine
guanine   thymine   cytosine   3' end   5' end   base pairing   double helix
complementary pairing   major groove   minor groove   intercalation

Nucleic acids and heredity:
replication   semiconservative   DNA polymerase   replication fork   DNA ligase
transcription   mRNA   rRNA   tRNA   sense (coding) strand   antisense (noncoding)
strand   promoter sites   exon   intron   translation   codon   anticodon

DNA technology:
DNA sequencing   Maxam-Gilbert method   restriction endonuclease   restriction fragment
palindrome   Sanger dideoxy method   DNA synthesis   DMT ether   phosphoramidite
phosphite polymerase chain reaction (PCR)

Metabolic pathways:
metabolism   anabolism   catabolism   digestion   phosphoric acid anhydride   ATP
NAD^+ $NADH/H^+$ β-oxidation pathway   glycolysis   Schiff base   pyruvate   acetyl
CoA   pyruvate   dehydrogenase complex   thiamine   lipoamide   citric acid cycle
electron-transport chain   transamination   oxidative deamination   gluconeogenesis   biotin

Types of Problems:

After studying these chapters you should be able to:

– Draw the structures of fats, oils, steroids and other lipids.

– Determine the structure of a fat.

– Predict the products of reactions of fats and steroids.

– Locate the five-carbon units in terpenoids.

– Understand the mechanism of terpenoid and steroid biosynthesis.

– Draw the structures and conformations of steroids and other fused-ring systems.

– Draw purines, pyrimidines, nucleosides, nucleotides, and representative segments of DNA and their complements.

– List the base sequence that codes for a given amino acid or peptide.

– Deduce an amino acid sequence from a given mRNA sequence (and *vice versa*).

– Draw the anticodon sequence of tRNA, given the mRNA sequence.

– Outline the process of DNA sequencing, and deduce a DNA sequence from an electrophoresis pattern.

– Outline the method of DNA synthesis, and formulate the mechanisms of synthetic steps.

–   Explain the basic concepts of metabolism, and understand the energy relationships of
     biochemical reactions.
–   Answer questions relating to the metabolic pathways of carbohydrates, fatty acids and
     amino acids.
–   Formulate mechanisms for metabolic pathways similar to those in the text.

Points to Remember:

*   When trying to locate the five-carbon units in a terpenoid, look for an isopropyl group first;
     at least one should be apparent. After finding it, count 5 carbons, and locate the second
     fivecarbon unit. If there are two possibilities for the second unit, choose the one that has the
     double bond in the correct location.

*   In general, the reactions of steroids that are presented in this book are familiar and
     uncomplicated. Keeping track of the stereochemistry of the tetracyclic ring system is
     somewhat more complicated.

*   In situations where base-pairing occurs, such as replication, transcription or translation, a
     polynucleotide chain (written with the 5' end on the left and the 3' end on the right) pairs
     with a second chain (written with the 3' end on the left and the 5' end on the right). Base
     pairing is complementary, and the two chains are always read in opposite directions.

*   Note the difference between transamination and oxidative deamination. Transamination is a
     reaction in which an amino group of an α-amino acid is transferred to α-ketoglutarate,
     yielding an α-keto acid and glutamate. In oxidative deamination, glutamate loses its amino
     group in an NAD^+-dependent reaction that regenerates α-ketoglutarate and produces NH_4^+.

*   Look at the steps of glycolysis, and then look at the steps of gluconeogenesis. Several steps
     in one pathway are the exact reverse of steps in the other pathway because the energy
     required for these steps is small. Other, high-energy transformations must occur by steps
     that are not the exact reverse and that require different enzymes. Gluconeogenesis is a
     metabolic pathway that takes place mainly during fasting and strenuous exercise because
     dietary sources of carbohydrates are usually available.

*   The conversion pyruvate $\rightarrow$ acetyl CoA is catalyzed by pyruvate dehydrogenase complex.
     The conversion acetyl CoA $\rightarrow$ carbohydrates doesn't occur in animals because they can
     obtain carbohydrates from food and don't usually need to synthesize carbohydrates. Only
     plants can, at times, use acetyl CoA to synthesize carbohydrates.

Self- test:

A

Lactaroviolin
(an antibiotic)

B

Toyocyamycin
(an antibiotic)

What type of terpenoid is **A**? Show the location of the five-carbon units.
Toyocamycin (**B**) is related to which nucleoside? What are the differences between **B** and
the nucleoside?

$$3' \qquad\qquad\qquad\qquad 5'$$
$$-ACG-CCT-TAG-GGC-TTA-GGA-$$

C

C represents a segment of the antisense strand of a molecule of DNA. Draw: (a) the sense
strand; (b) the mRNA that is synthesized from **C** during transcription; (c) the tRNA
anticodons that are complementary to the mRNA codons; (d) the amino acids (use one-
letter codes) that form the peptide that **C** codes for.

D   **E**

The above reaction is part of a metabolic pathway that occurs in plants. Identify **D** and **E**.
What type of reaction is taking place? Do think that NAD^+, FAD, or ATP are needed for
this reaction to occur?

Multiple choice:

1. Which type of molecule is most likely to be found in a lipid bilayer?
    (a) triacylglycerol
    (b) prostaglandin
    (c) sphingomyelin
    (d) triterpene

2. Which of the following terpenoids might have been formed by a tail-to-tail coupling?
    (a) monoterpenoid
    (b) sesquiterpenoid
    (c) diterpenoid
    (d) triterpenoid

3. Prostaglandins and related compounds have all of the following structural features in common except:
    (a) cis double bonds
    (b) a carboxylic acid group
    (c) a C_{20} chain
    (d) hydroxyl groups

4. Which of the following steps doesn't occur in the synthesis of isopentenyl diphosphate?
    (a) Claisen condensation
    (b) oxidation
    (c) aldol condensation
    (d) decarboxylation

5. Which nucleic acid has nonstandard bases, in addition to the usual bases?
    (a) DNA
    (b) mRNA
    (c) rRNA
    (d) tRNA

6. Which base doesn't need a protecting group in DNA synthesis?
    (a) Thymine
    (b) Cytosine
    (c) Adenine
    (d) Guanine

7. Which amino acid has only one codon?
    (a) Tyrosine
    (b) Arginine
    (c) Lysine
    (d) Tryptophan

8.   Which of the following enzyme cofactors is not involved in the conversion of pyruvate to acetyl CoA?
     (a)   Thiamine pyrophosphate
     (b)   Pyridoxal phosphate
     (c)   Lipoamide
     (d) NAD^+

9.   Which of the following steps of the citric acid cycle doesn't produce reduced coenzymes?
     (a)   Isocitrate $\rightarrow \alpha$-Ketoglutarate
     (b) α-Ketoglutarate $\rightarrow$ Succinyl CoA
     (c)   Fumarate $\rightarrow$ Malate
     (d)   Succinate $\rightarrow$ Fumarate

10. The amino acid aspartate can be metabolized as what citric acid cycle intermediate after transamination?
     (a)   Oxaloacetate
     (b)   Malate
     (c) α-Ketoglutarate
     (d)   Succinate

Chapter 30 – Orbitals and Organic Chemistry: Pericyclic Reactions

Chapter Outline

I. Molecular orbitals and pericyclic reactions of conjugated pi systems (Section 30.1).
   A. Molecular orbitals of conjugated π systems.
      1. The p orbitals of the sp^2-hybridized carbons of a polyene interact to form a set of π molecular orbitals.
      2. The energies of these orbitals depend on the number of nodes they have.
         a. The molecular orbitals with fewer nodes are bonding MOs.
         b. The molecular orbitals with more nodes are antibonding MOs.
      3. A molecular orbital description can be used for any conjugated π system.
         a. In the ground state, only the bonding orbitals are used.
         b. On irradiation with UV light, an electron is promoted to an antibonding orbital.
            i. This is known as an excited state.
   B. Molecular orbitals and pericyclic reactions.
      1. The mechanisms of pericyclic reactions can be explained by molecular orbital theory.
         a. A pericyclic reaction can take place only if the lobes of the reactant MOs have the correct algebraic sign in the transition state.
         b. If the symmetries of both reactant and product orbitals correlate, the reaction is symmetry-allowed.
         c. If the symmetries don't correlate, the reaction is symmetry-disallowed.
            i. The reaction may still take place, but only by a nonconcerted, high-energy pathway.
      2. A modification of MO theory states that only two MOs need be considered (frontier orbitals):
         a. The highest occupied molecular orbital (HOMO).
         b. The lowest unoccupied molecular orbital (LUMO).
II. Electrocyclic reactions (Sections 30.2–30.4).
   A. General description of electrocyclic reactions Section 30.2).
      1. Nature of electrocyclic reactions.
         a. An electrocyclic reaction involves the cyclization of a conjugated polyene.
            i. One π bond is broken, a new σ bond is formed and a cyclic compound results.
         b. Electrocyclic reactions are reversible.
            i. The triene–cyclohexadiene equilibrium favors the ring-closed product.
            ii. The diene–cyclobutene equilibrium favors the ring-opened product.
      2. Stereochemistry of electrocyclic reactions.
         a. A specific E,Z bond isomer yields a specific cyclic stereoisomer under thermal conditions.
         b. The stereochemical results are opposite when the reactions are carried out under photochemical conditions.

3. Orbital explanation for outcomes of electrocyclic reactions.
    a. The signs of the outermost lobes of the interacting orbitals explain these results.
        i. For a bond to form, the lobes must be of the same sign.
    b. The outermost π lobes of the polyene must rotate so that the lobes that form the bonds are of the same sign.
        i. If the lobes are on the same side of the molecule, the lobes must rotate in opposite directions – disrotatory motion.
        ii. If the lobes of the same sign are on opposite sides of the polyene, both lobes must rotate in the same direction – conrotatory motion.
B. Stereochemistry of thermal electrocyclic reactions (Section 30.3).
    1. The stereochemistry of an electrocyclic reaction is determined by the symmetry of the polyene HOMO.
    2. The ground-state electronic configuration is used to identify the HOMO for thermal reactions.
        a. For trienes, the HOMO has lobes of like sign on the same side of the molecule, and ring-closure is disrotatory.
        b. For dienes, ring closing is conrotatory.
    3. In general, polyenes with odd numbers of double bonds undergo disrotatory thermal electrocyclic reactions, and polyenes with even numbers of double bonds undergo conrotatory thermal electrocyclic reactions.
C. Stereochemistry of photochemical electrocyclic reactions (Section 30.4).
    1. UV irradiation of a polyene causes excitation of one electron from the ground-state HOMO to the ground-state LUMO.
    2. UV irradiation changes the symmetry of HOMO and LUMO and also changes the reaction stereochemistry.
        a. Photochemical electrocyclic reactions of trienes occur with conrotatory motion.
        b. Photochemical electrocyclic reactions of dienes occur with disrotatory motion.
    3. Thermal and photochemical electrocyclic reactions always take place with opposite stereochemistry.
III. Cycloaddition reactions (Sections 30.5–30.6).
A. General description of cycloaddition reactions (Section 30.5).
    1. A cycloaddition reaction is a reaction in which two unsaturated molecules add to give a cyclic product.
    2. Cycloadditions are controlled by the orbital symmetry of the reactants.
        a. Reactions that are symmetry-disallowed either don't take place or occur by a higher-energy nonconcerted pathway.
    3. The Diels–Alder cycloaddition is an example.
        a. Reaction occurs between a diene and a dienophile to yield a cyclic product.
        b. The products have a specific stereochemistry.
        c. The reaction is known as a [4 + 2] cycloaddition.
    4. Cycloadditions can only occur if the terminal π lobes have the correct stereochemistry.

    a. In suprafacial cycloadditions, a bonding interaction takes place between lobes on the same face of one reactant and lobes on the same face of the other reactant.

    b. Antarafacial cycloadditions occur between lobes on the same face of one reactant and lobes on opposite faces of the other reactant.

    c. Often, antarafacial cycloadditions are symmetry-allowed but geometrically constrained.

  B. Stereochemistry of cycloadditions (Section 30.6).

    1. A cycloaddition reaction takes place when a bonding interaction occurs between the HOMO of one reactant and the LUMO of the other reactant.

    2. The symmetries of the terminal lobes of the HOMO and LUMO of the reactants in a [4 + 2] thermal cycloaddition allow the reaction to proceed with suprafacial geometry.

    3. For [2 + 2] cycloadditions:

      a. Orbital symmetry shows that thermal cyclization must occur by an antarafacial pathway.

      b. Because of geometrical constraints, thermal [2 + 2] cycloadditions aren't seen.

      c. Photochemical [2 + 2] cycloadditions take place because the addition can occur by a suprafacial pathway.

    4. Thermal and photochemical cycloadditions always take place by opposite stereochemical pathways.

IV. Sigmatropic rearrangements (Sections 30.7–30.8).

  A. General description of sigmatropic rearrangements (Section 30.7).

    1. In a sigmatropic rearrangement, a σ-bonded atom or group migrates across a π electron system.

      a. A σ bond is broken, the π bonds move, and a new σ bond is formed in the product.

      b. The σ bonded group can be either at the end or in the middle of the π system.

      c. The notation [3,3] indicates the positions in the groups to which migration occurs.

    2. Sigmatropic rearrangements are controlled by orbital symmetry.

      a. Migration of a group across the same face of the π system is suprafacial rearrangement.

      b. Migration from one face to the other face is antarafacial rearrangement.

      c. Both types of rearrangements are symmetry-allowed, but suprafacial rearrangements are geometrically easier.

  B. Examples of sigmatropic rearrangements (Section 30.8).

    1. The [1,5] migration of a hydrogen atom across two double bonds of a π system is very common.

      a. Thermal [1,3] hydrogen shifts are unknown.

    2. The Cope rearrangement and the Claisen rearrangement involve reorganization of an odd number of electron pairs and proceed by suprafacial geometry.

V. A summary of rules for pericyclic reactions (Section 30.9).

    A. Thermal reactions with an even number of electron pairs are either conrotatory or antarafacial.

    B. A change from thermal to photochemical, or from even to odd, changes the outcome to disrotatory/suprafacial.

    C. A change of both thermal and even causes no change.

Solutions to Problems

30.1 **For ethylene:**

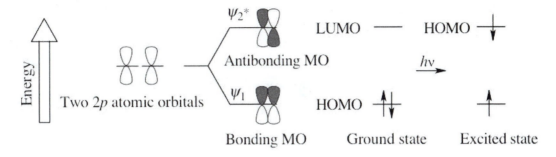

The two π electrons of ethylene occupy ψ_1 in the ground state, making ψ_1 the HOMO and ψ_2^* the LUMO. In the excited state, one electron occupies ψ_1 and the other occupies ψ_2^*, making ψ_2^* the HOMO. Since all orbitals are occupied in the excited state, there is no LUMO.

For 1,3-butadiene:

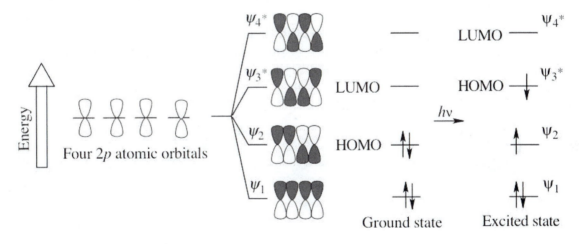

In the ground state, ψ_2 is the HOMO, and ψ_3^* is the LUMO. In the excited state, ψ_3^* is the HOMO, and ψ_4^* is the LUMO.

30.2

The symmetry of the octatriene HOMO predicts that ring closure occurs by a disrotatory path in the thermal reaction and that only *cis* product is formed.

30.3 Note: *Trans*-3,4-dimethylcyclobutene is chiral; the *S,S* enantiomer will be used for this argument.

Path A:

Path B:

Conrotatory ring opening of *trans*-3,4-dimethylcyclobutene can occur in either a clockwise or a counterclockwise manner. Clockwise opening (**path A**) yields the *E,E* isomer; counterclockwise opening (**path B**) yields the *Z,Z* isomer. Production of (2*Z*,4*Z*)-2,4-hexadiene is disfavored because of steric strain between the methyl groups in the transition state leading to ring-opened product.

30.4

(2*E*,4*Z*,6*E*)-
2,4-Octatriene
Ground state HOMO

Excited state HOMO

trans-5,6-Dimethyl-
1,3-cyclohexadiene

(2*E*,4*Z*,6*Z*)-
2,4-Octatriene
Ground state HOMO

Excited state HOMO

cis-5,6-Dimethyl-
1,3-cyclohexadiene

Photochemical electrocyclic reactions of 6 π electron systems always occur in a conrotatorymanner.

30.5

(2E,4E)-2,4-Hexadiene   (2E,4Z)-2,4-Hexadiene

The Diels–Alder reaction is a thermal [4 + 2] cycloaddition, which occurs with suprafacial geometry. The stereochemistry of the diene is maintained in the product.

30.6

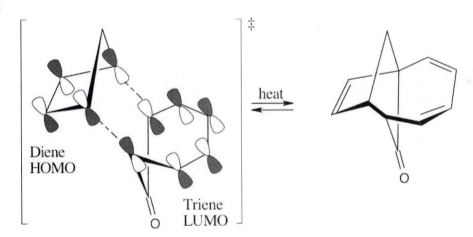

The reaction of cyclopentadiene and cycloheptatrienone is a [6 + 4] cycloaddition. This thermal cycloaddition proceeds with suprafacial geometry since five electron pairs are involved in the concerted process. The π electrons of the carbonyl group do not take part in the reaction.

30.7   This [1,7] sigmatropic reaction proceeds with antarafacial geometry because four electron pairs are involved in the rearrangement.

30.8   Scrambling of the deuterium label of 1-deuterioindene occurs by a series of [1,5] sigmatropic rearrangements. This thermal reaction involves three electron pairs – one pair of π electrons from the six-membered ring, the π electrons from the five-membered ring, and two electrons from a carbon-deuterium (or hydrogen) single bond – and proceeds with suprafacial geometry.

30.9

The Claisen rearrangement of an unsubstituted allyl phenyl ether is a [3,3] sigmatropic rearrangement in which the allyl group usually ends up in the position *ortho* to oxygen. In this problem both *ortho* positions are occupied by methyl groups. The Claisen intermediate undergoes a second [3,3] rearrangement, and the final product is *p*-allyl phenol.

30.10

Type of reaction	Number of electron pairs	Stereochemistry
(a) Thermal electrocyclic	four	conrotatory
(b) Photochemical electrocyclic	four	disrotatory
(c) Photochemical cycloaddition	four	suprafacial
(d) Thermal cycloaddition	four	antarafacial
(e) Photochemical sigmatropic rearrangement	four	suprafacial

Visualizing Chemistry

30.11

This reaction is a [3,3] sigmatropic rearrangement that yields 1,5-cyclodecadiene as a product.

30.12

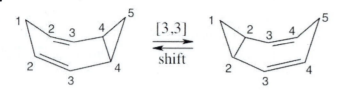

The ^{13}C NMR spectrum of homotropilidene would show five peaks if rearrangement were slow. In fact, rearrangement occurs at a rate that is too fast for NMR to detect. The ^{13}C NMR spectrum taken at room temperature is an average of the two equilibrating forms, in which positions 1 and 5 are equivalent, as are positions 2 and 4. Thus, only three distinct types of carbons are visible in the ^{13}C NMR spectrum of homotropilidene.

Mechanism Problems

30.13

30.14 (a)

(b)   The rearrangement product absorbs at a longer wavelength because it has a more extensive system of conjugated double bonds.

30.15

The first reaction is an electrocyclic opening of a cyclobutene ring.

Formation of estrone methyl ether occurs by a Diels–Alder [4 + 2] cycloaddition.

30.16

Reaction 1:   Reverse Diels–Alder [4 + 2] cycloaddition;

Reaction 2:   Conrotatory electrocyclic opening of a cyclobutene ring;

Reaction 3:   Diels–Alder [4 + 2] cycloaddition.

Treatment with base enolizes the ketone and changes the ring junction from *trans* to *cis*. A *cis* ring fusion is less strained when a six-membered ring is fused to a five-membered ring.

30.17

Each of the two electrocyclic reactions involves two pairs of electrons and proceeds in a conrotatory manner.

Additional Problems

Electrocyclic Reactions

30.18 (a)

Rotation of the orbitals in the 6 π electron system occurs in a disrotatory fashion. According to the rules in Table 30.1, the reaction should be carried out under thermal conditions.

(b)

Ground state HOMO

Excited state HOMO

For the hydrogens to be *trans* in the product, rotation must occur in a conrotatory manner. This can happen only if the HOMO has the symmetry pictured. For a 6 π electron system, this HOMO must arise from photochemical excitation of a π electron. To obtain a product having the correct stereochemistry, the reaction must be carried out under photochemical conditions.

30.19 The diene can cyclize by either of two conrotatory paths to form cyclobutenes **A** and **B**.

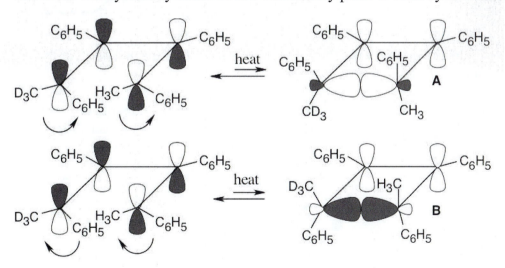

Opening of each cyclobutene ring can occur by either of two conrotatory routes to yield the isomeric dienes. Using **B** as an example:

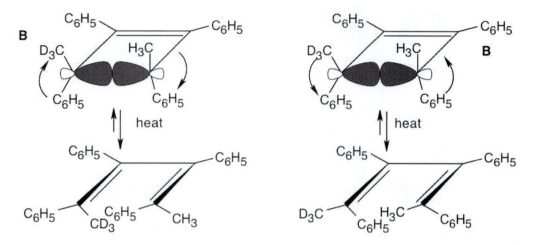

30.20   A photochemical electrocyclic reaction involving two electron pairs proceeds in a *disrotatory* manner (Table 30.1).

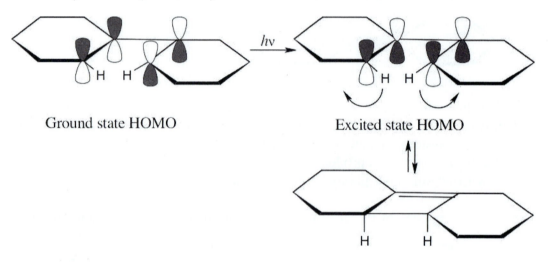

Ground state HOMO   Excited state HOMO

The two hydrogen atoms in the four-membered ring are cis to each other in the cyclobutene product.

30.21   The cyclononatriene is a 6 π electron system that cyclizes by a disrotatory route under thermal conditions. The two hydrogens at the ring junction have a cis relationship.

30.22

(2E,4Z,6Z,8E)-
2,4,6,8-Decatetraene

Four electron pairs undergo reorganization in this electrocyclic reaction. The thermal reaction occurs with conrotatory motion to yield a pair of enantiomeric *trans*-7,8-dimethyl- 1,3,5-cyclooctatrienes. The photochemical cyclization occurs with disrotatory motion to yield the *cis*-7,8-dimethyl isomer.

30.23

(2E,4Z,6Z,8Z)-
2,4,6,8-Decatetraene

30.24

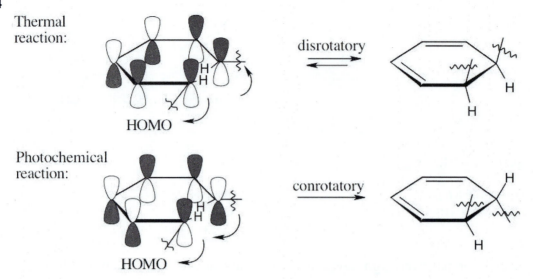

Thermal reaction: HOMO   disrotatory

Photochemical reaction: HOMO   conrotatory

Two electrocyclic reactions, involving three electron pairs each, occur in this isomerization. The thermal reaction is a disrotatory process that yields two *cis*-fused sixmembered rings. The photochemical reaction yields the *trans*-fused isomer. The two pairs of π electrons in the eight-membered ring do not take part in the electrocyclic reaction.

Carbohydrate Reactions

30.25

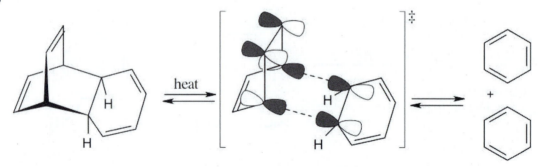

This reaction is a reverse [4 + 2] cycloaddition. The reacting orbitals have the correct symmetry for the reaction to take place by a favorable suprafacial process.

This [2 + 2] reverse cycloaddition is not likely to occur as a concerted process because the antarafacial geometry required for the thermal reaction is not possible for a four π-electron system.

30.26

Formation of the bicyclic ring system occurs by a suprafacial [4 + 2] Diels–Alder cycloaddition process. Only one pair of π electrons from the alkyne is involved in the reaction; the carbonyl π electrons are not involved.

30.27

The first reaction is a Diels–Alder [4 + 2] cycloaddition, which proceeds with suprafacial geometry.

The second reaction is a reverse Diels–Alder [4 + 2] cycloaddition.

Sigmatropic Rearrangements

30.28

This thermal sigmatropic rearrangement is a suprafacial process since five electron pairs are involved in the reaction.

30.29   The product of this [3,3] sigmatropic rearrangement is an enol that tautomerizes to a ketone.

30.30

Vinylcyclopropane   Cyclopentene

This reaction is a [1,3] sigmatropic rearrangement.

30.31

An allene is formed by a [3,3] sigmatropic rearrangement.

Acid catalyzes isomerization of the allene to a conjugated dienone *via* an intermediate enol.

30.32

Karahanaenone

Karahanaenone is formed by a [3,3] sigmatropic rearrangement (Claisen rearrangement).

General Problems

30.33 Tables 30.1–30.3 may be helpful. The first step is always to find the number of electron pairs involved in the reaction.

Type of reaction	Number of electron pairs	Stereochemistry
(a) Photochemical [1,5] sigmatropic rearrangement	3	antarafacial
(b) Thermal [4 + 6] cycloaddition	5	suprafacial
(c) Thermal [1,7] sigmatropic rearrangement	4	antarafacial
(d) Photochemical [2 + 6] cycloaddition	4	suprafacial

30.34 Ring opening of Dewar benzene is a process involving two electron pairs and, according to Table 30.1, should occur by a conrotatory pathway. However, if you look back to other ring openings of cis-fused cyclobutenes, you will see that conrotatory ring opening produces a diene in which one of the double bonds is trans. Since a trans double bond in a six-membered ring is not likely to be formed, ring opening occurs by a different, higher energy, nonconcerted pathway.

30.35

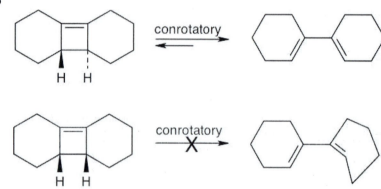

Ring opening of the *trans*-cyclobutene isomer proceeds by the expected conrotatory route to form the observed product. For the *cis*-cyclobutene isomer, the observed product can be formed by a four-electron *pericyclic* process only if the four-membered ring geometry is trans. Ring opening of the cis isomer by a concerted process would form a severely strained six-membered ring containing a trans double bond. Reaction of the cis isomer to yield the observed product occurs instead by a higher energy, nonconcerted path.

30.36

Both reactions are [2 + 2] photochemical electrocyclic reactions, which occur with disrotatory motion.

30.37

Bullvalene

Bullvalene can undergo [3,3] sigmatropic rearrangements in all directions. At 100 °C, the rate of rearrangement is fast enough to make all hydrogen atoms equivalent, and only one signal is seen in the ^{1}H NMR spectrum.

30.38

Suprafacial shift:

The observed products **A** and **B** result from a [1,5] sigmatropic hydrogen shift with suprafacial geometry, and they confirm the predictions of orbital symmetry. **C** and **D** are not formed.

30.39

This [2,3] sigmatropic rearrangement involves three electron pairs and should occur with suprafacial geometry.

30.40

Concerted thermal ring opening of a cis-fused cyclobutene is conrotatory and yields a product having one cis and one trans double bond. The ten-membered ring product of reaction 2 is large enough to accommodate a trans double bond, but a seven-membered ring containing a trans double bond is highly strained. Opening of the cyclobutene ring in reaction 1 occurs by a higher energy nonconcerted process to yield a seven-membered ring having two cis double bonds.

30.41

Thermal ring opening of the methylcyclobutene ring can occur by either of two symmetry-allowed conrotatory paths to yield the observed product mixture.

Chapter 31 – Synthetic Polymers

Chapter Outline

I. Chain-growth polymers (Sections 31.1–31.3).
  A. General features of chain-growth polymerization reactions (Section 31.1).
    1. How polymerization occurs.
      a. An initiator adds to a carbon–carbon double bond of a vinyl monomer.
      b. The reactive intermediate adds to a second molecule of monomer.
      c. The process is repeated.
    2. Types of polymerization.
      a. A radical initiator leads to radical polymerization.
      b. An acid causes cationic polymerization.
        i. Acid-catalyzed polymerization is effective only if the vinyl monomers contain electron-donating groups.
      c. Anionic polymerization can be brought about by anionic catalysts.
        i. Vinyl monomers in anionic catalysis must have electron-withdrawing groups.
        ii. Polymerization occurs by conjugate nucleophilic addition to the monomer.
        iii. Acrylonitrile, styrene and methyl methacrylate can be polymerized anionically.
        iv. "Super glue" is an example of an anionic polymer.
  B. Stereochemistry of polymerization (Section 31.2).
    1. There are three possible stereochemical outcomes of polymerization of a substituted vinyl monomer.
      a. If the substituents all lie on the same side of the polymer backbone, the polymer is isotactic.
      b. If the substituents alternate along the backbone, the polymer is syndiotactic.
      c. If the substituents are randomly oriented, the polymer is atactic.
    2. The three types of polymers have different properties.
    3. Although polymerization using radical initiators can't be control stereochemically, Ziegler–Natta catalysts can yield polymers of desired stereochemical orientation.
      a. Ziegler–Natta catalysts are organometallic-transition metal complexes.
        i. They are usually formed by treatment of an alkylaluminum with titanium tetrachloride.
      b. Ziegler–Natta polymers have very little chain-branching.
      c. Ziegler–Natta catalysts are stereochemically controllable.
      d. Polymerization occurs by coordination of the alkene monomer to the complex, followed by insertion into the polymer chain.
    4. Common Ziegler–Natta polymers.
      a. Polyethylene produced by the Ziegler–Natta process (high-density polyethylene) is linear, dense, strong, and heat-resistant.
      b. Other high-molecular-weight polyethylenes have specialty uses.

C   Copolymers (Section 31.3).

    1. Copolymers are formed when two different monomers polymerize together.

    2. The properties of copolymers are different from those of the corresponding monomers.

    3. Types of copolymers.

        a. Random copolymers.

        b. Alternating copolymers.

        c. Block copolymers.

            i. Block copolymers are formed when an excess of a second monomer is added to a still-active mix.

        d. Graft copolymers.

            i. Graft copolymers are made by gamma irradiation of a completed homopolymer to generate a new radical initiation site for further growth of a chain.

II. Step-growth polymers (Section 31.4).

  A. Step-growth polymer are formed by reactions in which each bond is formed independently of the others.

  B. Most step-growth polymers result from reaction of two difunctional compounds.

    1. Step-growth polymers can also result from polymerization of a single difunctional compound.

  C. Types of step-growth polymers.

    1. Polyamides and polyesters.

    2. Polycarbonates (formed from carbonates and alcohols or phenols).

    3. Polyurethanes.

        a. A urethane has a carbonyl group bonded to both an $-NR_2$ group and an $-OR$ group.

        b. Most polyurethanes are formed from the reaction of a diisocyanate and a diol.

        c. Polyurethanes are used as spandex fibers and insulating foam.

            i. Foaming occurs when a small amount of water is added during polymerization, producing bubbles of CO_2.

            ii. Polyurethane foams often use a polyol, to increase the amount of crosslinking.

III. Olefin metathesis polymerization (Section 31.5).

  A. General features.

    1. In an olefin metathesis reaction, two olefins (alkenes) exchange substituents.

    2. The catalysts contain a carbon–metal (usually ruthenium) double bond.

        a. They react reversibly with an alkene to form a 4-membered metallacyte.

        b. The metallacyte opens to give a different catalyst and a different alkene.

    3. The reaction is compatible with many olefin functional groups.

    4. The double bonds allow for further manipulations.

  B. Ring-opening metathesis polymerization (ROMP).

    1. The monomer is a moderately strained cycloalkene.

    2. The resulting polymer has double bonds spaced regularly along the chain.

C. Acyclic diene metathesis (ADMET).

   1. The monomer is a long-chain dialkene with double bonds at the end of the chain.

   2. As the reaction progresses, gaseous ethylene escapes, driving the reaction toward product.

IV. Polymer structure and physical properties (Section 31.6).

A. Physical properties of polymers.

   1. Because of their large size, polymers experience large van der Waals forces.

      a. These forces are strongest in linear polymers.

   2. Many polymers have regions held together by van der Waals forces; these regions are known as crystallites.

      a. Polymer crystallinity is affected by the substituents on the chains.

      b. T_m is the temperature at which the crystalline regions of a polymer melt.

   3. Some polymers have little ordering but are hard at room temperature.

      a. These polymers become soft at a temperature T_g (glass transition temperature).

B. Polymers can be classified by physical behavior.

   1. Thermoplastics.

      a. Thermoplastics have a high T_g and are hard at room temperature.

      b. Because they become soft at higher temperatures, they can be molded.

      c. Plasticizers such as dialkyl phthalates are often added to thermoplastics to keep them from becoming brittle at room temperature.

   2. Fibers.

      a. Fibers are produced by extrusion of a molten polymer.

      b. On cooling and drawing out, the crystallite regions orient along the axis of the fiber to add tensile strength.

   3. Elastomers.

      a. Elastomers are amorphous polymers that can stretch and return to their original shape.

      b. These polymers have a low T_g and a small amount of cross-linking.

      c. The randomly coiled chains straighten out in the direction of the pull, but they return to their random orientation when stretching is done.

      d. Natural rubber is an elastomer, but gutta percha is highly crystalline.

   4. Thermosetting resins.

      a. Thermosetting resins become highly cross-linked and solidify when heated.

      b. Bakelite, a phenolic resin formed from phenol and formaldehyde, is the most familiar example.

Solutions to Problems

31.1

Most reactive ⟶ *Least reactive*

$$H_2C{=}CHC_6H_5 \; > \; H_2C{=}CHCH_3 \; > \; H_2C{=}CHCl \; > \; H_2C{=}CHCO_2CH_3$$

The alkenes most reactive to cationic polymerization contain electron-donating functional groups that can stabilize the carbocation intermediate. The reactivity order of substituents in cationic polymerization is similar to the reactivity order of substituted benzenes in electrophilic aromatic substitution reactions.

31.2

Most reactive ⟶ *Least reactive*

$$H_2C{=}CHC{\equiv}N \; > \; H_2C{=}CHC_6H_5 \; > \; H_2C{=}CHCH_3$$

Anionic polymerization occurs most readily with alkenes having electron-withdrawing substituents.

31.3

The intermediate anion can be stabilized by resonance involving the phenyl ring.

31.4

$$n \; H_2C{=}CCl_2 \longrightarrow \left(CH_2{-}CCl_2 \right)_n$$

Vinylidene chloride doesn't polymerize in isotactic, syndiotactic or atactic forms because no asymmetric centers are formed during polymerization.

31.5 None of the polypropylenes rotate plane-polarized light. If an optically inactive reagent and an achiral compound react, the product must be optically inactive. For every chirality center generated, an enantiomeric chirality center is also generated, and the resulting polymer mixture is optically inactive.

31.6

2-Methyl-1,3-butadiene   2-Methylpropene

31.7

polybutadiene chain

polystyrene chain

Irradiation homolytically cleaves an allylic C–H bond because it has the lowest bond energy. The resulting radical adds to styrene to produce a polystyrene graft.

31.8

PET

31.9

a urethane

31.10

1,9-Decadiene

catalyst

Vestenamer   + n $H_2C{=}CH_2$

Vestenamer can be formed by an ADMET synthesis using 1,9-decadiene.

n catalyst

Vestenamer

Vestenamer is usually synthesized from cyclooctene by a ROMP polymerization.

n catalyst

Norbornene   Norsorex

Norbornene undergoes ROMP polymerization to yield Norsorex.

31.11

Natural rubber

$\xrightarrow[\text{Pd/C}]{H_2}$

In·

CH_3
$HC{=}CH_2$ + $H_2C{=}CH_2$

The product of hydrogenation of natural rubber is atactic. This product also results from the radical copolymerization of propene with ethylene.

31.12

This product, formed by electrophilic aromatic substitution, can react many times with additional formaldehyde and phenol to yield Bakelite. Reaction occurs at both *ortho* and *para* positions of phenol.

Visualizing Chemistry

31.13

The polymer is a polycarbonate synthesized from the above monomer units.

31.14

(a)

(b)

Both of these polymers are chain-growth polymers. To draw the polymer in (a), break the double bond, and draw its extensions, one on each side of the former double bond. In (b), break both double bonds and draw extensions at both ends of the former diene. The remaining double bond migrates to a position between the double bonds of the former diene.

Mechanism Problems

31.15

31.16 (a) The diamine is formed by an electrophilic aromatic substitution reaction of formaldehyde with two equivalents of aniline.

(b)   The diamine reacts with two equivalents of phosgene.

MDI

31.17

31.18 Step 1: Polystyrene and the phthalimide combine in an electrophilic aromatic substitution reaction

Step 2: The phthalimide is cleaved in a series of steps that involve nucleophilic acyl substitution reactions.

Let $\left(\!CH_2\!-\!CH\!\right)_n$ = R

attack of nucleophile

proton shift

ring opening

proton shift

attack of nucleophile

$$HN\!-\!NH \quad + \quad RNH_2 \quad = \quad \left(\!CH_2\!-\!CH\!\right)_n$$

31.19 (a)

Cyclopentadiene   Dicyclopentadiene

Dicyclopentadiene is formed by a Diels–Alder cycloaddition of two molecules of cyclopentadiene.

(b)

(c)

Crosslinking takes place when the remaining double bond of dicyclopentadiene is involved in the polymerization process.

Additional Problems

31.20

(a)

Chain-growth
polymer

(b)

$$n \quad \overset{F}{\underset{F}{\diagdown}}C=C\overset{Cl}{\underset{F}{\diagup}} \longrightarrow \quad -\!\!\left(\!CF_2-CFCl\!\right)\!_n$$

Chain-growth
polymer

(c)

$$n \quad \text{(pyrrolidinone)} \longrightarrow -\!\!\left(\!NHCH_2CH_2CH_2\overset{O}{\overset{\|}{C}}\!\right)\!_n \quad + \quad n\ H_2O$$

$$\left(n\ \ H_2NCH_2CH_2CH_2\overset{O}{\overset{\|}{C}}-OH\right)$$

Step-growth
polymer

(d)

$$n\ HO\!-\!\!\text{(benzene ring)}\!-\!\overset{O}{\overset{\|}{C}}-OH \longrightarrow -\!\!\left(\!O\!-\!\!\text{(benzene ring)}\!-\!\overset{O}{\overset{\|}{C}}\!\right)\!_n \quad + \quad n\ H_2O$$

Step-growth
polymer

(e)

$$n\ HO\!-\!\!\text{(ring)}\!-\!OH\ +\ n\ ROCOR \longrightarrow -\!\!\left(\!O\!-\!\!\text{(ring)}\!-\!O\!-\!\overset{O}{\overset{\|}{C}}\!\right)\!_n$$

Step-growth   + 2n ROH
polymer

31.21 Remember that isotactic polymers have identical groups on the same side of the polymer backbone. Syndiotactic polymers have alternating identical groups along the polymer backbone. Atactic polymers have a random orientation of groups.

(a)

Syndiotactic polyacrylonitrile

(b)

Atactic poly(methyl methacrylate)

(c)

Isotactic poly(vinyl chloride)

31.22

Kodel

31.23

Ring opening of the epoxide occurs by an S$_N$2 pathway at the less substituted epoxide carbon.

31.24

Nomex

31.25

$$n \; H_2N(CH_2)_{10}NH_2 \; + \; n \; HOC(CH_2)_8COH \; \longrightarrow \; \left(NH(CH_2)_{10}NH - C(CH_2)_8C \right)_n$$

Nylon 10,10 $\quad + 2n \; H_2O$

31.26

repeat many times

Polycyclopentadiene

31.27 *p*-Divinylbenzene is incorporated into the growing polystyrene chain.

$$CH_2CH \; + \; CH_2{=}CH \; \longrightarrow \; CH_2CH - CH_2CH \xrightarrow[\text{repeat many times}]{C_6H_5CH{=}CH_2}$$

CH=CH₂ CH=CH₂

Another growing polymer chain reacts with the second double bond of *p*-divinylbenzene.

$$CH_2CH - CH_2CH \; + \; {\cdot}CHCH_2 \; \longrightarrow \; CH_2CH - CH_2CH$$

CH=CH₂ ·CHCH₂—CHCH₂

The final product contains polystyrene chains cross-linked by *p*-divinylbenzene units.

31.28 The white coating on the distillation flask is due to the thermal polymerization of nitroethylene.

31.29 Poly(vinyl alcohol) is formed by chain-growth polymerization of vinyl acetate, followed by hydrolysis of the acetate groups.

Reaction of poly(vinyl alcohol) with butanal to form the poly(cyclic acetal) produces poly(vinyl butyral).

Poly(vinyl butyral)

31.30

The polymer is a polyester.

31.31

Use of glycerol as a monomer causes the cross-linking that gives Glyptal its strength.

31.32   Repeated nucleophilic acyl substitution reactions result in the formation of Melmac.

Melmac

31.33 (a)

The prepolymer contains epoxide rings and hydroxyl groups. Copolymerization with a triamine occurs at the epoxide ends of the prepolymer.

(b)

Cross-linking occurs when the triamine opens epoxide rings on two different chains of the prepolymer.

31.34

$$H_2N \overset{\overset{\displaystyle O}{\|}}{C} NH_2 \; + \; CH_2O \xrightarrow{\text{heat}}$$

31.35

$$2 \; CH_3CH_2CH_2CHO \xrightarrow[\text{2. heat}]{\text{1. NaOH, EtOH}} CH_3CH_2CH_2CH\!=\!\overset{\overset{\displaystyle CH_2CH_3}{|}}{C}CHO$$

1. NaBH$_4$
2. H$_3$O$^+$

$$\overset{\overset{\displaystyle CH_2CH_3}{|}}{CH_3CH_2CH_2CH_2CHCH_2OH} \xleftarrow[\text{Pd/C}]{H_2} CH_3CH_2CH_2CH\!=\!\overset{\overset{\displaystyle CH_2CH_3}{|}}{C}CH_2OH$$

2-Ethyl-1-hexanol

Review Unit 12: Pericyclic Reactions, Synthetic Polymers

Major Topics Covered (with vocabulary):

Pericyclic reactions:
pericyclic reaction   concerted reaction   symmetry-allowed   symmetry-disallowed   frontier orbitals   HOMO   LUMO   electrocyclic reaction   disrotatory motion   conrotatory motion   cycloaddition reaction   suprafacial cycloaddition   antarafacial cycloaddition   sigmatropic rearrangement   suprafacial rearrangement   antarafacial rearrangement   Cope rearrangement   Claisen rearrangement

Synthetic polymers:
chain-growth polymer   Ziegler–Natta catalyst   isotactic   syndiotactic   atactic   homopolymer   copolymer   block copolymer   graft copolymer   step-growth polymer   polycarbonate   polyurethane   olefin metathesis   Grubbs catalyst   ADMET   ROMP   crystallite   melt transition temperature   glass transition   thermoplastic   fiber   elastomer   thermosetting resin   plasticizer

Types of Problems:

After studying these chapters, you should be able to:

– Understand the principles of molecular orbitals, and locate the HOMO and LUMO of conjugated π systems.
– Predict the stereochemistry of thermal and photochemical electrocyclic reactions.
– Know the stereochemical requirements for cycloaddition reactions, and predict the products of cycloadditions.
– Classify sigmatropic reactions by order and predict their products.
– Know the selection rules for pericyclic reactions.
– Locate the monomer units of a polymer; predict the structure of a polymer, given its monomer units.
– Formulate the mechanisms of radical, cationic, anionic, and step-growth polymerizations.
– Understand the stereochemistry of polymerization, and draw structures of atactic, isotactic, and syndiotactic polymers.
– Understand copolymerization, graft polymerization and block polymerization.
– Selsct monomers to form products by olefin metathesis.

Points to Remember:

* Just because a reaction is symmetry-disallowed doesn't mean that it can't occur. Reactions that are symmetry-allowed occur by relatively low-energy, concerted pathways. Reactions that are symmetry-disallowed must take place by higher energy, nonconcerted routes.

* To predict if a reaction is symmetry-allowed, it is only necessary to be concerned with the signs of the outermost lobes.

* The notations in brackets in a sigmatropic rearrangement refer to the positions in the migrating groups to which migration occurs.

*   The stereochemical outcome of a concerted reaction run under thermal conditions is always opposite to the stereochemical outcome of the same reaction run under photochemical conditions.

*   To show the monomer unit of a chain-growth polymer, find the smallest repeating unit, break the polymer bonds, and draw the monomer with its original double bond in place. To show the monomer unit of a step-growth polymer, find the smallest repeating unit, break the polymer bonds, and draw the monomer unit or units with the small molecules that were displaced by polymerization added to the monomer units.

*   Fishhook arrows are used to show movement of single electrons.

Self-test:

$$H_2C=CCO_2CH_2CHCH_3$$

with CN and CH3 substituents, labeled **B**

$$H_2C-CH-CH-CH_2$$ with epoxide oxygens, labeled **C**

What type of reaction is occurring in **A**? Describe it by order and type. If the reaction of the stereoisomer shown proceeds readily, is the reaction being carried out under thermal or photochemical conditions?

Under what conditions would you expect monomer **B** to polymerize? (Actually it polymerizes well under all conditions). Is the polymer a chain-growth or a step-growth polymer? Draw a representative segment of the polymer.

Suggest a use for **C** in polymerizations.

Multiple choice:

1. In which orbitals do the outermost lobes have opposite signs on the same side of the
    π system?
    (a) HOMO in the ground state of a 2 π electron system
    (b) HOMO in the excited state of a 4 π electron system
    (c) LUMO in the excited state of a 6 π electron system
    (d) HOMO in the ground state of a 4 π electron system

2. Which reaction is symmetry-disallowed?
    (a) conrotatory photochemical ring-opening of a 6 π electron system
    (b) suprafacial thermal cycloaddition of a 6 π electron system
    (c) antarafacial thermal sigmatropic rearrangement of a 4 π electron system
    (d) antarafacial photochemical cycloaddition of a 6 π electron system

3. Which of the following reactions is symmetry-allowed but geometrically constrained?
    (a) thermal electrocyclic reaction of a 4 π electron system
    (b) photochemical cycloaddition of a 4 π electron system
    (c) thermal sigmatropic rearrangement of a 4 π electron system
    (d) photochemical electrocyclic reaction of a 4 π electron system

4. All of the following sigmatropic rearrangements involve 6 π electrons except:
    (a) rearrangement of allyl phenyl ether to o-allyl phenol
    (b) rearrangement of 1,5-heptadiene to 3-methyl-1,5-hexadiene
    (c) rearrangement of 1,3,5-heptatriene in which a hydrogen atom migrates across the π
    system
    (d) rearrangement of homotropilidene

5. Consider the 4 π electron thermal electrocyclic reactions of two double-bond stereoisomers.
    All of the following are true except:
    (a) one reaction is concerted and one isn't
    (b) the equilibrium lies on the side of the ringopened product
    (c) the reaction proceeds with conrotatory motion
    (d) The ring-closed products are stereoisomers

6. Which of the following monomers is most likely to undergo cationic polymerization?
    (a) $H_2C=CF_2$
    (b) $H_2C=CH_2$
    (c) formaldehyde
    (d) $H_2C=C(CH_3)_2$

7. Which of the following is not a copolymer?
    (a) Saran
    (b) Nylon 6
    (c) Dacron
    (d) Lexan

8. In which step-growth polymer is an alcohol the by product?
    (a) polyester
    (b) polyamide
    (c) polyurethane
    (d) polycarbonate

9. Which type of polymer has large regions of oriented crystallites and little or no crosslinking?
    (a) a thermoplastic
    (b) a fiber
    (c) an elastomer
    (d) a thermosetting resin

10. A copolymer formed by irradiating a homopolymer in the presence of a second monomer is called a:
    (a) random copolymer
    (b) alternating copolymer
    (c) graft copolymer
    (d) block copolymer

Answers to Practice Your Scientific Reasoning I - VII

Practice Your Scientific Analysis and Reasoning I
The Chiral Drug Thalidomide

1. B
2. D
3 A
4. A
5. A

Practice Your Scientific Analysis and Reasoning II
From Mustard Gas to Alkylating Anticancer Drugs

1. B
2. B
3. A
4. C

Practice Your Scientific Analysis and Reasoning III
Photodynamic Therapy (PDT)

1. A
2. D
3. B
4. B

Practice Your Scientific Analysis and Reasoning IV
Selective Serotonin Reuptake Inhibitors (SSRIs)

1. C
2. C
3. B
4. B
5. B

Practice Your Scientific Analysis and Reasoning V

Thymine in DNA

1.   B
2.   C
3.   A
4.   D
5.   C
6.   C

Practice Your Scientific Analysis and Reasoning VI

Melatonin and Serotonin

1.   C
2.   B
3.   A
4.   A
5.   B
6.   C

Practice Your Scientific Analysis and Reasoning VII

The Potent Antibiotic of Endiandric Acid C

1.   D
2.   A
3.   C
4.   B

Functional-Group Synthesis

The following table summarizes the synthetic methods by which important functional groups can be prepared. The functional groups are listed alphabetically, followed by reference to the appropriate text section and a brief description of each synthetic method.

Acetals, $R_2C(OR')_2$
(Sec. 19.10)   from ketones and aldehydes by acid-catalyzed reaction with alcohols

Acid anhydrides, RCO_2COR'
(Sec. 21.3)   from dicarboxylic acids by heating
(Sec. 21.5)   from acid chlorides by reaction with carboxylate salts

Acid bromides, RCOBr
(Sec. 21.4)   from carboxylic acids by reaction with PBr_3

Acid chlorides, RCOCl
(Sec. 21.3)   from carboxylic acids by reaction with $SOCl_2$

Alcohols, ROH
(Sec. 8.4)   from alkenes by oxymercuration/demercuration
(Sec. 8.5)   from alkenes by hydroboration/oxidation
(Sec. 8.7)   from alkenes by hydroxylation with OsO_4
(Sec. 11.2, 11.3)   from alkyl halides and tosylates by S_N2 reaction with hydroxide ion
(Sec. 18.3)   from ethers by acid-induced cleavage
(Sec. 18.6)   from epoxides by acid-catalyzed ring opening with either H_2O or HX
(Sec. 18.6)   from epoxides by base-induced ring opening
(Sec. 17.4, 19.7)   from ketones and aldehydes by reduction with $NaBH_4$ or $LiAlH_4$
(Sec. 17.5, 19.7)   from ketones and aldehydes by addition of Grignard reagents
(Sec. 21.3)   from carboxylic acids by reduction with either $LiAlH_4$ or BH_3
(Sec. 21.4)   from acid chlorides by reduction with $LiAlH_4$
(Sec. 21.4)   from acid chlorides by reaction with Grignard reagents
(Sec. 21.5)   from acid anhydrides by reduction with $LiAlH_4$
(Sec. 17.4, 21.6)   from esters by reduction with $LiAlH_4$
(Sec. 17.5, 21.6)   from esters by reaction with Grignard reagents

Aldehydes, RCHO
(Sec. 8.8)   from disubstituted alkenes by ozonolysis
(Sec. 8.8)   from 1,2-diols by cleavage with sodium periodate
(Sec. 9.4)   from terminal alkynes by hydroboration followed by oxidation
(Sec. 17.7, 19.2)   from primary alcohols by oxidation
(Sec. 19.2, 21.6)   from esters by reduction with DIBAH [$HAl(i\text{-}Bu)_2$]

Alkanes, RH
(Sec. 8.6)   from alkenes by catalytic hydrogenation
(Sec. 10.5)   from alkyl halides by protonolysis of Grignard reagents
(Sec. 10.7)   from alkyl halides by coupling with Gilman reagents
(Sec. 19.9)   from ketones and aldehydes by Wolff–Kishner reaction

Alkenes, $R_2C=CR_2$

(Sec. 8.1, 11.8)	from alkyl halides by treatment with strong base (E2 reaction)
(Sec. 8.1, 17.6)	from alcohols by dehydration
(Sec. 9.5)	from alkynes by catalytic hydrogenation using the Lindlar catalyst
(Sec. 9.5)	from alkynes by reduction with lithium in liquid ammonia
(Sec. 19.11)	from ketones and aldehydes by treatment with alkylidenetriphenylphosphoranes (Wittig reaction)
(Sec. 22.3)	from α-bromo ketones by heating with pyridine
(Sec. 24.7)	from amines by methylation and Hofmann elimination

Alkynes, $RC\equiv CR$

(Sec. 9.2)	from dihalides by base-induced double dehydrohalogenation
(Sec. 9.8)	from terminal alkynes by alkylation of acetylide anions

Amides, $RCONH_2$

(Sec. 21.3)	from carboxylic acids by heating with ammonia
(Sec. 21.4)	from acid chlorides by treatment with an amine or ammonia
(Sec. 21.5)	from acid anhydrides by treatment with an amine or ammonia
(Sec. 21.6)	from esters by treatment with an amine or ammonia
(Sec. 20.7)	from nitriles by partial hydrolysis with either acid or base
(Sec. 26.7)	from a carboxylic acid and an amine by treatment with dicyclohexylcarbodiimide (DCC)

Amines, RNH_2

(Sec. 19.13)	from conjugated enones by addition of primary or secondary amines
(Sec. 21.7, 24.6)	from amides by reduction with $LiAlH_4$
(Sec. 20.7, 24.6)	from nitriles by reduction with $LiAlH_4$
(Sec. 24.6)	from primary alkyl halides by treatment with ammonia
(Sec. 24.6)	from primary alkyl halides by Gabriel synthesis
(Sec. 24.6)	from primary alkyl azides by reduction with $LiAlH_4$
(Sec. 24.6)	from acid chlorides by Curtius rearrangement of acyl azides
(Sec. 24.6)	from primary amides by Hofmann rearrangement
(Sec. 24.6)	from ketones and aldehydes by reductive amination with an amine and $NaBH_3CN$

Amino Acids, $RCH(NH_2)CO_2H$

(Sec. 26.3)	from α-bromo acids by S_N2 reaction with ammonia
(Sec. 26.3)	from α-keto acids by reductive amination
(Sec. 26.3)	from primary alkyl halides by alkylation with diethyl acetamidomalonate
(Sec. 26.3)	from (Z)-amido acids by enantioselective hydrogenation

Arenes, Ar–R

(Sec. 16.3)	from arenes by Friedel–Crafts alkylation with an alkyl halide
(Sec. 16.10)	from aryl alkyl ketones by catalytic reduction of the keto group
(Sec. 24.8)	from arenediazonium salts by treatment with hypophosphorous acid

Arylamines, Ar–NH$_2$

(Sec. 16.2, 24.6)   from nitroarenes by reduction with either Fe, Sn, or H$_2$/Pd.

Arenediazonium salts, Ar–N$_2^+$ X$^-$

(Sec. 24.8)   from arylamines by reaction with nitrous acid

Arenesulfonic acids Ar–SO$_3$H

(Sec. 16.2)   from arenes by electrophilic aromatic substitution with SO$_3$/H$_2$SO$_4$

Azides, R–N$_3$

(Sec. 11.2, 24.6)   from primary alkyl halides by S$_N$2 reaction with azide ion

Carboxylic acids, RCO$_2$H

(Sec. 8.8)   from mono- and 1,2-disubstituted alkenes by ozonolysis
(Sec. 16.9)   from arenes by side-chain oxidation with Na$_2$Cr$_2$O$_7$ or KMnO$_4$
(Sec. 19.3)   from aldehydes by oxidation
(Sec. 20.5)   from alkyl halides by conversion into Grignard reagents followed by
                    reaction with CO$_2$
(Sec. 20.5, 20.7)   from nitriles by acid or base hydrolysis
(Sec. 21.4)   from acid chlorides by reaction with aqueous base
(Sec. 21.5)   from acid anhydrides by reaction with aqueous base
(Sec. 21.6)   from esters by hydrolysis with aqueous base
(Sec. 21.7)   from amides by hydrolysis with aqueous base

Cyanohydrins, RCH(OH)CN

(Sec. 19.6)   from aldehydes and ketones by reaction with HCN

Cycloalkanes

(Sec. 8.9)   from alkenes by addition of dichlorocarbene
(Sec. 8.9)   from alkenes by reaction with CH$_2$I$_2$ and Zn/Cu (Simmons–Smith
                    reaction)
(Sec. 16.10)   from arenes by hydrogenation

Disulfides, RS–SR'

(Sec. 18.8)   from thiols by oxidation with bromine

Enamines, RCH=CRNR$_2$

(Sec. 19.8)   from ketones or aldehydes by reaction with secondary amines

$$R_2C \overset{O}{\overset{/\backslash}{-}} CR_2$$

Epoxide s ,

(Sec. 8.9, 18.5)   from alkenes by treatment with a peroxyacid
(Sec. 18.5)   from halohydrins by treatment with base

Esters, RCO$_2$R'

(Sec. 21.3)	from carboxylic acid salts by S$_N$2 reaction with primary alkyl halides
(Sec. 21.3)	from carboxylic acids by acid-catalyzed reaction with an alcohol (Fischer esterification)
(Sec. 21.4)	from acid chlorides by base-induced reaction with an alcohol
(Sec. 21.5)	from acid anhydrides by base-induced reaction with an alcohol
(Sec. 22.7)	from alkyl halides by alkylation with diethyl malonate
(Sec. 22.7)	from esters by treatment of their enolate ions with alkyl halides

Ethers, R–O–R'

(Sec. 16.7)	from activated haloarenes by reaction with alkoxide ions
(Sec. 16.8)	from unactivated haloarenes by reaction with alkoxide ions via benzyne intermediates
(Sec. 18.2)	from primary alkyl halides by S$_N$2 reaction with alkoxide ions (Williamson ether synthesis)
(Sec. 18.2)	from alkenes by alkoxymercuration/demercuration
(Sec. 18.5)	from alkenes by epoxidation with peroxyacids

Halides, alkyl, R$_3$C–X

(Sec. 7.7)	from alkenes by electrophilic addition of HX
(Sec. 8.2)	from alkenes by addition of halogen
(Sec. 8.3)	from alkenes by electrophilic addition of hypohalous acid (HOX) to yield halohydrins
(Sec. 9.3)	from alkynes by addition of halogen
(Sec. 9.3)	from alkynes by addition of HX
(Sec. 10.3)	from alkenes by allylic bromination with *N*-bromosuccinimide (NBS)
(Sec. 10.5)	from alcohols by reaction with HX
(Sec. 10.5)	from alcohols by reaction with SOCl$_2$
(Sec. 10.5)	from alcohols by reaction with PBr$_3$
(Sec. 11.2, 11.3)	from alkyl tosylates by S$_N$2 reaction with halide ions
(Sec. 16.9)	from arenes by benzylic bromination with *N*-bromosuccinimide (NBS)
(Sec. 18.3)	from ethers by cleavage with either HX
(Sec. 22.3)	from ketones by α-halogenation with bromine
(Sec. 22.4)	from carboxylic acids by α-halogenation with phosphorus and PBr$_3$ (Hell–Volhard–Zelinskii reaction)

Halides, aryl, Ar–X

(Sec. 16.1, 16.2)	from arenes by electrophilic aromatic substitution with halogen
(Sec. 24.8)	from arenediazonium salts by reaction with cuprous halides (Sandmeyer reaction)

Halohydrins, R$_2$CXC(OH)R$_2$

(Sec. 8.3)	from alkenes by electrophilic addition of hypohalous acid (HOX)
(Sec. 18.6)	from epoxides by acid-induced ring opening with HX

Imines. R₂C=NR'

$\text{Imines. } R_2C=NR'$

(Sec. 19.8)   from ketones or aldehydes by reaction with primary amines

Ketones, $R_2C=O$

(Sec. 8.8)   from alkenes by ozonolysis
(Sec. 8.7)   from 1,2-diols by cleavage reaction with sodium periodate
(Sec. 9.4)   from alkynes by mercuric-ion-catalyzed hydration
(Sec. 9.4)   from alkynes by hydroboration/oxidation
(Sec. 16.3)   from arenes by Friedel–Crafts acylation reaction with an acid chloride
(Sec. 17.7, 19.2) from secondary alcohols by oxidation
(Sec. 19.2, 21.4) from acid chlorides by reaction with lithium diorganocopper (Gilman) reagents
(Sec. 19.13)   from conjugated enones by addition of lithium diorganocopper reagents
(Sec. 20.7)   from nitriles by reaction with Grignard reagents
(Sec. 22.7)   from primary alkyl halides by alkylation with ethyl acetoacetate
(Sec. 22.7)   from ketones by alkylation of their enolate ions with primary alkyl halides

Nitriles, $R–C\equiv N$

(Sec. 11.3, 20.7) from primary alkyl halides by S_N2 reaction with cyanide ion
(Sec. 20.7)   from primary amides by dehydration with $SOCl_2$
(Sec. 22.7)   from nitriles by alkylation of their α-anions with primary alkyl halides
(Sec. 24.8)   from arenediazonium ions by treatment with CuCN

Nitroarenes, $Ar–NO_2$

(Sec. 16.2)   from arenes by electrophilic aromatic substitution with nitric/sulfuric acids

Organometallics, R–M

(Sec. 10.6)   formation of Grignard reagents from organohalides by treatment with magnesium
(Sec. 10.7)   formation of organolithium reagents from organohalides by treatment with lithium
(Sec. 10.7)   formation of lithium diorganocopper reagents (Gilman reagents) from organolithium reagents by treatment with cuprous halides

Phenols, Ar–OH

(Sec. 24.8)   from arenediazonium salts by reaction with Cu_2O and $Cu(NO_3)_2$
(Sec. 16.7)   from aryl halides by nucleophilic aromatic substitution with hydroxide ion

Quinones,

(Sec. 17.10)   from phenols by oxidation with Fremy's salt $[(KSO_3)_2NO]$

Sulfides, R–S–R'

(Sec. 18.8)   from thiols by S_N2 reaction of thiolate ions with primary alkyl halides

Sulfones, R–SO$_2$–R'

(Sec. 18.8)   from sulfides or sulfoxides by oxidation with peroxyacids

Sulfoxides, R–SO–R'

(Sec. 18.8)   from sulfides by oxidation with H$_2$O$_2$

Thiols, R–SH

(Sec. 11.3)   from primary alkyl halides by S$_N$2 reaction with hydrosulfide anion

(Sec. 18.8)   from primary alkyl halides by S$_N$2 reaction with thiourea, followed by
                     hydrolysis

Functional-Group Reactions

The following table summarizes the reactions of important functional groups. The functional groups are listed alphabetically, followed by a reference to the appropriate text section.

Acetal
1. Hydrolysis to yield a ketone or aldehyde plus alcohol (Sec. 19.10)

Acid anhydride
1. Hydrolysis to yield a carboxylic acid (Sec. 21.5)
2. Alcoholysis to yield an ester (Sec. 21.5)
3. Aminolysis to yield an amide (Sec. 21.5)
4. Reduction to yield a primary alcohol (Sec. 21.5)

Acid chloride
1. Friedel–Crafts reaction with an aromatic compound to yield an aryl ketone (Sec. 16.3)
2. Hydrolysis to yield a carboxylic acid (Sec. 21.4)
3. Alcoholysis to yield an ester (Sec. 21.4)
4. Aminolysis to yield an amide (Sec. 21.4)
5. Reduction to yield a primary alcohol (Sec. 21.4)
6. Grignard reaction to yield a tertiary alcohol (Sec. 21.4)
7. Reaction with a lithium diorganocopper reagent to yield a ketone (Sec. 21.4)

Alcohol
1. Acidity (Sec. 17.2)
2. Oxidation (Sec. 17.7)
   a. Reaction of a primary alcohol to yield an aldehyde or acid
   b. Reaction of a secondary alcohol to yield a ketone
3. Reaction with a carboxylic acid to yield an ester (Sec. 21.3)
4. Reaction with an acid chloride to yield an ester (Sec. 21.4)
5. Reaction with an acid anhydride to yield an ester (Sec. 21.5)
6. Dehydration to yield an alkene (Sec. 17.6)
7. Reaction with a primary alkyl halide to yield an ether (Sec. 18.2)
8. Conversion into an alkyl halide (Sec. 17.6)
   a. Reaction of a tertiary alcohol with HX
   b. Reaction of a primary or secondary alcohol with $SOCl_2$
   c. Reaction of a primary or secondary alcohol with PBr_3

Aldehyde
1. Oxidation to yield a carboxylic acid (Sec. 19.3)
2. Nucleophilic addition reactions
   a. Reduction to yield a primary alcohol (Secs. 17.4, 19.7)
   b. Reaction with a Grignard reagent to yield a secondary alcohol (Secs. 17.5, 19.7)
   c. Grignard reaction of formaldehyde to yield a primary alcohol (Sec. 17.5)
   d. Reaction with HCN to yield a cyanohydrin (Sec. 19.6)
   e. Wolff–Kishner reaction with hydrazine to yield an alkane (Sec. 19.9)
   f. Reaction with an alcohol to yield an acetal (Sec. 19.10)
   g. Wittig reaction to yield an alkene (Sec. 19.11)

h.   Reaction with an amine to yield an imine or enamine (Sec. 19.8)
3. Aldol reaction to yield a β-hydroxy aldehyde (Sec. 23.1)
4. Alpha bromination of an aldehyde (Sec. 22.3)

Alkane

1. Radical halogenation to yield an alkyl halide (Secs. 6.3, 10.3)

Alkene

1. Electrophilic addition of HX to yield an alkyl halide (Secs. 7.7–7.11)
    Markovnikov regiochemistry is observed.
2. Electrophilic addition of halogen to yield a 1,2-dihalide (Sec. 8.2)
3. Oxymercuration/demercuration to yield an alcohol (Sec. 8.4)
    Markovnikov regiochemistry is observed, yielding the more highly substituted alcohol.
4. Hydroboration/oxidation to yield an alcohol (Section 8.5)
5. Hydrogenation to yield an alkane (Sec. 8.6)
6. Hydroxylation to yield a 1,2-diol (Sec. 8.7)
7. Oxidative cleavage to yield carbonyl compounds (Sec. 8.8)
8. Simmons–Smith reaction with CH_2I_2 to yield a cyclopropane (Sec. 8.9)
9. Reaction with dichlorocarbene to yield a dichlorocyclopropane (Sec. 8.9)
10. Allylic bromination with NBS (Sec. 10.4)
11. Alkoxymercuration to yield an ether (Sec. 18.2)
12. Reaction with a peroxyacid to yield an epoxide (Secs. 8.7, 18.5)

Alkyne

1. Electrophilic addition of HX to yield a vinylic halide (Sec. 9.3)
2. Electrophilic addition of halogen to yield a dihalide (Sec. 9.3)
3. Mercuric-sulfate-catalyzed hydration to yield a methyl ketone (Sec. 9.4)
4. Hydroboration/oxidation to yield an aldehyde (Sec. 9.4)
5. Alkylation of an alkyne anion (Sec. 9.8)
6. Reduction (Sec. 9.5)
    a.   Hydrogenation over Lindlar catalyst to yield a cis alkene
    b.   Reduction with Li/NH_3 to yield a trans alkene

Amide

1. Hydrolysis to yield a carboxylic acid (Sec. 21.7)
2. Reduction with $LiAlH_4$ to yield an amine (Sec. 21.7)
3. Dehydration to yield a nitrile (Section 20.7)

Amine

1. Basicity (Sec. 24.3)
2. S_N2 alkylation of an alkyl halide to yield an amine (Sec. 24.6)
3. Nucleophilic acyl substitution reactions
    a.   Reaction with an acid chloride to yield an amide (Sec. 21.4)
    b.   Reaction with an acid anhydride to yield an amide (Sec. 21.5)
4. Hofmann elimination to yield an alkene (Sec. 24.7)
5. Formation of an arenediazonium salt (Sec. 24.8)

Arene

1. Oxidation of an alkylbenzene side chain to yield a benzoic acid (Sec. 16.9)
2. Catalytic reduction to yield a cyclohexane (Sec. 16.10)
3. Reduction of an aryl alkyl ketone to yield an arene (Sec. 16.10)
4. Electrophilic aromatic substitution (Secs. 16.1–16.3)
   a. Halogenation (Secs. 16.1–16.2)
   b. Nitration (Sec. 16.2)
   c. Sulfonation (Sec. 16.2)
   d. Friedel–Crafts alkylation (Sec. 16.3)
      Aromatic ring must be at least as reactive as a halobenzene
   e. Friedel–Crafts acylation (Sec. 16.3)

Arenediazonium salt

1. Conversion into an aryl chloride (Sec. 24.8)
2. Conversion into an aryl bromide (Sec. 24.8)
3. Conversion into an aryl iodide (Sec. 24.8)
4. Conversion into an aryl cyanide (Sec. 24.8)
5. Conversion into a phenol (Sec. 24.8)
6. Conversion into an arene (Sec. 24.8)

Carboxylic acid

1. Acidity (Secs. 20.2–20.4)
2. Reduction to yield a primary alcohol (Secs. 17.4, 21.3)
   a. Reduction with $LiAlH_4$
   b. Reduction with BH_3
3. Nucleophilic acyl substitution reactions (Sec. 21.3)
   a. Conversion into an acid chloride
   b. Conversion into an acid anhydride
   c. Conversion into an ester
      (1) Fischer esterification
      (2) S_N2 reaction with an alkyl halide
4. Alpha bromination (Hell–Volhard–Zelinskii reaction) (Sec. 22.4)

Diene

1. Conjugate addition of HX and X_2 (Sec. 14.2)
2. Diels–Alder reaction (Secs. 14.4, 14.5, 30.5)

Epoxide

1. Acid-catalyzed ring opening with HX to yield a halohydrin (Sec. 18.6)
2. Ring opening with aqueous acid to yield a 1,2-diol (Sec. 18.6)

Ester

1. Hydrolysis to yield a carboxylic acid (Sec. 21.6)
2. Aminolysis to yield an amide (Sec. 21.6)
3. Reduction to yield a primary alcohol (Secs. 17.4, 21.6)
4. Partial reduction with DIBAH to yield an aldehyde (Sec. 21.6)

5. Grignard reaction to yield a tertiary alcohol (Secs. 17.5, 21.6)
6. Claisen condensation to yield a β-keto ester (Sec. 23.7)

Ether

1. Acid-induced cleavage to yield an alcohol and an alkyl halide (Sec. 18.3)
2. Claisen rearrangement of an allyl aryl ether to yield an o-allyl phenol (Secs. 18.4, 30.8)

Halide, alkyl

1. Reaction with magnesium to form a Grignard reagent (Sec. 10.6)
2. Reduction to yield an alkane (Sec. 10.6)
3. Coupling with a diorganocopper reagent to yield an alkane (Sec. 10.7)
4. Reaction with an alcohol to yield an ether (Sec. 18.2)
5. Nucleophilic substitution (S_N1 or S_N2) (Secs. 11.1–11.5)
6. Dehydrohalogenation to yield an alkene (E1 or E2) (Secs. 11.7–11.10)

Halohydrin

1. Conversion into an epoxide (Sec. 18.5)

Ketone

1. Nucleophilic addition reactions
   a. Reduction to yield a secondary alcohol (Secs. 17.4, 19.7)
   b. Reaction with a Grignard reagent to yield a tertiary alcohol (Secs. 17.5, 19.7)
   c. Wolff–Kishner reaction with hydrazine to yield an alkane (Sec. 19.9)
   d. Reaction with HCN to yield a cyanohydrin (Sec. 19.6)
   e. Reaction with an alcohol to yield an acetal (Sec. 19.10)
   f. Wittig reaction to yield an alkene (Sec. 19.11)
   g. Reaction with an amine to yield an imine or enamine (Sec. 19.8)
2. Aldol reaction to yield a β-hydroxy ketone (Sec. 23.1)
3. Alpha bromination (Sec. 22.3)

Nitrile

1. Hydrolysis to yield a carboxylic acid (Secs. 20.5, 20.7)
2. Reduction to yield a primary amine (Sec. 20.7)
3. Reaction with a Grignard reagent to yield a ketone (Sec. 20.7)

Nitroarene

1. Reduction to yield an arylamine (Secs. 16.2, 24.6)

Organometallic reagent

1. Reduction by treatment with acid to yield an alkane (Sec. 10.6)
2. Nucleophilic addition to a carbonyl compound to yield an alcohol (Secs. 17.5, 19.7)
3. Conjugate addition of a lithium diorganocopper to an α,β-unsaturated ketone (Sec. 19.13)
4. Coupling reaction of a lithium diorganocopper reagent with an alkyl halide to yield analkane (Sec. 10.7)

5. Coupling reaction of a lithium diorganocopper with an acid chloride to yield a ketone (Sec. 21.4)
6. Reaction with carbon dioxide to yield a carboxylic acid (Sec. 20.5)

Phenol
1. Acidity (Sec. 17.2)
2. Reaction with an acid chloride to yield an ester (Sec. 21.4)
3. Reaction with an alkyl halide to yield an ether (Sec. 18.2)
4. Oxidation to yield a quinone (Sec. 17.10)

Quinone
1. Reduction to yield a hydroquinone (Sec. 17.10)

Sulfide
1 Reaction with an alkyl halide to yield a sulfonium salt (Sec. 18.8)
2. Oxidation to yield a sulfoxide (Sec. 18.8)
3. Oxidation to yield a sulfone (Sec. 18.8)

Thiol
1. Reaction with an alkyl halide to yield a sulfide (Sec. 18.8)
2. Oxidation to yield a disulfide (Sec. 18.8)

Reagents in Organic Chemistry

The following table summarizes the uses of some important reagents in organic chemistry. The reagents are listed alphabetically, followed by a brief description of the uses of each and references to the appropriate text sections.

Acetic acid, CH_3CO_2H: Used as a solvent for the reduction of ozonides with zinc (Section 8.8) and the α-bromination of ketones and aldehydes with Br_2 (Section 22.3).

Acetic anhydride, $(CH_3CO)_2O$: Reacts with alcohols to yield acetate esters (Sections 21.5 and 25.6) and with amines to yield acetamides (Section 21.5).

Aluminum chloride, $AlCl_3$: Acts as a Lewis acid catalyst in Friedel–Crafts alkylation and acylation reactions of aromatic compounds (Section 16.3).

Ammonia, NH_3: Used as a solvent for the reduction of alkynes by lithium metal to yield trans alkenes (Section 9.5).
- Reacts with acid chlorides and acid anhydrides to yield amides (Sections 21.4 and 21.5).

Borane, BH_3: Adds to alkenes, giving alkylboranes that can be oxidized with alkaline H_2O_2 to yield alcohols (Section 8.5).
- Adds to alkynes, giving vinylic organoboranes that can be oxidized with H_2O_2 to yield aldehydes (Section 9.4).
- Reduces carboxylic acids to yield primary alcohols (Section 21.3).

Bromine, Br_2: Adds to alkenes, yielding 1,2-dibromides (Sections 8.2, 14.2).
- Adds to alkynes yielding either 1,2-dibromoalkenes or 1,1,2,2-tetrabromoalkanes (Section 9.3).
- Reacts with arenes in the presence of $FeBr_3$ catalyst to yield bromoarenes (Section 16.1).
- Reacts with ketones in acetic acid solvent to yield α-bromo ketones (Section 22.3).
- Reacts with carboxylic acids in the presence of PBr_3 to yield α-bromo carboxylic acids (Hell–Volhard–Zelinskii reaction; Section 22.4).
- Oxidizes aldoses to yield aldonic acids (Section 25.6).

N-Bromosuccinimide (NBS), $(CH_2CO)_2NBr$: Reacts with alkenes in the presence of aqueous dimethylsulfoxide to yield bromohydrins (Section 8.3).
- Reacts with alkenes in the presence of light to yield allylic bromides (Section 10.3).
- Reacts with alkylbenzenes in the presence of light to yield benzylic bromides; (Section 16.9).

Di-*tert*-butoxy dicarbonate, $(\underline{t}\text{-BuOCO})_2O$: Reacts with amino acids to give *t*-Boc protected amino acids suitable for use in peptide synthesis (Section 26.7).

Butyllithium, CH₃CH₂CH₂CH₂Li: A strong base; reacts with alkynes to yield acetylide anions, which can be alkylated (Section 9.8).
- Reacts with dialkylamines to yield lithium dialkylamide bases such as LDA [lithium diisopropylamide] (Section 22.5).
- Reacts with alkyltriphenylphosphonium salts to yield alkylidenephosphoranes (Wittig reagents (Section 19.11).

Carbon dioxide, CO₂: Reacts with Grignard reagents to yield carboxylic acids (Section 20.5).

Chlorine, Cl₂: Adds to alkenes to yield 1,2-dichlorides (Sections 8.2 and 14.2).
- Reacts with alkanes in the presence of light to yield chloroalkanes by a radical chain reaction pathway (Section 10.2).
- Reacts with arenes in the presence of FeCl₃ catalyst to yield chloroarenes (Section 16.2).

***m*-Chloroperoxybenzoic acid, *m*-ClC₆H₄CO₃H:** Reacts with alkenes to yield epoxides (Sections 8.7, 18.5).

Chlorotrimethylsilane, (CH₃)₃SiCl: Reacts with alcohols to add the trimethylsilyl protecting group (Section 17.8).

Chromium trioxide, CrO₃: Oxidizes alcohols in aqueous acid to yield carbonyl-containing products. Primary alcohols yield carboxylic acids, and secondary alcohols yield ketones (Sections 17.7 and 19.3).

Cuprous bromide, CuBr: Reacts with arenediazonium salts to yield bromoarenes (Sandmeyer reaction; Section 24.8).

Cuprous chloride, CuCl: Reacts with arenediazonium salts to yield chloroarenes (Sandmeyer reaction; Section 24.8).

Cuprous cyanide, CuCN: Reacts with arenediazonium salts to yield substituted benzonitriles (Sandmeyer reaction; Section 24.8).

Cuprous iodide, CuI: Reacts with organolithiums to yield lithium diorganocopper reagents (Gilman reagents; Section 10.7).

Cuprous oxide, Cu₂O: Reacts with arenediazonium salts to yield phenols (Section 24.8).

Dess–Martin periodinane, C₇H₄IO₂(OAc)₄: Oxidizes primary alcohols to aldehydes (Sections 17.7 and 19.2)

Dichloroacetic acid, Cl₂CHCO₂H: Cleaves DMT protecting groups in DNA synthesis (Section 28.7).

Dicyclohexylcarbodiimide (DCC), C_6H_{11}–N=C=N–C_6H_{11}:   Couples an amine with a carboxylic acid to yield an amide. DCC is often used in peptide synthesis (Section 26.7).

Diethyl acetamidomalonate, $CH_3CONHCH(CO_2Et)_2$:   Reacts with alkyl halides in a common method of α-amino acid synthesis (Section 26.3).

Diiodomethane, CH_2I_2: Reacts with alkenes in the presence of zinc–copper couple to yield cyclopropanes (Simmons–Smith reaction; Section 8.9).

Diisobutylaluminum hydride (DIBAH), (i-Bu)$_2$AlH: Reduces esters to yield aldehydes (Sections 19.2 and 21.6).

2,4-Dinitrophenylhydrazine, 2,4-$(NO_2)_2C_6H_3NHNH_2$:   Reacts with aldehydes and ketones to yield 2,4-DNPs that serve as useful crystalline derivatives (Section 19.8).

Ethylene glycol, $HOCH_2CH_2OH$:   Reacts with ketones or aldehydes in the presence of an acid catalyst to yield acetals that serve as useful carbonyl protecting groups (Section 19.10).

F-TEDA-BF$_4$, $C_7H_{14}B_2ClF_9N_2$: Fluorinates an aromatic ring (Section 16.2).

Ferric bromide, $FeBr_3$: Acts as a catalyst for the reaction of arenes with Br_2 to yield bromoarenes (Section 16.1).

Ferric chloride, $FeCl_3$: Acts as a catalyst for the reaction of arenes with Cl_2 to yield chloroarenes (Section 16.2).

Grignard reagent, RMgX: Reacts with acids to yield alkanes (Section 10.6).
– Adds to carbonyl-containing compounds (ketones, aldehydes, esters) to yield alcohols (Sections 17.6 and 19.7).
– Adds to nitriles to yield ketones (Section 20.9).

Hydrazine, H_2NNH_2:   Reacts with ketones or aldehydes in the presence of KOH to yield the corresponding alkanes (Wolff–Kishner reaction; Section 19.19).

Hydrogen bromide, HBr: Adds to alkenes with Markovnikov regiochemistry to yield alkyl bromides (Sections 7.7 and 14.2).
– Adds to alkynes to yield either bromoalkenes or 1,1-dibromoalkanes (Section 9.3).
– Reacts with alcohols to yield alkyl bromides (Sections 10.5 and 17.6).
– Cleaves ethers to yield alcohols and alkyl bromides (Section 18.3).

Hydrogen chloride, HCl: Adds to alkenes with Markovnikov regiochemistry to yield alkyl chlorides (Sections 6.7 and 14.2).
– Adds to alkynes to yield either chloroalkenes or 1,1-dichloroalkanes (Section 8.3).
– Reacts with alcohols to yield alkyl chlorides (Sections 10.6 and 17.6).

Hydrogen cyanide, HCN: Adds to ketones and aldehydes to yield cyanohydrins (Section 19.6).

Hydrogen iodide, HI: Reacts with alcohols to yield alkyl iodides (Section 17.6).
− Cleaves ethers to yield alcohols and alkyl iodides (Section 18.3).

Hydrogen peroxide, H_2O_2: Oxidizes organoboranes to yield alcohols. Used in conjunction with addition of borane to alkenes, the overall transformation effects syn Markovnikov addition of water to an alkene (Section 8.5).
− Oxidizes vinylic boranes to yield aldehydes.(Section 9.4).
− Oxidizes sulfides to yield sulfoxides (Section 18.8).

Hydroxylamine, NH_2OH: Reacts with ketones and aldehydes to yield oximes (Section 19.8).
− Reacts with aldoses to yield oximes as the first step in the Wohl degradation of aldoses (Section 25.6).

Hypophosphorous acid, H_3PO_2:   Reacts with arenediazonium salts to yield arenes (Section 24.8).

Iodine, I_2: Reacts with arenes in the presence of CuCl or H_2O_2 to yield iodoarenes (Section 16.2).

Iodomethane, CH_3I: Reacts with alkoxide anions to yield methyl ethers (Section 18.2).
− Reacts with carboxylate anions to yield methyl esters (Section 21.6).
− Reacts with enolate ions to yield α-methylated carbonyl compounds (Section 22.7).
− Reacts with amines to yield methylated amines (Section 24.6).

Iron, Fe: Reacts with nitroarenes in the presence of aqueous acid to yield anilines (Section 24.6).

Lindlar catalyst: Acts as a catalyst for the partial hydrogenation of alkynes to yield cis alkenes (Section 9.5).

Lithium, Li: Reduces alkynes in liquid ammonia solvent to yield trans alkenes (Section 9.5).
− Reacts with organohalides to yield organolithium compounds (Section 10.7).

Lithium aluminum hydride, $LiAlH_4$: Reduces ketones, aldehydes, esters, and carboxylic acids to yield alcohols (Sections 17.4, 19.7, and 20.6).
− Reduces amides to yield amines (Section 21.7).
− Reduces alkyl azides to yield amines (Section 24.6).
− Reduces nitriles to yield amines (Sections 20.7 and 24.6).

Lithium diisopropylamide (LDA), $LiN(i\text{-}Pr)_2$:   Reacts with carbonyl compounds (aldehydes, ketones, esters) to yield enolate ions (Sections 22.5 and 22.7).

Lithium diorganocopper reagent (Gilman reagent), LiR₂Cu: Couples with alkyl halides to yield alkanes (Section 10.7).
– Adds to α,β-unsaturated ketones to give 1,4-addition products (Section 19.13).
– Reacts with acid chlorides to give ketones (Section 21.4).

Magnesium, Mg: Reacts with organohalides to yield Grignard reagents (Section 10.6).

Mercuric acetate, Hg(O₂CCH₃)₂: Adds to alkenes in the presence of water, giving α-hydroxy organomercury compounds that can be reduced with NaBH₄ to yield alcohols. The overall effect is the Markovnikov hydration of an alkene (Section 8.4).

Mercuric sulfate, HgSO₄: Acts as a catalyst for the addition of water to alkynes in the presence of aqueous sulfuric acid, yielding ketones (Section 9.4).

Mercuric trifluoroacetate, Hg(O₂CCF₃)₂: Adds to alkenes in the presence of alcohol, giving α-alkoxy organomercury compounds that can be reduced with NaBH₄ to yield ethers. The overall reaction effects a net addition of an alcohol to an alkene (Section 18.2).

Nitric acid, HNO₃: Reacts with arenes in the presence of sulfuric acid to yield nitroarenes (Section 16.2).
– Oxidizes aldoses to yield aldaric acids (Section 25.6).

Nitrous acid, HNO₂: Reacts with amines to yield diazonium salts (Section 24.8).

Osmium tetroxide, OsO₄: Adds to alkenes to yield 1,2-diols (Section 8.7).
– Reacts with alkenes in the presence of periodic acid to cleave the carbon–carbon double bond, yielding ketone or aldehyde fragments (Section 8.8).

Ozone, O₃: Adds to alkenes to cleave the carbon–carbon double bond and give ozonides, which can be reduced with zinc in acetic acid to yield carbonyl compounds (Section 8.8).

Palladium on barium sulfate, Pd/BaSO₄: Acts as a hydrogenation catalyst for nitriles in the Kiliani–Fischer chain-lengthening reaction of carbohydrates (Section 25.6).

Palladium on carbon, Pd/C: Acts as a hydrogenation catalyst for reducing carbon–carbon multiple bonds. Alkenes and alkynes are reduced to yield alkanes (Sections 8.6 and 9.5).
– Acts as a hydrogenation catalyst for reducing aryl ketones to yield alkylbenzenes (Section 16.10).
– Acts as a hydrogenation catalyst for reducing nitroarenes to yield anilines (Section 24.6).

Periodic acid, HIO₄: Reacts with 1,2-diols to yield carbonyl-containing cleavage products (Section 8.8).

Peroxyacetic acid, CH₃CO₃H: Oxidizes sulfoxides to yield sulfones (Section 18.8)

Phenylisothiocyanate, C₆H₅–N=C=S: Used in the Edman degradation of peptides to identify N-terminal amino acids (Section 26.6).

Phosphorus oxychloride, POCl₃: Reacts with secondary and tertiary and alcohols to yield alkene dehydration products (Section 17.6).

Phosphorus tribromide, PBr₃: Reacts with alcohols to yield alkyl bromides (Section 10.5).
- Reacts with carboxylic acids to yield acid bromides (Section 21.4).
- Reacts with carboxylic acids in the presence of bromine to yield α-bromo carboxylic acids (Hell–Volhard–Zelinskii reaction; Section 22.4).

Platinum oxide (Adams catalyst), PtO₂: Acts as a hydrogenation catalyst in the reduction of alkenes and alkynes to yield alkanes (Sections 8.6 and 9.5).

Potassium hydroxide, KOH: Reacts with alkyl halides to yield alkenes by an elimination reaction (Sections 8.1 and 11.7).
- Reacts with 1,1- or 1,2-dihaloalkanes to yield alkynes by a twofold elimination reaction (Section 9.2).

Potassium nitrosodisulfonate (Fremy's salt), K(SO₃)₂NO: Oxidizes phenols to yield quinones (Section 17.10).

Potassium permanganate, KMnO₄: Oxidizes alkenes under neutral or acidic conditions to give carboxylic acid double-bond cleavage products (Sections 8.9).
- Oxidizes alkynes to give carboxylic acid triple-bond cleavage products (Section 9.6).
- Oxidizes aromatic side chains to yield benzoic acids (Section 16.9).

Potassium phthalimide, C₆H₄(CO)₂NK: Reacts with alkyl halides to yield *N*-alkylphthalimides, which are hydrolyzed by aqueous sodium hydroxide to yield amines (Gabriel amine synthesis; Section 24.6).

Potassium *tert*-butoxide, KO-*t*-Bu: Reacts with alkyl halides to yield alkenes (Sections 11.7 and 11.8).
- Reacts with allylic halides to yield conjugated dienes (Section 14.1).
- Reacts with chloroform in the presence of an alkene to yield a dichlorocyclopropane (Section 8.9).

Pyridine, C₅H₅N: Acts as a basic catalyst for the reaction of alcohols with acid chlorides to yield esters (Section 21.4).
- Acts as a basic catalyst for the reaction of alcohols with acetic anhydride to yield acetate esters (Section 21.5).
- Reacts with α-bromo ketones to yield α,β-unsaturated ketones (Section 22.3).

Pyrrolidine, C₄H₈N: Reacts with ketones to yield enamines for use in the Stork enamine reaction (Sections 19.8 and 23.11).

Rhodium on carbon, Rh/C: Acts as a hydrogenation catalyst in the reduction of benzene rings to yield cyclohexanes (Section 16.10).

Silver oxide, Ag$_2$O: Catalyzes the reaction of monosaccharides with alkyl halides to yield ethers (Section 25.6).

– Reacts with tetraalkylammonium salts to yield alkenes (Hofmann elimination; Section 24.7).

Sodium amide, NaNH$_2$: Reacts with terminal alkynes to yield acetylide anions (Section 9.7).

– Reacts with 1,1- or 1,2-dihalides to yield alkynes by a twofold elimination reaction (Section 9.2).

– Reacts with aryl halides to yield anilines by a benzyne aromatic substitution mechanism (Section 16.8).

Sodium azide, NaN$_3$:   Reacts with alkyl halides to yield alkyl azides (Section 24.6).

– Reacts with acid chlorides to yield acyl azides. On heating in the presence of water, acyl azides yield amines and carbon dioxide (Section 24.6).

Sodium bisulfite, NaHSO$_3$: Reduces osmate esters, prepared by treatment of an alkene with osmium tetroxide, to yield 1,2-diols (Section 8.7).

Sodium borohydride, NaBH$_4$:   Reduces organomercury compounds, prepared by oxymercuration of alkenes, to convert the C–Hg bond to C–H (Section 8.5).

– Reduces ketones and aldehydes to yield alcohols (Sections 17.4 and 19.7).

– Reduces quinones to yield hydroquinones (Section 17.10).

Sodium cyanide, NaCN: Reacts with alkyl halides to yield alkanenitriles (Sections 20.6 and 20.7).

Sodium dichromate, Na$_2$Cr$_2$O$_7$:   Oxidizes primary alcohols to yield carboxylic acids and secondary alcohols to yield ketones (Sections 17.7 and 19.2).

– Oxidizes alkylbenzenes to yield benzoic acids (Section 16.9).

Sodium hydride, NaH: Reacts with alcohols to yield alkoxide anions (Section 17.2).

Sodium hydroxide, NaOH: Reacts with aryl halides to yield phenols by a benzyne aromatic substitution mechanism (Section 16.8).

Sodium iodide, NaI: Reacts with arenediazonium salts to yield aryl iodides (Section 24.8).

Stannous chloride, SnCl$_2$:   Reduces nitroarenes to yield anilines (Sections 16.2 and 24.6).

– Reduces quinones to yield hydroquinones (Section 17.10).

Sulfur trioxide, SO$_3$:   Reacts with arenes in sulfuric acid solution to yield arenesulfonic acids (Section 16.2).

Sulfuric acid, H$_2$SO$_4$:   Reacts with alcohols and water to yield alkenes (Section 8.4).
– Reacts with alkynes in the presence of water and mercuric sulfate to yield ketones (Section 9.4).
– Catalyzes the reaction of nitric acid with aromatic rings to yield nitroarenes (Section 16.2).
– Catalyzes the reaction of SO$_3$ with aromatic rings to yield arenesulfonic acids (Section 16.2).

Tetrazole: Acts as a coupling reagent for use in DNA synthesis (Section 28.7).

Thionyl chloride, SOCl$_2$:   Reacts with primary and secondary alcohols to yield alkyl chlorides (Section 10.5).
– Reacts with carboxylic acids to yield acid chlorides (Section 21.4).

Thiourea, H$_2$NCSNH$_2$:   Reacts with primary alkyl halides to yield thiols (Section 18.8).

p-Toluenesulfonyl chloride, p-CH$_3$C$_6$H$_4$SO$_2$Cl: Reacts with alcohols to yield tosylates (Sections 11.1 and 17.6).

Trifluoroacetic acid, CF$_3$CO$_2$H: Acts as a catalyst for cleaving *tert*-butyl ethers, yielding alcohols and 2-methylpropene (Section 18.3).
– Acts as a catalyst for cleaving the *t*-Boc protecting group from amino acids in peptide synthesis (Section 26.7).

Triphenylphosphine, (C$_6$H$_5$)$_3$P: Reacts with primary alkyl halides to yield the alkyltriphenylphosphonium salts used in Wittig reactions (Section 19.11).

Zinc, Zn: Reduces ozonides, produced by addition of ozone to alkenes, to yield ketones and aldehydes (Section 8.8).
– Reduces disulfides to yield thiols (Section 18.8).

Zinc–copper couple, Zn–Cu: Reacts with diiodomethane in the presence of alkenes to yield cyclopropanes (Simmons–Smith reaction; Section 8.9).

Name Reactions

Acetoacetic ester synthesis (Section 22.7): a multistep reaction sequence for converting a primary alkyl halide into a methyl ketone having three more carbon atoms in the chain.

$$RCH_2X + CH_3\!-\!\overset{O}{\overset{\|}{C}}\!-\!\overset{..}{\underset{}{C}}H\!-\!\overset{O}{\overset{\|}{C}}\!-\!OCH_3 \quad \xrightarrow[\text{2. }H_3O^+,\text{ heat}]{\text{1. Heat}} \quad RCH_2\!-\!CH_2\overset{O}{\overset{\|}{C}}CH_3 + CO_2 + CH_3OH$$

Adams' catalyst (Section 8.6): PtO_2, a catalyst used for the hydrogenation of carbon–carbon double bonds.

Aldol condensation reaction (Section 23.1): the nucleophilic addition of an enol or enolate ion to a ketone or aldehyde, yielding a β-hydroxy ketone.

$$2\ R\!-\!\overset{O}{\overset{\|}{C}}\!-\!\overset{|}{\underset{|}{C}}\!-\!H \quad \xrightarrow{\text{NaOH}} \quad R\!-\!\overset{O}{\overset{\|}{C}}\!-\!\overset{|}{\underset{R}{C}}\!-\!\overset{OH}{\overset{|}{C}}\!-\!\overset{|}{\underset{|}{C}}\!-\!H$$

Amidomalonate amino acid synthesis (Section 26.3): a multistep reaction sequence, similar to the malonic ester synthesis, for converting a primary alkyl halide into an amino acid.

$$RCH_2X + \ ^-\!\!:\!\!\underset{NHAc}{\overset{|}{C}}(CO_2Et)_2 \quad \xrightarrow[\text{2. }H_3O^+,\text{ heat}]{\text{1. mix}} \quad RCH_2\!-\!\underset{NH_2}{\overset{|}{C}}H\overset{O}{\overset{\|}{C}}OH + CO_2 + 2\ EtOH$$

Benedict's test (Section 25.6): a chemical test for aldehydes, involving treatment with cupric ion in aqueous sodium citrate.

Cannizzaro reaction (Section 19.12): the disproportionation reaction that occurs when a nonenolizable aldehyde is treated with base.

$$2\ R_3C\overset{O}{\overset{\|}{C}}H \quad \xrightarrow[\text{2. }H_3O^+]{\text{1. }HO^-} \quad R_3C\overset{O}{\overset{\|}{C}}OH + R_3C\overset{}{C}H_2OH$$

Claisen condensation reaction (Section 23.7): a nucleophilic acyl substitution reaction that occurs when an ester enolate ion attacks the carbonyl group of a second ester molecule. The product is a β-keto ester.

$$2\ R\!-\!CH_2\!-\!\overset{O}{\overset{\|}{C}}\!-\!OCH_3 \quad \xrightarrow[\text{2. }H_3O^+]{\text{1. }HO^-} \quad R\!-\!CH_2\!-\!\overset{O}{\overset{\|}{C}}\!-\!\underset{R}{\overset{|}{C}}H\!-\!\overset{O}{\overset{\|}{C}}\!-\!OCH_3 + CH_3OH$$

Claisen rearrangement (Sections 18.4 and 30.8): the thermal [3.3] sigmatropic rearrangement of an allyl vinyl ether or an allyl phenyl ether.

Cope rearrangement (Section 30.8): the thermal [3.3] sigmatropic rearrangement of a 1,5-diene to a new 1,5-diene.

Curtius rearrangement (Section 24.6): the thermal rearrangement of an acyl azide to an isocyanate, followed by hydrolysis to yield an amine.

$$R-\overset{\overset{\displaystyle O}{\|}}{C}-N=\overset{+}{N}=\overset{-}{N} \quad \xrightarrow[\text{2. } H_2O]{\text{1. heat}} \quad RNH_2 + CO_2 + N_2$$

Diazonium coupling reaction (Section 24.8): the coupling reaction between an aromatic diazonium salt and a phenol or aniline.

Dieckmann reaction (Section 23.9): the intramolecular Claisen condensation reaction of a 1,6- or 1,7-diester, yielding a cyclic β-keto ester.

Diels–Alder cycloaddition reaction (Sections 14.4–14.5 and 30.5): the reaction between a diene and a dienophile to yield a cyclohexene ring.

Edman degradation (Section 26.6): a method for cleaving the N-terminal amino acid from a peptide by treatment of the peptide with *N*-phenylisothiocyanate.

Fehling's test (Section 25.6): a chemical test for aldehydes, involving treatment with cupric ion in aqueous sodium tartrate.

Fischer esterification reaction (Section 21.3): the acid-catalyzed reaction between a carboxylic acid and an alcohol, yielding the ester.

Friedel–Crafts reaction (Section 16.3): the alkylation or acylation of an aromatic ring by treatment with an alkyl- or acyl chloride in the presence of a Lewis-acid catalyst.

Gabriel amine synthesis (Section 24.6): a multistep sequence for converting a primary alkyl halide into a primary amine by alkylation with potassium phthalimide, followed by hydrolysis.

Gilman reagent (Section 10.7): a lithium dialkylcopper reagent, R_2CuLi, prepared by treatment of a cuprous salt with an alkyllithium. Gilman reagents undergo a coupling reaction with alkyl halides, a 1,4-addition reaction with α,β-unsaturated ketones, and a coupling reaction with acid chlorides to yield ketones.

Glycal assembly method (Section 25.9): a method of polysaccharide synthesis in which a glycal is converted into its epoxide, which is then opened by reaction with an alcohol.

Grignard reaction (Section 19.7): the nucleophilic addition reaction of an alkylmagnesium halide to a ketone, aldehyde, or ester carbonyl group.

Grignard reagent (Section 10.6): an organomagnesium halide, RMgX, prepared by reaction between an organohalide and magnesium metal. Grignard reagents add to carbonyl compounds to yield alcohols.

Hell–Volhard–Zelinskii reaction (Section 22.4): the α-bromination of a carboxylic acid by treatment with bromine and phosphorus tribromide.

Hofmann elimination (Section 24.7): a method for effecting the elimination reaction of an amine to yield an alkene. The amine is first treated with excess iodomethane, and the resultant quaternary ammonium salt is heated with silver oxide.

Hofmann rearrangement (Section 24.6): the rearrangement of an N-bromoamide to a primary amine by treatment with aqueous base.

Kiliani–Fischer synthesis (Section 25.6): a multistep sequence for chain-lengthening an aldose into the next higher homolog.

Knowles amino acid synthesis (Section 26.3): an enantioselective method of amino acid synthesis by hydrogenation of a Z enamido acid with a chiral hydrogenation catalyst.

Koenigs–Knorr reaction (Section 25.6): a method for synthesizing a glycoside by reaction of a pyranosyl bromide with an alcohol and Ag_2O.

Malonic ester synthesis (Section 22.7): a multistep sequence for converting an alkyl halide into a carboxylic acid with the addition of two carbon atoms to the chain.

McLafferty rearrangement (Section 19.14): a mass spectral fragmentation pathway for carbonyl compounds having a hydrogen three carbon atoms away from the carbonyl carbon.

Meisenheimer complex (Section 16.7): an intermediate formed in the nucleophilic aryl substitution reaction of a base with a nitro-substituted aromatic ring.

Merrifield solid-phase peptide synthesis (Section 26.8): a rapid and efficient means of peptide synthesis in which the growing peptide chain is attached to an insoluble polymer support.

Michael reaction (Section 23.10): the 1,4-addition reaction of a stabilized enolate anion such as that from a 1,3-diketone to an α,β-unsaturated carbonyl compound.

Robinson annulation reaction (Section 23.12): a multistep sequence for building a new cyclohexenone ring onto a ketone. The sequence involves an initial Michael reaction of the ketone followed by an internal aldol cyclization.

Sandmeyer reaction (Section 24.8): a method for converting an aryldiazonium salt into an aryl halide by treatment with a cuprous halide.

Sanger dideoxy method (Section 28.6): an enzymatic method for DNA sequencing.

Simmons–Smith reaction (Section 8.9): a method for preparing a cyclopropane by treating an alkene with diiodomethane and zinc–copper.

Stork enamine reaction (Section 23.11): a multistep sequence whereby a ketone is converted into an enamine by treatment with a secondary amine, and the enamine is then used in Michael reactions.

Suzuki–Miyaura reaction (Section 10.7): an organometallic coupling of an aromatic or vinyl substituted boronic acid with an aryl or vinyl substituted organohalide in the presence of a base and a palladium catalyst.

$$\text{Ar}-\text{B(OH)}_2 \ \ + \ \ \text{I}-\text{Ar'} \ \xrightarrow[\substack{\text{CaCO}_3 \\ \text{THF}}]{\text{Pd(Ph}_3)_4} \ \ \text{Ar}-\text{Ar'}$$

(or vinyl)

Tollens' test (Section 25.6): a chemical test for detecting aldehydes by treatment with ammoniacal silver nitrate. A positive test is signaled by formation of a silver mirror on the walls of the reaction vessel.

Walden inversion (Section 11.1): the inversion of stereochemistry at a chirality center during an S_N2 reaction.

Williamson ether synthesis (Section 18.2): a method for preparing an ether by treatment of a primary alkyl halide with an alkoxide ion.

$$\text{R–O}^- \ \text{Na}^+ + \text{R'CH}_2\text{Br} \longrightarrow \text{R}-\text{O}-\text{CH}_2\text{R'} + \text{NaBr}$$

Wittig reaction (Section 19.11): a general method of alkene synthesis by treatment of a ketone or aldehyde with an alkylidenetriphenylphosphorane.

Wohl degradation (Section 25.6): a multistep reaction sequence for degrading an aldose into the next lower homolog.

Wolff–Kishner reaction (Section 19.9): a method for converting a ketone or aldehyde into the corresponding hydrocarbon by treatment with hydrazine and strong base.

$$\underset{R}{\overset{O}{\underset{}{\parallel}}}\overset{}{\underset{}{C}}\underset{R'}{} \quad \xrightarrow{\text{N}_2\text{H}_4,\ \text{KOH}} \quad \text{R–CH}_2\text{–R'}$$

Woodward–Hoffmann orbital symmetry rules (Section 30.9): a series of rules for predicting the stereochemistry of pericyclic reactions. Even-electron species react thermally through either antarafacial or conrotatory pathways, whereas odd-electron species react thermally through either suprafacial or disrotatory pathways.

Abbreviations

Å	symbol for Angstrom unit (10^{-8} cm $= 10^{-10}$ m)
ADMET	acyclic diene metathesis, a method of polymerization
Ac–	Acetyl group, $CH_3\overset{\displaystyle O}{\overset{\|}{C}}-$
Ar–	aryl group
at. no.	atomic number
at. wt.	atomic weight
$[\alpha]_D$	specific rotation
Boc	*tert*-butoxycarbonyl group, $(CH_3)_3\overset{\displaystyle O}{\overset{\|}{C}}OC-$
bp	boiling point
n-Bu	*n*-butyl group, $CH_3CH_2CH_2CH_2-$
sec-Bu	*sec*-butyl group, $CH_3CH_2CH(CH_3)-$
t-Bu	*tert*-butyl group, $(CH_3)_3C-$
cm	centimeter
cm^{-1}	wavenumber, or reciprocal centimeter
D	stereochemical designation of carbohydrates and amino acids
DCC	dicyclohexylcarbodiimide, $C_6H_{11}-N{=}C{=}N-C_6H_{11}$
δ	chemical shift in ppm downfield from TMS
Δ	symbol for heat; also symbol for change
ΔH	heat of reaction
dm	decimeter (0.1 m)
DMF	dimethylformamide, $(CH_3)_2NCHO$
DMSO	dimethyl sulfoxide, $(CH_3)_2SO$
DNA	deoxyribonucleic acid
DNP	dinitrophenyl group, as in 2,4-DNP (2,4-dinitrophenylhydrazone)
(E)	entgegen, stereochemical designation of double bond geometry
E_{act}	activation energy
E1	unimolecular elimination reaction
E1cB	unimolecular elimination that takes place through a carbanion intermediate
E2	bimolecular elimination reaction
Et	ethyl group, CH_3CH_2-
g	gram

hv	symbol for light
Hz	Hertz, or cycles per second (s^{-1})
i-	iso
IR	infrared
J	Joule
J	symbol for coupling constant
K	Kelvin temperature
K_a	acid dissociation constant
kJ	kilojoule
L	stereochemical designation of carbohydrates and amino acids
LAH	lithium aluminum hydride, $LiAlH_4$
Me	methyl group, CH_3-
Mg	milligram (0.001 g)
MHz	megahertz (10^6 s^{-1})
mL	milliliter (0.001 L)
mm	millimeter (0.001 m)
mp	melting point
μg	microgram (10^{-6} g)
mμ	millimicron (nanometer, 10^{-9} m)
MW	molecular weight
n-	normal, straight-chain alkane or alkyl group
ng	nanogram (10^{-9} gram)
nm	nanometer (10^{-9} meter)
NMR	nuclear magnetic resonance
–OAc	acetate group, $-OCCH_3$ with O double-bonded
Ph	phenyl group, $-C_6H_5$
pH	measure of acidity of aqueous solution
pK_a	measure of acid strength ($= -\log K_a$)
pm	picometer (10^{-12} m)
ppm	parts per million
*n-*Pr	*n*-propyl group, $CH_3CH_2CH_2-$

i-Pr	isopropyl group, $(CH_3)_2CH–$
pro-R	designation of a prochirality center
pro-S	designation of a prochirality center
R–	symbol for a generalized alkyl group
(R)	*rectus*, designation of chirality center
Re face	a face of a planar, sp^2-hybridized carbon atom
RNA	ribonucleic acid
ROMP	ring-opening metathesis polymerization
(S)	*sinister*, designation of chirality center
sec-	secondary
Si face	a face of a planar, sp^2-hybridized carbon atom
S_N1	unimolecular substitution reaction
S_N2	bimolecular substitution reaction
tert-	tertiary
THF	tetrahydrofuran
TMS	tetramethylsilane nmr standard, $(CH_3)_4Si$

Tos	tosylate group,
UV	ultraviolet
X–	halogen group (–F, –Cl, –Br, –I)
(Z)	zusammen, stereochemical designation of double bond geometry
⟶	chemical reaction in direction indicated
⇌	reversible chemical reaction
⟷	resonance symbol
⌒➤	curved arrow indicating direction of electron flow
≡	is equivalent to
>	greater than
<	less than
≈	approximately equal to
R⌇	indicates that the organic fragment shown is a part of a larger molecule
◀	single bond coming out of the plane of the paper

-----	single bond receding into the plane of the paper
......	partial bond
$\delta+$, $\delta-$	partial charge
‡	denoting the transition state

Infrared Absorption Frequencies

Functional Group		Frequency (cm^{-1})	Text Section
Alcohol	$-$O$-$H	3300–3600 (s)	17.11
	$\diagdown$C$-$O$-$	1050 (s)	
Aldehyde	$-$CO$-$H	2720, 2820 (m)	19.14
aliphatic	$\diagdown$C$=$O	1725 (s)	
aromatic		1705 (s)	
Alkane			12.8
	$\diagdown$C$-$H	2850–2960 (s)	
	$\diagdown$C$-$C$\diagup$	800–1300 (m)	
Alkene			12.8
	$=$C$\diagdown^{H}$	3020–3100 (s)	
	$\diagdown$C$=$C$\diagdown$	1650–1670 (m)	
	RCH$=$CH$_2$	910, 990 (m)	
	R$_2$C$=$CH$_2$	890 (m)	
Alkyne	$\equiv$C$-$H	3300 (s)	12.8
	$-$C$\equiv$C$-$	2100–2260 (m)	
Alkyl bromide			12.8
	$\diagdown$C$-$Br	500–600 (s)	
Alkyl chloride			12.8
	$\diagdown$C$-$Cl	600–800 (s)	
Amine, *primary*			24.10
	$-$N$\diagdown^{H}_{H}$	3400, 3500 (s)	
secondary			
	$\diagdown$N$-$H	3350 (s)	

Ammonium salt			24.10
	$\overset{|}{\underset{|}{-N}}\!{}^{+}\!-H$	2200–3000 (broad)	
Aromatic ring	Ar–H	3030 (m)	15.8
monosubstituted	Ar–R	690–710 (s)	
		730–770 (s)	
o-disubstituted		735–770 (s)	
m-disubstituted		690–710 (s)	
		810–850 (s)	
p-disubstituted		810–840 (s)	
Carboxylic acid	–O–H	2500–3300 (broad)	20.8
associated	C=O	1710 (s)	
free		1760 (s)	
Acid anhydride			21.10
	C=O	1760, 1820 (s)	
Acid chloride			21.10
aliphatic	C=O	1810 (s)	
aromatic		1770 (s)	
Amide			21.10
aliphatic	C=O	1810 (s)	
aromatic		1770 (s)	
N-substituted		1680 (s)	
N,N-disubstituted		1650 (s)	
Ester			21.10
aliphatic	C=O	1735 (s)	
aromatic		1720 (s)	
Ether			18.9
	–O–C	1050–1150 (s)	

Ketone   19.14

    aliphatic $\diagup^{\diagdown}C=O$ 1715 (s)

    aromatic   1690 (s)

    6-memb. ring   1715 (s)

    5-memb. ring   1750 (s)

Nitrile   20.8

    aliphatic $-C\equiv N$ 2250 (m)

    aromatic   2230 (m)

Phenol $-O-H$ 3500 (s)   17.11

(s) = strong; (m) = medium intensity

Proton NMR Chemical Shifts

Type of Proton		Chemical Shift (δ)	Text Section
Alkyl, primary	$R–CH_3$	0.7–1.3	13.9
Alkyl, secondary	$R–CH_2–R$	1.2–1.4	13.9
Alkyl tertiary	$R_3C–H$	1.4–1.7	13.9
Allylic	$-\overset{\mid}{C}=\overset{\mid}{C}-\overset{\mid}{C}-H$	1.6–1.9	13.9
α to carbonyl	$-\overset{O}{\overset{\parallel}{C}}-\overset{\mid}{C}-H$	2.0–2.3	19.14
Benzylic	$Ar–\overset{\mid}{C}-H$	2.3–3.0	15.8
Acetylenic	$R–C\equiv C–H$	2.5–2.7	13.9
Alkyl chloride	$Cl–\overset{\mid}{C}-H$	3.0–4.0	13.9
Alkyl bromide	$Br–\overset{\mid}{C}-H$	2.5–4.0	13.9
Alkyl iodide	$I–\overset{\mid}{C}-H$	2.0–4.0	13.9
Amine	$N–\overset{\mid}{C}-H$	2.2–2.6	24.10
Epoxide	$\overset{O}{\overset{}{C-C}}\ H$	2.5–3.5	18.9
Alcohol	$HO–\overset{\mid}{C}-H$	3.5–4.5	17.11
Ether	$RO–\overset{\mid}{C}-H$	3.5–4.5	18.9
Vinylic	$-\overset{\mid}{C}=\overset{\mid}{C}-H$	5.0–6.0	13.9
Aromatic	$Ar–H$	6.5–8.0	15.8
Aldehyde	$R–\overset{O}{\overset{\parallel}{C}}-H$	9.7–10.0	19.14
Carboxylic acid	$R–\overset{O}{\overset{\parallel}{C}}-O–H$	11.0–12.0	20.8
Alcohol	$R–O–H$	3.5–4.5	17.11
Phenol	$Ar–O–H$	2.5–6.0	17.11

Nobel Prizes in Chemistry

1901 **Jacobus H. van't Hoff** (The Netherlands):
"for the discovery of laws of chemical dynamics and of osmotic pressure"

1902 **Emil Fischer** (Germany):
"for syntheses in the groups of sugars and purines"

1903 **Svante A. Arrhenius** (Sweden):
"for his theory of electrolytic dissociation"

1904 **Sir William Ramsey** (Britain):
"for the discovery of gases in different elements in the air and for the determination of their place in the periodic system"

1905 **Adolf von Baeyer** (Germany):
"for his researches on organic dyestuffs and hydroaromatic compounds"

1906 **Henri Moissan** (France):
"for his research on the isolation of the element fluorine and for placing at the service of science the electric furnace that bears his name"

1907 **Eduard Buchner** (Germany):
"for his biochemical researches and his discovery of cell-less formation"

1908 **Ernest Rutherford** (Britain):
"for his investigation into the disintegration of the elements and the chemistry of radioactive substances"

1909 **Wilhelm Ostwald** (Germany):
"for his work on catalysis and on the conditions of chemical equilibrium and velocities of chemical reactions"

1910 **Otto Wallach** (Germany):
"for his services to organic chemistry and the chemical industry by his pioneer work in the field of alicyclic substances"

1911 **Marie Curie** (France):
"for her services to the advancement of chemistry by the discovery of the elements radium and polonium"

1912 **Victor Grignard** (France):
"for the discovery of the so-called Grignard reagent, which has greatly helped in the development of organic chemistry"

Paul Sabatier (France):
"for his method of hydrogenating organic compounds in the presence of finely divided metals"

1913   **Alfred Werner** (Switzerland):
"for his work on the linkage of atoms in molecules by which he has thrown new light on earlier investigations and opened up new fields of research especially in inorganic chemistry"

1914   **Theodore W. Richards** (U.S.):
"for his accurate determinations of the atomic weights of a great number of chemical elements"

1915   **Richard M. Willstätter** (Germany):
"for his research on plant pigments, principally on chlorophyll"

1916 - 17   No award

1918   **Fritz Haber** (Germany):
"for the synthesis of ammonia from its elements, nitrogen and hydrogen"

1919   No award

1920   **Walther H. Nernst** (Germany):
"for his thermochemical work"

1921   **Frederick Soddy** (Britain):
"for his contributions to the chemistry of radioactive substances and his investigations into the origin and nature of isotopes"

1922   **Francis W. Aston** (Britain):
"for his discovery, by means of his mass spectrograph, of the isotopes of a large number of nonradioactive elements, as well as for his discovery of the whole-number rule"

1923   **Fritz Pregl** (Austria):
"for his invention of the method of microanalysis of organic substances"

1924   No award

1925   **Richard A. Zsigmondy** (Germany):
for his demonstration of the heterogeneous nature of colloid solutions, and for the methods he used, which have since become fundamental in modern colloid chemistry"

1926   **Theodor Svedberg** (Sweden):
"for his work on disperse systems"

1927   **Heinrich O. Wieland** (Germany):
"for his research on bile acids and related substances"

1928   **Adolf O. R. Windaus** (Germany):
"for his studies on the constitution of the sterols and their connection with the vitamins"

1929   **Arthur Harden** (Britain):
Hans von Euler-Chelpin (Sweden):
"for their investigation on the fermentation of sugar and of fermentative enzymes"

1930   **Hans Fischer** (Germany):
"for his researches into the constitution of hemin and chlorophyll, and especially for his synthesis of hemin"

1931   **Frederich Bergius** (Germany):
Carl Bosch (Germany):
"for their contributions to the invention and development of chemical high-pressure methods"

1932   **Irving Langmuir** (U.S.):
"for his discoveries and investigations in surface chemistry"

1933   No award

1934   **Harold C. Urey** (U.S.):
"for his discovery of heavy hydrogen"

1935   **Frederic Joliot** (France):
Irene Joliot-Curie (France):
"for their synthesis of new radioactive elements"

1936   **Peter J. W. Debye** (Netherlands/U.S.):
"for his contributions our knowledge of molecular structure through his investigations on dipole moments and on the diffraction of X rays and electrons in gases"

1937   **Walter N. Haworth** (Britain):
"for his researches into the constitution of carbohydrates and vitamin C"

Paul Karrer (Switzerland):
"for his researches into the constitution of carotenoids, flavins, and vitamins A and B"

1938   **Richard Kuhn** (Germany):
"for his work on carotenoids and vitamins"

1939   **Adolf F. J. Butenandt** (Germany):
"for his work on sex hormones"

Leopold Ruzicka (Switzerland):
"for his work on polymethylenes and higher terpenes"

1940 - 42   No award

1943   **Georg de Hevesy** (Hungary):
"for his work on the use of isotopes as tracer elements in researches on chemical processes"

1944   **Otto Hahn** (Germany):
"for his discovery of the fission of heavy nuclei"

1945   **Artturi I. Virtanen** (Finland):
"for his researches and inventions in agricultural and nutritive chemistry, especially for his fodder preservation method"

1946   **James B. Sumner** (U.S.):
"for his discovery that enzymes can be crystallized"

John H. Northrop (U.S.):
Wendell M. Stanley (U.S.):
for their preparation of enzymes and virus proteins in a pure form"

1947   **Sir Robert Robinson** (Britain):
"for his investigations on plant products of biological importance, particularly the alkaloids"

1948   **Arne W. K. Tiselius** (Sweden):
"for his researches on electrophoresis and adsorption analysis, especially for his discoveries concerning the complex nature of the serum proteins"

1949   **William F. Giauque** (U.S.):
"for his contributions in the field of chemical thermodynamics, particularly concerning the behavior of substances at extremely low temperatures"

1950   **Kurt Alder** (Germany):
Otto P. H. Diels (Germany):
"for their discovery and development of the diene synthesis"

1951   **Edwin M. McMillan** (U.S.):
Glenn T. Seaborg (U.S.):
"for their discoveries in the chemistry of the transuranium elements"

1952   **Archer J. P. Martin** (Britain):
Richard L. M. Synge (Britain):
"for their development of partition chromatography"

1953   **Hermann Staudinger** (Germany):
"for his discoveries in the field of macromolecular chemistry"

1954   **Linus C. Pauling** (U.S.):
"for his research into the nature of the chemical bond and its application to the elucidation of the structure of complex substances"

1955   **Vincent du Vigneaud** (U.S.):
"for his work on biochemically important sulfur compounds, especially for the first synthesis of a polypeptide hormone"

1956   **Sir Cyril N. Hinshelwood** (Britain):
Nikolai N. Semenov (U.S.S.R.):
"for their research in clarifying the mechanisms of chemical reactions in gases"

1957   **Sir Alexander R. Todd** (Britain):
"for his work on nucleotides and nucleotide coenzymes"

1958   **Frederick Sanger** (Britain):
"for his work on the structure of proteins, particularly insulin"

1959   **Jaroslav Heyrovsky** (Czechoslovakia):
"for his discovery and development of the polarographic method of analysis"

1960   **Willard F. Libby** (U.S.):
"for his method to use carbon-14 for age determination in archaeology, geology, geophysics, and other branches of science"

1961   **Melvin Calvin** (U.S.):
"for his research on the carbon dioxide assimilation in plants"

1962   **John C. Kendrew** (Britain):
Max F. Perutz (Britain):
"for their studies of the structures of globular proteins"

1963   **Giulio Natta** (Italy):
Karl Ziegler (Germany):
"for their work in the controlled polymerization of hydrocarbons through the use of organometallic catalysts"

1964   **Dorothy C. Hodgkin** (Britain):
"for her determinations by X-ray techniques of the structures of important biochemical substances, particularly vitamin B_{12} and penicillin"

1965   **Robert B. Woodward** (U.S.):
"for his outstanding achievements in the 'art' of organic synthesis"

1966   **Robert S. Mulliken** (U.S.):
"for his fundamental work concerning chemical bonds and the electronic structure of molecules by the molecular orbital method"

1967   **Manfred Eigen** (Germany):
Ronald G. W. Norrish (Britain):
George Porter (Britain):
"for their studies of extremely fast chemical reactions, effected by disturbing the equilibrium with very short pulses of energy"

1968   **Lars Onsager** (U.S.):
"for his discovery of the reciprocal relations bearing his name, which are fundamental for the thermodynamics of irreversible processes"

1969   **Sir Derek H. R. Barton** (Britain):
Odd Hassel (Norway):
"for their contributions to the development of the concept of conformation and its application in chemistry"

1970   **Luis F. Leloir** (Argentina):
"for his discovery of sugar nucleotides and their role in the biosynthesis of carbohydrates"

1971   **Gerhard Herzberg** (Canada):
"for his contributions to the knowledge of electronic structure and geometry of molecules, particularly free radicals"

1972   **Christian B. Anfinsen** (U.S.):
"for his work on ribonuclease, especially concerning the connection between the amino acid sequence and the biologically active conformation"

Stanford Moore (U.S.):
William H. Stein (U.S.):
"for their contribution to the understanding of the connection between chemical structure and catalytic activity of the active center of the ribonuclease molecule"

1973   **Ernst Otto Fischer** (Germany):
Geoffrey Wilkinson (Britain):
"for their pioneering work, performed independently, on the chemistry of the organometallic sandwich compounds"

1974   **Paul J. Flory** (U.S.):
"for his fundamental achievements, both theoretical and experimental, in the physical chemistry of macromolecules"

1975 **John Cornforth** (Australia/Britain):
"for his work on the stereochemistry of enzyme-catalyzed reactions"

Vladimir Prelog (Yugoslavia/Switzerland):
"for his work on the stereochemistry of organic molecules and reactions"

1976 **William N. Lipscomb** (U.S.):
"for his studies on the structures of boranes illuminating problems of chemical bonding"

1977 **Ilya Pregogine** (Belgium):
"for his contributions to nonequilibrium thermodynamics, particularly the theory of dissipative structures"

1978 **Peter Mitchell** (Britain):
"for his contribution to the understanding of biological energy transfer through the formulation of the chemiosmotic theory"

1979 **Herbert C. Brown** (U.S.):
"for his application of boron compounds to synthetic organic chemistry"

Georg Wittig (Germany):
"for developing phosphorus reagents, presently bearing his name"

1980 **Paul Berg** (U.S.):
"for his fundamental studies of the biochemistry of nucleic acids, with particular regard to recombinant DNA"

Walter Gilbert (U.S.)
Frederick Sanger (Britain):
"for their contributions concerning the determination of base sequences in nucleic acids"

1981 **Kenichi Fukui** (Japan)
Roald Hoffmann (U.S.):
for their theories, developed independently, concerning the course of chemical reactions"

1982 **Aaron Klug** (Britain):
"for his development of crystallographic electron microscopy and his structural elucidation of biologically important nucleic acid – protein complexes"

1983 **Henry Taube** (U.S.):
"for his work on the mechanisms of electron transfer reactions, especially in metal complexes"

1984 **R. Bruce Merrifield** (U.S.):
"for his development of methodology for chemical synthesis on a solid matrix"

1985   **Herbert A. Hauptman** (U.S.):
Jerome Karle (U.S.):
"for their outstanding achievements in the development of direct methods for the determination of crystal structures"

1986   **John C. Polanyi** (Canada):
"for his pioneering work in the use of infrared chemiluminescence in studying the dynamics of chemical reactions"

Dudley R. Herschbach (U.S.):
Yuan T. Lee (U.S.):
"for their contributions concerning the dynamics of chemical elementary processes"

1987   **Donald J. Cram** (U.S.):
Jean-Marie Lehn (France):
Charles J. Pedersen (U.S.):
"for their development and use of molecules with structure-specific interactions of high selectivity"

1988   **Johann Deisenhofer** (Germany):
Robert Huber (Germany):
Hartmut Michel (Germany):
"for their determination of the structure of the photosynthetic reaction center of bacteria"

1989   **Sidney Altman** (U.S.):
Thomas R. Cech (U.S.):
"for their discovery of catalytic properties of RNA"

1990   **Elias J. Corey** (U.S.):
"for his development of the theory and methodology of organic synthesis"

1991   **Richard R. Ernst** (Switzerland):
"for his contributions to the development of the methodology of high resolution NMR spectroscopy"

1992   **Rudolph A. Marcus** (U.S.):
"for his contributions to the theory of electron-transfer reactions in chemical systems"

1993   **Kary B. Mullis** (U.S.):
"for his development of the polymerase chain reaction"
Michael Smith (Canada):
"for his fundamental contributions to the establishment of oligonucleotide-based sitedirected mutagenesis and its development for protein studies"

1994   **George A Olah** (U.S.):
"for pioneering research on carbocations and their role in the chemical reactions of hydrocarbons"

1995   **F. Sherwood Rowland** (U.S.)
Mario Molina (U.S.)
Paul Crutzen (Germany)
"for their work in atmospheric chemistry, particularly concerning the formation and decomposition of ozone"

1996   **Robert F. Curl, Jr.** (U.S.)
Harold W. Kroto (U.K.)
Richard E. Smalley (U.S.)
"for their discovery of carbon atoms bound in the form of a ball (fullerenes)."

1997   **Paul D. Boyer** (U.S.)
John E. Walker (U.K.)
"for having elucidated the mechanism by which ATP synthase catalyzes the synthesis of adenosine triphosphate, the energy currency of living cells"

Jens C. Skou (Denmark)
"for his discovery of the ion-transporting enzyme Na^+–K^+ ATPase, the first molecular pump"

1998   **Walter Kohn** (U.S.)
John A. Pople (U.S.)
"to Walter Kohn for his development of the density-functional theory and to John Pople for his development of computational methods in quantum chemistry"

1999   **Ahmed H. Zewail** (Egypt, U.S.)
"for his studies of the transition states of chemical reactions using femtosecond spectroscopy."

2000   **Alan J. Heeger** (U.S.)
Alan G. MacDiarmid (U.S.)
Hideki Shirakawa (Japan)
"for opening and developing the important new field of electrically conductive polymers"

2001   **William S. Knowles** (U.S.)
Ryoji Noyori (Japan)
K. Barry Sharpless (U.S.)
"for their work on chirally catalysed hydrogenation and oxidation reactions"

2002   **John B. Fenn** (U.S.)
Koichi Tanaka (Japan)
"for their development of soft desorption ionisation methods for mass spectrometric analyses of biological macromolecules"

Kurt Wüthrich (Switzerland)
"for his development of nuclear magnetic resonance spectroscopy for determining the threedimensional structure of biological macromolecules in solution"

2003   **Peter Agre** (U.S.)
"for the discovery of water channels in cell membranes"

Roderick MacKinnon (U.S.)
"for structural and mechanistic studies of ion channels in cell membranes"

2004   **Aaron Ciechanover** (Israel)
Avram Hershko (Israel)
Irwin Rose (U.S.)
"for the discovery of ubiquitin-mediated protein degradation"

2005   **Yves Chauvin** (France)
Robert H. Grubbs (U.S.)
Richard R.Schrock (U.S.)
"for the development of the metathesis method in organic synthesis"

2006   **Roger D. Kornberg** (U.S.)
"for his studies of the molecular basis of eukaryotic transcription"

2007   **Gerhard Ertl** (Germany)
"for his studies of chemical processes on solid surfaces"

2008   **Osamu Shimomura** (U.S.)
Martin Chalfie (U.S.)
Roger Y. Tsien (U.S.)
"for the discovery and development of the green fluorescent protein"

2009   **Venkatraman Ramakrishnan** (U.S.)
Thomas A. Steitz (U.S.)
Ada E. Yonath (Israel)
"for studies of the structure and function of the ribosome"

2010   **Richard F. Heck** (U.S.)
Ei-ichi Negishi (U.S.)
Akira Suzuki (Japan)
"for palladium-catalyzed cross couplings in organic synthesis"

2011   **Dan Shechtman** (Israel)
"for the discovery of quasicrystals"

2012   **Robert J. Lefkowitz** (U.S)
Brian K. Kobilka
"for studies of G-protein-coupled receptors"

2013 **Martin Karplus** (U.S)
      Michael Levitt (U.S)
      Arieh Warshel (U.S)
      "for the development of multiscale models for complex chemical systems".

2014 **Eric Betzig** (U.S)
      Stefan W. Hell (Germany)
      William E. Moerner (U.S.)
      "for the development of super-resolved fluorescence microscopy".

Answers to Multiple-Choice Questions in Review Units 1–12

Review Unit 1:	1. d	2. b	3. a	4. c	5. d	6. b	7. a	8. d	9. a	10. c
Review Unit 2:	1. c	2. a	3. d	4. c	5. d	6. b	7. a	8. d	9. b	10. c
Review Unit 3:	1. a	2. b	3. c	4. b	5. c	6. b	7. b	8. a	9. d	10. d
Review Unit 4:	1. d	2. b	3. b	4. c	5. b	6. a	7. d	8. c	9. a	10. d
	11. c									
Review Unit 5:	1. b	2. b	3. c	4. a	5. d	6. d	7. c	8. a	9. b	10. c
	11. c									
Review Unit 6:	1. c	2. a	3. b	4. d	5. c	6. c	7. a	8. d	9. d	10. a
Review Unit 7:	1. d	2. c	3. b	4. b	5. a	6. d	7. b	8. a	9. c	10. c
Review Unit 8:	1. a	2. b	3. a	4. b	5. c	6. a	7. c	8. d	9. c	10. b
Review Unit 9:	1. b	2. d	3. a	4. a	5. c	6. d	7. b	8. c	9. b	10. d
Review Unit 10:	1. b	2. a	3. d	4. b	5. c	6. a	7. d	8. b	9. a	10. d
Review Unit 11:	1. c	2. d	3. a	4. b	5. d	6. a	7. d	8. b	9. c	10. a
Review Unit 12:	1. d	2. a	3. c	4. c	5. a	6. d	7. b	8. a	9. b	10. c